ENVIRONMENTAL SCIENCE
A Study of Interrelationships

Fourth Edition

Eldon D. Enger
Delta College

Bradley F. Smith
Delta College

 WCB Wm. C. Brown Publishers

Book Team

Editor *Kevin Kane*
Developmental Editor *Margaret J. Manders*
Visuals/Design Consultant *Marilyn Phelps*
Production Editor *Debra DeBord*
Designer *David C. Lansdon*
Art Editor *Carla Marie Heathcote*
Permissions Editor *Vicki Krug*
Visuals Processor *Joyce Watters*

 Wm. C. Brown Publishers

President *G. Franklin Lewis*
Vice President, Publisher *George Wm. Bergquist*
Vice President, Operations and Production *Beverly Kolz*
National Sales Manager *Virginia S. Moffat*
Group Sales Manager *Vincent R. Di Blasi*
Vice President, Editor in Chief *Edward G. Jaffe*
Marketing Manager *Paul Ducham*
Advertising Manager *Amy Schmitz*
Managing Editor, Production *Colleen A. Yonda*
Manager of Visuals and Design *Faye M. Schilling*
Production Editorial Manager *Julie A. Kennedy*
Production Editorial Manager *Ann Fuerste*
Publishing Services Manager *Karen J. Slaght*

WCB Group

President and Chief Executive Officer *Mark C. Falb*
Chairman of the Board *Wm. C. Brown*

Cover photo © 1991 David Muench, birch trees on the Isle Royale Park, Michigan.

Copyeditor *Mary M. Monner*

Photo Researcher *Toni Michaels*

The credits section for this book begins on page 505, and is considered an extension of the copyright page.

BRIEF CONTENTS

CONTENTS

PREFACE

The study of environmental science is interdisciplinary. Because it is a broad area of study, it is important to develop its many aspects simultaneously, so that the reader may acquire a unified understanding of environmental issues and concerns.

Most environmental issues are best understood when historical background is provided and political, economic, and social implications are integrated with purely scientific information. This text thoroughly explores all facets of environmental issues and provides a comprehensive, readable, and objective treatment of this complex field. The authors have diligently avoided the inclusion of personal biases and fashionable philosophies.

Environmental Science: The Study of Interrelationships is designed for a one-semester, introductory course. Students from a wide range of disciplines will find it informative and interesting. The text is not a catalog of facts; appropriate facts are used to develop concepts, and concepts are carefully and effectively exemplified. The concept of interrelatedness is central to the text; understanding this concept will enable students to grow in their capacity for intelligent environmental decision making.

Organization and Content

This book is divided into five parts and twenty chapters. It is organized to provide an even and logical flow of concepts and to treat this complex field in a clear and thorough manner.

Part 1 establishes the theme of the book in chapter one by presenting the complexity of interrelations inherent in dealing with the human use of the Great Lakes. Many different constituencies use this tremendous resource, and in so doing often infringe on the desires of others. Chapter two broadens the philosophical base needed to examine environmental issues by discussing various ethical and moral stands that people take regarding environmental issues.

Part 2 develops the ecological principles that are basic to an understanding of environmental interactions and the flow of matter and energy in ecosystems. All living systems can be described in terms of the flow of energy through them. The efficiency of energy flow and the intricacies of interorganism interaction are developed, as is the creative process of natural selection. Principles of population structure and organization are also developed in this section, with particular attention being paid to the implications of these principles to human population problems.

Part 3 focuses on energy. A major emphasis is on the historically important fossil fuels that have been responsible for the development of the major economies of the world. Because of the current concerns about the role nuclear power should play in the energy equation of developed nations, the topic of nuclear power is covered in detail.

Part 4 focuses on how human use of the natural world has resulted in changes in natural ecosystems. The historical development of human effects on major ecosystems and types of organisms is presented. Land, water, and soil use practices are major topics. Because of the heavy use of pesticides in agriculture, this topic is also covered in this section.

Part 5 deals with major types of pollution and the economic, political, and risk analyses involved in determining government policy. Air pollution, solid waste, and hazardous and toxic wastes are covered in this section.

New to This Edition

Several organizational and content changes have been made in this edition.

1. Chapter 2, **Environmental Ethics,** has been substantially rewritten to include a more global perspective and to clarify material in previous editions of this text.
2. Chapter 16, **Risk and Cost: Elements of Decision Making,** is a new chapter that highlights risk analysis and cost-benefit analysis as it applies to environmental problems. It also discusses the difference between real and perceived risk and the importance of the perceptions of the public in formulating environmental policy.
3. Chapter 18 is a new chapter on solid waste, which deals with methods of disposal, source reduction, and recycling.
4. Chapter 20, **Environmental Policy and Decision Making,** is completely rewritten to emphasize changes in environmental policy between the two Earth Days in North America and other parts of the world, and the changes that seem to forecast the nineties as an environmental decade.
5. Two chapters on land-use planning issues have been consolidated into one chapter.
6. Two chapters on water-use issues have been consolidated into one chapter.

7. Fifteen new boxes have been added including: Chief Seattle's lament, the periodic table of elements, the northern spotted owl, the Greens, and the alar controversy.
8. Since the text has been produced in a full-color format, all illustrations have been reevaluated and most have been changed to reflect the addition of color.
9. All data have been updated to include the most recent information available, considering the rapid rate of change in this field.

Special Features and Learning Aids

1. Each of the five parts of the text begins with an **introduction** that places the upcoming chapters in context for the reader by recalling previously discussed material and by describing the organization of the chapters to come.
2. Each chapter begins with a set of learning **objectives,** an **outline,** a list of **key terms,** and a **conceptual diagram**—all of which give the student a broad overview of the interrelated forces that are involved in the material to be discussed. The student is encouraged to refer to these resources while reading and reviewing the chapter.
3. Chapters conclude with a **case study, summary,** and **review questions.** The case studies have been specifically selected to allow the reader to apply the chapter concepts to actual situations. Review questions are related to the chapter objectives, and thus serve to reinforce understanding of basic concepts and principles.
4. To dramatize and clarify text material, each chapter includes a number of **tables, charts, graphs, maps, drawings,** or **photographs.** Each illustration has been carefully chosen to provide a pictorial image or an organized format for showing detailed information, which helps the reader comprehend the chapter material.
5. Each chapter also includes one or more **boxed items.** These point out the far-reaching effects of environmental forces and exemplify the interrelationships

that are central to understanding environmental issues. They are intended to interest the reader and reflect current concerns related to chapter content.
6. The text concludes with a **list of active environmental organizations,** a **metric conversion chart, suggested readings,** a thorough **glossary,** and an **index.**

Useful Ancillaries

An **instructor's manual** accompanies the text. It includes chapter outlines, objectives, and key terms; a range of test and discussion questions; suggestions for demonstrations; and suggestions for audiovisual materials and other teaching aids. The Instructor's Manual also provides **additional case studies** for instructors who wish to use additional concrete examples of how the concepts in the chapter can be applied to the real world.
2. A set of **fifty-one transparencies** is also available to users of the text. The transparencies duplicate text figures that clarify essential ecological, political, economic, social, and historical concepts.
3. A completely revised **Laboratory Manual** is available that features thirty exercises on environmental topics. The manual can be adapted to a variety of situations, including rural and urban, northern and southern, desert and forestland, coastal and inland. A **Laboratory Resource Guide** can also be ordered that specifies procedures, objectives, and equipment for each exercise. For courses using computers, a software program on the chi-square is also available in IBM (ISBN 14176–01).
4. *You Can Make a Difference* by Judith Getis is a practical guide for students and consumers to heighten environmental awareness and to encourage participation as a traveler on the Spaceship Earth.
5. **WCB TestPak** rounds out the supplementary materials. It is a computerized testing service available to adopters of this text. It provides either a call-in/mail-in testing service, or, if you have access to an Apple® IIe or IIc, or to an IBM® PC, you can receive a complete test-item file on a microcomputer diskette.

Acknowledgments

We wish to thank our colleagues who read all or part of this text as we developed the fourth edition. Their suggestions and comments contributed greatly to the final product.

Fourth Edition Reviewers

Terry C. Allison, University of Texas–Pan American
Loretta M. Bates, Concordia College at Moorhead
John H. Green, Nicholls St. University
Wayne E. Kiefer, Central Michigan University
R. G. Litchford, University of Tennessee at Chattanooga
David McCalley, University of Northern Iowa
Jane Maler, Austin Community College
Eric J. Meyer, Iowa Western Community College
Muthena Naseri, Moorpark College–California
Robert M. Shealy, University of South Carolina–Spartanburg
James M. Willard, Cleveland St. University
Ray E. Williams, Rio Hondo College

The authors would like to thank Carolyn Russell for her contributions to this project.

Market Research Respondents

We would like to thank the following adopters of the second edition for their help in preparing the third edition. Each contributed greatly to our understanding of the relative strengths and weaknesses of the second edition by responding to a user's survey of the text.

John W. Adams, University of Texas at San Antonio
Mark Aronson, Scott Community College
Glenn Bellah, Bethany College
Jeffrey H. Black, Oklahoma Baptist University
Larry G. Blackburn, Davidson County Community College
Marion E. Bontrager, Hesston College
Dominic Calvetti, National College
Jay P. Clymer III, Marywood College
Richard D. Coleman, Volunteer State Community College
Sam Crowley, National College
Darlene Coleman Deecher, Mohawk Valley Community College
Sara B. Dodd, Davidson County Community College

Lowell L. Getz, University of Illinois
William H. Gilbert, Simpson College
John Lee Griffis, Southwest State University
Charles R. Hart, University of Wisconsin Center–Manitowoc County
Harry L. Holloway, Jr., University of North Dakota
Jerry F. Howell, Jr., Morehead State University
John A. Jones, Miami-Dade Community College
Ronald R. Keiper, Penn State University–Mont Alto
Eric Meyer, Iowa Western Community College
Robert Miller, Moorpark College
Mercedes C. Mondecar, Morehouse College
Muthena Naseri, Moorpark College
Lisa H. Newton, Fairfield University
Victor J. Newton, Fairfield University
Jarl Roine, Northern Michigan University
Anna E. Ross, Christian Brothers College
Lynette Rushton, Centralia College
Mel Seifert, Sheldon Jackson College
Wayne L. Smith, University of Tampa
Philip C. Shelton, University of Virginia–Clinch Valley College
Susan P. Speece, Anderson University
David A. Stewart, Ferris State University
John B. Topp, Gaston College
Jack A. Turner, University of South Carolina, Spartanburg
Deborah F. Verfaillie, Grossmont College
Linda L. Wallace, University of Oklahoma
Phillip L. Watson, Ferris State University
Clint Westervelt, Chapman College
Ray E. Williams, Rio Hondo College
Jim Yoder, Hesston College

Third Edition Reviewers

Dr. John W. Adams
University of Texas,
San Antonio

Dr. Theodore J. Crovello
California State University,
Los Angeles

Dr. Deborah S. Temperley
Northwood Institute

Dr. Linda Wallace
University of Oklahoma,
Norman

Dr. Phillip L. Watson
Ferris State University

Dr. Jim Yoder
Hesston College

PUBLISHER'S NOTE TO INSTRUCTORS AND STUDENTS

Both the paperback and hardcover versions of this text are printed on recycled paper. All of the book's ancillaries, as well as all advertising pieces for the book, will also be printed on recycled paper, subject to market availability.

Our goal in offering the text and its ancillary package on *recycled paper* is to take an important first step toward minimizing the environmental impact of our products. If you have any questions about recycled paper use, the text, its package, or any of our other biology texts, feel free to call us at 1–800–331–2111. Thank you.

Kevin Kane
Senior Editor
Biology

Paul Ducham
Marketing Manager
Biology

PART ONE
Interrelatedness

Environmental science has evolved as an interdisciplinary study that seeks to describe problems caused by our use of the natural world. In addition, it seeks some of the remedies for these problems. An understanding of three major areas of information is necessary to deal with this complex topic. First, it is important to understand the natural processes (both physical and biological) that operate in the world. Second, it is important to appreciate the role that technology plays in our society and its capacity to alter natural processes as well as solve problems caused by human impact. Third, the complex social processes that are characteristic of human populations must be

understood and integrated with knowledge of technology and natural processes to fully appreciate the role of the human animal in the natural world.

Chapter 1 deals with a specific issue that illustrates how all of these areas are interrelated. This chapter introduces the central theme of interrelatedness by analyzing the competing demands placed on a unique natural resource—the Great Lakes. Chapter 2 discusses the differences that can exist between individuals in a society and the different behaviors exhibited, depending on whether the person is acting as an individual, as a part of a corporation, or as a part of government.

CHAPTER ONE
The Great Lakes:
A Study of Interrelationships

Objectives

After reading this chapter, you should be able to:

Understand why environmental problems are complex and interrelated.

Realize that environmental problems are tied to social, political, and economic issues.

Understand that acceptable solutions to environmental problems are not often easy to achieve.

Understand that all organisms have an impact on their surroundings.

Appreciate that individuals view environmental problems from different perspectives.

Understand what is meant by an ecosystem approach to environmental problem solving.

Chapter Outline

Key Terms

comprehensive water
  management
comprehensive water
  planning

dredging
ecosystem
eutrophication

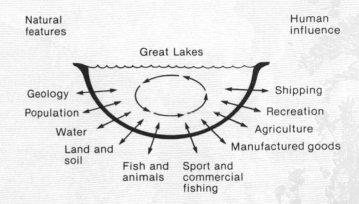

Natural
features

Human
influence

Great Lakes

Geology

Population

Water

Land and
soil

Fish and
animals

Sport and
commercial
fishing

Shipping

Recreation

Agriculture

Manufactured goods

The Ecosystem Approach

We often hear it said that there are no simple solutions to complex problems. Environmental problems are often particularly difficult to resolve because they are interrelated with other societal issues, such as employment levels, profits, and environmental quality. Since by their nature environmental problems are complex and interrelated, arriving at solutions may be a difficult process. Often, solutions to one problem create other problems just as difficult to resolve. Solutions frequently involve compromise.

As one example, consider what happens to the products people buy and use. Nearly everything eventually becomes garbage or trash. Garbage may be carried to a landfill, but if a landfill is inadequate or operated improperly, some harmful wastes may seep into groundwater supplies or drain directly to nearby streams, which flow into lakes used for sources of drinking water. Also, liquid wastes may be dumped accidentally or intentionally into a storm water drainage system, thus polluting water sources.

As with other complex organizations, biological interactions are often referred to as systems. A word-processing unit is a system. It includes many parts that interrelate to store, manipulate, and print information. In the living world, this kind of a system is called an **ecosystem** (ecological system). To better understand the interrelatedness of problems, complexity of competing uses, and compromises involved in environmental decision making, let us analyze a real-world situation: the issue of competing demands upon a natural resource, the Great Lakes.

When we think about ecological disasters, what probably comes to mind are oil tanker accidents fouling beaches with crude oil, the failure of a nuclear reactor at a power-generating facility resulting in massive radioactive exposure to nearby inhabitants, or air-pollution warnings. However, in the case of the Great Lakes, the first evidence of an environmental disaster was of a much smaller nature. It was signaled by a mayfly nymph.

In 1953, scientists taking routine samples of bottom clay from Lake Erie were puzzled. They were checking for the presence of mayfly nymphs, insect larvae that were an important link in the Lake Erie food chain. Samples from an area about one half-meter square produced 485 dead nymphs and no live ones. In another area, which normally produced 1,087 live nymphs per square meter, not a single living nymph was found. Mayfly nymphs grow up to be adult mayflies: short-lived, delicate-looking insects with translucent wings and a long, streamerlike tail. Fish find both the immature nymphs and the adult mayflies to be extremely tasty, and anglers have long designed artificial flies and lures to look like these insects. (See figure 1.1.)

Mayflies once swarmed in great clouds during certain times of the year on Lake Erie. Perch, walleye, and trout would feed on both mayfly and nymph. When nymphs would hatch into mayflies, even bottom-feeding fish such as carp would come to the surface to dine. Hungry bass would flash out of the water to snap at them. Mayflies also provided an important dietary staple for many birds in the Lake Erie region, including robins, swallows, and pheasants. This seemingly insignificant insect played an ecological role much greater than its small size might suggest, and its disappearance was a catastrophe to Lake Erie.

What happened? The answer provides a good example of an ecosystem in action and demonstrates that factors that affect one element within nature can set off a far-reaching chain of events that might have destructive consequences—even in areas far removed from the original point of impact.

As might be expected, the mayfly disappeared as a direct result of the pollution of Lake Erie. The shallow western end of the lake trapped tons

Interrelatedness

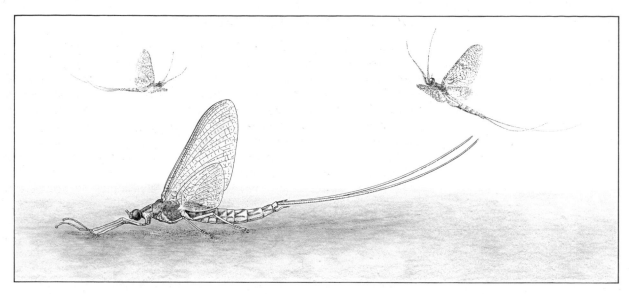

FIGURE 1.1
Water Pollution and Mayfly Populations. The mayfly is an important food source for many types of fish. When pollution reduces the number of mayflies, the fish also decline.

of nutrients, which poured in as stream runoff from both the United States and Canada, and as waste from major urban centers. (See figure 1.2.) Fertilizer and soil washed from surrounding farmlands eventually found their way into the lake as well. This mixture of agricultural runoff and municipal effluent killed the mayfly by the seemingly harmless but lethal process of nutrient enrichment, which stimulated the growth of algae and plants. Organic and inorganic nutrients, particularly phosphorus compounds present as a result of human activity, served as fertilizers for algae—tiny, green aquatic organisms—and species of larger plants, which proliferated far beyond what would have normally occurred in nature. When the plants and algae died, they sank to the bottom, where they were broken down by bacteria. (See figure 1.3.) Bacteria need oxygen to do their work, and the more they took from the bottom water to break down dead organic matter, the less was left for bottom-dwelling mayfly nymphs. When a long spell of unusually hot and still weather prevented mixing of the water, oxygen was depleted from the lower depths of the lake and the mayflies and nymphs died from a lack of oxygen. The chemical balance of Lake Erie had changed beyond the point where the bottom-dwelling nymph could survive and hence the adult mayfly disappeared. And with the mayfly gone, other species were affected as well. Perch, walleye, and bass became less and less common, and carp and suckers became more common.

Persons concerned about the changing species of fish in Lake Erie had to take a look at a long and complex chain of events; the species changed because they were part of a complicated, interrelated ecosystem. When one part of the delicate balance of life within the lake was affected, the reverberations were felt throughout the system.

It was recognized that the nutrients entering the lake were a major contributor to the decline in the mayfly population. High levels of phosphorus and other nutrients stimulated growth of algae and plants. This resulted in poorer water quality and led to a decline in many commercial and sport fish. Consequently, a major effort was undertaken to reduce phosphorus by improving sewage treatment and eliminating phosphate-containing detergents. In response, the amount of phosphorus, a nutrient required by algae and plants, entering Lake Erie has decreased by about 10,000 metric tons a year since 1972, and the ecosystem is responding to this change in its

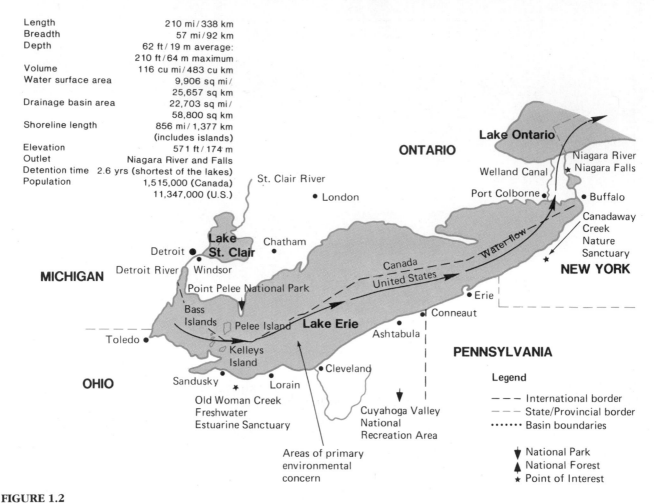

Length	210 mi / 338 km
Breadth	57 mi / 92 km
Depth	62 ft / 19 m average:
	210 ft / 64 m maximum
Volume	116 cu mi / 483 cu km
Water surface area	9,906 sq mi /
	25,657 sq km
Drainage basin area	22,703 sq mi /
	58,800 sq km
Shoreline length	856 mi / 1,377 km
	(includes islands)
Elevation	571 ft / 174 m
Outlet	Niagara River and Falls
Detention time	2.6 yrs (shortest of the lakes)
Population	1,515,000 (Canada)
	11,347,000 (U.S.)

FIGURE 1.2

Lake Erie Facts. The western portion of Lake Erie was the area of greatest environmental concern. The shallow western end trapped the nutrients that flowed into the lake.

balance. This time the change is positive. With improvements in water quality, the mayfly has returned. Fish species associated with cleaner water, such as pickerel and bass, have made a significant comeback in Lake Erie. At one time, freshwater perch had declined to the point of being rare. Now, perch make up a majority of the catch in an expanding fishery in Lake Erie's western basin.

The story of the mayfly and the fall and rise of the Lake Erie fishery offers hope, and a good example of what can be done to restore and preserve the Great Lakes, if we respect and work within the framework of their ecosystem.

In an ecosystem, all components, including humans, interact in complex and often subtle, yet extremely important ways. An ecosystem can be used to refer to a body of water, an area of land with plants and a surrounding envelope of air, or even the entire earth. Actions that might have an impact on the ecosystem cannot be considered in isolation, but in all their far-reaching implications. Humans do not exist outside an ecosystem, but rather as an integral part of it.

Recognition of this principle is exemplified by the Great Lakes Water Quality Agreement between Canada and the United States, which was updated in 1978. The Great Lakes ecosystem was defined as "the interacting components of air, land, and water, and living organisms, including humans, within the drainage basin of the St. Lawrence River at or upstream from the point at which this river becomes the international boundary between

FIGURE 1.3
Floating Mats of Algae. By providing nutrients, agricultural runoff and municipal effluent stimulate the growth of algae and plants.

Canada and the United States." Although the language is formal and legalistic, the agreement recognized humans as part of the ecosystem and stated that ecological problems could no longer be considered in isolation: ". . . restoration and enhancement of the boundary waters cannot be achieved independently of other parts of the Great Lakes basin ecosystem with which these waters interact." Managing the Great Lakes, which are composed of many interrelated elements and processes, means adopting what is called an ecosystem approach.

We have often failed to see, let alone anticipate, the "ripple effect" that a single change in environmental balance can have. All this is indicative of one general problem: our failure to develop an ecosystem approach. This is why we must always view ourselves as part of the ecosystem and carefully weigh the consequences of our actions. Those consequences will be around far longer than we will be.

Setting the Stage

In the past 175 years, the population of the Great Lakes basin has increased more than a hundredfold, from 300,000 to 40 million. With all these people living in, working in, and polluting the area, it is little wonder that the lakes have experienced ecological imbalances. (See figure 1.4.)

The Great Lakes' influence on the people, history, and development of North America has been profound. The lakes and their surrounding land area were rich in natural resources for native people and later for Europeans who explored and settled the continent. The lakes and their river systems provided the means of transportation that made access to these resources possible. Since the development of industrialized urban society, however, we have been exploiting the lakes in increasingly harmful ways. It took society almost 150 years to realize that it could not go on abusing and neglecting the Great Lakes forever; that, even though they are huge, unchecked pollution and unplanned growth could destroy this seemingly limitless resource.

The physical dimensions of the Superior, Huron, Michigan, Erie, and Ontario lakes, which make up the largest freshwater system in the world,

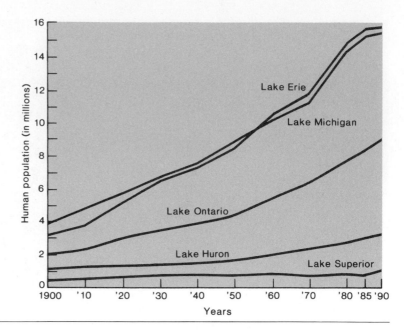

FIGURE 1.4

Population Growth in the Great Lakes Basin since 1900. The population of the Great Lakes Basin continues to increase, placing ever-growing demands on the lakes and surrounding natural resources.

are impressive. Stretching from Minnesota to Montreal, the lakes cover an area in excess of 250,000 square kilometers. One-fifth of the world's fresh surface water, 22.7 quadrillion liters, is contained in the lakes. Spread evenly across the continental United States, the Great Lakes would submerge the country under 2.7 to 3.7 meters of water. Over 15,000 kilometers of Great Lakes coastline are shared between Canada and the United States. The U.S. Great Lakes coastline alone, over 7,000 kilometers, is longer than the U.S. Atlantic and Gulf of Mexico coasts combined. (See figure 1.5.)

Lake Superior has the largest surface area of any freshwater lake in the world. Lake Huron, the second largest of the Great Lakes, is the fifth largest lake in the world. Lake Michigan, the only Great Lake encompassed in the United States, is the world's sixth-largest lake. In almost every aspect, the lakes look and behave much like oceans and are often referred to as the "inland seas."

The Formation of the Great Lakes

The Great Lakes were formed by a series of geologic processes. These processes included the erosion of the earth's surface by immense sheets of ice known as glaciers, and the deposition of large amounts of silt, clay, sand, gravel, and boulders. The glaciers that carved the Great Lakes developed when the climate in North America changed. Short, cool summers followed by long winters and heavy snowfall caused the formation of thick accumulations of ice in northern Canada.

As the thickness of the ice grew, the glaciers slowly moved southward over North America, scouring rock surfaces and plucking up bedrock as they progressed. As a glacier travels over a region, it usually takes the path of least resistance. When an area of nonresistant surface rock is adjacent to an area of resistent rock, the glacier will generally carve much deeper into the yielding material and leave the more resistant rock relatively unchanged.

Several periods of glaciation greatly affected the Great Lakes region. Massive walls of ice, often in excess of a kilometer in height, first covered the land and then melted, only to be followed by a later glacier. The initial

Great Lakes Basin

Canada

Lake Huron   Lake Ontario

Lake Superior

Lake Michigan

Lake Erie

United States

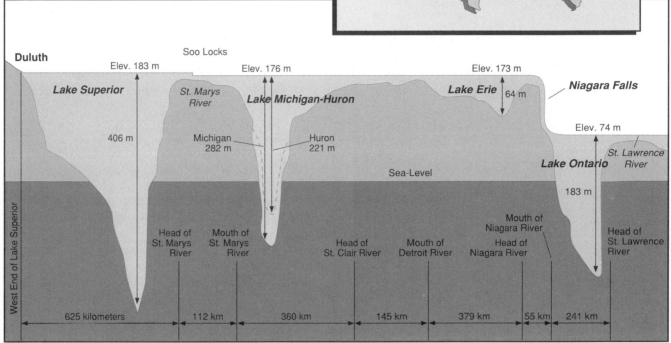

Duluth

Soo Locks

Elev. 183 m   Elev. 176 m   Elev. 173 m   Niagara Falls

Lake Superior

St. Marys
River

Lake Michigan-Huron

Lake Erie 64 m

Elev. 74 m

406 m

Michigan
282 m

Huron
221 m

St. Lawrence
River

Lake Ontario

Sea-Level

183 m

West End of Lake Superior

Head of
St. Marys
River

Mouth of
St. Marys
River

Head of
St. Clair River

Mouth of
Detroit River

Mouth of
Niagara River

Head of
Niagara River

Head of
St. Lawrence
River

625 kilometers   112 km   360 km   145 km   379 km   55 km   241 km

a.

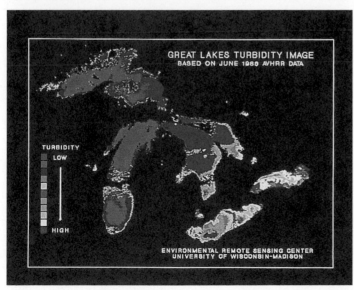

GREAT LAKES TURBIDITY IMAGE
BASED ON JUNE 1988 AVHRR DATA

TURBIDITY
LOW

HIGH

ENVIRONMENTAL REMOTE SENSING CENTER
UNIVERSITY OF WISCONSIN-MADISON

b.

FIGURE 1.5

Great Lakes Profile. (*a*) The Great
Lakes cover an area in excess of 250,000
square kilometers. The size and depth of
the lakes vary considerably. (*b*) The
Great Lakes coastline forms an aquatic
ecosystem almost equal in size to North
America's Atlantic, Pacific, and Arctic
combined shorelines. The lakes contain
one-fifth of all the fresh surface water on
earth. This photo was taken from a
weather satellite and shows ranges of
turbidity, which is influenced by
sediments, river inputs, and algal growth.
Eutrophication, the natural aging
process of lakes, can be accelerated by
nutrient runoff from agriculture and
effluent from municipal sewage plants.
The process is evident where orange and
red indicate unusually high algal growth.

lakes were formed as the ice retreated and the melting ice filled the basins. Each succeeding period of glaciation contributed to shaping the lakes as we currently know them. The freshwater resource created by glaciers thousands of years ago would come to be utilized by humans for many different purposes, not all of which were compatible.

Competing Uses and Demands on the Great Lakes

For centuries, people have used the Great Lakes for many purposes—drinking, boating, swimming, industrial processing, agricultural irrigation, and waste disposal. Millions of Americans and Canadians now depend on the lakes for economic, recreational, and aesthetic benefits. If the water quality of the lakes is not maintained, many of these benefits will be lost.

The multiple uses of the Great Lakes are often highly incompatible. Manufacturing processes, power generation, shipping, dredging, and other human activities often harm water quality. If the lakes become contaminated, fish, other aquatic wildlife, and plants die. Humans also suffer because we eventually pay the price. When we cannot safely use the water for swimming, fishing, sailing, or drinking, tax dollars must be used for complicated water and sewage treatment processes.

Both industry and the public have a right to use the lakes, yet often one use interferes with another use. The lakes are a shared resource, not only between boaters and anglers, and commercial and industrial users, but between citizens of two different countries and several provinces and states as well. The water of the lakes not only serves as a pleasant, natural boundary between the two countries but also allows pollutants to be shared between neighbors. This binds basin states and provinces together, giving each a vested interest in the activities of the other. The management of such a complex resource is not a simple task.

Society's use of the Great Lakes is so great that they are threatened from many quarters. The natural aging process of the lakes is vastly speeded up by sewage and agricultural runoff. Acid rain and the hazards posed by toxic chemicals pose perhaps an even greater danger. Other threats like urban sprawl and habitat destruction are more subtle, though also important. Where society has recognized problems and determined to do something about them, results have been encouraging. Over the past decade, the water quality of the lakes has actually improved dramatically in some areas. Tremendous strides have been made in treating sewage and reducing phosphate pollution. (See figure 1.6.)

Unfortunately, in the past, the environment was allowed to deteriorate badly before humans became fully aware that action was required. Frequently, this resulted in harm to wildlife, flora, and water quality. The late 1960s and early 1970s saw a great heightening of public consciousness about environmental matters. This, combined with an increasing awareness that the lakes seemed to be dying, awakened people, and many cleanup measures were taken. However, an entirely new set of problems came to the fore in the 1980s and 1990s, and some of the old difficulties proved to be more persistent than was initially anticipated.

Eutrophication, the accelerated aging process of the lakes, continues, though at a vastly reduced rate. Acid rain does not yet threaten the Great Lakes themselves, but is a problem in northern regions of the Great Lakes watershed. The tragic events at Love Canal in New York were followed by the discovery that toxic chemicals were leaching into the Niagara River.

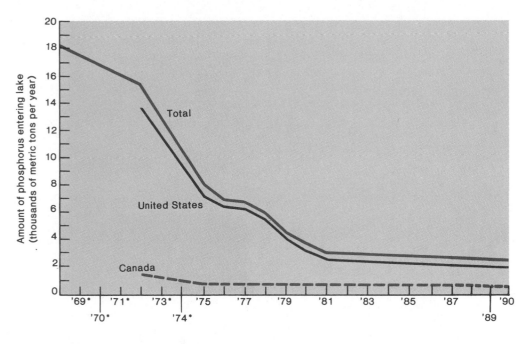

*Estimate only—no data available   Years

FIGURE 1.6
Phosphorus Entering Lake Erie from Municipal Sources. The amount of phosphorus entering Lake Erie from municipal sources has declined steadily over the past two decades. This is primarily due to improved municipal wastewater treatment.

These events, combined with scientists' newly enhanced ability to detect minute quantities of chemical substances in water and fish, shook public confidence in the safety of its drinking water, and advisories against eating fish were published. At the same time, the population of the Great Lakes region makes increasing demands on the lakes. Furthermore, the economic problems of the heavily industrialized Great Lakes region during the 1980s caused delays in the installation of necessary pollution control equipment. In the minds of some, the economy is forcing harsh choices to be made between economic necessities and ecological benefits.

There are also discussions about the feasibility of diverting Great Lakes water to the southwestern United States. As the population of the southwestern states increases, the pressure will mount to divert the water. With any significant water diversion, there will be ecological consequences. The political forces in the Great Lakes states are strongly opposed to water diversion. Regional political rivalry between the Great Lakes states and the southwestern states could be a growing problem.

While the abundance and quality of water serve as powerful incentives for investment in the Great Lakes region, development that limits or degrades water resources will discourage further prosperity. Although the amenities of water still abound, the unique nature of the Great Lakes environment may be transformed over time into something far less desirable.

As our knowledge increases, complex interrelationships between water and related land resources become more apparent. The institutional structure for managing water resources is also complex. Various governmental agencies have important roles at the local, substate, state, federal, and international levels. Numerous organizations, interest groups, and universities are involved in research, policy development, and related activities. Citizens influence the decision-making process throughout government and are directly affected by the water-management programs. To better understand the issue, it is important to see what is meant by the terms *water management* and *water planning*.

Comprehensive water management is the process of protecting and developing all the water resources in a state, region, or province in a broad, integrated, and foresighted manner. In practice, this is a complicated endeavor, since comprehensive water management involves a number of functions that are closely related but that are carried out by different agencies and organizations. These functions include water law and policy making, regulation, technical assistance and coordination, monitering and evaluation, administration and financing, and public education and involvement. Collectively, they form the water-management structure for the Great Lakes.

Comprehensive water planning is the process used to integrate the diverse functions necessary for comprehensive water management. The purpose of this process is to identify alternative courses of action to protect and develop water resources. In the course of comprehensive water planning, problems are identified, data are collected and analyzed, and projections are made. Laws, policies, and institutions are also examined. Alternative solutions are then developed and analyzed, and a selection is made. Finally, an implementation and monitoring program is put into place. This process provides a basis for integrating all of the functional components of comprehensive water management.

To more fully appreciate the demands placed on the Great Lakes, let us analyze how the resource is utilized and the problems that arise from conflicting uses.

Industrial Uses

The development of accessible, efficient water transportation after harbor improvements and other modifications also encouraged the development of industry. The Great Lakes region is a primary center for industrial production. One-fifth of U.S. industry is located around the lakes. Many of these industries were first developed here because of the availability of abundant, free water and accessible, efficient transportation. Over 50 percent of U.S. and 62 percent of Canadian steel production is in the basin, along with the major auto-producing areas in North America. The power-generating capacity of the Great Lakes states and Ontario is estimated at 77,640 megawatts per year.

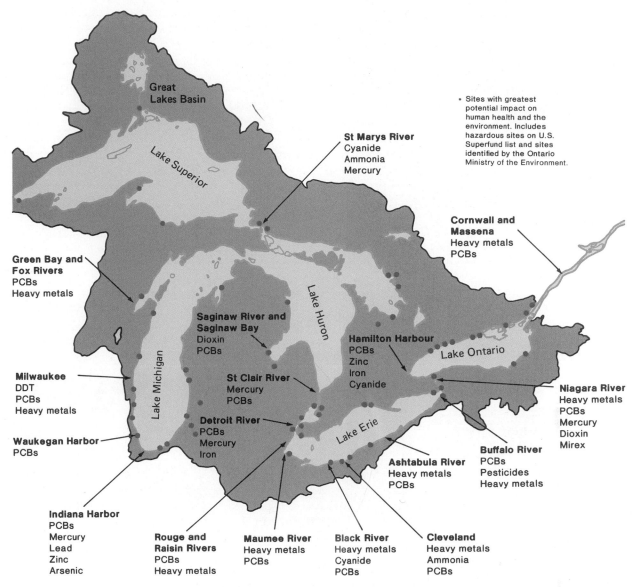

Great Lakes Basin

Lake Superior

St Marys River
Cyanide
Ammonia
Mercury

• Sites with greatest
potential impact on
human health and the
environment. Includes
hazardous sites on U.S.
Superfund list and sites
identified by the Ontario
Ministry of the Environment.

**Cornwall and
Massena**
Heavy metals
PCBs

**Green Bay and
Fox Rivers**
PCBs
Heavy metals

Lake Huron

**Saginaw River and
Saginaw Bay**
Dioxin
PCBs

Hamilton Harbour
PCBs
Zinc
Iron
Cyanide

Lake Ontario

Milwaukee
DDT
PCBs
Heavy metals

Lake Michigan

St Clair River
Mercury
PCBs

Niagara River
Heavy metals
PCBs
Mercury
Dioxin
Mirex

Detroit River
PCBs
Mercury
Iron

Lake Erie

Waukegan Harbor
PCBs

Buffalo River
PCBs
Pesticides
Heavy metals

Ashtabula River
Heavy metals
PCBs

Indiana Harbor
PCBs
Mercury
Lead
Zinc
Arsenic

**Rouge and
Raisin Rivers**
PCBs
Heavy metals

Maumee River
Heavy metals
PCBs

Black River
Heavy metals
Cyanide
PCBs

Cleveland
Heavy metals
Ammonia
PCBs

FIGURE 1.7

Toxic Substances in the Great Lakes.
Perhaps the greatest problem associated
with industrial uses of the Great Lakes is
contamination from toxic waste. There
are many sources of toxic substances
throughout the Great Lakes. Included
are hazardous sites on the U.S.
Superfund Priorities list and sites
identified by the Ontario Ministry of the
Environment.

One of the greatest problems associated with industrial uses of the Great Lakes is the problem of contamination from toxic waste. (See figure 1.7.) The issue of toxic contaminants in the Great Lakes is subtle because the effects generally cannot be seen directly. Toxic contaminants do not cause observable effects, such as algae mats and obnoxious odors or turbid water. The levels of many of the contaminants in water are difficult to detect, even with the best scientific instruments.

A major concern about these pollutants is the effects they have on aquatic life and fish-eating birds as they bioaccumulate up the food chain. There is also concern about the accumulative effects of toxic contaminants on aquatic life, birds, and humans. The concentrations of some chemicals in the fat tissue of top predators, such as lake trout and salmon, can be a million times higher than the levels in the water.

High levels of contaminants in some species of fish have led to the issuance of fish consumption advisories. The larger, older fish in a species are most likely to be restricted. Nursing mothers, pregnant women, women

Public health agencies have issued advisories for consumption of certain Great Lakes fish		
	Restrict consumption[a]	Do not eat
Lake Michigan	Lake trout 20 to 23 inches, coho salmon over 26 inches, chinook salmon 21 to 32 inches, and brown trout up to 23 inches	Lake trout over 23 inches, chinook salmon over 32 inches, brown trout over 23 inches, carp, and catfish
Lake Superior	Lake trout up to 30 inches, walleye up to 26 inches	Lake trout over 30 inches, walleye over 26 inches
Lake Huron	Lake trout, rainbow trout, and brown trout	
Lake Erie	Carp and catfish (New York State waters—eat no more than one meal per month)	Carp and catfish
Lake Ontario	White perch, coho salmon up to 21 inches, rainbow trout up to 18 inches (eat no more than one meal per month)	American eel, channel catfish, lake trout, chinook salmon, coho salmon over 21 inches, rainbow trout over 25 inches, brown trout over 18 inches

[a] Nursing mothers, pregnant women, women who anticipate bearing children, and children age 15 and under should not eat the fish listed in any of these categories.

a.

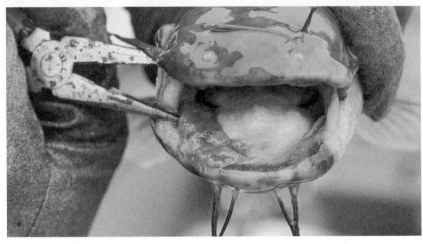

b.

FIGURE 1.8

Fish Consumption Advisory and Fish Cancer. (*a*) Warnings against humans consuming certain species of fish caught in areas of the Great Lakes are now common. (*b*) The lip cancer on this bullhead indicates carcinogens in Wisconsin's Fox River. In some tributaries of the Great Lakes, fish cancer rates may reach 84 percent.
(*b*) Source: U.S. Department of Commerce.

who anticipate bearing children, and children under age fifteen are advised not to eat lake trout over a certain size from Lakes Michigan, Superior, Huron, and Ontario. In addition, men and women of all ages are advised against eating very large lake trout from all five lakes. (See figure 1.8.)

Contamination also has led to waterfowl advisories. The New York Department of Environmental Conservation has tested seven species of waterfowl for polychlorinated biphenyls (PCBs) and organochlorine pesticides. Birds nesting near Lake Ontario were found to have significantly higher concentrations of PCBs than birds in other areas, except for the Hudson River. High levels of PCBs have caused New York and Wisconsin to issue advisories against eating certain species of waterfowl.

Industrial toxic wastes affect almost all other uses of the Great Lakes, from safe drinking water to secure food sources. While industry is a primary source of toxic wastes, other users of the lakes, such as agricultural and municipal users, also contribute hazardous substances.

BOX 1.1
Atmospheric Deposition Loadings of Toxic Chemicals to the Great Lakes

Atmospheric deposition is an important, if not dominant, source of many toxic organic chemicals, heavy metals, and pesticides to the Great Lakes. Atmospheric deposition of PCBs to the Great Lakes ranges from about 94 percent of total inputs for Lake Superior to about 11 percent of total inputs for Lake Ontario. In humans, PCBs produce liver ailments and skin lesions. In high concentrations, they can damage the nervous system, and they are suspected carcinogens. In fish, PCBs can interfere with reproduction. Atmospheric deposition supplies significant quantities (over 10 percent) of lead, cadmium, mercury, and arsenic relative to total inputs to the Great Lakes. Estimates of atmospheric deposition to all of the lakes suggest that 50 to 90 percent of lead and 13 to 75 percent of cadmium loadings derive from the atmospheric pathway. Atmospheric deposition of lead is decreasing due to limits on lead usage in gasoline.

Atmospheric deposition of toxic chemicals occurs over a large geographical area.

Source: Courtesy of Environment Canada, Atmospheric Environment Service (AES).

Airborne emissions of critical pollutants in the United States.

	Metric tons per year
Lead	*21,305*
Gasoline combustion	15,400
Municipal waste incineration	2,800
Industrial processes	2,300
Coal combustion	778
Other sources	27
Arsenic	*3,332*
Use of arsenical pesticides	1,500
Coal combustion	1,390
Primary copper smelters	312
Other sources	130
Mercury	*1,668*
Natural	1,019
Use of consumer goods	352
Coal combustion	132
Municipal waste incineration	68
Copper smelting	41
Sewage sludge incineration	36
Other sources	20

Source: International Joint Commission.

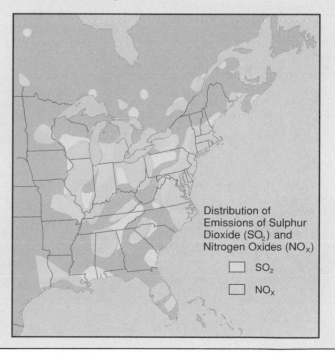

2. Transport
3. Transformation
4. Deposition
1. Emissions
5. Sensitive ecosystems
Long distance

Distribution of Emissions of Sulphur Dioxide (SO_2) and Nitrogen Oxides (NO_x)

☐ SO_2
☐ NO_x

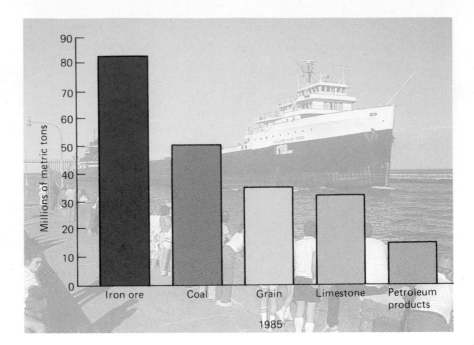

FIGURE 1.9

Great Lakes Shipping. While most of the shipping conducted on the Great Lakes is by Canada and the United States, the lakes are also used extensively by many other nations. The economic benefits of shipping are great, but not without environmental costs. The shipping data show the major commodities transported on the Great Lakes.

Shipping

For hundreds of years, the Great Lakes have been used for shipping everything from fur pelts to iron ore. Today, the lakes are an extremely important inland transportation route for coal, limestone, sand and gravel, heavy machinery, iron ore, foodstuffs, grain, and other agricultural products. (See figure 1.9.) Shipping activities extend from the western shore of Lake Superior through the St. Lawrence River and out through the Atlantic Ocean on an almost year-round basis. The majority of all shipping is carried out by Canadian- and U.S.-owned and operated vessels.

The economic benefits of shipping to the Great Lakes region are enormous but are not without environmental costs. Some cargos are toxic, and others may consist of oil or other hazardous materials. Any of these can be spilled accidentally. Vessels and crews also produce wastes, which can pose a potential pollution problem. Problems stemming from shipping are sometimes complicated because each state and province has its own regulations to control vessels and their wastes in its portion of the lakes.

A problem directly related to shipping is that of **dredging,** the mechanical removal of accumulated sediments from a watercourse. Since materials from natural processes, such as erosion, and human activities, such as waste disposal and construction, gradually fill the bottoms of rivers, harbors, and lakes, dredging is necessary to maintain navigation channels at proper depths and to excavate new channels or harbor facilities. The dredging process was once perceived to be a simple matter of scooping up the sediments with a special boat and transporting the dredged materials to a deeper part of the water for disposal or depositing them on nearby shores.

However, since many of the sediments near industrial centers are contaminated, dealing with contaminated dredged materials is not so simple. First, there is the problem that toxic substances or other hazardous polluting materials buried in the bottom sediments could be resuspended and recontaminate the water during dredging. A second major problem lies in

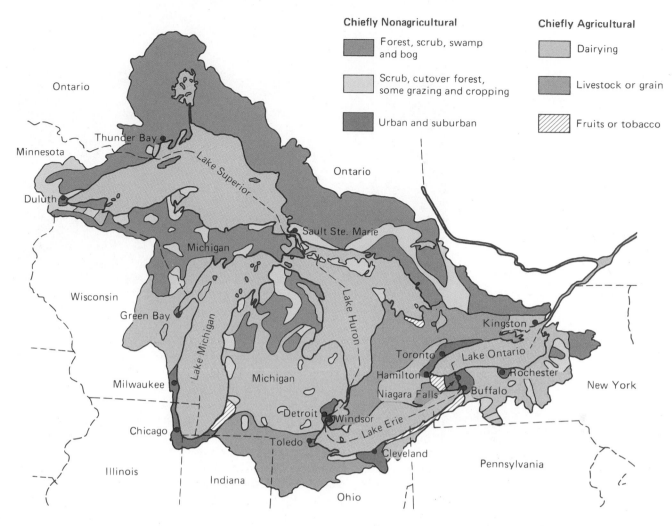

Chiefly Nonagricultural

- Forest, scrub, swamp and bog
- Scrub, cutover forest, some grazing and cropping
- Urban and suburban

Chiefly Agricultural

- Dairying
- Livestock or grain
- Fruits or tobacco

Ontario
Minnesota
Thunder Bay
Duluth
Lake Superior
Ontario
Sault Ste. Marie
Michigan
Wisconsin
Green Bay
Lake Michigan
Lake Huron
Kingston
Toronto
Lake Ontario
Hamilton
Rochester
Niagara Falls
Buffalo
New York
Milwaukee
Michigan
Chicago
Detroit
Windsor
Lake Erie
Illinois
Toledo
Cleveland
Indiana
Pennsylvania
Ohio

FIGURE 1.10

Land Use in the Great Lakes Basin.
Land use in the Great Lakes Basin is closely interrelated with water use and water quality. Study of this map shows why lower Lake Michigan, Lake Erie, and Lake Ontario have been areas of environmental concern.

the difficulty of finding a safe and acceptable disposal site for polluted dredged materials. While the problems resulting from dredging are acute in certain areas along the Great Lakes, environmental concerns resulting from agricultural practices affect a majority of the basin.

Agricultural and Other Land-Use Activities

Agriculture is a major industry in the Great Lakes region. The moderating effect of the Great Lakes climate and the ample water supplies support an agricultural industry that ranks among the top three contributors to the region's economy. Great Lakes states and provinces produce much of the U.S. and Canadian beans, sunflower seeds, and dairy products. Orchards, particularly apple and cherry, lie parallel to the shores of several lakes. Canneries, breweries, and cheese factories are also major users of Great Lakes water.

Land-use activities along the basin are closely interrelated with water quality. (See figure 1.10.) In recent years, there has been a growing concern about the amount of Great Lakes pollution that may come from diffuse or widespread sources such as farming. Farmers use pesticides and fertilizers, which wash off their fields into adjoining streams or lakes. Raising livestock creates wastes, which often receive no special treatment and drain off the

FIGURE 1.11

Fishing. Recreational fishing is a major economic activity in the Great Lakes region, providing both jobs and tourist dollars for the area.

land into the water. The destruction of vegetated buffer strips along stream and ditch banks accelerates erosion. Poor plowing practices rob the rural areas of valuable topsoil and create channels for pollution to run off the land.

It is difficult to estimate how important each of these agricultural activities is to the polluting of the Great Lakes. Even more difficult is finding ways to control some of these problems, since they are diffuse problems, which are difficult to police. From a pragmatic point of view, however, it is less expensive to control sources of pollution than to capture and treat pollutants after they get into the Great Lakes system.

Recreation

Great Lakes recreation consists of activities on or near the shore, as well as on, in, or under the water. Rock collecting, sunbathing, hiking, sand-dune exploration, birdwatching, and picnicking are just a few forms of shore-bound leisure activities. Sailing, powerboating, swimming, fishing, and scuba diving attract people into or onto the water. Great Lakes states spend approximately $7 million annually promoting tourism. In Michigan, 33 percent of all boating activity is on the Great Lakes. In 1990, almost 87,000 boat owners in Ohio reported that they used their vessels primarily on Lake Erie. Surveys have indicated that one of every three families in Ontario owns or uses a boat—many of them on the Great Lakes. The Great Lakes states sold almost $10 million in fishing licenses in 1990, many of them for use in the Great Lakes or their tributaries. (See figure 1.11.)

Fishing on the Great Lakes is divided between sport and commercial uses. Sport fishing attracts about five million anglers yearly, with a regional economic benefit exceeding $1.2 billion annually. The dockside value of the Great Lakes commercial fishing industry is about $48 million a year and has a regional economic impact more than four times that. As you can imagine, water quality and the health of the Great Lakes fishing industry are closely interrelated.

Managing a resource as diverse as the Great Lakes is not a simple task. As we have seen, a variety of demands are placed on the lakes, and many users of the resource are frequently in conflict with one another. The resource cannot be left unregulated. Policy decisions regarding the use and future of the Great Lakes must be based on the conflicting demands for their use and must also recognize the economic and political problems of the region.

The Decision Makers

A vast array of organizations share in managing the Great Lakes. They range from local community associations to international commissions. The institutions and groups involved in Great Lakes management change as the issues change. The future of the lakes depends on a multitude of complex decision-making interactions between governmental units, each of which plays a different role in responding to the problems and opportunities the lakes present.

The U.S. federal government has traditionally set environmental protection goals, funded enforcement efforts, and performed research. In Canada, the primary federal role is defined in an agreement with Ontario. The Canada–U.S. Great Lakes Water Quality Agreement of 1978 sets clean water goals for both federal governments. Each country implements this agreement according to its own laws and customs.

The Federal Clean Water Act sets overall water policy in the United States. The U.S. Environmental Protection Agency (EPA) implements that act's provisions through fines and sanctions, and by providing money for cleanup and repair processes.

A unique organization, the International Joint Commission (IJC) plays a major role in providing direction on many issues, especially water quality. The IJC is composed of six members, three from the United States and three from Canada. The commission was established in 1909 when Canada and the United States signed the Boundary Waters Treaty, which concerned all the waters that form or cross the border between the two countries. Article IV of the treaty provided that "boundary waters and waters flowing across the boundary shall not be polluted on either side to the injury of health or property on the other." The treaty created the International Joint Commission to assist the parties in carrying out their treaty obligations. Over the years, the IJC has studied Great Lakes pollution many times at the request of the two governments. While IJC recommendations do not carry the force of law, they are likely to guide policy makers.

Other organizations and agencies also influence different Great Lakes issues. The U.S. Army Corps of Engineers, the Maritime Administration, and the St. Lawrence Seaway Development Corporation, all representing the U.S. federal government, play important roles in commercial navigation or flood control. Many state agencies and the Great Lakes Fishery Commission are active in fisheries management.

BOX 1.2
Saginaw Bay Remedial Action Plan

Saginaw Bay is a southwestern extension of Lake Huron, located in the east central portion of Michigan's Lower Peninsula. The bay has a large surface area of 2,960 square kilometers, and its drainage basin of 20,907 square kilometers includes 13 percent of Michigan's total land area.

Saginaw Bay is one of fourteen Great Lakes Areas of Concern (AOC) identified by the International Joint Commission (IJC). Areas of Concern are defined as areas where severe environmental quality problems result in the inability to use the resource for one or more designated uses.

The Saginaw Bay AOC is presently assessed as a category 3 on the IJC classification list. In a category 3 area, the causes of the environmental problems are known, but a Remedial Action Plan (RAP) has not been developed to abate the problem. Eutrophication and toxic contaminants have been identified as the primary water-quality issues in Saginaw Bay. These two problems have produced several environmental effects, including contaminated sediments and fish consumption health advisories.

Work began on the Saginaw Bay RAP in July 1986 with the assignment of a RAP site coordinator from the Surface Water Quality Division of the Michigan Department of Natural Resources (MDNR). During September, a contract was developed between the MDNR and the East Central Michigan Planning and Development Region (ECMPDR) to retain ECMPDR to produce the initial draft RAP document. ECMPDR is a regional planning agency that serves fourteen of the counties that surround Saginaw Bay. ECMPDR then subcontracted with the Great Lakes Natural Resource Center of the National Wildlife Federation (NWF) to have NWF prepare several sections of the draft RAP. NWF, in turn, subcontracted with the school of Natural Resources at the University of Michigan for input.

In November 1986, a Technical Work Group (TWG) was formed to review drafts of the RAP for completeness and technical content. This group included approximately thirty representatives, including the NWF, MDNR, IJC, U.S. Environmental Protection Agency, U.S. Geological Survey, U.S. Fish and Wildlife Service, National Oceanographic and Atmospheric Administration, U.S. Army Corps of Engineers, U.S. Soil Conservation Service, Michigan Department of Public Health, as well as several independent consulting firms.

Saginaw Bay is an extension of Lake Huron and designated Area of Environmental Concern by the International Joint Commission.

In the spring of 1987, ECMPDR conducted a series of meetings to solicit input to the RAP process. A separate meeting was held with groups representing industry, agriculture, commerce, conservation/recreation, and municipal/local governments. The ECMPDR also formed a public group called the Saginaw Basin Natural Resource Steering Committee. This body coordinated public input to the RAP process and provided public review during the RAP's developmental stages.

Copies of the draft RAP were provided to both the Technical Work Group and Steering Committee. It then went to the Michigan Water Resources Commission (WRC). In 1990, the IJC sent its comments to the MDNR that the RAP needed to address the issues of contaminants and remediation more fully. By early 1991, the IJC had not accepted the RAP.

If this process appears somewhat confusing, then perhaps you can appreciate the complexity involved in environmental decision making. It is important that the public be involved in decision making. Confusion surfaces, however, when many different "publics" want to be heard and represented.

BOX 1.3
Foreign Organisms Invading the Great Lakes

Scientists have discovered that three aquatic animal species from Europe have invaded the waters of the Great Lakes in recent years—the zebra mussel, a waterflea, and a small perchlike fish called the ruffe. Available information suggests that they may have some highly undesirable effects on the Great Lakes ecosystem.

Of greatest concern is the zebra mussel, *Dressenia polymorpha*. Biologists believe that the zebra mussel came to North America from the Soviet Union in the ballast water of an ocean-going freighter about 1985. When the freighter dumped its ballast water, the zebra mussel was dumped with it. Zebra mussels breed in warm water, feed on plankton, and multiply extremely rapidly. The most frequent problem they cause is blocking the intake pipes of municipal water supplies because of the mussels' large colonies. They also attach themselves to docks and boat hulls. Densities of more than twenty-seven thousand zebra mussels per square meter have been documented in Lake Erie. In Europe, where the zebra mussel has been a problem for many years, kilometers of water pipe must frequently be replaced at great cost when mussels cannot be removed. Great Lakes environmental officials say that virtually all of the hard surfaces, such as rocks, docks, and even old car bodies, in Lake St. Clair and Lake Erie are now coated at least 2 centimeters thick with zebra mussel beds. There are indications that the mussels could devastate Lake Erie's population of walleye, one of the lake's most important commercial fish, by destroying breeding grounds.

The waterflea, *Bythotrephes cedestroemi,* is a small shrimplike crustacean found throughout Great Lakes waters. Like the zebra mussel, the waterflea is probably a by-product of ballast discharge from European freighters. The effect it will have on the Great Lakes ecosystem is still not known. On the negative side, the waterfleas are eating similar native species, which threatens to disrupt the food webs of the Great Lakes. On the positive side, yellow perch and walleye, popular commercial and sport fish, have found the waterflea a welcome addition to their diet.

A third foreigner that is making inroads in the lakes is the river ruffe, *Gynnocephilus cernuum.* This 8-to-12-centimeter-long Baltic relative of the yellow perch was first discovered in 1987. In Europe, the ruffe has been known to rapidly dominate yellow perch populations in some locations. In some ruffe-infested Soviet lakes, whitefish production has been reduced by

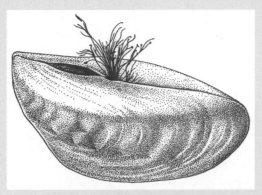

Zebra mussel

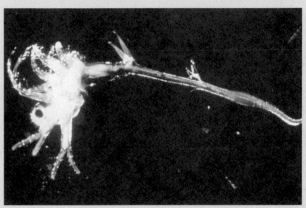

River ruffe

Spiny waterflea

50 percent. Such a situation in the Great Lakes would become a severe economic problem. Transport of small ruffe throughout the Great Lakes is being monitored. On the positive side, the ruffe does not seem to possess the explosive population characteristics of the waterflea or zebra mussel, nor does it readily swim to other areas.

Research pertaining to the Great Lakes is undertaken by various institutions, including the U.S. Fish and Wildlife Service, the National Oceanic and Atmospheric Administration, and the Great Lakes Environmental Research Laboratory. Several comparable institutions exist in the provinces of Ontario and Quebec and in the Canadian federal government.

Only a few of the Great Lakes research agencies, organizations, and institutions have been mentioned; there are many more. Literally hundreds of institutions affect the future of the Great Lakes. One thing is becoming clear—Great Lakes decision makers are increasingly aware of the importance of protecting and managing the lakes. Research institutions will continue to develop as the region's investigators. Federal governments will provide the basic legal framework for water quality. Private citizens and organizations will continue to influence government activity by bringing attention to specific issues and actions. Finally, the states, acting both individually and through their regional institutions, will provide the leadership and direct action on Great Lakes issues.

Summary

The future of the Great Lakes is not easy to predict. The problems are complicated, as are the potential solutions. Toxic wastes, competing water-use demands, agricultural runoff pollution, fluctuating water levels that cause economic damage, and diversion of water from the Great Lakes to the southwestern United States are issues that are only just beginning to surface. If and when solutions to such problems are developed, they will be the result of hard work by a multitude of professionals and organizations.

Policymakers respond to what they think the public wants. The ultimate question, however, is which "public" will be heard. As we have seen, the Great Lakes serve many publics, and each has a right to use the resource. Industry, agriculture, transportation, commerce, tourism, and recreation all want something different from the Great Lakes, and all place their own demands on the resource. Solutions to the problems faced by the Great Lakes by their very nature must be interrelated and present an ecosystem approach. The Great Lakes, however, are not unique in terms of environmental and societal questions. The issues faced are representative of similar problems throughout the world.

In the long run, society must determine what it really wants. Because society cannot control all pollutant sources completely, choices will have to be made to decide the degree of control that is most desirable. We will have to strike a balance between environmental protection and our requirements for resources, energy, and the luxuries and necessities that our present life-style demands. This is essentially a moral and ethical question that is related to, yet goes far beyond, the issue of the Great Lakes.

Review Questions

1. Why do you think finding a solution to an environmental conflict is so complex? Explain your answer.
2. Describe what is meant by an ecosystem approach to environmental problem solving.
3. What role did the mayfly play in the ecosystem of Lake Erie?
4. Define comprehensive water management and planning as it applies to the Great Lakes.
5. Describe the relationship between land use in the Great Lakes watershed and water quality problems in the lakes.
6. Why do you think so many organizations are involved in managing the Great Lakes?
7. What do you think the future of the Great Lakes is? Why?

CHAPTER TWO
Environmental Ethics

Objectives

After reading this chapter, you should be able to:

Differentiate between ethics and morals.

Define personal ethics.

Explain the connection between material wealth and resource exploitation.

Describe how industry must exploit resources and consume energy to produce goods.

Explain how corporate behavior is determined.

Describe the tremendous power that corporations wield because of their size.

Explain why governmental action was necessary to force all companies to meet environmental standards.

Describe what has been the general attitude of society toward the environment.

Explain the relationship between economic growth and environmental degradation.

List three conflicting attitudes toward nature.

Chapter Outline

Key Terms

conservation ethic
corporation
development ethic
economic growth
ethics

morals
preservation ethic
profitability
resource exploitation

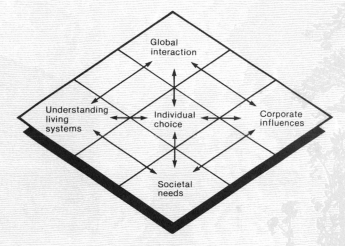

FIGURE 2.1

The Earth as Seen from Space.
Political, geographical, and nationalistic differences among humans do not seem as important from this perspective. In reality, we all share the same "home."

Views of Nature

The most beautiful object I have ever seen in a photograph in all my life, is the planet Earth seen from the distance of the moon, hanging in space, obviously alive. Although it seems at first glance to be made up of innumerable separate species of living things, on closer examination every one of its things, working parts, including us, is interdependently connected to all the other working parts. It is, to put it one way, the only truly closed ecosystem any of us know about.

Lewis Thomas

One of the marvels of technology of the past few decades is that we can see the Earth from the perspective of space, a blue sphere unique among all of the planets in our solar system. (See figure 2.1.) Looking at ourselves from space, it becomes obvious, says William Clark of Harvard University, that only as a global species, ". . . pooling our knowledge, coordinating our actions, and sharing what the planet has to offer—do we have any prospect for managing the planet's transformation along pathways of sustainable development."

BOX 2.1
What Is Ethical?

Ethics is one branch of philosophy. Ethics seeks to define fundamentally what is right and what is wrong, regardless of cultural differences. For example, most cultures have a reverence for life and feel that all individuals have a right to live. It is considered unethical to deprive an individual of life.

Morals differ somewhat from ethics because morals reflect the predominant feelings of a culture about ethical issues. For example, in almost all cultures it is certainly unethical to kill someone. However, when a country declares war, most of its people accept the necessity for killing the enemy. Therefore, it is a moral thing to do even though ethics says that killing is wrong. No nation has ever declared an immoral war.

Environmental issues also involve a consideration of ethics and morals. For example, because there is currently enough food in the world to feed everyone adequately, it is unethical to allow some people to starve while others have more than enough food. However, the predominant mood of the developed world is one of indifference. They don't feel morally bound to share what they have with others. In reality, this indifference says that it is permissible to allow people to starve. This moral stand is not consistent with a purely ethical one.

As we can see, ethics and morals are not always the same; thus, it is often difficult to clearly define what is right and what is wrong. Some individuals view the world's energy situation as serious and have reduced their own consumption. Others do not believe there is a problem, and, therefore, have not modified their energy use. Still others do not care what the situation is. They will use energy as long as it is available. Other similar issues are the population problem and pollution. Is it ethical to have more than two children when the world faces overpopulation? Should an industry try to persuade the public to vote "no" on a particular issue because it might reduce their profits, even though its passage would improve the environment? The stand taken on such issues often revolves around the position of the person making the statement. An industrial leader, for example, would probably not look upon pollution as negatively as someone who is active outdoors. In fact, many business leaders view the behavior of active preservationists as somewhat immoral because it restricts growth and, in some cases, causes unemployment.

Most ethical questions are very complex. Ethical issues dealing with the environment are no different. Therefore, it is important to explore environmental issues from several points of view before taking a stand. One point to consider is the difference between short-term and long-term effects of a course of action.

When we take an ethical stand, we become open to attack from those who disagree with our stand. Often, individuals are portrayed as villains for pursuing a course of action they consider righteous.

Many people see little value in an undeveloped river and feel it is unreasonable to leave a river flowing in a natural state. Many rivers throughout the world have been "controlled" to provide power, irrigation, and navigation for the people at the expense of the natural world. To not use these resources would be wasteful in the minds of many.

In the Pacific Northwest of the United States, there is a conflict over the value of old-growth forests. Many economic interests want to use the forests for timber production and feel that to not use these resources would bring economic hardship to many. The argument is that the trees are going to die anyway and they might as well be used for the betterment of the community. Others feel that all the living things that make up the forest have a value we do not yet appreciate. Removing the trees would destroy something that took hundreds of years to develop and may never be replaced.

BOX 2.2
"The Land Is Sacred to Us": Chief Seattle's Lament

Chief Sealth of the Duwamish League, known to us as Chief Seattle, delivered this speech in 1854—one year before a great treaty-making council between Indian tribes and the U.S. government. The government proposed that reservations be established, and although several tribes opposed this, treaties were signed: each of the tribes was to select its favorite home valley as its reservation. Three months later, war broke out.

The Great Chief in Washington sends word that he wishes to buy our land. The Great Chief also sends us words of friendship and goodwill. This is kind of him, since we know he has little need for our friendship in return. But we will consider your offer. For we know that if we do not sell, the white man may come with guns and take our land.

How can you buy or sell the sky, the warmth of the land? The idea is strange to us.

If we do not own the freshness of the air and the sparkle of the water, how can you buy them?

Every part of this earth is sacred to my people. Every shining pine needle, every sandy shore, every mist in the dark woods, every clearing, and humming insect is holy in the memory and experience of my people. The sap which courses through the trees carries the memories of the red man.

The white man's dead forget the country of their birth when they go to walk among the stars. Our dead never forget this beautiful earth, for it is the mother of the red man. We are part of the earth and it is part of us. The perfumed flowers are our sisters; the deer, the horse, the great eagle, these are our brothers. The rocky crests, the juices in the meadows, the body heat of the pony, and man—all belong to the same family.

So, when the Great Chief in Washington sends word that he wishes to buy our land, he asks much of us. The Great Chief sends word he will reserve us a place so that we can live comfortably to ourselves. He will be our father and we will be his children. So we will consider your offer to buy our land. But it will not be easy. For this land is sacred to us.

The red man has always retreated before the advancing white man, as the mist of the mountain runs before the morning sun. But the ashes of our fathers are sacred. Their graves are holy ground, and so these hills, these trees, this portion of the earth is consecrated to us. We know that the white man does not understand our ways. One portion of land is the same to him as the next, for he is a stranger who comes in the night and takes from the land whatever he needs. The earth is not his brother, but his enemy, and when he has conquered it, he moves on. He leaves his fathers' graves behind, and he does not care. He kidnaps the earth from his children. He does not care. His fathers' graves and his children's birthright are forgotten. He treats his mother, the earth, and his brother, the sky, as things to be bought, plundered, sold like sheep or bright beads. His appetite will devour the earth and leave behind only a desert.

I do not know. Our ways are different from your ways. The sight of your cities pains the eyes of the red man. But perhaps it is because the red man is a savage and does not understand.

There is no quiet place in the white man's cities. No place to hear the unfurling of leaves in spring or the rustle of insect's wings. But perhaps it is because I am a savage and do not understand. The clatter only seems to insult the ears. And what is there to life if a man cannot hear the lonely cry of the whippoorwill or the arguments of the frogs around a pond at night? I am a red man and do not understand. The Indian prefers the soft sound of the wind darting over the face of a pond and the smell of the wind itself, cleansed by a midday rain or scented with the pinon pine.

The air is precious to the red man, for all things share the same breath—the beast, the tree, the man, they all share the same breath. The white man does not seem to notice the air he breathes. Like a man dying for many days, he is numb to the stench. But if we sell you our land, you must remember that the air is precious to us, that the air shares its spirit with all the life it supports. The wind that gave our grandfather his first breath also receives his last sigh. And the wind must also give our children the spirit of life. And if we sell you our land, you must keep it apart and sacred, as a place where even the white man can go to taste the wind that is sweetened by the meadow's flowers.

So we will consider your offer to buy our land. If we decide to accept, I will make one condition: The white man must treat the beasts of this land as his brothers.

I am a savage and I do not understand any other way. I have seen a thousand rotting buffalos on the prairie, left by the white man who shot them from a passing train. I am a savage and I do not understand how the smoking iron horse

---

Although efforts to manage the interactions between people and their environment are as old as human civilization, the management problem has been transformed today by the unprecedented increases in the rate, scale, and complexity of those interactions. At one time, pollution was a local, temporary event, but today, pollution problems may involve several countries, as with the concern over acid deposition in Europe and in North America, and they affect multiple generations. The debates over chemical and radioactive waste disposal are examples of the increasingly international nature of pollution. For example, many European countries are con-

can be more important than the buffalo that we kill only to stay alive.

What is man without the beasts? If all the beasts were gone, men would die from a great loneliness of spirit. For whatever happens to the beasts, soon happens to man. All things are connected.

So we will consider your offer to buy our land. If we agree, it will be to secure the reservation you have promised. There, perhaps, we may live out our brief days as we wish. When the last red man has vanished from this earth, and his memory is only the shadow of a cloud moving across the prairie, these shores and forests will still hold the spirits of my people. For they love this earth as the newborn loves its mother's heartbeat. So if we sell you our land, love it as we've loved it. Care for it as we've cared for it. Hold in your mind the memory of the land as it is when you take it. And with all your strength, with all your mind, with all your heart, preserve it for your children, and love it . . . as God loves us all.

One thing we know, Our God is the same God. This earth is precious to Him. Even the white man cannot be exempt from the common destiny. We may be brothers after all. We shall see.

This we know: The earth does not belong to man; man belongs to the earth. This we know: All things are connected like the blood which unites one family. All things are connected.

Whatever befalls the earth befalls the sons of the earth. Man did not weave the web of life; he is merely a strand in it. Whatever he does to the web, he does to himself.

But we will consider your offer to go to the reservation you have for my people. We will live apart, and in peace. It matters little where we spend the rest of our days. Our children have seen their fathers humbled in defeat. Our warriors have felt shame, and after defeat they turn their days in idleness and contaminate their bodies with sweet foods and strong drink. It matters little where we pass the rest of our days. Tribes are made of men, nothing more. Men come and go like waves of the sea.

Even the white man, whose God walks and talks with him as friend to friend, cannot be exempt from the common destiny. We may be brothers after all; we shall see. One thing we know, which the white man may one day discover—our God is the same God. You may think now that you own Him as you wish to own our land; but you cannot. He is the God of man, and His compassion is equal for the red man and the white. This earth is precious to Him, and to harm the earth is to heap contempt on its Creator. The whites too shall pass; perhaps sooner than all other tribes. Continue to contaminate your bed, and you will one night suffocate in your own waste.

But in your perishing you will shine brightly, fired by the strength of the God who brought you to this land and for some special purpose gave you dominion over this land and over the red man. Your destiny is a mystery to us, for we do not understand when the buffalo are all slaughtered, the wild horses are tamed, the secret corners of the forest heavy with the scent of many men, and the view of the ripe hills blotted by talking wires. Where is the thicket? Gone. Where is the eagle? Gone. And what is it to say goodbye to the swift pony and the hunt? The end of living and the beginning of survival.

Source: From "The Land Is Sacred to Us: Chief Seattle's Lament."

cerned about the transportation of radioactive and toxic wastes across their borders. What were once straightforward confrontations between ecological preservation and economic growth now involve multiple linkages that blur the distinction between right and wrong. For example, the greenhouse effect is thought to result from energy consumption, agricultural practices, and climatic change.

Many believe that we have entered an era characterized by global change that stems from the interdependence between human development and the environment. It is increasingly argued that self-conscious,

BOX 2.3
Naturalist Philosophers

The philosophy behind the reemerging environmental movement actually had its roots in the last century. Although there have been many notable conservationist philosophers, several stand out. Some of these landmark thinkers are Ralph Waldo Emerson, Henry David Thoreau, John Muir, Aldo Leopold, and Rachel Carson.

In Emerson's first essay, *Nature,* published in 1836, he claimed that "behind nature, throughout nature, spirit is present." Emerson was an early critic of rampant economic development, and he sought to correct what he considered to be the social and spiritual errors of his time. In his *Journals,* published in 1840, Emerson stated that "a question which well deserves examination now is the Dangers of Commerce. This invasion of Nature by Trade with its Money, its Credit, its Steam, its Railroads, threatens to upset the balance of Man and Nature."

Another naturalist who held beliefs similar to Emerson's was Henry David Thoreau. Thoreau's bias fell on the side of "Truth in nature and wilderness over the deceits of urban civilization." The countryside around Concord, Massachusetts, fascinated and exhilarated him as much as the commercialism of the city depressed him. It was near Concord that Thoreau wrote his classic *Walden Pond,* which describes a year in which he lived in the country to have direct contact with nature's "essential facts of Life." In his later writings and journals (1861), Thoreau summarized his feelings toward nature with prophetic vision:

> But most men, it seems to me, do not care for Nature and would sell their share in all her beauty, as long as they may live, for a stated sum—many for a glass of rum. Thank God, men cannot as yet fly, and lay waste the sky as well as the earth! We are safe on that side for the present. It is for the very reason that some do not care for these things that we need to continue to protect all from the vandalism of a few.

John Muir combined the intellectual ponderings of a philosopher with the hard-core, pragmatic characteristics of a leader. Muir believed that

"wilderness mirrors divinity, nourishes humanity, and vivifies the spirit." Muir tried to convince people to leave the cities for a while to enjoy the wilderness. However, Muir felt that the wilderness was threatened. In the 1876 article entitled "God's First Temples: How Shall We Preserve Our Forests," published in the Sacramento *Record Union,* Muir argued that only government control could save California's finest sequoia groves from the "ravages of fools." In the early 1890s, Muir organized the Sierra Club to "explore, enjoy, and render accessible the mountain regions of the Pacific Coast" and to enlist the support of the government in preserving these areas. Muir's actions in the West convinced the federal government to restrict development in the Yosemite Valley, which preserved its beauty for generations to come.

Another thinker as well as a doer in the early conservation field was Aldo Leopold. As a philosopher, Leopold summed up his feelings in *A Sand County Almanac:*

> Wilderness is the raw material out of which man has hammered the artifact called civilization. No living man will see again the long grass prairie, where a sea of prairie flowers lapped at the stirrups of the pioneer. No living man will see again the virgin pineries of the Lake States, or the flatwoods of the coastal plain, or the giant hardwoods.

Leopold was the founder of the field of game management. In the 1920s, while serving in the Forest Service, he worked for the development of a wilderness policy and pioneered his concepts of game management. He wrote extensively in the *Bulletin* of the American Game Association and stated that the amount of space and the type of forage of a wildlife habitat determine the number of animals that can be supported in an area. Furthermore, he said that regulated hunting can maintain a proper balance of wildlife in a habitat.

While most people talk about the difficulties in changing the way things are, few actually have gone ahead and changed it. Rachel Carson ranks among those

intelligent management of the earth is one of the greatest challenges facing humanity as we approach the twenty-first century. Many argue that to meet this challenge a new environmental ethic must evolve.

Ethical issues dealing with the environment are no different from other kinds of problems. Depending on your perspective, the concept of an environmental ethic could encompass differing principles and beliefs. Perhaps ecologist/writer Aldo Leopold summed up an environmental ethic best in his essay "The Land Ethic" from *A Sand County Almanac and Sketches Here and There* (1949):

Ralph Waldo Emerson

Henry David Thoreau

John Muir

Aldo Leopold

Rachel Carson

few. A distinguished naturalist and best-selling nature writer, Rachel Carson published a series of articles in the *New Yorker* in 1960, which generated widespread discussion about pesticides. In 1962, she published *Silent Spring,* which dramatized the potential dangers of pesticides to food, wildlife, and humans.

Although some technical details of her book have been shown to be in error by later research, its basic thesis that pesticides can contaminate and cause widespread damage to the ecosystem has been established. Unfortunately, Carson's early death from cancer came before her book was recognized as one of the most important events in the history of environmental awareness and action in this century.

All ethics so far evolved rest upon a single premise: that the individual is a member of a community of interdependent parts. The land ethic simply enlarges the boundaries of the community to include soils, waters, plants, and animals, or collectively, the land . . . a land ethic changes the role of *Homo sapiens* from conqueror of the land-community to plain member and citizen of it. . . . It implies respect for his fellow-members, and also respect for the community as such.

Leopold also wrote that ". . . a thing is right when it tends to preserve the integrity, stability, and beauty of the biotic community. It is wrong when it tends otherwise. . . . We abuse land because we regard it as a commodity belonging to us. When we see land as a community to which we belong, we may begin to use it with love and respect."

Environmental ethicists argue that to consider environmental protection as a "right" of the planet is a natural extension of concepts of human rights. Many would also argue that an environmental ethic considers one's actions toward the environment as a matter of right and wrong, rather than one of self-interest.

Environmental Attitudes

Although there are many different attitudes about the environment, most of these attitudes would fall under one of three headings. These three positions can be labeled the development ethic, the preservation ethic, and the conservation ethic. Each of these ethical positions has its own appropriate code of conduct against which ecological morality may be measured. (See figure 2.2.)

The **development ethic** is based on action. It assumes that the human race is and should be the master of nature and that the earth and its resources exist for our benefit and pleasure. This view is reinforced by the work ethic, which dictates that humans should be busy creating continual change and that things that are bigger, better, and faster represent "progress," which itself is good. This philosophy is strengthened by the idea that, "if it can be done, it should be done," or that our actions and energies are best harnessed in creative work.

Examples of the development ethic abound. The notion that bigger is better is certainly not new to us, nor is the belief that if something can be done or built it should be. The dream of upward mobility is embodied in this ethic. In some circles, questioning growth is considered almost unpatriotic. Only in the last fifty to one hundred years have the by-products and waste associated with development been analyzed.

The **preservation ethic** considers nature special in itself. Different preservationists have different reasons for wanting to preserve nature. Some preservationists have an almost religious belief regarding nature. These individuals hold a reverence for life and respect the right of all creatures to live, no matter what the social and economic costs. Preservationists also include those whose interest in nature is primarily aesthetic or recreational. They believe that nature is beautiful and refreshing and should be available for picnics, hiking, camping, fishing, or just peace and quiet.

In addition to the semireligious and recreational preservationists, there are also preservationists whose reasons are essentially scientific. These individuals would argue that the human species depends on and has much to learn from nature. Rare and endangered species and ecosystems, as well as the more common ones, must be preserved because of their known or assumed long-range, practical utility. In this view, natural diversity, variety, complexity, and wildness is thought to be superior to humanized uniformity, simplicity, and domesticity.

The third environmental ethic is referred to as the **conservation ethic.** The conservation ethic is related to the scientific preservationist view, but extends the rational consideration to the entire earth and for all time. It recognizes the desirability of decent living standards, but it works toward a balance of resource use and resource availability. The conservation ethic stresses a balance between total development and absolute preservation. It further stresses that rapid and uncontrolled growth in population and economics is self-defeating in the long run. The goal of the conservation ethic is one people living together in one world, indefinitely.

Interrelatedness

FIGURE 2.2
The Views of Nature. Each of these individuals envisions the same resources used differently.

Societal Environmental Ethics

Society is composed of a great variety of people with differing viewpoints. However, this variety can be distilled into a set of ideas that reflect the prevailing attitudes of society. The collective attitudes of society can also be analyzed from an ethical point of view. Western, developed societies have long acted as if the earth has unlimited reserves of natural resources, an unlimited ability to assimilate wastes, and a limitless ability to accommodate unchecked growth.

The economic direction and rationale of developed nations have been one of continual growth. Unfortunately, this growth has not always been carefully planned or even desired. This "growthmania" has resulted in the use of our nonrenewable resources for comfortable homes, well-equipped hospitals, convenient transportation, as well as fast-food outlets, VCRs, home computers, and battery-operated toys. In economic statistics, such "growth" measures out as "productivity," and all looks rosy. The question then arises, "What is enough?" Many poor societies have too little, but a rich society never says, "Halt! We have enough." As the Indian philosopher and statesman Mahatma Gandhi said, "The earth provides enough to satisfy every person's need, but not every person's greed."

Growth, expansion, and domination remain the central sociocultural objectives of most advanced societies. **Economic growth** and **resource exploitation** are attitudes shared by developing societies. As a society, we

BOX 2.4
Pigeon River

In 1968, oil was discovered in a remote area of northern Michigan known as the Pigeon River Forest. The Pigeon River Forest is unique in many ways. Its 37,600 hectares is regarded as the wildest country in the state's Lower Peninsula. Because of its wilderness qualities, grouse, black bear, bobcats, deer, beaver, and many other animals inhabit the region. The streams in the area provide excellent conditions for healthy populations of native brook trout. In addition to hunting and fishing, other major forms of recreation in the forest include cross-country skiing, camping, and hiking. The forest is also the home of the largest North American elk herd east of the Mississippi.

Five successful oil wells were drilled in the area in the early 1970s. However, a series of lawsuits filed by environmental groups halted further exploration until 1981. A compromise was eventually reached between the oil companies, the environmental groups, and the state to restrict and closely monitor future oil drilling. The agreement also limited drilling to the southern 11,700 hectares of the forest, protecting what then was the elk herd's primary range. The argument in 1980 was that the oil from the Pigeon River area was needed to combat the energy crisis and that the state of Michigan would get several hundred million dollars in royalties from it, which could then be used to buy other recreational lands. By the early 1980s, it was clear that most of the glowing promises that resulted in the compromise to allow drilling had not been kept.

Much less oil had been found beneath the forest than was originally predicted, so royalties were a fraction of what had been estimated. More important than the loss of income was that the Pigeon River Forest had changed.

Even with all of the safeguards in the compromise, pollution problems did arise. Groundwater became contaminated with chloride from the drilling when pits meant to store brine began to leak. Part of the forest's most prized commodity—solitude—had been lost, despite the best efforts of state officials, citizen watchdogs, and the oil companies to minimize the disruption.

Even an untrained observer notices how a winding two-track road changes into a straight, flat, wide gravel-bed highway when it reaches the southernmost facility where gas and oil from wells in the forest are pumped and stored. At the forest's northern pumping site, a heavy odor of natural gas fumes is evident.

The Pigeon River was not true wilderness, but it was the closest thing to it, the largest single roadless area in the Lower Peninsula of Michigan, and the legacy of a half-century of careful management aimed at restoring what the early loggers had ruined. Trade-offs are a way of life in the world today. The case of the Pigeon River Forest was also a trade-off—solitude versus oil. A barrel of oil has a straightforward economic value. What value do we place on a day of solitude in nature?

a.

b.

(*a*) In economic terms, it is easy to set a value on a barrel of oil. (*b*) There is no good way to place a value on some things, such as scenery or solitude. We recognize that they have value; however, we often recognize this value only after it is lost.

continue to consume natural resources as if the supplies were never-ending. All of this is reflected in our increasingly unstable relationship with the environment, which grows out of our tendency to take from the "common good" without regard for the future.

This attitude is deeply embedded in the fabric of our society. Since the first settlers arrived in North America, nature has been considered an enemy. Frequently, the pioneers expressed their relation to the wilderness in military terms. They viewed nature as an enemy to be "conquered," "subdued," or "vanquished" by a pioneer "army." Any qualms the pioneers may have felt about invading and exploiting the wilderness were justified by religious beliefs. This attitude toward nature is still extremely popular today. Many view wilderness solely as underdeveloped land and see value in land only if it is farmed, built upon, or in some way developed. The notion that land and wilderness should be preserved is incomprehensible to some. The thought of purposely opting to not develop a resource is considered almost a sin by many.

Corporate Environmental Ethics

Many facets of industry, such as procuring raw materials, manufacturing and marketing, and waste disposal, are in large part responsible for pollution. This is not because any industry or company has adopted pollution as a corporate policy. Industry is naturally dirty because all industries consume energy and resources. In addition, when raw materials are processed, some waste is inevitable. It is usually not possible to completely control the dispersal of all by-products of a manufacturing process. Also, some of the waste material may simply be useless.

For example, the food-service industry uses energy in the preparation of meals. Much of this energy is lost as waste heat. In addition, smoke and odors are produced and released into the atmosphere. Finally, bone, fat, and discolored food items become wastes (useless materials) that must be discarded.

The cost of controlling waste can be very important in determining a company's profit margin. **Corporations** are legal entities designed to operate at a profit, which is not in itself harmful. The corporation has no ethics, but the people who make up the corporation are faced with ethical decisions. Ethics are involved, however, when a corporation cuts corners in production quality or waste disposal to maximize profit. The cheaper it is to produce an item, the greater the possible profit. It is cheaper to dump wastes into a river than to install a wastewater treatment facility, and it is cheaper to release wastes into the air than it is to trap them in filters. Many would consider such pollution unethical and immoral, but to the corporation, it is just one of the factors that determines **profitability.** (See figure 2.3.) Because stockholders expect an immediate return on their investment, corporations often make decisions based on short-term profitability rather than long-term benefit to society.

The amount of profit a corporation realizes determines the amount of expansion possible. To expand continually, a corporation creates an increased demand for its products through advertising. The more it expands, the more power it attains. The more power it has, the greater its influence over decision makers who can create conditions favorable to its expansion plans. Therefore, it becomes a seemingly never-ending spiral.

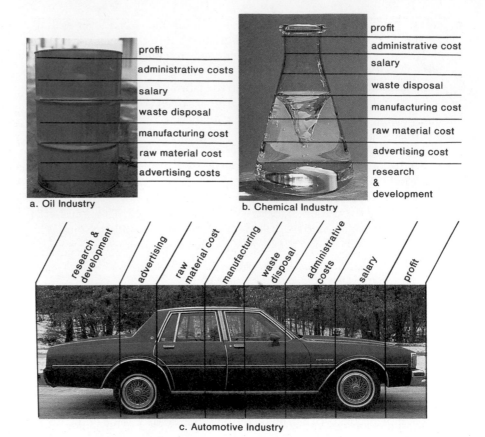

a. Oil Industry

b. Chemical Industry

c. Automotive Industry

FIGURE 2.3
Corporate Decision Making.
Corporations must make a profit. When they look at pollution control, they view its cost like any other cost: any reductions in cost increase profits.

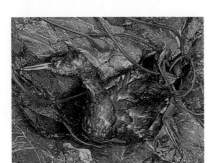

a.

b.

FIGURE 2.4
Valdez Principles. The 1989 oil spill in Alaska led to the development of the Valdez Principles. (*a*) The duck was a victim of the spill. (*b*) The *Exxon Valdez*.

Nations of the world must confront the problem of corporate irresponsibility toward the environment. In business, incorporation allows for the organization and concentration of wealth and power far surpassing the capacity of individuals or partnerships. Some of the most important decisions affecting our environment are made by individuals who wield massive corporate power rather than by governments or the public. Often, these corporate decisions involve only minimal concessions to the public interest, while every effort is exerted to maximize profits.

Business decisions and technological developments have increased the intensity of exploitation of natural resources. In addition, many political and legal institutions have generally supported the development of private enterprise. They have also promoted and defended private property rights rather than social and environmental concerns. Businesses and individuals typically use loopholes, political pressure, and the time-consuming nature of legal action to circumvent or delay compliance with established criteria for social or environmental improvement.

Is industry becoming more environmentally concerned? Corporations have certainly made increasing reference to worldwide environmental issues over the past several years, especially leading up to Earth Day 1990. Is such concern only rhetoric, or is it the beginning of a new corporate ethic? The "Valdez Principles," a tool that industry and the public can use to evaluate corporate environmental responsibility, were developed as a result of the 1989 oil spill in Alaska. (See figure 2.4.)

The Valdez Principles are a set of codes that businesses may voluntarily adopt. The codes provide environmental standards against which all companies can be assessed and compared. The ten principles encompass a wide range of goals that include minimizing pollutants, making sustainable use

Chico Mendes, whose real name was Francisco Alves Mendes Filho, was a Brazilian rubber tapper active in the rubber tappers' union. He and many other peasants made a living by extracting the latex from the rubber tree and selling the balls of raw latex. Rubber tappers also collected and sold other natural products of the forest, such as Brazil nuts, fruits, and native medicines. He was interested in preserving the portion of the Brazilian rain forest that provided their livelihood and supported a concept of "extractive reserves."

Extractive reserves involve setting aside land for the use of rubber tappers, who would continue their traditional life-styles and use the rain forest in its natural state for generations to come. This put Mendes in conflict with powerful people interested in clearing the rainforest so that cattle could be raised in the area. Most cattle-ranching operations show short-term economic gains but ultimately become uneconomical.

On 22 December 1988, Chico Mendes was shot by members of a vigilante group who supported local ranchers as Mendes walked from his home. In 1990, a rancher, Darli Alves da Silva, and his son were convicted of the murder.

Before his death, Mendes had said, "I want to live to defend the Amazon." It appears that his life and death will make a difference in the way the Brazilian rain forest is being used. Due to the circumstances that surrounded his death and his role as a leader in the rubber tappers' union, his death received international notice and caused many people to begin to ask if the natural rain forest perhaps has as much to offer as ranches.

During 1990, the Brazilian government established four refuges. The first, named the Chico Mendes Reserve in the Juruá River valley, covers about 6 percent of the state of Acre in northwest Brazil.

of renewable resources, reducing health and safety risks to employees and communities, and representing environmental interests on corporate boards. The Valdez Principles are looked upon as a guide for corporate environmentalism. The goal, some argue, should be to make compliance with the Valdez Principles a prerequisite for doing business.

Practicing an environmental ethic should not interfere with corporate and other social responsibilities or obligations. It must be integrated into overall systems of belief and coordinated with economic systems. Environmental advocates, in turn, need to consider others' objectives just as they demand of others the consideration of the environmental consequences in decision making. It makes little sense to preserve the environment if that objective produces national economic collapse. Nor does it make sense to maintain stable industrial productivity at the cost of depriving a country of breathable air, drinkable water, wildlife species, parks, and wilderness. But for business to maintain its profitability, influence, and freedom, it must be sensitive to its impact on current and future citizens, not just in terms of the price and quality of the goods it produces but also in terms of public approval of its social and political influence. A 1989 Harris Poll, for example, found that 68 percent of Americans wanted increased government spending to control acid rain and toxic waste dumping, even if they had to pay higher taxes. In another 1989 poll, by the Michael Peters Group, eight of ten Americans said they would be willing to pay extra for a product packaged with recyclable materials.

Individual Environmental Ethics

The environmental movement has been effective in influencing public opinion and in moving the business community toward an environmental ethic. The result of this changing view of business's responsibilities will complicate business decision making in the remainder of the twentieth

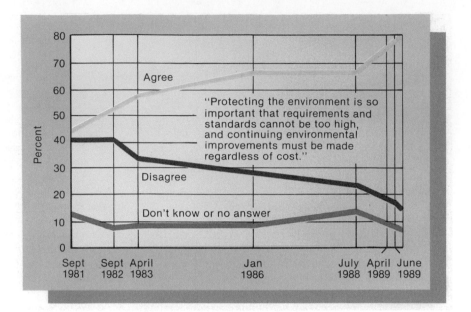

FIGURE 2.5

Environmental Values. Environmental values have drawn increasing support in the United States. In *New York Times/*CBS news polls taken since 1981, respondents were asked to react to this statement: "Protecting the environment is so important that requirements and standards cannot be too high and continuing environmental improvements must be made regardless of cost." The two latest polls were taken after the *Exxon Valdez* spill.

century. More complex demands by the public and a broadening of horizons on the part of business will be the dominant theme of corporate life during the next few years. As human populations and economic activity continue to grow, we are facing a number of environmental problems that threaten not only human health and the productivity of ecosystems, but in some cases the very habitability of the globe. (See figure 2.5.)

If we are to respond to those problems successfully, then our environmental ethic must express itself in broader and more fundamental ways. We have to recognize that each of us is individually responsible for the quality of the environment we live in and that our personal actions affect environmental quality, for better or worse. The recognition of individual responsibility must then lead to real changes in individual behavior. In other words, our environmental ethic must begin to express itself not only in national laws, but also in subtle but profound changes in the ways we all live our daily lives. (See figure 2.6.)

A Roper Poll in 1990 indicated that Americans think environmental problems can often be given a quick technological fix. Says the Roper organization, "They believe that cars, not drivers, pollute, so business should invent pollution-free autos. Coal utilities, not electricity consumers, pollute, so less environmentally dangerous generation methods should be found." It appears that many individuals want the environment cleaned up, but they do not want to make major life-style changes to make that happen.

Decisions and actions by individuals faced with ethical choices collectively determine the hopes and quality of life for everyone. As ecological knowledge and awareness begin to catch up with good intentions, people in all walks of life will need to live by an environmental ethic.

Global Environmental Ethics

In 1990, Noel Brown, the Director of the United Nations Environmental Program, stated:

> Suddenly and rather uniquely the world appears to be saying the same thing. We are approaching what I have termed a consensual moment in history,

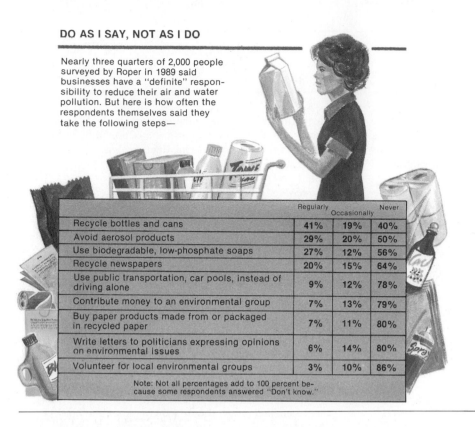

DO AS I SAY, NOT AS I DO

Nearly three quarters of 2,000 people surveyed by Roper in 1989 said businesses have a "definite" responsibility to reduce their air and water pollution. But here is how often the respondents themselves said they take the following steps—

	Regularly	Occasionally	Never
Recycle bottles and cans	41%	19%	40%
Avoid aerosol products	29%	20%	50%
Use biodegradable, low-phosphate soaps	27%	12%	56%
Recycle newspapers	20%	15%	64%
Use public transportation, car pools, instead of driving alone	9%	12%	78%
Contribute money to an environmental group	7%	13%	79%
Buy paper products made from or packaged in recycled paper	7%	11%	80%
Write letters to politicians expressing opinions on environmental issues	6%	14%	80%
Volunteer for local environmental groups	3%	10%	86%

Note: Not all percentages add to 100 percent because some respondents answered "Don't know."

FIGURE 2.6
Do As I Say, Not As I Do.
Source: Data from The Roper Corporation, Inc.

where suddenly from most quarters we get a sense that the world community is now agreeing that the environment has to become a matter of global priority and action.

This new sense of urgency and common cause about the environment is leading to unprecedented cooperation in some areas. Despite their political differences, the Arab, Israeli, Soviet, and American environmental professionals have been working together for several years. Ecological degradation in any nation almost inevitably impinges on the quality of life in others. For years, acid rain has been a major irritant in relations between the United States and Canada. Drought in Africa and deforestation in Haiti have resulted in waves of refugees. From the Nile to the Rio Grande, conflicts flare over water rights. The growing megacities of the Third World are time bombs of civil unrest.

Much of the current environmental crisis is rooted in and exacerbated by the widening gap between rich and poor nations. Industrialized countries contain only 23 percent of the world's population, yet they control 80 percent of the world's goods and are also responsible for a majority of its pollution. On the other hand, the developing countries are hardest hit by overpopulation, malnutrition, and disease. As these nations struggle to catch up with the developed world, a vicious circle begins: Their efforts at rapid industrialization poison their cities, while their attempts to boost agricultural production often result in the destruction of their forests and the depletion of their soils.

Perhaps one of the most important questions of the future is, "Will the nations of the world be able to put aside their political differences to work toward a global environmental course of action?" The United Nations Conference on Human Environment held in Stockholm, Sweden, in 1972 was a step in the right direction. Out of that international conference was born

BOX 2.6
Social Versus Deep Ecology

The debate about why and how we should protect the environment is not new. It goes back to the beginnings of the conservation movement in the United States during the nineteenth century, when the utilitarian conservationists (wise use) led by Grifford Pinchot fought with the altruistic preservationists (strict preservation) led by John Muir. The conservationists were progressives who sought to work within the existing system to reform it. Their approach to resource management was scientific, professional, activist, and technological. Husbanding and stewardship of resources were key concepts in this view. As humanists and populists, they considered the goals of conservation to be preventing waste and saving resources for prudent, constructive, and efficient human use. Their motto was, "For the greatest good, for the greatest number, for the longest time." Pinchot said, "There may be just as much waste in neglecting the development and use of certain natural resources as there is in their destruction." In contrast, the preservationists were antimodernists who distrusted technology and civilization. They were visionary, zealous, and uncompromising in their defense of nature; egalitarian in regarding humans as just ordinary members of the biological community; and appreciative of nature as a source of beauty, knowledge, and power.

Today, the conservationist and preservationist themes have reemerged as the social ecology and deep ecology movements, respectively. The social ecologists advocate a political solution modeled on democratic socialism or recent Green Party platforms. They believe that only humankind can solve the problems we have created and see us as custodians and stewards of nature. The attack, they say, must be upon the capitalist, the consumerist, or the growth-and-progress-oriented system that began with the Industrial Revolution and is now threatening the planet's survival. According to the social ecologists, the problem is human ignorance and greed and can be solved only by reform of human institutions. Right-thinking people, they say, should redistribute wealth and food resources.

Deep ecologists, on the other hand, see humankind itself as the main problem. They believe that the earth is a complex organism with its own needs, metabolism, and immune system and that humankind's relationship with the earth is increasingly parasitic. In the book *Deep Ecology: Living Nature As If Nature Mattered,* proponents Bill Devall and George Sessions clearly state their principles: (1) Humans have no right to reduce the richness and diversity of life except to satisfy vital needs; (2) the quality of human life and culture is compatible with a substantial *decrease* in the human population; and (3) the flourishing of nonhuman life requires such a decrease. One deep ecology organization called Earth First publishes a manual called *Ecodefense: A Field Guide to Monkey-wrenching.* In the manual are instructions on how to spike trees so as to damage saws, "harvest" billboards, sabotage heavy machinery, and "disrupt" land development. Earth First states that since nobody asked "our" permission, to put billboards in "our" forests, "we" need not ask permission to burn them. Earth First argues that, since the Earth's exploiters often operate in closed rooms, under the cover of darkness, and at the margins of the law, why can't her defenders?

At the base of the controversy between these radical and conservative environmentalists are some deep differences in opinion about society and human nature. Which of these philosophies we ascribe to makes a great deal of difference in how we might attain a sustainable future. Anyone planning to join an environmental group should examine their feelings about which kind of group they would feel most comfortable with by considering the following questions:

Is it best to be progressive and reformist, or radical and revolutionary? Should we work to improve existing structures, or should we tear them down and start over?

Are you anthropocentric or biocentric? Do humans have a special importance in the world, or are we "plain members of a biotic community"?

Do you believe in conservation, preservation, or some combination of the views? Is the purpose of saving resources to preserve them for use by later generations or simply to save them for their own sake?

Do you favor professionalism (with representative democracy) or participatory democracy? Are efficiency and effectiveness the measures of success, or is direct participation by everyone more important?

Would you prefer integration of environmentalism into a broadly agreed-upon national agenda, or do you find yourself fundamentally opposed to the predominant view?

Can technology offer solutions to our problems, or will it inevitably make them worse?

Is environmentalism a science, religion, or social movement?

Is further growth possible? Even necessary? Impossible?

CONSIDER THIS CASE STUDY
Antarctica—Resource or Refuge?

Few places on earth have not been exploited by humans. One such place is Antarctica. While Antarctica is as close to an unpolluted environment as there is on earth, it is not without its problems.

Seals and whales brought the earliest exploitation to Antarctica. There was money to be made in Antarctica, and this "opportunity" resulted in the near extinction of the southern fur seal, the elephant seal, and the blue whale.

By the 1950s, aboveground nuclear testing had spread radioactive particles around the planet, including Antarctica. Pesticides like DDT were also turning up in the tissues and blood of certain Antarctic bird and marine mammal species. A growing hole in the ozone layer above the Antarctic continent is caused by the use of chlorofluorocarbons. Fossil-fuel combustion contributes to the greenhouse effect, which, in turn, threatens to melt the ice in Antarctica's Western Peninsula.

During the past several decades, Antarctica has witnessed extensive scientific exploration. Much of this exploration has been economically motivated rather than scientific. For example, there are government scientists who, with the aid of satellites, are advising oil and mineral prospectors. In addition, much of the so-called scientific research is conducted with geopolitical or military objectives in mind.

Antarctica is also being proposed as a tourist attraction. Australia has suggested building a hotel, while Argentina is considering chartering a vessel to transport six hundred tourists from South America seven times a year. Several sites are also being viewed as potential ski resorts.

From an ecological perspective, Antarctica is fragile. The thin layer on the surface of the ocean, nourished by the sun, supports the tiny shrimplike krill, which, in turn, sustain fish, whales, seals, and penguins. These short, simple food chains are extremely sensitive to environmental insults.

In the mid-1970s, New Zealand proposed making the continent into an Antarctic World Park. This would turn Antarctica into an international wilderness area, a region on earth where we recognize that humanity does not belong.

Should we turn a continent into a world park?
Would humanity be better served by developing the natural resources of Antarctica, such as oil and minerals?
Should the natural beauty of Antarctica be opened up to tourism so it can be enjoyed by many?
Is it possible to strike a balance between preservation and development in a fragile ecosystem? Can you give examples?

---

the U.N. Environment Programme, a separate department of the United Nations that deals with environmental issues. Through organizations such as this, nations can work together to solve common environmental problems.

At the individual level, people have begun to respond to increased awareness of global environmental change by altering their values, beliefs, and actions. Changes in individual behavior are necessary but are not enough. As a global species, we are changing the planet. By pooling our knowledge, coordinating our actions, and sharing what the planet has to offer, we may achieve a global environmental ethic.

Summary

People of different cultures view their place in the world from different perspectives. Among the things that shape their views are religious understandings, economic pressures, and fundamental knowledge of nature. Because of this diversity of background, different cultures put different values on the natural world and the individual organisms that comprise it.

Three prevailing attitudes toward nature are the development ethic, which assumes that nature is for people to use for their own purposes, the preservation ethic, which assumes that nature has a value of itself and should not be disturbed, and the conservation ethic, which recognizes that we will use nature but that it should be used wisely.

Ethical issues can be examined at several different levels. Growth and exploitation have been the prevailing attitude of our society for generations. This does not mean that everyone in society has the same opinions, but the general attitude has been one of development rather than preservation. Most environmental decisions have really been economic decisions, and the prevailing attitude has been: If a resource is available for use, it should be used.

Corporate ethics are even more strongly influenced by economics. Corporations are in existence to make a profit. Any way that they can reduce costs makes them more profitable. Unfortunately, pollution and exploitation of rare resources may be costly to individuals or society while being profitable to corporations. In addition, corporations wield tremendous economic power and can sway public opinion and political will. Recently, many corporations have begun to openly acknowledge their responsibility to carefully examine their impact on the natural world.

Society and corporations are composed of individuals. An increasing sensitivity of individual citizens to environmental concerns can change the political and economic climate for society and corporations. However, people often do not have a clear idea of what should be done and often do not act in a way that supports their stated beliefs.

Global environmental concerns have become more important. The world is getting "smaller" and more interrelated. As more people are added each year, there is increasing competition for the resources needed for a decent life. An environmental disaster is no longer a local problem, but affects us globally. The increasing economic difference between rich and poor nations is also related to the global environment, since the poor aspire to have what the rich take for granted. We need to work together to solve the problem.

Review Questions

1. How does personal wealth relate to ethics?
2. Why do industries pollute?
3. Why would normal economic forces work against pollution control?
4. Is it reasonable to expect a totally unpolluted environment? Why or why not?
5. What has been the prominent societal attitude toward resource use?
6. Describe the differences between development, preservation, and conservation ethics.
7. What is a corporation's major motivating force?
8. Why do decision makers view the actions of corporations differently than they view the actions of individuals?

PART TWO
Ecological Principles and Their Application

Many of the environmental problems that have surfaced over the last twenty years can be better understood if one has a clear understanding of how organisms interact with their physical environment and each other. Chapter 3 deals with energy principles and the structure of matter. Chapters 4, 5, and 6 present principles of ecology and population biology that can be applied to specific problems we face as a result of a growing world population putting demands on a shrinking set of resources. Chapter 7 focuses on human population issues. It specifically relates population growth to social and economic problems.

CHAPTER THREE
Interrelated Scientific Principles: Matter, Energy, and Environment

Objectives

After reading this chapter, you should be able to:

Understand that science is exact because information is gathered in a manner that requires evaluation and revision.

Understand that environmental science is a new discipline that includes both applied and theoretical aspects of traditional science, and that social, economic, and political aspects are also involved.

Understand that matter is made up of atoms that have a specific subatomic structure made up of protons, neutrons, and electrons.

Recognize that different elements have different atomic structures and that different atoms of the same element may differ in the number of neutrons present.

Recognize that atoms may be combined and held together by chemical bonds to produce molecules.

Understand that rearranging of chemical bonds results in a chemical reaction and that this is associated with energy changes.

Recognize that matter may be solid, liquid, or gas, depending on the amount of kinetic energy contained in the molecules.

Realize that energy can be neither created nor destroyed, but that in converting energy from one form to another, some energy is converted into a useless form.

Understand that energy can be of different qualities.

Chapter Outline

Key Terms

acid
activation energy
atom
base
catalyst
cellular respiration
chemical bond
combustion
compound
controlled experiment
electron
element
energy
environmental science
experiment
first law of
   thermodynamics
hydroxyl ion
hypothesis

isotope
kinetic energy
law
matter
molecule
neutron
nucleus
observation
pH
photosynthesis
potential energy
proton
repeatability
science
scientific method
second law of
   thermodynamics
theory

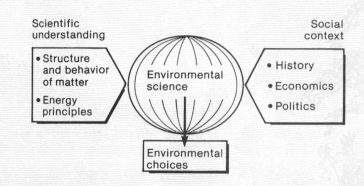

Scientific Thinking

Since environmental science requires the analysis of scientific input, it is appropriate to understand how scientists gather and evaluate information. It is also important to have a basic understanding of some chemical and physical principles as a background for making environmental decisions. An understanding of these concepts is also necessary to appreciate the ecological concepts in the chapters that follow.

The word *science* creates a variety of images in the mind. Some people feel that it is a powerful word and are threatened by it. Others are baffled by topics that are scientific and, therefore, have developed an idea that scientists are brilliant individuals who can solve any problem that comes along. Neither of these images accurately reflects what science is really like. **Science** is a body of knowledge characterized by the requirement that information be gathered and evaluated by impartial testing of hypotheses and that information be shared so that it can be evaluated by others in the field. The **scientific method** of gathering information generally involves the following elements: observation, hypothesis formation, hypothesis testing, critical evaluation of results, and the publishing of findings. (See figure 3.1.)

Underlying all of these activities is constant attention to accuracy and freedom from bias. **Observation** simply means the ability to notice something. Sometimes, the observation is made with the unaided senses—we see, feel, or smell something. Often, special machines, such as microscopes, chemical analyzers, or radiation detectors, may be used to extend our senses. Because these machines are often complicated, we often get the feeling that science is incredibly complex, when in reality the scientific questions being asked might be relatively easy to understand.

When a question has been formulated and needs scientific investigation, the first step is the formation of a hypothesis. A **hypothesis** is a logical guess that explains an event or answers a question. A good hypothesis should be as simple as possible, while taking all of the known facts into account. Furthermore, a hypothesis must be testable. In other words, you must be able to prove it correct or incorrect. The construction and testing of hypotheses is one of the most difficult (creative) aspects of the scientific method. Often, artificial situations must be constructed to test hypotheses. These are called **experiments.** A standard kind of experiment is one called a controlled experiment. A **controlled experiment** is one in which two groups are created that are identical in all respects except one. The control group has nothing done to it that is out of the ordinary. The experimental group has one thing different. If the experimental group gives different results from the control group, it must be the result of the single difference (variable) between the two groups.

The results of a well-designed experiment should be able to prove or disprove a hypothesis. However, this does not always occur. Sometimes, the results of an experiment are inconclusive. This means that a new experiment needs to be conducted or that more information needs to be collected. Often, it is necessary to have large amounts of information before a decision can be made about the validity of a hypothesis. The public often finds it difficult to understand why it is necessary to perform experiments with large numbers being involved, or why it is necessary to repeat experiments again and again. The concept of **repeatability** is important to the scientific method. Because it is often not easy to eliminate unconscious bias, it is useful to have independent investigators repeat the same experiment to see if they get the same results. To do this, they must have a complete and accurate written document to work from. This process of

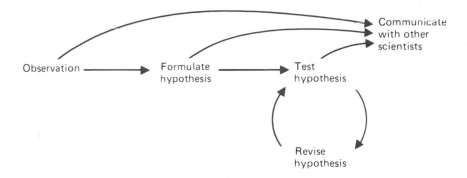

FIGURE 3.1

Steps in the Scientific Method. The scientific method consists of a series of steps. Observation of a natural phenomenon is usually the first step. This is followed by the construction of a hypothesis that explains why the phenomenon occurred. The hypothesis is then tested to see if it is correct. Often, this involves experimentation. If the hypothesis is not substantiated, it is modified and tested in its new form. It is important at all steps that others in the scientific community be informed by publishing observations of unusual events, their probable cause, and the results of experiments that test the hypotheses.

publishing results for others to examine and criticize is one of the most important steps in the process of scientific discovery. If a hypothesis is supported by many experiments and by different investigators, it is considered reliable. A hypothesis that has survived repeated examination by many investigators may become known as a **law.** Broadly written statements that cover large bodies of scientific knowledge are often called **theories.** A theory is generally accepted by scientists to be true, but it cannot be proven true in *every* case because it is impossible to test *every* case.

Now that we have some idea of how the scientific method works, let's look at an example. In many rivers in industrial parts of the world, it is possible to notice tumors of the skin and liver in the fish that live in the rivers (observation). Many people feel that the tumors are caused by the toxic chemicals that have been released into the rivers by industrial plants (hypothesis). Now, how could an experiment be conducted that would test the hypothesis? If a specific industrial plant or group of plants are suspected of releasing toxic chemicals that cause tumors, a simple experiment can be conducted to test the hypothesis. Resident species of fish, who do not migrate, can be collected upstream and downstream from the wastewater discharge pipes (outfall) of the industries. Those collected above the outfall constitute the control group, and those collected below the out-fall constitute the experimental group. Large numbers of fish would need to be collected and examined. If the fish below the outfall are found to have a significantly higher number of tumors than those above the outfall, then it is because of where they live in the river and the industrial plants are probably the cause of the tumors. After the data have been evaluated, the results of the experiment would be published. Certainly, the industrial plants might want to look at the data and perhaps repeat the experiment to see if they get the same results.

Limitations of Science

Science is a powerful tool for developing an understanding of the natural world, but it cannot analyze international politics, decide if family-planning programs should be instituted, nor evaluate the significance of a beautiful landscape. These are issues that are beyond the scope of scientific investigation. This does not mean that scientists cannot comment on these issues. They often do. But they should not be regarded as being more knowledgeable on these issues just because they are scientists. Scientists may be more knowledgeable on the scientific aspects of these issues but must struggle with the same moral and ethical questions that face all people, and their judgments on these matters can be just as faulty as anyone else's.

Because scientists tend to be methodical and follow established rules for gathering and evaluating information, there is a tendency to assume they are correct. However, scientists form and state opinions that may or may not be supported by fact, just as other people do. Equally reputable scientists commonly state opinions that are in direct contradiction. This is especially true in environmental science, where predictions about the future must be based on inadequate or fragmentary data.

It is important to recognize that knowledge and science are not synonymous and that some knowledge can be used in both a scientific and a nonscientific manner. For example, it is a fact that many of the kinds of chemicals used in modern agriculture are toxic to humans and other animals. This does not mean that foods grown with the use of these chemicals are less nutritious or that "organically grown" foods are necessarily more nutritious because they have been grown without the use of agricultural chemicals. The idea that something that is artificial is necessarily bad and that which is natural is necessarily good is false. After all, tobacco, poison ivy, and the prickly cactus are natural, while chemical fertilizers are responsible for most of the food grown in the world. It is often very easy to jump to conclusions or confuse fact with hypothesis. This is particularly true when we generalize too much.

Environmental Science

Environmental science is an interdisciplinary area of study that includes both applied and theoretical aspects of humanity's impact on the world. Since humans are generally organized into groups, environmental science must also deal with the areas of politics, social organization, economics, ethics, and philosophy. Thus, environmental science is a mixture of traditional science, societal values, and political awareness. In chapter 1, we looked at the Great Lakes and the variety of factors that influence how a person looks at them. To one person, they are waterways, to another fisheries, to a third a source of water for industry. Further complicating the situation are the international boundary and several state boundaries that divide this resource into smaller parts.

Environmental science is a field of study that is still in the process of evolving, but its beginnings are rooted in the early history of civilization. The thoughts expressed in the following quote from Henry David Thoreau (1817–1862) are consistent with current environmental philosophy:

> I wish to speak a word for Nature, for absolute freedom and wildness, as contrasted with a freedom and culture merely civil . . . to regard man as an inhabitant, or a part and parcel of Nature, rather than a member of society.

The current interest in the state of the environment began with people like Thoreau and other philosophers but received an additional push by the organization of the first Earth Day, 22 April, 1970. This committment was reaffirmed on 22 April, 1990, when the second Earth Day was held. As a result of this continuing interest in the state of the world and how people affect it, environmental science is now a standard course on many college campuses and is also a part of high school course offerings. Most of the concepts covered by environmental science courses had been previously taught in ecology, conservation, or geography courses. These physical and biological ideas were incorporated, along with input from the social sciences, such as economics, sociology, and political science, into a new interdisciplinary field of environmental science. (See figure 3.2.)

Since environmental science requires an understanding of the physical world, it is important to have an appreciation for the fundamental subunits of matter that make up the physical world.

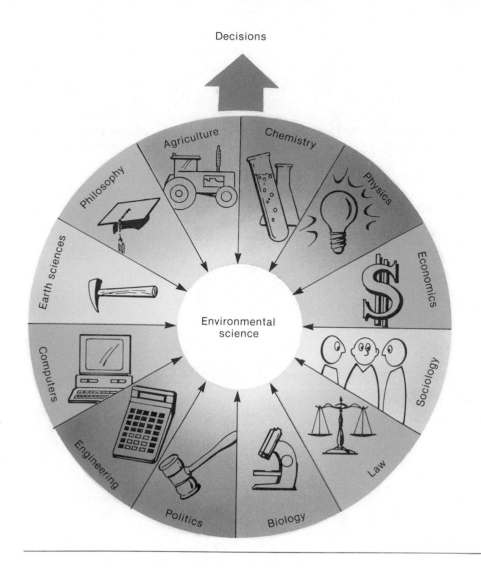

Decisions

FIGURE 3.2
Environmental Science.
Environmental science draws from a variety of subject areas that influence decisions related to the environment. Some are political, some are biological, some are economic, and some are sociological. All are interrelated in the discipline of environmental science.

The Structure of Matter

The science that deals with the composition and properties of matter is called chemistry. **Matter** is anything that takes up space and has weight. Therefore, air, water, trees, cement, and gold are all examples of matter. Although each of these kinds of matter is different, they are similar in one fundamental way. They are all made up of moving subunits called **atoms.** Ninety-two different kinds of atoms are found in nature. Each one forms a specific kind of matter known as an **element.** Gold, oxygen, and mercury are examples of elements. Each has a different kind of atom making up its structure.

Atomic Structure

All atoms are composed of a central region called a **nucleus,** which is composed of two kinds of relatively heavy particles. Positively charged particles are called **protons** and uncharged particles are called **neutrons.** Surrounding the nucleus are clouds of relatively lightweight, fast-moving, negatively charged particles called **electrons.** As mentioned earlier, each kind of element is composed of a different kind of atom. They differ from one

BOX 3.1

The Periodic Table of Elements

Traditionally, elements are represented in a shorthand form by letters. For example, the symbol for water, H_2O, shows that a molecule of water consists of two atoms of hydrogen and one atom of oxygen. These chemical symbols can be found on any periodic table of elements. Using the periodic table on page 51, we can determine the number and position of the various parts of atoms. Notice that atoms number 3, 11, 19, and so on, are in column one. The atoms in this column act in a similar way since they all have one electron in their outermost layer. In the next column,

Be, Mg, Ca, and so on act alike because these metals all have two electrons in their outermost electron layer. Similarly, atoms number 9, 17, 35, and so on all have seven electrons in their outer layer. Knowing how fluorine, chlorine, and bromine act, you can probably predict how iodine will act under similar conditions. At the far right in the last column, argon, neon, and so on all act alike. They all have eight electrons in their outer electron layer. Atoms with eight electrons in their outer electron layer seldom form bonds with other atoms.

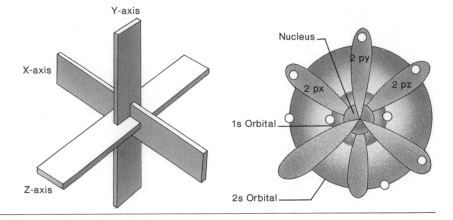

FIGURE 3.3

Diagrammatic Oxygen Atom. Most oxygen atoms are composed of a nucleus containing eight positively charged protons and eight neutrons without charges. In addition, there are eight negatively charged electrons that spin around the nucleus.

another in the number of protons, neutrons, and electrons present. For example, a typical atom of mercury contains eighty protons and eighty electrons, while gold has seventy-nine and oxygen only eight. (See figure 3.3.) While an atom always has the same number of protons and electrons, the number of neutrons may vary from one atom to the next. Atoms of the same element that differ from one another in the number of neutrons they contain are called **isotopes.**

Molecular Structure and Chemical Reactions

When two or more atoms combine with one another, they form a new, stable unit known as a **molecule.** While there are only ninety-two kinds of atoms, there can be millions of different kinds of molecules. Many of the kinds of matter that you are familiar with are composed of molecules. The water molecule is composed of an oxygen atom and two hydrogen atoms. (See figure 3.4.) The types of matter that are composed of only one kind of molecule are called **compounds.** Just as atoms are in constant motion, molecules also move. The atoms within a molecule are held together by chemical bonds. **Chemical bonds** are physical attractions between atoms

Periodic Table of the Elements

Key:
- Atomic number
- Symbol and name
- Atomic weight

Group	Element data
IA	1 H Hydrogen 1.008
IIA	4 Be Beryllium 9.012; 12 Mg Magnesium 24.31

Period 1
- 1 H Hydrogen 1.008
- 2 He Helium 4.003

Period 2
- 3 Li Lithium 6.941
- 4 Be Beryllium 9.012
- 5 B Boron 10.81
- 6 C Carbon 12.01
- 7 N Nitrogen 14.01
- 8 O Oxygen 16.00
- 9 F Flourine 19.00
- 10 Ne Neon 20.18

Period 3
- 11 Na Sodium 22.99
- 12 Mg Magnesium 24.31
- 13 Al Aluminum 26.98
- 14 Si Silicon 28.09
- 15 P Phosphorus 30.97
- 16 S Sulfur 32.08
- 17 Cl Chlorine 35.45
- 18 Ar Argon 39.95

Period 4
- 19 K Potassium 39.10
- 20 Ca Calcium 40.08
- 21 Sc Scandium 44.95
- 22 Ti Titanium 47.90
- 23 V Vanadium 50.94
- 24 Cr Chromium 52.00
- 25 Mn Manganese 54.94
- 26 Fe Iron 55.85
- 27 Co Cobalt 58.93
- 28 Ni Nickel 58.71
- 29 Cu Copper 63.55
- 30 Zn Zinc 65.38
- 31 Ga Gallium 69.72
- 32 Ge Germanium 72.59
- 33 As Arsenic 74.92
- 34 Se Selenium 78.96
- 35 Br Bromine 79.90
- 36 Kr Krypton 83.80

Period 5
- 37 Rb Rubidium 85.47
- 38 Sr Strontium 87.62
- 39 Y Yttrium 89.91
- 40 Zr Zirconium 91.22
- 41 Nb Niobium 92.91
- 42 Mo Molybdenum 95.94
- 43 Tc Technetium 98.91
- 44 Ru Ruthenium 101.1
- 45 Rh Rhodium 102.9
- 46 Pd Palladium 106.4
- 47 Ag Silver 107.9
- 48 Cd Cadmium 112.4
- 49 In Indium 114.8
- 50 Sn Tin 116.7
- 51 Sb Antimony 121.8
- 52 Te Tellurium 127.6
- 53 I Iodine 126.9
- 54 Xe Xenon 131.3

Period 6
- 55 Cs Cesium 132.9
- 56 Ba Barium 137.3
- 57 La¹ Lanthanum 138.9
- 72 Hf Hafnium 178.5
- 73 Ta Tantalum 180.9
- 74 W Tungsten 183.9
- 75 Re Rhenium 186.2
- 76 Os Osmium 190.2
- 77 Ir Iridium 192.2
- 78 Pt Platinum 195.1
- 79 Au Gold 197.0
- 80 Hg Mercury 200.6
- 81 Tl Thallium 204.4
- 82 Pb Lead 207.2
- 83 Bi Bismuth 209.0
- 84 Po Polonium (210)*
- 85 At Astatine (210)*
- 86 Rn Radon (222)*

Period 7
- 87 Fr Francium (223)*
- 88 Ra Radium 226.0
- 89 Ac² Actinium (227)*
- 104 [Rf]† Rutherfordium (257)*
- 105 [Ha]† Hahnium (262)*
- 106 [ ]† (263)*

¹Lanthanide series
- 58 Ce Cerium 140.1
- 59 Pr Praseodymium 140.9
- 60 Nd Neodymium 144.2
- 61 Pm Promethium (145)*
- 62 Sm Samarium 150.4
- 63 Eu Europium 152.0
- 64 Gd Gadolinium 157.2
- 65 Tb Terbium 158.9
- 66 Dy Dysprosium 162.5
- 67 Ho Holmium 164.9
- 68 Er Erbium 167.3
- 69 Tm Thulium 168.9
- 70 Yb Ytterbium 173.0
- 71 Lu Lutetium 175.0

²Actinide series
- 90 Th Thorium 232.0
- 91 Pa Protactinium 231.0
- 92 U Uranium 238.0
- 93 Np Neptunium 237.0
- 94 Pu Plutonium (244)*
- 95 Am Americium (243)*
- 96 Cm Curium (247)*
- 97 Bk Berkelium (247)*
- 98 Cf Californium (251)*
- 99 Es Einsteinium (254)*
- 100 Fm Fermium (257)*
- 101 Md Mendelevium (258)*
- 102 No Nobelium (259)*
- 103 Lr Lawrencium (260)*

Legend:
- Metals
- Metalloids
- Nonmetals
- Noble gases

*Mass of most stable isotope
†Brackets surround names that are not officially accepted

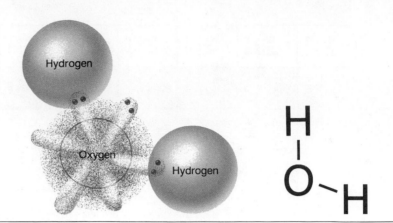

FIGURE 3.4

Water Molecule. Water molecules consist of an atom of oxygen bonded to two atoms of hydrogen, with the hydrogen atoms located on one side of the oxygen.

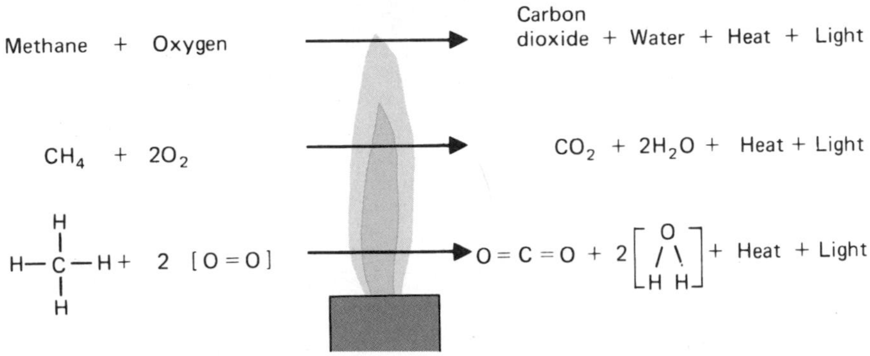

FIGURE 3.5

A Chemical Reaction. When methane is burned, chemical bonds are changed, and the excess chemical bond energy is released as light and heat. The same atoms are present, but they are bonded in different ways, resulting in different molecules.

Methane + Oxygen ⟶ Carbon dioxide + Water + Heat + Light

CH_4 + $2O_2$ ⟶ CO_2 + $2H_2O$ + Heat + Light

$$H-\overset{\displaystyle H}{\underset{\displaystyle H}{C}}-H + 2\ [O=O] \longrightarrow O=C=O + 2\left[\begin{matrix}O\\ /\ \backslash\\ H\ \ H\end{matrix}\right] + \text{Heat} + \text{Light}$$

resulting from the interaction of their electrons. When chemical bonds are broken or formed, a chemical reaction occurs. When a chemical reaction occurs, the amount of energy within chemical bonds changes, and some of the energy may be released as heat and light. A common example of a chemical reaction is the burning of natural gas. The primary ingredient in natural gas is the compound methane. When methane and oxygen are mixed together and a small amount of energy is used to start the reaction, the chemical bonds in the methane and oxygen (reactants) will be rearranged to form the two new compounds of carbon dioxide and water (products). In this kind of reaction, there is some chemical bond energy left over, which is released as light and heat. (See figure 3.5.)

Other kinds of chemical reactions require an input of energy to enable them to occur. The process of food formation in plants known as **photosynthesis** requires an input of sunlight energy. Light energy is required to enable the lower-energy reactants (carbon dioxide and water) to combine to form higher-energy products (sugar and oxygen). (See figure 3.6.)

In every reaction, the amount of energy in the reactants and in the products can be compared and the differences accounted for by energy loss or gain. However, even with energy-yielding reactions, it is usually necessary to have an input of energy to get the reaction started. This initial input of energy is called **activation energy.** In certain cases, the amount of activation energy required to start the reaction can be reduced by the use of a catalyst. A **catalyst** is a substance that alters the rate of a reaction, but it is not altered in the process. Catalysts are used in catalytic converters,

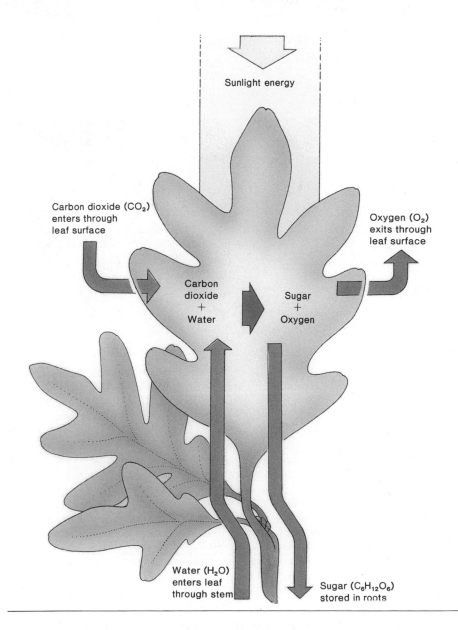

Sunlight energy

Carbon dioxide (CO_2) enters through leaf surface

Oxygen (O_2) exits through leaf surface

Carbon dioxide + Water

Sugar + Oxygen

Water (H_2O) enters leaf through stem

Sugar ($C_6H_{12}O_6$) stored in roots

FIGURE 3.6

Photosynthesis. This reaction is an example of one that requires an input of energy (sunlight) to combine low-energy molecules (CO_2 and H_2O) to form sugar with a greater amount of chemical bond energy.

which are attached to automobile exhaust systems. The purpose of the catalytic converter is to bring about more complete burning of the fuel, thus resulting in less air pollution. Most of the materials that are not completely burned by the engine require high temperatures to react further; with the presence of catalysts, these reactions can occur at lower temperatures.

Living things also contain catalysts, which are composed of proteins and are known as enzymes. Most chemical reactions that occur in organisms are assisted by enzymes that reduce the activation energy needed to start the reactions. This is important since the high temperatures required to start these reactions without enzymes would destroy living organisms. Many of these enzymes are arranged in such a way that they cooperate in the controlling of a chain of reactions, as in photosynthesis and respiration.

Modern society makes extensive use of many different kinds of chemicals. A survey of a typical household would yield many of the following kinds of inorganic chemicals:

Common name	Chemical name	Use
Table salt	Sodium chloride, NaCl	Flavor
Saltpeter	Potassium nitrate, KNO$_3$	Preservative
Baking soda	Sodium bicarbonate, NaHCO$_3$	Leavening agent
Ammonia	Ammonia, NH$_3$	Disinfectant
Bleach	Sodium hypochlorite, NaHClO	Bleaching
Caustic soda	Sodium hydroxide, NaOH	Drain cleaner

Other kinds of products we use contain mixtures of inorganic chemicals. Fertilizers are good examples. They usually contain a nitrate such as ammonium nitrate, NH$_4$NO$_3$, a phosphate such as phosphoric acid, P$_2$O$_5$, and potash, which is potassium oxide, K$_2$O.

In addition, we use a vast array of different kinds of organic chemicals: ethyl alcohol in alcoholic beverages, acetic acid in vinegar, methyl alcohol for fuel, and cream of tartar (tartaric acid) for flavoring. We also use a large number of complex mixtures of organic molecules in flavorings, pesticides, cleaners, and many other applications.

Most of us know very little about the activities of the molecules we use. Many of them can be very dangerous if used improperly. Fertilizer is poisonous, caustic soda can cause severe burns, and bleach or ammonia can damage skin or other tissues if present in high enough concentrations. Furthermore, the disposal of unused or unwanted household chemicals is a problem. Many of them should not just be dumped down the sink but should be disposed of in such a way that the material is converted to a harmless product or stored in a secure place. Unfortunately, most citizens do not know how to dispose of unwanted chemicals.

Acids, Bases, and pH

Acids and bases are two classes of biologically important compounds. Their characteristics are determined by the nature of their chemical bonds. When acids are dissolved in water, hydrogen ions (H$^+$) are set free. The hydrogen ion is positive because it has lost its electron and now has only the positive charge of the proton. An **acid** is any ionic compound that releases a hydrogen ion in a solution. One other way of thinking of an acid is that it is a substance able to donate a proton to a solution. This is only part of the definition of an acid. We also think of acids as compounds that act like the hydrogen ion—they attract negatively charged particles. An example of a common acid is the sulfuric acid (H$_2$SO$_4$) in our automobile batteries.

A **base** is the opposite of an acid in that it is an ionic compound that releases a group known as a **hydroxyl ion,** or OH$^-$ group. This group is composed of an oxygen atom and a hydrogen atom bonded together but with an additional electron. The hydroxyl ion is negatively charged. It is a base because it is able to donate electrons to the solution. A base can also be thought of as any substance able to attract positively charged particles. A very strong base used in oven cleaners is NaOH, sodium hydroxide.

The strength of an acid or base is represented by a number called its **pH** number. The pH scale is a measure of hydrogen ion concentration. A

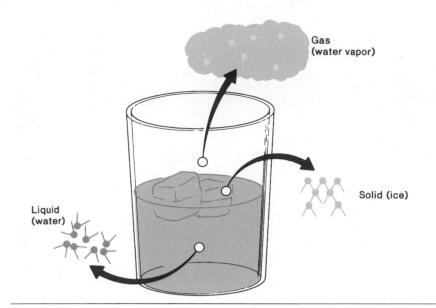

Gas
(water vapor)

Solid (ice)

Liquid
(water)

FIGURE 3.7
States of Matter. Matter exists in one of
three states, depending on the amount
of kinetic energy the molecules have.
The higher the amount of energy, the
greater the distance between molecules
and the greater the degree of freedom of
movement of the molecules.

pH of seven indicates that the solution is neutral and has an equal number
of H^+ ions and OH^- ions to balance each other. As the pH number gets
smaller, the number of hydrogen ions in the solution increases. The lower
the number, the stronger the acid. A number higher than seven indicates
that the solution has more OH^- than H^+. As the pH number gets larger, the
number of hydroxyl ions increases. The higher the number, the stronger
the base.

Energy Principles

Although the previous section started out with a description of matter, it
was necessary to use energy concepts to describe the movement of atoms
and molecules, chemical bonds, and chemical reactions. Energy and matter
are inseparable. It is difficult to describe one without the other. **Energy** is
the ability to move matter over a distance. This even occurs at the molecular
level.

Depending on the amount of energy present, matter can occur in three
different states: solid, liquid, or gas. The physical nature of matter changes
because of changes in the amount of kinetic energy a molecule contains,
but the chemical nature of matter remains the same. For example, water
vapor, liquid water, and ice all have the same chemical composition, but
differ in the arrangement and activities of the molecules. The amount of
kinetic energy contained by molecules determines how rapidly they move.
(See figure 3.7.) In solids, the molecules have low amounts of energy, and
they vibrate in place very close to one another. Higher-energy molecules
are farther apart and will roll, tumble, and flow over each other as is typical
of liquids. The molecules of gases move very rapidly, and are very far apart.
So all that is necessary to change the physical nature of a type of matter is
an energy change. Heat must be added or removed.

There are several kinds of energy. Heat, light, electricity, and chemical
energy are common forms. The energy contained by moving objects is called
kinetic energy. The moving molecules in air have kinetic energy, as does
water running downhill or a dog chasing a ball. **Potential energy** is in a
special category because it is the energy matter has because of its position.

FIGURE 3.8
Kinetic and Potential Energy. Kinetic and potential energy are interconvertible. The potential energy possessed by the water behind a dam is converted to kinetic energy as the water flows to a lower level.

The water behind a dam has potential energy by virtue of its elevated position. (See figure 3.8.) An electron at some distance from the nucleus has potential energy due to the distance between the electron and the nucleus.

First and Second Laws of Thermodynamics

Energy can exist in several different forms. However, the total amount of energy remains constant. The **first law of thermodynamics** states that energy can neither be created nor destroyed; it can only be changed from one form into another. Some forms of energy are more useful than others from a human perspective. We tend to make extensive use of electrical energy for a variety of purposes, but there is very little electrical energy present in nature. Therefore, we convert other forms of energy into electrical energy. When converting energy from one form to another, some of the useful energy is lost. This is the **second law of thermodynamics.** There is no loss of *total* energy, but there is a loss of *useful* energy. For example, coal can be burned in a power plant to produce electrical energy. However, large amounts of useless heat energy are also produced. Therefore, the amount of useful energy (electricity) is much less than the total amount of chemical energy present in the coal. (See figure 3.9.)

Energy is being converted from one form to another continuously within the universe. Stars are converting nuclear energy into heat and light. Animals are converting the chemical energy found in food into the kinetic energy of motion. Plants are converting sunlight energy into the chemical bond energy of sugar molecules. In each of these cases, some useless energy is produced, generally in the form of heat.

Environmental Implications of Energy Flow

The heat produced when energy conversions occur is dissipated throughout the universe. This is a common experience. Valuable things always seem to disintegrate unless we work to prevent it. Houses require constant maintenance, automobiles rust, and appliances wear out. In reality, all of these phenomena involve the loss of heat. The organisms that decompose the wood in our houses release heat. The chemical reaction that causes rust releases heat. Friction, caused by the movement of parts of a machine against each other, generates heat and causes the parts to wear.

Loss
of
heat

Loss
of
heat

Power
plant

Loss
of
heat

FIGURE 3.9
Second Law of Thermodynamics.
Whenever energy is converted from one
form to another, some of the useful
energy is lost, usually in the form of
heat. The conversion of fuel to electricity
results in the production of heat, which
is lost to the atmosphere. As the
electricity moves through the wire,
resistance generates some additional
heat. When the electricity is converted to
light in a light bulb, heat is produced as
well. All of these steps result in the
production of useless heat as a result of
the second law of thermodynamics.

Orderly arrangements of matter always tend to become disordered by
the constant flow of energy toward a dilute form of heat. This dissipated
form of energy is low quality heat and, therefore, has little value to us. It
is important to understand that different energy forms can be of different
quality. Some are of high quality, such as electrical energy, which can be
easily used to perform a variety of useful actions, while some are of low
quality, such as the heat in the water of the ocean. Although the total *quantity* of heat energy in the ocean is much greater than the total amount of
electrical energy in the world, there is little that can be done with the energy
in the ocean because it is of low *quality*. Therefore, it is not as valuable as
some other forms of energy, which can be used to do work for us. When
two objects differ in heat content, heat will flow to the cooler object. The
greater the temperature difference, the more useful the work that can be
done. Because of this, fossil-fuel power plants use cold water to provide a
steep heat gradient. Because the average temperature of the ocean is not
high, and it is difficult to find another object that has a greatly lower temperature than the ocean, the tremendously large heat content of the ocean
cannot do useful work for us.

These quantitative/qualitative factors are also evident in the energy expended by a stream as the water runs downhill. The steeper the slope, the
greater the amount of energy expended per kilometer of its length. The
stream has low-quality energy, because the energy is dissipated along the
entire length of the stream, if there is no point along the stream where the
slope is very steep. To make this a high-quality (concentrated) source of
energy, the water must be dammed so that it will drop a long distance at
one point. This means that it will give up much of its energy over a short
distance. With damming, the *quantity* of energy has not changed; the *quality*
has.

Because of the second law of thermodynamics, all organisms, including humans, are in the process of converting high-quality energy into

CONSIDER THIS CASE STUDY
Alcohol Production in Brazil

During the 1970s, when the price of oil increased tremendously, many countries developed new energy policies. Many instituted energy reforms that reduced energy consumption or increased the production of oil from their own resources. Brazil was in a particularly difficult position since it has very small oil reserves within its borders. Since liquid fuels are preferred for automobiles, Brazilians sought an alternative fuel and embarked on a massive alcohol production program. They had several factors working in their favor. They had large amounts of land suitable for the growing of sugar cane, which is a well-understood crop in Brazil; they had many small distilleries that could be converted to the production of alcohol for fuel; and they had a large population of unemployed workers, who could be put to useful work growing more sugar cane. In many respects, the project was a success. Automobile engines were developed that used alcohol effectively, many unemployed workers were employed, and Brazil's dependence on foreign sources of oil for liquid fuel decreased. Today 50 percent of automobile fuel in Brazil is alcohol, and Brazil imports significantly less oil today than in the past.

But how does this work out from an energy point of view? Energy is used to plant, fertilize, care for, and harvest the sugar cane. Energy is used to haul the cane to the still. Energy is used in the distillation process. And finally, energy is used in the distribution of the alcohol fuel to the place where it is consumed. A total energy analysis shows that there is less energy in the alcohol than in the energy expended to produce and market that fuel. Yet, many consider this to be a successful project from the point of view of the economy and welfare of the country. Less money is being given to other countries to buy oil. This same money is being paid as wages to Brazilians in the production process. Nevertheless, Brazil is still an economically unstable country since it owes more than $100 billion to foreign banks who have loaned it money to develop its alcohol fuel project and other important industrial developments.

Is alcohol fuel really efficient?
What special circumstances made alcohol fuel feasible in Brazil?

TABLE 3.1
The efficiency of some energy conversion systems.

Energy conversion system	% Efficiency*
Electric generator	99
Home oil furnace	65
Steam-power plant	40
High-intensity lamp	32
Automobile engine	25
Fluorescent lamp	20
Incandescent lamp	4

*Efficiency in this case means the efficiency with which the energy of the power or fuel source is converted to a useful form.

Source: Robert H. Romer, *Energy: An Introduction to Physics,* W. H. Freeman and Company, 1976.

low-quality energy. Waste heat is produced when the chemical-bond energy in food is converted into the energy needed to move, grow, or respond. The process of releasing chemical-bond energy from food by organisms is known as **cellular respiration.** From an energy point of view, it is similar to the process of **combustion,** which is the burning of fuel to obtain heat, light, or some other form of useful energy. The efficiency of cellular respiration is relatively high. About 40 percent of the energy contained in food is released in a useful form. The rest is dissipated as low-quality heat. Table 3.1 lists the efficiencies of many common energy conversion systems.

The amount of energy in the universe is limited. Only a small amount of that energy is of high quality. The use of high-quality energy decreases the amount of useful energy available, as more low-quality heat is generated. All life and all activities are subject to these important physical principles known as the first and second laws of thermodynamics.

Summary

Science is a method of gathering and organizing information, and involves observation, hypothesis formation, and experimentation. A hypothesis is a logical guess about how things work. The process of science attempts to be careful, unbiased, and reliable in the way information is collected and evaluated. This usually involves conducting experiments to test the validity of hypotheses. If a hypothesis is used for a long period of time and is continually supported by new facts, it is often stated as a law. A theory is a broadly written statement that cannot be proven true in every case because it is impossible to test every case.

Environmental science is a combination of theoretical science and applied areas of science, such as engineering. In addition, it incorporates many components from such fields as economics, philosophy, and political science.

The fundamental unit of matter is the atom, which is made up of protons and neutrons in the nucleus and electrons circling the nucleus. The number of protons for any one type of atom is constant, but the number of neutrons may vary. Protons have a positive charge, neutrons lack a charge, and electrons have a negative charge.

When two or more atoms combine with one another, they form stable units known as molecules. Chemical bonds are physical attractions between atoms resulting from the interaction of their electrons. When chemical bonds are broken or formed, a chemical reaction occurs, and the amount of energy within the chemical bonds is changed. Chemical reactions require activation energy to get the reaction started.

Matter can occur in three different states: solid, liquid, and gas. These three differ in the amount of energy the molecules contain and the distance between the molecules. Kinetic energy is the energy contained by moving objects. Potential energy is the energy an object has because of its position.

The first law of thermodynamics states that the amount of energy is constant, that energy can neither be created nor destroyed. The second law of thermodynamics states that, when energy is converted from one form to another, some of the useful energy is lost. Some forms of energy are more useful than others. The quality of the energy determines how much useful work can be accomplished by expending the energy. Low-temperature heat sources are of poor quality, since they cannot be used to do useful work.

Review Questions

1. How do scientific disciplines differ from nonscientific disciplines?
2. What is a hypothesis? Why is it an important part of the way scientists think?
3. Why are events that happen only once difficult to analyze from a scientific point of view?
4. What is the scientific method, and what steps are involved in it?
5. How does environmental science differ from traditional science?
6. Diagram an atom of oxygen and label its parts.
7. What happens to atoms during a chemical reaction?
8. State the first and second laws of thermodynamics.
9. How do solids, liquids, and gases differ from one another at the molecular level?
10. List five kinds of energy.
11. Are all kinds of energy equal in their capacity to bring about changes? Why or why not?

CHAPTER FOUR
Interactions: Environment and Organisms

Objectives

After reading this chapter, you should be able to:
Identify the abiotic and biotic factors in an ecosystem.
Define *niche*.
Describe the process of natural selection as it operates to refine the fit between organism, habitat, and niche.
Describe predator-prey, parasite-host, competitive, mutualistic, and commensalistic relationships.
Differentiate between a community and an ecosystem.
List some of the components of an ecosystem.
Define the role of producer, herbivore, carnivore, omnivore, scavenger, parasite, and decomposer.
Describe energy flow in an ecosystem.
Relate the concept of food web and food chain to trophic levels.
Explain the cycling of nutrients, such as nitrogen, carbon, and phosphorus, through an ecosystem.

Chapter Outline

Key Terms

abiotic factor
biomass
biotic factor
carbon cycle
carnivore
commensalism
community
competition
consumer
decomposer
denitrifying bacteria
detritus
ecology
ecosystem
ectoparasite
endoparasite
environment
evolution
food chain
food web
free-living nitrogen-
    fixing bacteria
habitat
herbivore

host
interspecific competition
intraspecific competition
limiting factor
mutualism
natural selection
niche
nitrogen cycle
nitrogen-fixing bacteria
omnivore
parasite
parasitism
predator
prey
primary consumer
producer
range of tolerance
secondary consumer
speciation
species
symbiosis
symbiotic nitrogen-fixing
    bacteria
trophic level

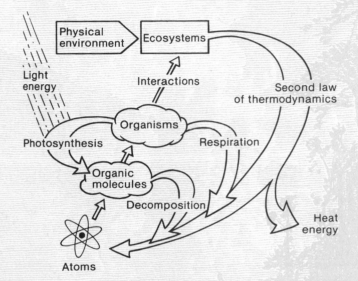

FIGURE 4.1
Ecology. The study of ecology is pursued at various levels, ranging from atoms and molecules to individual organisms to groups of interacting organisms to ecosystems.

Ecosystem

Community

Population

Organism

Atoms

Ecological Concepts

Energy and matter are interrelated. Matter contains energy, which, in turn, determines the physical properties of matter. Living systems can also be regarded as having matter and energy interrelationships. Living things require a constant flow of energy to assure their survival. If energy flow ceases, the organism dies. All organisms are dependent on other organisms. Sometimes, one organism eats another and uses it for energy. In other cases, one organism may just make temporary use of another without harming it. The large number of ways that organisms and their surroundings interact is the basic concern of the branch of science known as **ecology.** It deals with the ways in which organisms are molded by their surroundings, how they make use of these surroundings, and how an area is altered by the presence and activities of organisms. The study of ecology can be divided into many specialties and be looked at from several levels of organization. (See figure 4.1.) Before we can explore the field of ecology in greater depth, we must become familiar with some of the standard vocabulary used in this field.

Environment

Everything that affects an organism during its lifetime is collectively known as its **environment.** (See figure 4.2.) This idea of environment is very broad. For example, an animal is likely to interact with hundreds of other organisms, drink many liters of water, breathe a lot of air, and respond to many changes in temperature and humidity during its lifetime, and this does not begin to detail everything that makes up its environment. Because of this complexity, it is convenient to classify the environmental factors that influence an organism into **abiotic** (nonliving) and **biotic** (living) **factors.** The abiotic factors include the flow of energy necessary to maintain the

FIGURE 4.2

Environment. Everything that affects an organism during its lifetime is collectively called its environment. Physical factors, such as weather, soil type, altitude, living space, and the amount of sunlight, have a significant effect on the kinds of organisms that can live in an area. Likewise, how the organism interacts with other organisms, such as the types of plants used for food and shelter, parasites, and predators, is part of the environment.

organism, the supply of molecules required for the organism's various life functions, and the physical factors that affect the organism.

The ultimate source of energy for almost all organisms is the sun; in the case of plants, the sun directly supplies the energy necessary for the organisms to maintain themselves. Animals obtain their energy by eating plants or other animals who eat plants. Ultimately, the amount of living material that can exist in an area is determined by the plants and the amount of sunlight they can trap.

All forms of life require carbon, nitrogen, phosphorus, water, and other atoms and molecules. Organisms constantly remove these from the environment, use them for a period of time, and then they are returned to the environment.

The abiotic or physical factors include such things as climate; temperature; type, amount, and seasonal distribution of precipitation; soil type; and even the amount and three-dimensional shape of the space the organism inhabits.

The biotic factors influencing an organism include all forms of life in its environment. Plants that carry on photosynthesis, animals that eat other organisms, fungi that cause decay, and bacteria that cause disease are all part of an organism's biotic environment. Both the types of living forms and the number of each kind are important in characterizing the environment. Organisms must respond to all of the individual factors that constitute their environment, but some are of primary consideration and determine the success of a species more than any other factor. These are called **limiting factors.**

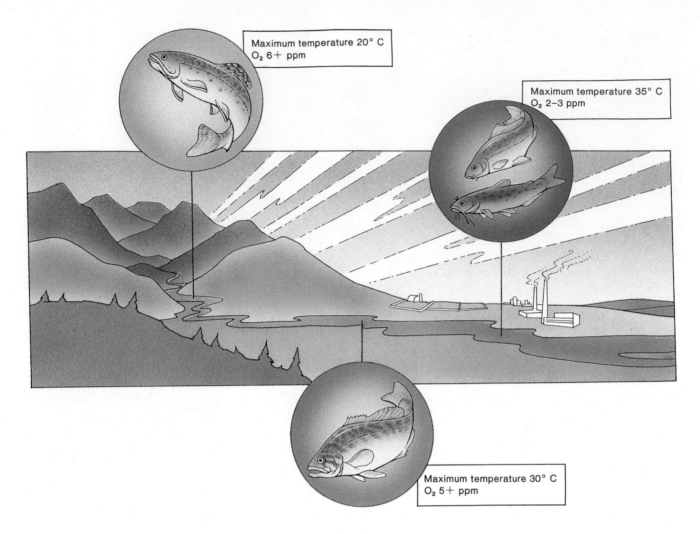

Maximum temperature 20° C
O$_2$ 6+ ppm

Maximum temperature 35° C
O$_2$ 2–3 ppm

Maximum temperature 30° C
O$_2$ 5+ ppm

FIGURE 4.3

Limiting Factors. In aquatic habitats oxygen can often be a limiting factor for many species of fish. Cool, highly oxygenated water, which is typical of the rapidly flowing upper sections of river systems, supports trout, but warmer, less oxygenated water is unsuited for trout. Other fish, which are more tolerant to low levels of oxygen, such as bass and carp, occupy the lower sections of the river, where the water is warmer and there is less oxygen.

Limiting Factors

The limiting factor for many species of fish is the amount of dissolved oxygen present in the water. In a swiftly flowing, tree-lined mountain stream, the level of dissolved oxygen is high and provides a favorable environment for trout. (See figure 4.3.) As the stream continues down the mountain, the steepness of the slope decreases, which results in fewer rapids to oxygenate the water. In addition, the canopy of trees over the stream usually is thinner, allowing more sunlight to reach the stream and warm the water. Warm water cannot hold as much dissolved oxygen as cool water. Therefore, decreasing turbulence and increasing temperature result in a decreased level of dissolved oxygen to a level below that required by trout. Fish such as black bass and walleye replace the trout, since they are able to tolerate lower oxygen concentrations and higher water temperatures. These species have a greater **range of tolerance** to oxygen concentration and water temperature. Thus, low levels of oxygen and high water temperatures are limiting factors for the distribution of trout.

In addition to temperature and oxygen concentration, other factors may influence the ability of water to support certain species of fish. If we continue to follow our stream, it will eventually become a broad, slowly flowing river. Much of the land along the river probably will be devoted to agriculture. This will result in significant amounts of silt and other soil particles

Ecological Principles and Their Application

entering the river. Most of the surface of the river will be exposed to the warming effects of the sun, resulting in additional temperature increases and a reduced oxygen concentration. Under these conditions, we may see the bass and walleye replaced by such species as carp and catfish, which have an even greater range of tolerance to high temperatures and low oxygen concentrations than do the bass and walleye.

Habitat and Niche

As we have just seen, it is impossible to understand an organism apart from its environment. The environment influences the organism, and organisms affect the environment. To focus attention on specific elements of this interaction, ecologists have developed two concepts that need careful explanation: habitat and niche.

The **habitat** of an organism is the space that the organism inhabits, the place where it lives. The place where an organism lives has specific characteristics, some biotic and some abiotic, and the specific mixture of these characteristics determines whether or not it is a good habitat for the organism. We tend to identify a habitat with particular physical environmental characteristics like soil type, availability of water, or climatic conditions, or we identify it with the predominant plant species that exist in the area. For example, mosses are small plants that dry out and die if they are exposed to sunlight, wind, and drought. Therefore, the habitat of moss is likely to be cool, moist, and shady. (See figure 4.4.) Likewise, a rapidly flowing, cool, well-oxygenated stream with many bottom-dwelling insects is good trout habitat, while open prairie with lots of grass is preferred by such animals as bison, prairie dogs, and many kinds of hawks and falcons. Similarly, elm bark beetles will only reside in areas where elm trees are found. The particular biological requirements of an organism will determine the kind of habitat in which it is likely to be found.

A second concept related to habitat is niche. The **niche** of an organism is the functional role it has in its surroundings. A niche is everything that both affects and is affected by an organism during the organism's lifetime. For example, beavers frequently build dams of mud and sticks, which cause the flooding of surrounding areas. (See figure 4.5.) The flooding has several effects. It provides beavers with a larger area of deep water, which they need for protection; it provides a pond habitat that many other species of animals, like ducks and fish, find to their liking; and it brings about the death of trees. Furthermore, after a period of time, the beavers will have eaten all the suitable food, like aspen, and they will abandon the pond and migrate to other areas along the stream and begin the whole process over again. Although most beavers only leave the pond to obtain food during the late evening or night, which provides them with some protection from predators, they are still used as a food source by several kinds of animals.

In this recitation of beaver characteristics, we have listed several effects that the animal has. It changes the physical environment by flooding, it kills trees, it enhances the environment for other animals, and it avoids predators by being active at night. These factors are all part of the beaver's niche. This is not, however, all there is to the beaver's niche and serves as only a glimpse of the many aspects of the beaver's interaction with its environment. A complete catalog of all aspects of its niche would make up a separate book.

Another familiar organism is the dandelion. Since it is a plant, one of the major aspects of its niche is to carry on photosynthesis and grow. Many kinds of animals use dandelions as food, including humans. The leaves

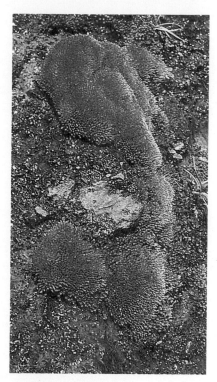

FIGURE 4.4
Moss Habitat. The habitat of mosses must be cool and shady, since mosses tend to dry out and die if they are exposed to the drying action of the sun.

FIGURE 4.5

Ecological Niche. The ecological niche of an organism involves everything that affects it during its lifetime as well as all the effects it has. The ecological niche of beaver includes: the need for streams with aspen trees nearby, the building of dams and flooding of forested areas, the killing of trees, the providing of ponds for ducks and other animals, and many other effects.

may be eaten in a salad, and the blossoms can be used to make dandelion wine. Bees also frequently visit the flowers to obtain nectar. This plant is an opportunist in that it has the ability to produce, in a few days, large numbers of parachutelike seeds that can be easily carried by the wind over long distances. Furthermore, it often produces several sets of flowers per year. Since there are so many seeds and they are so easily distributed, the plant can easily establish itself in any sunny area that is disturbed, including lawns. The long, tapering root can store nutrients, so that even if the top of the plant is ripped away by an animal, it can regenerate from the root. However, dandelions need a large amount of sunlight to grow successfully and do not grow well in shady areas. Mowing of lawns helps to provide just the right kind of conditions for dandelions since the vegetation is never allowed to get so tall that the dandelions are shaded.

The Role of Natural Selection

As we have just seen, each organism is finely tuned to a particular habitat. Each organism plays a very specific role and also influences the physical surroundings of itself and other organisms. But how is it that each plant,

Ecological Principles and Their Application

animal, fungus, and bacterium seems to fit into its environment in such a precise way? The process that causes organisms to adapt to their environment is known as **natural selection.**

In the 1800s, Charles Darwin, a British naturalist, explained the process of natural selection in his book *The Origin of Species*. In this work, he suggested the following:

1. All populations of organisms show variations in characteristics that are determined by the organisms' genetic makeup. No organisms have exactly the same genetic makeup.
2. Organisms typically produce more offspring than can survive. This means that there is not enough suitable habitat for all of the offspring to grow to maturity.
3. There is a struggle for survival. The organisms must compete with each other for food, space, mates, or other requirements in limited supply.
4. Because they have different genetic characteristics, certain individuals of a species have a better chance for survival and reproduction than others. Characteristics such as blindness in lions or lack of chlorophyll in plants would obviously result in a high mortality rate for the organisms with the characteristic.
5. Because not all organisms have the same likelihood of survival and reproduction, some genetic characteristics will become more rare and others will become more common from one generation to the next. Those individuals with particularly valuable combinations of genes have a better chance of reproducing and passing their genes on to their offspring.

In a classic experiment, H. B. D. Kettlewell studied the processes that resulted in some areas of England having mostly dark-colored forms of the peppered moth, while other areas had mostly light-colored forms. He hypothesized that the color of tree trunks was a major influence. Peppered moths normally rest on tree trunks during the daytime. If they are noticed by birds, they are eaten. In industrialized regions of England, the trees and other objects were blackened by the presence of soot from the smokestacks of coal-burning power plants. Kettlewell discovered that, in these areas, the light-colored moths were more conspicuous and were more frequently located and eaten by the local bird population. (See figure 4.6.) However, in areas where the trees were light colored because of lower air pollution and were covered with light-colored lichens, the birds had greater difficulty locating the light-colored moths, and the dark-colored moth was rare. In these two different locations, natural selection, through the act of birds killing and eating moths, was selecting for those moths that were dark in the one case and for those that were light in the other. In both cases, the moth population that remained was better adapted to the environment than that which was previously present.

When natural selection occurs over long periods of time (thousands to millions of years), a whole new species of organisms can result. Recall that Darwin wrote about natural selection, but the title of the book is *The Origin of Species*. A **species** is a group of organisms that can interbreed and produce offspring capable of reproduction. Hybrids such as the mule, which is a cross between two species (horse and donkey), are generally sterile and therefore are not considered species. When one species is divided into physically separated subpopulations, it is possible that, as each subpopulation is subjected to natural selection, the differences between the

FIGURE 4.6
Natural Selection in Peppered Moths.
Air pollution in industrial regions of England resulted in blackened tree trunks. The dark form of peppered moths was less conspicuous and, therefore, they were not seen as readily by their bird predators. The light form was more easily seen and more frequently eaten. The number of dark moths increased and the number of light moths decreased, because the predators ate more of the light form of the moth, preventing them from breeding and passing on their genes for light color.

FIGURE 4.7
Predator/Prey Relationship. Lions are predators on zebras. The quicker lions are more likely to get food, and the slower, sickly, or weaker zebras are more likely to become prey.

two populations become so great that they eventually become incapable of interbreeding with each other. This process of developing new species is called **speciation.**

In a similar fashion, it is possible to imagine the gradual change of organisms over millions of years as their environment changes and they are continually subjected to the action of natural selection. This process is known as **evolution.** The key idea in all of this discussion is that species of organisms change and that these changes result in organisms that are well adapted to the environment in which they live.

Kinds of Organism Interactions

The concept of natural selection allows us to see how interacting organisms can result in populations that are better adapted to their environment. We can also recognize that the way organisms interact can be organized into several broad categories.

Predation

One common kind of interaction occurs when one animal, known as a **predator,** kills and eats another, known as the **prey.** (See figure 4.7.) Different predator-prey relationships include lions and zebras, birds and worms, wolves and moose, and frogs and insects. Predator-prey relationships are often thought to be one-sided. The benefits seem to be totally with the predator. The predator has the meal and goes on living after killing and eating the prey. However, this relationship can also be seen as having value for the prey as well. Prey species have a higher reproductive rate than predator species. For example, field mice may have ten to twenty offspring per year, while hawks typically have two to three. Because of this high reproductive rate, prey species can endure a high mortality rate. Certainly, the individual organism that has been killed and eaten did not benefit, but the species did, since the prey organisms that die are likely to be the old, the slow, the sick, and the less fit members of the population. The healthier, quicker, and more fit individuals are more likely to survive. When these survivors reproduce, their offspring are more likely to be better adapted to their environment. At the same time, a similar process is taking place in the predator population, since poorly adapted individuals are less likely to

FIGURE 4.8

Competition. Whenever a needed
resource is in limited supply, organisms
compete for it. This competition may be
between members of the same species
and, therefore, is called intraspecific
competition or it may be between
different species and is called
interspecific competition. This
photograph shows several vultures
competing for a food source.

capture prey. The predator is a participant in the natural selection process
and so is the prey. This dynamic relationship between predator and prey
species is a complex one that continues to intrigue ecologists.

Competition

A second type of interaction between species is competition. **Competition**
is a kind of interaction in which two organisms strive to obtain the same
limited resource, and in the process both organisms are harmed to some
extent. (See figure 4.8.) However, this does not mean that there is no winner.
If two robins are competing for the same worm, only one gets it. Both organisms were harmed because they had to expend energy in fighting for
the worm, but one got some food and was harmed less than the one that
fought but got nothing. This example of competition, which is between
members of the same species, is known as **intraspecific competition.** This
kind of competition might involve competing for food, mates, sunlight, soil
nutrients, den sites, or a great variety of other environmental factors that
are limited.

Competition among members of the same species does not only have
a negative impact. As was the case with predation, competition can have
its good side when seen from the point of view of the species as a whole.
When resources are limited, the less fit organisms are more likely to die or
be denied mating privileges, and consequently, the next generation of organisms will contain more of the genetic characteristics that are favorable
for survival in that particular environment. Since all individuals of the same
species have very similar requirements, competitive pressure among individuals of the same species is usually very intense. A slight advantage may
mean the difference between survival and death. Natural selection is responsible for maintaining a well-adapted population.

Competition may also occur among organisms of different species. This
is called **interspecific competition.** In a forest, very little light reaches
the forest floor. Therefore, the different species of plants present must compete for the available light. Mosses and ferns can tolerate shade while grasses
cannot; therefore, the grasses lose in this competitive interaction.

The more similar two kinds of individuals are, the more intense the
competition between them. If one of the two competing species is better

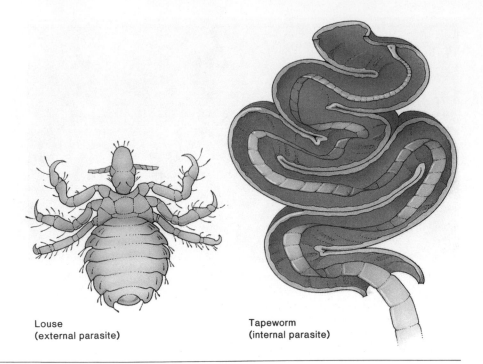

FIGURE 4.9

Parasitism. Lice are small insects that live in the feathers of birds or the fur of mammals, where they bite their hosts to obtain blood. Since they live on the outside of their hosts, they are called ectoparasites. Tapeworms live inside the intestines of their hosts, where they absorb food from their hosts' intestines. Since they live inside their hosts, they are called endoparasites.

Louse
(external parasite)

Tapeworm
(internal parasite)

adapted to live in the area than the other, the less fit organism must either evolve into a slightly different niche, migrate to a different geographic area, or become extinct. As with intraspecific competition, one of the effects of competition is that organisms emerge from the interaction better adapted to their environment.

Symbiotic Relationships

Symbiosis is a close, long-lasting, physical relationship between two different species of organisms. In other words, the two organisms are usually in physical contact and at least one of the organisms derives some sort of benefit from this contact. There are three different categories of symbiotic relationships: parasitism, commensalism, and mutualism.

Parasitism

Parasitism is a relationship in which one organism, known as the **parasite,** lives in or on another organism, known as the **host,** from which it derives nourishment. Generally, the parasite is much smaller than the host. Although the host is harmed by the interaction, it is generally not killed by the parasite. Because the evolution of a parasitic way of life involves a long-standing interaction between two species—the parasite and the host—they have generally evolved in such a way that they can accommodate one another. It is not in the parasite's best interest to kill its host. If it does, it must find another. Likewise, the host has been evolving defenses against the parasite, often reducing the harm done by the parasite to something the host can tolerate.

Those parasites that live on the surface of their host are known as **ectoparasites.** Fleas, lice, and some molds and mildews are examples of parasites that live on the surface of their hosts. (See figure 4.9.) Many other parasites, like tapeworms, malaria parasites, many kinds of bacteria, and some fungi, are called **endoparasites,** which live inside the bodies of their hosts. The tapeworm lives in the intestines of its host, where not only is it able to resist being digested, but it also makes use of the nutrients in the intestine. If the host has only one or two tapeworms, it can live for some time with little discomfort, supporting itself and its parasites. However, if the number of parasites is large, it may result in the death of the host. Mistletoe is a flowering plant that is parasitic on trees. It gets established on the surface of a tree when a bird transfers the seed to the tree. It then grows down into the tissues of the tree and uses the tree as a source of nutrients. Parasitism is a very common kind of niche. If we were to categorize all the organisms in the world, we would find that there are many more parasitic species than there are nonparasitic species. Each organism, including you, has many others that use it as a host.

Commensalism

If the relationship between organisms is one in which one organism benefits while the other is not affected, it is called **commensalism.** It is possible to visualize a parasitic relationship evolving into a commensal one. Since parasites generally evolve to do as little harm to their host as possible and the host is combating the negative effects of the parasite, they might eventually evolve to the point that the host is not harmed at all. There are many examples of commensal relationships. Many orchids use trees as a surface upon which to grow. The tree is not harmed or helped, but the orchid needs a surface upon which to establish itself and also benefits by being close to the top of the tree, where it can get more sunlight and rain. Some mosses, ferns, and many vines make use of the surfaces of trees in this way.

In the ocean, many sharks have a smaller fish known as a remora attached to them. Remoras have a sucker on the top of their heads that they can use to attach to the shark. In this way, they can hitchhike a ride as the shark swims along. When the shark feeds, the remora is able to obtain small bits of food that the shark misses. The shark does not appear to be positively or negatively affected by the remoras. (See figure 4.10.)

FIGURE 4.10
Commensalism. Remoras hitchhike a ride on sharks and feed on the scraps of food lost by the sharks. This is a benefit to the remoras. The sharks do not appear to be affected by the presence of the remoras.

Mutualism

Some symbiotic relationships are actually beneficial to both species of organisms involved. This kind of a relationship is called **mutualism.** In many mutualistic relationships, the relationship is obligatory. Flowers of the yucca plant are pollinated by a specific insect known as the yucca moth. The moth, in turn, lays it eggs in the flower, where the immature moth larvae feed on the developing seeds. The yucca is dependent on the moth for pollination, and the moth is dependent on the seeds as a source of food for its larvae. Neither organism can exist without the other. The number of eggs laid by the moth is small enough that some of the yucca seeds grow to maturity to provide new yucca plants for future generations of yucca moths. Many other flowering plants are also dependent on specific insects as pollinators.

One of the nutrients that plants need is nitrogen, which is usually in short supply in the soil. Many kinds of plants, such as beans, clover, and alder trees, have bacteria that live in their roots in little nodules. These nodules are formed by the roots when they are infected with certain kinds of bacteria. However, the bacteria do not cause disease, but provide the plants with nitrogen-containing molecules that the plants can use for growth. The nitrogen-fixing bacteria benefit from the living site and nutrients that the plants provide, and the plants benefit from the nitrogen they receive. (See figure 4.11.)

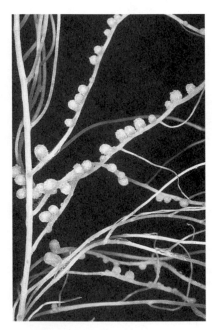

FIGURE 4.11
Mutualism. The growth on the root of this plant contains bacteria that are beneficial to the plant because they make nitrogen available to it. The relationship is also beneficial to the bacteria because the plant makes necessary raw materials available to the bacteria. It is a mutually beneficial relationship.

Community and Ecosystem Interactions

Thus far, we have discussed specific ways in which individual organisms interact with one another and with their physical surroundings. This does not accurately depict the complexity of interactions that occur among various assemblages of organisms that affect one another in varied and intricate ways. We have come to think of interacting groups of organisms as **communities** in which each organism has a specific niche or role to play. The removal of any role would result in a change in the community. Some organisms play minor roles while others play major roles, but all are part of the community. For example, the grasses of the prairie have a major role since, without them, there would be no prairie, but a meadowlark, while it is a conspicuous and colorful part of the prairie scene, has little to do with maintaining a prairie community.

Communities are generally thought of as consisting of interacting species, but these species do not interact in a vacuum. These same organisms must interact with the physical world around them as well. The physical world has a major impact on what kinds of plants and animals can live in an area. We do not expect to see a banana tree in the arctic or a walrus in the Mississippi River. Banana trees are adapted to warm, moist, tropical areas, and walruses require cold ocean waters. However, organisms can also have an impact on their physical surroundings. Trees break the force of the wind, grazing animals form paths, and earthworms create holes in the soil. This system of interacting organisms and their nonliving surroundings is frequently referred to as an **ecosystem.** While the two ideas of community and ecosystem are closely related, an ecosystem is a broader concept because it involves the physical as well as the biological realms.

Every system has parts that are related to one another in very specific ways. A bicycle has wheels, a frame, handlebars, brakes, pedals, and a seat. They must be organized in a very specific way or the system known as a bicycle does not function. Similarly, an ecosystem has parts that must be organized in a specific way or the system will not operate. We can look at ecosystems from three different points of view: the major kinds of roles played by organisms, the way energy is utilized within ecosystems, and the way atoms are cycled from one organism to another.

Major Roles of Organisms

Several different categories of organisms are found in any ecosystem. **Producers** are able to make new, complex, organic material from the atoms in their environment. To do this, they must have a source of energy. In nearly all ecosystems, this energy is supplied by the sun, and the organisms that utilize this energy are plants that carry on photosynthesis. Since producers are the only organisms in an ecosystem that can trap energy and make new organic material from inorganic material, all other organisms must rely on producers as a source of food, either directly or indirectly. These other organisms are called **consumers** because they consume organic matter to provide themselves with energy and the organic molecules necessary to build their own bodies. **Primary consumers** are those that eat producers (plants) as a source of food. They are also known as **herbivores. Secondary consumers** or **carnivores** eat other animals. Some carnivores eat herbivores, while others may eat other carnivores. In addition, many kinds of animals have mixed diets that include both plants and animals. These kinds of animals are usually called **omnivores.**

A final category of consumer is the decomposer. **Decomposers** use nonliving organic matter as a source of food. Many small animals, fungi, and bacteria would fill this niche. (See table 4.1.)

Energy Flow through Ecosystems

An ecosystem is a stable, self-regulating unit. To maintain itself, it must have a continuous input of energy. The only significant source of energy for most ecosystems is sunlight energy. Producers are the only organisms that are capable of trapping this energy through the process of photosynthesis and making it available to the ecosystem. The energy is stored in the

TABLE 4.1
Roles in an ecosystem.

Category	Major role or action	Examples
Producer	Converts simple inorganic molecules into organic molecules by the process of photosythesis	Trees, flowers, grasses, ferns, mosses, algae
Consumer	Uses organic matter as a source of food	Animals, fungi, bacteria
Herbivore	Eats plants directly	Grasshopper, elk, human vegetarian
Carnivore	Kills and eats animals	Wolf, pike, dragonfly
Omnivore	Eats both plants and animals	Rats, raccoons, most humans
Scavenger	Eats meat, but often gets it from animals that died by accident or illness, or that were killed by other animals	Coyote, vulture, blowflies
Parasite	Lives in or on another living organism and gets food from it	Tapeworm, many bacteria, some insects
Decomposer	Returns organic material to inorganic material; completes recycling of atoms	Fungi, bacteria, some insects and worms

form of chemical bonds in large organic molecules such as carbohydrates (sugars, starches), fats, and protein. The energy in these molecules can be transferred to other organisms when the organisms consume the plants. Each step in the flow of energy through an ecosystem is known as a **trophic level.** Producers (plants) constitute the first trophic level, and herbivores constitute the second trophic level. Carnivores that eat herbivores would be the third trophic level, and carnivores that eat other carnivores would be a fourth trophic level. Omnivores, parasites, and scavengers occupy different trophic levels, depending on what they happen to be eating at the time. If we eat a piece of steak, we are at the third trophic level; if we eat celery, we are at the second trophic level. (See figure 4.12.)

As energy passes from one trophic level to the next, some of the useful energy is lost due to the second law of thermodynamics. Much of this loss is in the form of low-quality heat, which is dissipated to the surroundings to warm the air, water, or soil. In addition to this loss of heat, organisms must expend energy to maintain their own life processes. It takes energy to chew food, defend nests, walk to waterholes, or raise offspring. Therefore, the amount of energy contained in higher trophic levels is considerably less than that at lower trophic levels. Approximately 90 percent of the useful energy is lost with each transfer to the next highest trophic level. So in any ecosystem, the amount of energy contained in the herbivore's trophic level is only about 10 percent of the energy contained in the producer's trophic level. The amount of energy at the third trophic level would be approximately 1 percent of that found in the first trophic level.

Because it is technically difficult to actually measure the amount of energy contained in each trophic level, ecologists often use other measures to approximate the relationship between the amounts of energy at each trophic level. One of these is to measure the biomass present. The **biomass** is the weight of living material in a trophic level. It is often possible in a simple ecosystem to collect and weigh all the producers, herbivores, and carnivores. When this is done, the weights often show the same 90 percent loss as one passes from one trophic level to the next as happens with the amount of energy present.

Food Chains

The passage of energy from one trophic level to the next as a result of one organism consuming another is known as a **food chain.** For example,

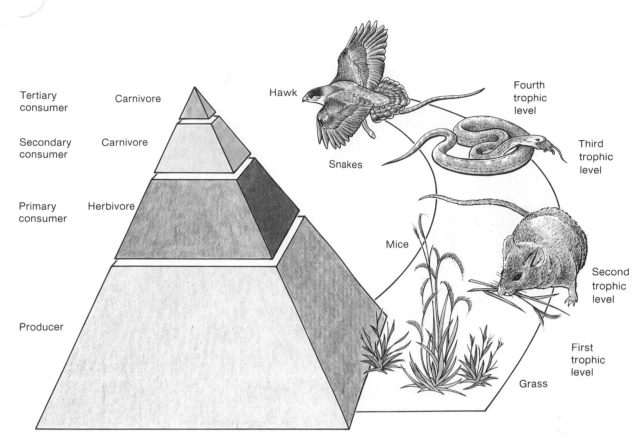

FIGURE 4.12

Energy Flow through an Ecosystem.
As energy flows through an ecosystem, it passes through several levels known as trophic levels. Each time energy moves to a new trophic level, approximately 90 percent of the useful energy is lost. Therefore, high trophic levels contain less energy and fewer organisms in most ecosystems.

willow trees grow well in very moist soil that might be present near a stream, river, or pond. The plant uses its leaves to capture sunlight and convert carbon dioxide and water into sugars and other organic molecules. The leaves of the willow serve as a food source for insects, such as caterpillars and leaf beetles, with chewing mouth parts and a digestive system adapted to plant food. Some of these insects fall from the tree into the pond below, where they are consumed by a frog. As the frog swims from one lily pad to another, a large bass consumes the frog. Occasionally, a human may use an artificial frog as a lure to entice the bass from its hiding place. A fish dinner is the final step in this long chain of events that began with the leaves of a willow tree. (See figure 4.13.) This food chain has five trophic levels. Each organism occupies a specific niche and has special abilities that fit it for the niche, and each organism in the food chain is involved in converting energy and matter from one form to another.

When a plant or animal dies, the chemical energy contained within its body is ultimately released as heat by organisms that decompose the body into smaller molecules, such as water and carbon dioxide. These decomposers—bacteria, fungi, and some small consumers, such as insects and worms—play a significant role in the ultimate recycling of the atoms that make up the bodies of living things. They are involved in the breakdown of small bits of organic material called **detritus.** Detritus food chains are found in a variety of situations. The bottoms of the deep lakes and oceans are too dark for photosynthesis. The animals and decomposers that live there rely on a steady rain of small bits of organic matter from the upper layers of the water where photosynthesis does take place. Similarly, in most streams, leaves and other organic debris serve as the major source of organic material and energy. A sewage treatment plant is also a detritus food

FIGURE 4.13
Food Chain. As one organism feeds upon another organism, there is a flow of energy through the series. This is called a food chain.

chain in which particles and dissolved organic matter are constantly supplied to a series of bacteria and protozoa that use this material for food.

Food Webs

In another example, the soil on a forest floor receives leaves, which also fuel a detritus food chain. In detritus food chains, a mixture of insects, crustaceans, worms, bacteria, and fungi cooperate in the breakdown of the large pieces of organic matter, while at the same time feeding on one another. For example, when a leaf dies and falls to the forest floor, it is colonized by bacteria and fungi, which begin the breakdown process. An earthworm will also feed on the leaf and at the same time consume the bacteria and fungi. If that earthworm is eaten by a bird, it becomes part of a larger food chain that includes both material from a detritus and from a photosynthesis-driven food chain. When several food chains overlap and intersect, they make up a **food web.** (See figure 4.14.)

Notice in the upper-left-hand corner of figure 4.14 that the Coopers and sharpshinned hawks use many different kinds of birds as a source of food. These hawks fit into several different food chains. If one source of prey is in short supply, they can switch to something else without too much trouble. These kinds of complex food webs tend to be more stable than simple food chains with few cross-links.

Nutrient Cycles in Ecosystems

Organisms are composed of molecules and atoms that must be cycled from one living organism to another. Some atoms are more common in living things than others. Carbon, nitrogen, oxygen, hydrogen, and phosphorus are found in all living things and are recycled when an organism dies.

Carbon Cycle

Carbon and oxygen combine to form the molecule carbon dioxide, which is a gas present in the atmosphere in small quantities. During photosynthesis, carbon dioxide from the atomosphere is taken into a plant leaf and combined with hydrogen from water molecules, which were absorbed from the soil by the roots and transported to the leaf, to form complex organic molecules. Oxygen molecules are released from the leaf into the atmosphere when water molecules are split to provide hydrogen atoms for the organic molecules. In this total process, light energy is converted to chemical-bond energy in organic molecules such as sugar. These sugars are used by plants for growth and to provide energy for other processes.

Herbivores can use these complex organic molecules as food. When an herbivore eats a plant, it breaks down the complex organic molecules into simpler molecules, which can be incorporated into the chemical structure of the animal. The carbon atom, which was once part of a plant, is now part of an herbivore. All organisms also carry on the process of respiration, in which oxygen from the atmosphere is used to break down large organic molecules into carbon dioxide and water. Much of the chemical-bond energy is released by respiration and is lost as heat, but the remainder is used by the herbivore for movement, growth, and other activities.

When an herbivore is eaten by a carnivore, the carbon-containing molecules of the herbivore become incorporated into the body of the carnivore. Carnivores also carry on respiration and release carbon dioxide and water in the process of obtaining energy from organic molecules.

The waste products of all kinds of organisms, and any organisms that die, are acted upon by decomposers so that even the naturally occurring

Ecological Principles and Their Application

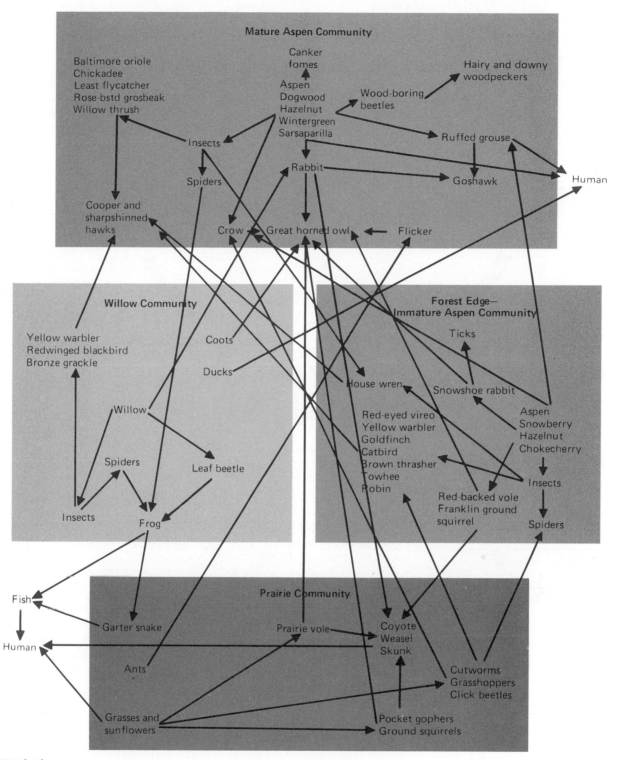

FIGURE 4.14

Food Web. The many kinds of interactions among organisms in an ecosystem constitute a food web. In this network of interactions, several organisms would be affected if one key organism were reduced in number. Look at the rabbit in the mature aspen community and note how many organisms use it as food.

Source: Adapted from Ralph D. Bird, "Biotic Communities of the Aspen Parkland of Central Canada." *Ecology* 11 (April 1930): 410.

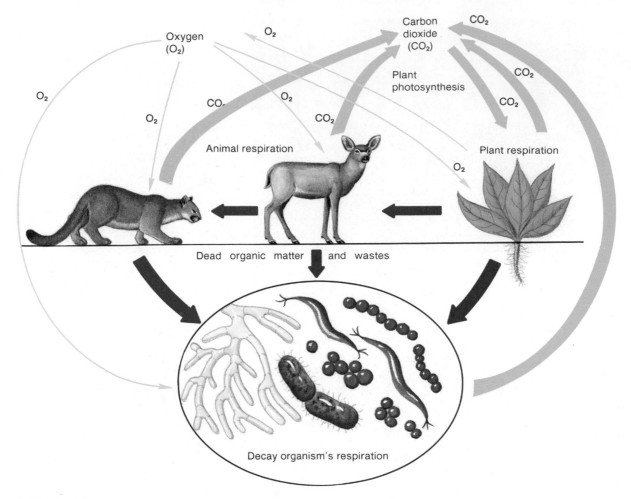

FIGURE 4.15

Carbon Cycle. Carbon atoms are cycled through ecosystems. Plants can incorporate carbon atoms from carbon dioxide into organic molecules when they carry on photosynthesis. These carbon-containing organic molecules are passed to animals when they eat plants or other animals. Organic wastes or dead organisms are consumed by decay organisms. All organisms, plants, animals, and decomposers return carbon atoms to the atmosphere when they carry on respiration. Carbon atoms are being cycled at the same time that oxygen atoms are being cycled.

organic molecules in nonliving matter are used by decomposers to provide energy. Carbon dioxide is produced in the process. The carbon atom began as part of a carbon dioxide molecule in the atmosphere, became part of an organic molecule in one or more organisms, and then was released back into the atmosphere by the process of respiration. The mechanism of flow of carbon atoms is known as the **carbon cycle.** (See figure 4.15.)

Nitrogen Cycle

Another very important cycle, the **nitrogen cycle,** involves the flow of nitrogen atoms through organisms in an ecosystem. Seventy-eight percent of the gas in the air we breathe is made up of molecules of nitrogen gas. However, very few organisms are able to use it in this form. Since plants are at the base of nearly all food chains, they must make new nitrogen-containing molecules, such as proteins and DNA. Plants are unable to use the nitrogen in the atmosphere and must get it in the form of nitrate (NO_3^-) or ammonia (NH_3). The amount of nitrogen available to plants controls their growth and is often a limiting factor. The primary way in which plants obtain nitrogen compounds they can use is with the help of bacteria.

Some bacteria are able to convert the nitrogen gas (N_2) in the atmosphere into forms that plants can use. These bacteria are called **nitrogen-fixing bacteria.** Some kinds of these bacteria live in the soil and are called **free-living nitrogen-fixing bacteria,** while others live in nodules in the roots of plants known as legumes (peas, beans, and clover) and are known

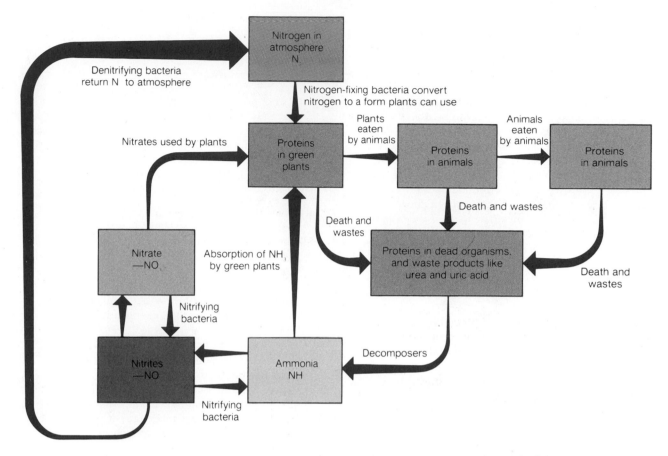

Denitrifying bacteria
return N to atmosphere

Nitrogen in
atmosphere
N

Nitrogen-fixing bacteria convert
nitrogen to a form plants can use

Nitrates used by plants

Plants
eaten
by animals

Animals
eaten
by animals

Proteins
in green
plants

Proteins
in animals

Proteins
in animals

Nitrate
—NO

Absorption of NH
by green plants

Death and
wastes

Death and
wastes

Proteins in dead organisms
and waste products like
urea and uric acid

Death and
wastes

Nitrifying
bacteria

Nitrites
—NO

Ammonia
NH

Decomposers

Nitrifying
bacteria

FIGURE 4.16

Nitrogen Cycle. Nitrogen atoms are
cycled in ecosystems. Atmospheric
nitrogen is converted by nitrogen-fixing
bacteria to a form that plants can use
to make protein and other compounds.
Proteins are passed to other organisms
when one organism is eaten by another.
Dead organisms and waste products are
acted upon by decay organisms to form
ammonia, which may be reused by
plants or converted to other nitrogen
compounds by other kinds of bacteria.
Denitrifying bacteria are able to convert
inorganic nitrogen compounds into
atmospheric nitrogen.

as **symbiotic nitrogen-fixing bacteria.** Some grasses and evergreen trees
appear to have a similar relationship with certain root fungi that seem to
improve the nitrogen-fixing capacity of the plant.

Other kinds of bacteria can also contribute to making nitrogen available
to plants. Dead organisms and their waste products contain molecules
that contain nitrogen. Many kinds of soil bacteria decompose these large
nitrogen-containing organic molecules, releasing ammonia, which can be
used directly by many kinds of plants or be converted to nitrate by other
kinds of bacteria. An additional kind of bacteria known as **denitrifying
bacteria** is, under certain conditions, able to convert a molecule known
as nitrite to atmospheric nitrogen gas where the nitrogen atoms can enter
the cycle again with the aid of nitrogen-fixing bacteria. (See figure 4.16.)

In naturally occurring soil, nitrogen is often a limiting factor of plant
growth. To increase yields, farmers often provide extra sources of nitrogen
in several different ways. Inorganic fertilizers are a primary method of increasing
the nitrogen available. These fertilizers may contain ammonia, nitrate,
or both.

Since the manufacture of nitrogen fertilizer takes a lot of energy, it means
that fertilizer is expensive, and farmers have looked for alternative methods
to reduce this cost. A farmer might alternate nitrogen-demanding crops like
corn with nitrogen-yielding crops like soybeans. Soybeans are legumes that
have symbiotic nitrogen-fixing bacteria in their roots. If soybeans are planted
one year, then the excess nitrogen left in the soil can be used by the corn
plants grown the next year. A slightly different technique involves the
growing of a nitrogen-fixing crop for a short period of time and then plowing

the crop under and letting the organic matter decompose. Farmers can also add nitrogen to the soil by spreading manure on the field and relying on the soil bacteria to decompose the organic matter and release the nitrogen for plant use.

Phosphorus Cycle

Phosphorus is another kind of atom common in the structure of living things. It is present in many important biological molecules such as DNA and in the membrane structure of cells. In addition, the bones and teeth of animals contain significant quantities of phosphorus. The ultimate source of phosphorus atoms is rock. In nature, new phosphorus compounds are released by the erosion of rock. Plants use the dissolved phosphorus compounds to construct the molecules they need. Animals consume plants or other animals and obtain needed phosphorus. When an organism dies or excretes waste products, decomposer organisms recycle the phosphorus compounds back into the soil. (See figure 4.17.) In many soils, phosphorus is in short supply and must be provided to plants under agricultural conditions to get maximum yields.

Fertilizers usually contain nitrogen, phosphorus, and potassium compounds. The numbers on a fertilizer bag indicate the percentage of each

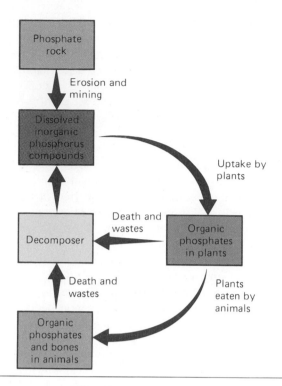

FIGURE 4.17
Phosphorus Cycle. The source of phosphorus is rock that, when dissolved, provides a source of phosphate to be used by plants and animals.

in the fertilizer. For example, a 6-24-24 fertilizer has 6 percent nitrogen, 24 percent phosphorus, and 24 percent potassium compounds. In addition to carbon, nitrogen, and phosphorus, potassium and other elements are cycled within ecosystems. In an agricultural ecosystem, these elements are removed when the crop is harvested. Therefore, the farmer must return these elements to the soil with fertilizer.

CONSIDER THIS CASE STUDY
The Reintroduction of the Moose into Michigan

At one time, much of the American continent was covered by forests. The Upper Penninsula of Michigan was the southern portion of a spruce and fir forest, which originally supported such animals as wolves, moose, the martin, and the fisher. During the 1800s, mining was the main occupation in the region, and trees were cut for lumber, to heat homes, to provide mining timbers, and to provide charcoal for smelting. This destroyed much of the wilderness character of the forest, and wolves, moose, and many other kinds of animals disappeared due to the combination of reduced habitat and hunting. Since moose are browsing animals, they require large amounts of small trees as a source of food and travel over large areas to get what they need.

Over many years, the importance of the mining industry diminished, people were less dependent on fuel wood for heating, and the population declined, allowing many areas to regrow to something approximating the original forest. In 1986, the state of Michigan and the province of Ontario worked out a swap of animals. Michigan would provide wild turkeys for a restocking effort in Ontario, and Ontario would provide moose for restocking in the Upper Penninsula of Michigan. The moose were checked by veterinarians, tagged with radio collars so they could be followed, and then released. Currently, the animals seem to be establishing a viable population in the area.

What value does such a reintroduction have?
What kinds of relationships existed between moose and humans?
What made reintroduction possible?

Summary

Everything that affects an organism during its lifetime is collectively known as its environment. The environment of an organism can be divided into biotic (living) and abiotic (nonliving or physical) components.

The space an organism occupies is known as its habitat, and the role it plays in its environment is known as its niche. The niche of an organism is the result of natural selection directing the adaptation of the organism to a specific set of environmental conditions.

Organisms interact with one another in a variety of ways. Organisms that have the same needs compete with one another and do mutual harm, but one is usually harmed less and survives. Symbiotic relationships are those in which organisms live in physical contact with one another. Parasites live in or on another organism and derive benefit from the relationship, harming the host in the process. Commensal organisms derive benefit from another organism but do not harm the host. Mutualistic organisms both derive benefit from their relationship.

A community is a set of interacting groups of organisms. Those organisms and their abiotic environment constitute an ecosystem. In an ecosystem, energy flows from producers through various trophic levels of consumers (herbivores, carnivores, omnivores, and decomposers). About 90 percent of the energy is lost as it passes from one trophic level to the next. This means that the amount of biomass at higher trophic levels is usually much less than that at lower trophic levels. The sequence of organisms through which energy flows is known as a food chain. Several interconnecting food chains constitute a food web.

The flow of atoms through an ecosystem involves all the organisms in the community. The carbon, nitrogen, and phosphorus cycles are examples of how these materials are cycled in ecosystems.

Review Questions

1. Define *environment.*
2. Describe, in detail, the niche of a human.
3. How is natural selection related to the concept of niche?
4. List five predators and their prey organisms.
5. How is an ecosystem different from a community?
6. Humans raising cattle for food is what kind of relationship?
7. Give examples of organisms that are herbivores, carnivores, omnivores.
8. What are some different trophic levels in an ecosystem?
9. Describe the carbon cycle, the nitrogen cycle, and the phosphorus cycle.
10. Analyze an aquarium as an ecosystem. Identify the major abiotic and biotic factors. List members of the producer, primary consumer, secondary consumer, and decomposer trophic levels.

CHAPTER FIVE
Kinds of Ecosystems and Communities

Objectives

After reading this chapter, you should be able to:

Recognize the difference between primary and secondary succession.

Describe the process of succession from pioneer to climax community in both terrestrial and aquatic situations.

Associate typical plants and animals with the various terrestrial biomes.

Recognize the physical environmental factors that determine the kind of climax community that will develop.

Differentiate the forest biomes that develop based on temperature and rainfall.

Describe the various kinds of aquatic ecosystems and the factors that determine their characteristics.

Chapter Outline

Key Terms

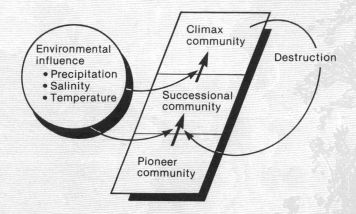

abyssal ecosystem
alpine tundra
benthic
benthic ecosystem
biochemical oxygen
  demand (BOD)
biome
boreal forest
climax community
coral reef ecosystem
desert
emergent plants
estuary
euphotic zone
eutrophic lake
freshwater ecosystem
grassland
limnetic zone
littoral zone
mangrove swamp
  ecosystem
marine ecosystem
marsh
northern coniferous
  forest

oligotrophic lake
pelagic
pelagic ecosystem
periphyton
permafrost
phytoplankton
pioneer community
prairie
primary succession
savanna
secondary succession
seral stage
sere
steppe
submerged plants
succession
successional stage
swamp
taiga
temperate deciduous
  forest
tropical rain forest
tundra
zooplankton

FIGURE 5.1
Pioneer Organism. The lichen growing on this rock is able to accumulate bits of debris, carry on photosynthesis, and aid in breaking down the rock. All of these activities contribute to the formation of a thin layer of soil, which is necessary for the success of plants in the early stages of succession.

Succession

Ecosystems are dynamic changing units. On a daily basis, plants grow and die, animals feed on plants and one another, and ultimately, decomposers recycle the chemical elements that make up the biotic portion of any ecosystem. Since all organisms are linked together in a community, any change in the community is going to affect a large number of different kinds of organisms within it. Certain conditions within a community are keys to the kinds of organisms that develop in association with one another. Grasshoppers need grass for food, robins need trees to build nests, and herons need shallow water to find food. Each organism has specific requirements that must be met in a community, or it will not survive.

Over long periods of time, it is possible to see trends in the way the structure of a community changes. These regular, predictable changes in the structure of a community over time are called **succession.** Succession occurs because organisms cause changes in their surroundings that make the environment less suitable for themselves and more suitable for other kinds of organisms. One community of organisms is replaced by a slightly changed community, which itself is replaced by a subsequent collection of organisms. Ecologists recognize two different kinds of succession: **primary succession,** which begins with bare mineral surfaces or water, and **secondary succession,** which begins with the destruction or disturbance of an existing ecosystem.

Primary Succession

Primary succession can begin on a bare rock surface, pure sand, or standing water. Since succession on rock and sand is somewhat different from that which occurs with watery situations, we deal with them separately. We discuss terrestrial succession first.

Terrestrial Primary Succession

Bare rock is a very inhospitable place for organisms to live. The temperature changes drastically, there is little moisture, the organisms are exposed to the damaging effect of the wind, few nutrients are available, and few places are available for organisms to attach themselves or hide. Primarily what is lacking is soil. However, a few kinds of organisms can survive in even this kind of environment. These are known as the **pioneer community** because they are the first to colonize bare rock. (See figure 5.1.)

The dominant organism in this initial community is something called a lichen. Lichens are actually mutualistic relationships between two kinds of organisms: algae that are green and carry on photosynthesis and fungi that retain moisture and attach to the rock surface. The growth and development of lichens is a slow process. It may take lichens one hundred years to grow as large as a dinner plate. A lichen serves as a producer in this simple ecosytem, and many tiny organisms may be found associated with a lichen. They may feed on it and use it as a place of shelter since even a drizzle is like a torrential rain for a microscopic animal. Since lichens are firmly attached to rock surfaces, they also tend to accumulate bits of airborne debris and store small amounts of water that would otherwise blow away or run off the rock surface. Furthermore, acids produced by the lichen tend to cause the breakdown of the rock substrate into smaller particles. This fragmentation of rock, aided by physical and chemical weathering processes, along with the trapping of debris and the contribution of organic matter by the death of lichens and other organisms, ultimately leads to the accumulation of a very thin layer of soil.

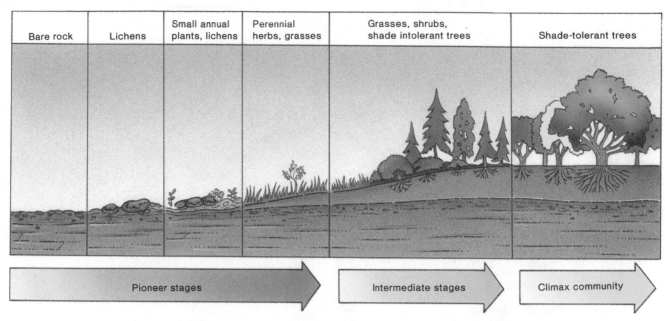

| Bare rock | Lichens | Small annual plants, lichens | Perennial herbs, grasses | Grasses, shrubs, shade intolerant trees | Shade-tolerant trees |

Pioneer stages → Intermediate stages → Climax community →

Hundreds of years

FIGURE 5.2

Primary Succession on Land. The formation of soil is a major step in primary succession. Until soil is formed, the area is unable to support large amounts of vegetation, which modify the harsh environment. Once soil formation begins, the site proceeds through an orderly series of stages toward a climax community.

This thin layer of soil is the key to the next stage in the successional process. The thin layer of soil can support some fungi, certain small worms, insects, bacteria, protozoa, and perhaps a few tiny annual plants. As these organisms grow, reproduce, and die, they contribute additional organic material for the soil-building process. This stage eliminates the lichen community and in return is replaced by a community of small perennial plants. The perennial grasses and herbs are eventually replaced by shrubs, which are often replaced by trees that require lots of sunlight, which are replaced by trees that can tolerate shade. Eventually, a relatively stable, long-lasting, more complex, and interrelated community of plants, animals, fungi, and bacteria is produced, known as the **climax community.** Each step in this process is called a **successional stage,** or **seral stage,** and the entire sequence of stages—from pioneer community to climax community—is called a **sere.** (See figure 5.2.) The specific kind of climax community produced depends on such things as climate and soil type, which are discussed in greater detail later in this chapter. However, certain characteristics are typical of climax communities regardless of the specific type: (1) Climax communities are able to reproduce themselves; (2) Climax communities are in energy balance. While successional communities tend to accumulate large amounts of new material (gain energy), climax communities tend to have just as much material and energy leave the system as enter it; (3) Climax communities also tend to have larger numbers of kinds of organisms and kinds of interactions among organisms than does a successional community. The general trend is toward increasing complexity and energy efficiency in climax communities, as compared to the successional communities that preceeded them.

Aquatic Primary Succession

The principal concepts of land succession can also be applied to aquatic ecosystems. Except for the oceans, most aquatic ecosystems are considered temporary. Certainly, some are going to be around for thousands of years, but eventually, they will disappear and be replaced by terrestrial ecosystems as a result of normal successional processes. All aquatic ecosystems

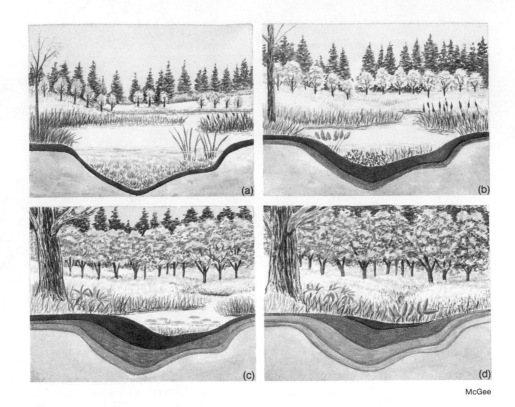

McGee

FIGURE 5.3

Primary Succession from Water.
(*a*) Lakes are in the process of being filled in by sediment from the surrounding land and the organic matter produced by plants in the water. (*b*) As more material accumulates in the lake the shallower portions can eventually support emergent vegetation, (*c*) which is followed by grasses and woody plants as the area becomes dryer. (*d*) Eventually a climax community is produced.

have a continuous input of soil particles and organic matter from surrounding land, which results in the gradual filling in of shallow bodies of water like ponds and lakes.

In deep portions of lakes and ponds, only floating plants and algae can exist, but as the amount of sediment accumulates, it becomes possible for submerged plants to become established on the bottom of shallow bodies of water. They carry on photosynthesis, resulting in a further accumulation of organic matter, along with the trapping of sediments that flow into the pond or lake from the streams or rivers. Eventually, as the water becomes shallower, emergent plants become established. They have leaves that float on the surface or project into the air. The network of roots and stems below the surface of the water results in the accumulation of more material, and the water becomes even shallower as material accumulates on the bottom. As the process continues, more sediment accumulates, the wet soil thus formed begins to dry out, and grasses and other plants that can live in wet soil become established. This is often called a wet meadow. Once this occurs, the stage is set for a typical terrestrial successional series of changes, eventually resulting in a climax community. (See figure 5.3.)

In many northern ponds and lakes, where sphagnum moss is a common organism, a floating mat of vegetation may become established and support a new community of organisms composed of terrestrial species that can tolerate wet soil. (See figure 5.4.)

Since the shallower portions of most lakes and ponds are at the shore, it is often possible to see the various stages in aquatic succession from the shore. In the central, deeper portions of the lake, there are only floating plants and algae. As we approach the shore, we first find submerged plants like *Elodea* and algal mats, then emergent vegetation like water lilies and cattails, then grasses and sedges that can tolerate wet soil, and on the shore the beginnings of a typical terrestrial succession resulting in the climax community typical for the area.

FIGURE 5.4

Floating Bog. In many northern regions, sphagnum moss forms a floating mat that can support the growth of other plants. If a person were to walk on this mat, it would bounce up and down, because it is floating on water.

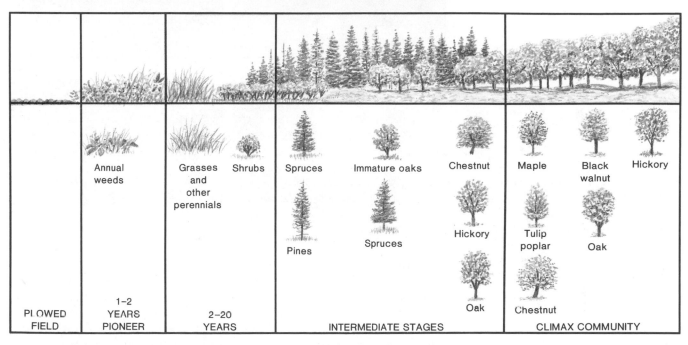

	Annual weeds	Grasses and other perennials Shrubs	Spruces Immature oaks Chestnut Pines Spruces Hickory Oak	Maple Black walnut Hickory Tulip poplar Oak Chestnut
PLOWED FIELD	1–2 YEARS PIONEER	2–20 YEARS	INTERMEDIATE STAGES	CLIMAX COMMUNITY

⟵————————————— 200 YEARS (VARIABLE) —————————————⟶

McGee

FIGURE 5.5

Secondary Succession on Land. A plowed field in the southeastern United States shows a parade of changes involving plant and animal associations over a period of years. The general pattern is for annual weeds to be replaced by grasses and other perennial herbs, which are replaced by shrubs, which are replaced by trees. As the plant species change, so do the animal species.

Secondary Succession

The same processes and activities that drive primary succession result in secondary succession. The major difference is that secondary succession occurs when an existing community is destroyed in some way. A forest fire, the flooding of an area, or the destruction of a natural ecosystem to convert the land to agricultural use may be causes. Usually, the destroyed ecosystem is not completely returned to bare rock. Much of the soil may remain, and many of the nutrients necessary for plant growth may be available for the reestablishment of the previously existing ecosystem. Consequently, secondary succession tends to be a more rapid process than primary succession. Figure 5.5 shows the typical secondary succession found on abandoned farmland in the southeastern United States.

FIGURE 5.6
**Beaver Pond Succession—Secondary
Succession from Water.** A colony of
beavers can dam up streams and kill
trees by the flooding that occurs and by
using trees for food. Once the site is
abandoned, it will slowly return to the
original forest community by a process
of succession.

Similarly, when beavers flood an area, the existing trees die and an
aquatic ecosystem is established. As the area behind the dam fills in with
sediment and organic matter, it goes through a typical series of stages that
eventually return the area to the typical climax community. (See figure 5.6.)

Major Types of Climax Communities—Biomes

Several major types of climax communities, often known as **biomes,** can
be found throughout the world. The species involved may not be identical
in various places, but the general structure of the ecosystem and the kinds
of niches present are similar in broad terms. (See figure 5.7.) Two primary
nonbiological parameters have major impacts on the kind of climax community that develops: precipitation and temperature. Several aspects of precipitation are important: the total amount of precipitation per year, the form
in which it arrives (rain, snow, sleet), and the seasonal distribution of the
precipitation. Is it evenly spaced throughout the year? Are there wet and
dry seasons?

Ecological Principles and Their Application

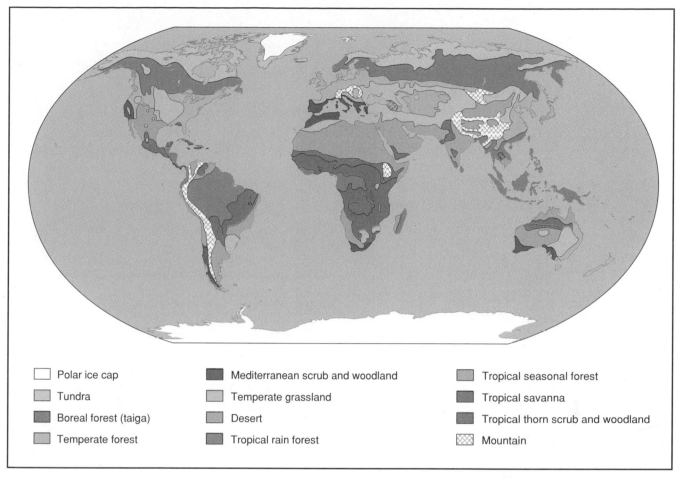

☐ Polar ice cap	■ Mediterranean scrub and woodland	▨ Tropical seasonal forest
☐ Tundra	▨ Temperate grassland	■ Tropical savanna
■ Boreal forest (taiga)	▨ Desert	■ Tropical thorn scrub and woodland
▨ Temperate forest	■ Tropical rain forest	▨ Mountain

FIGURE 5.7

Biomes of the World. Although most biomes are named for a major type of vegetation, each includes a specialized group of animals adapted to the plants and the biome's climatic conditions.

The temperature also can vary considerably. Many tropical areas have warm, relatively unchanging temperatures throughout the year. Many areas near the poles have long winters with extremely cold temperatures and relatively short, cool summers. Many other areas have the year more evenly divided between cold and warm periods of the year. The interplay between these two major parameters can account for many of the differences seen in climax communities. (See figure 5.8.)

In addition to temperature and precipitation are several other factors that may influence the kind of climax community present. Some climax communities seem to rely on periodic fires to prevent the establishment of larger, woody species. Similarly, severe wind may prevent the establishment of trees and cause rapid drying of the soil. Sandy soils tend to dry out quickly and may not allow the establishment of more water-demanding species like trees, while extremely wet soils may only allow certain species of trees to grow. Obviously, the kinds of organisms currently living in the area are also important, since their offspring will be the ones available to colonize a new area. Let's look at some of the major types of biomes and the factors that shape their character.

Desert

Deserts are areas that generally receive fewer than 25 centimeters of precipitation per year. A lack of water is the primary factor that determines that

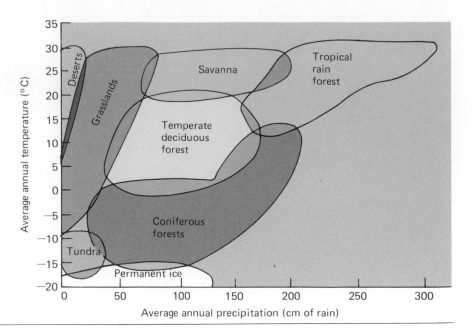

FIGURE 5.8

Influence of Precipitation and Temperature on Vegetation.
Temperature and moisture are two major factors that influence the kind of vegetation that can occur in an area. Areas with low moisture and low temperatures produce tundra; areas with high moisture and freezing temperatures during part of the year produce deciduous or coniferous forests; dry areas produce deserts; moderate amounts of rainfall or seasonal rainfall support grasslands or savannas; and areas with high rainfall and high temperatures support tropical rain forests.

an area will be a desert. (See figure 5.9.) Rain comes in the form of thundershowers at infrequent intervals. Much of the water runs off into gullies and, since the rate of evaporation is high, the rains provide only brief periods of rapid growth. Deserts are also likely to be windy. We often think of deserts as hot, dry wastelands devoid of life. However, many deserts are quite cool during a major part of the year. Certainly, the Sahara Desert and the deserts of the southwestern United States are hot during much of the year, but the desert areas of the northwestern United States and the Gobi Desert in Central Asia can be extremely cold during winter months and have relatively cool summers. Furthermore, the temperature can vary greatly during a twenty-four-hour period. Since deserts receive little rainfall, it is logical that they have infrequent cloud cover. Since no clouds block out the sun, the earth tends to heat up rapidly. Therefore, throughout the day the soil surface and the air above it tend to heat up rapidly. However, after the sun has set, the absence of clouds allows heat energy to be reradiated from the earth, and the area cools off rapidly. Cool to cold nights are typical even in "hot" deserts, especially during the winter months.

Another misconception about deserts is that they have a low diversity of organisms inhabiting them. Many types of organisms live in deserts, but they have special adaptations that allow them to survive in dry, often hot environments. For example, since water evaporates from the surfaces of leaves, many plants have very small leaves that allow them to conserve water. Some even lose their leaves entirely during the driest part of the year. Many desert plants are spiny. Some plants like cactus have the ability to store water in their spongy bodies for use during drier periods. Many other plants have parts or seeds that lie dormant until the rains come. Then they grow rapidly, reproduce, and die, or become dormant until the next rains. Even the perennial plants are tied to the infrequent rains. During these times, the plants are most likely to produce flowers and reproduce.

The desert has many different kinds of animals, often inconspicuous and overlooked because many are small or are inactive during the hot part of the day. In addition they don't occur in large, conspicuous groups. Many kinds of lizards, snakes, small mammals, grazing mammals, carnivorous

Ecological Principles and Their Application

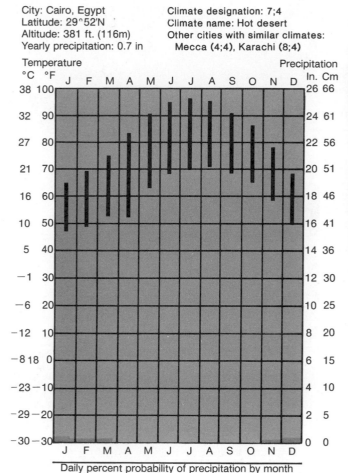

City: Cairo, Egypt
Latitude: 29°52'N
Altitude: 381 ft. (116m)
Yearly precipitation: 0.7 in

Climate designation: 7;4
Climate name: Hot desert
Other cities with similar climates:
Mecca (4;4), Karachi (8;4)

Temperature
°C °F

Precipitation
In. Cm

Daily percent probability of precipitation by month

a. 3 3 3 0 0 0 0 0 0 0 3 3

b.

FIGURE 5.9

Desert. (*a*) Climagraph for Cairo, Egypt. (*b*) The desert receives less than 25 centimeters of precipitation per year, yet it teems with life. Cactus, sagebrush, lichens, snakes, small mammals, birds, and insects inhabit the desert. Because daytime temperatures are often high, most animals are only active at night, when the air temperature drops significantly. Cool deserts also exist in many parts of the world, where rainfall is low but temperatures are not high.

mammals, and birds are very common in desert areas. All of the animals that live in desert areas have an ability to survive with a minimal amount of water. Some get nearly all of their water from the moisture in the food they eat. They generally have an outer skin or cuticle that is waterproof; therefore, they lose little water by evaporation. In addition, animals often limit their activities to the cooler part of the day (the evening) and may spend considerable amounts of time in underground burrows during the day, which allows them to avoid extreme temperatures and to conserve water.

Grassland

Grasslands, also known as **prairies** or **steppes,** are widely distributed over the world. As with deserts, the major factor that contributes to the establishment of a grassland is the amount of available moisture. Grasslands generally receive between 25 and 75 centimeters of precipitation per year. These areas are windy with hot summers and cold to mild winters. Fire is an important force in this biome. Trees that generally require greater amounts of water are rare in these areas except along watercourses. (See figure 5.10.) Although grasses are the dominant vegetation and make up 60–90 percent of the vegetation, many other kinds of flowering plants are found interspersed with the grasses.

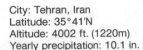

City: Tehran, Iran
Latitude: 35°41′N
Altitude: 4002 ft. (1220m)
Yearly precipitation: 10.1 in.

Climate designation: 10;4
Climate name: Midlatitude dryland
Other cities with similar climates:
    Salt Lake City (12;4), Ankara (10;4)

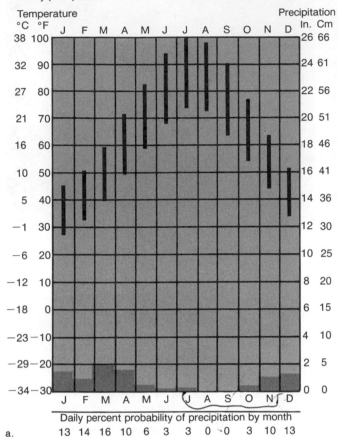

Daily percent probability of precipitation by month

J	F	M	A	M	J	J	A	S	O	N	D
13	14	16	10	6	3	3	0	0	3	10	13

a.

b.

FIGURE 5.10

Grassland. (*a*) Climagraph for Tehran, Iran. (*b*) Grasses are better able to withstand low water levels than are trees. Therefore, in areas that have moderate rainfall, grasses are the dominant plants.

Characteristic of these areas are large herds of migratory, grazing mammals like bison, wildebeests, wild horses, and various kinds of sheep, cattle, and goats. Most of the grasslands of the world have been converted to agriculture, since the rich, deep soil that developed as a result of the activities of centuries of soil building is useful for growing cultivated grasses like corn (maize) and wheat. The drier grasslands have been converted to the raising of domesticated grazers like cattle, sheep, and horses.

In addition to grazing mammals, many kinds of insects, including grasshoppers and other herbivorous insects, dung beetles (which feed on the dung of grazing animals), and flies (which bite the large mammals), are common. Many kinds of small herbivorous mammals, such as mice and ground squirrels, are also common. Birds are often associated with grazing mammals where they eat the insects stirred up by the large mammals or feed on the insects that bite the large grazing animals. Other birds feed on the seeds and other plant parts available. The reptiles (snakes and lizards) present often feed on small mammals and insects.

Savanna

In parts of Africa, South America, and Australia are extensive grasslands spotted with occasional trees or patches of trees. (See figure 5.11.) This kind of a biome is often called a **savanna.** These areas of the world are

Ecological Principles and Their Application

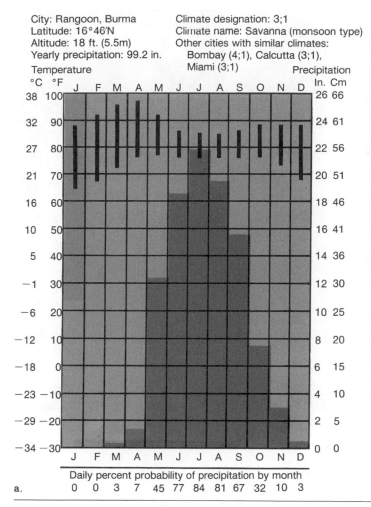

City: Rangoon, Burma
Latitude: 16°46′N
Altitude: 18 ft. (5.5m)
Yearly precipitation: 99.2 in.

Climate designation: 3;1
Climate name: Savanna (monsoon type)
Other cities with similar climates:
 Bombay (4;1), Calcutta (3;1),
 Miami (3;1)

Daily percent probability of precipitation by month

J	F	M	A	M	J	J	A	S	O	N	D
0	0	3	7	45	77	84	81	67	32	10	3

a.

b.

FIGURE 5.11

Savanna. (*a*) Climagraph for Rangoon, Burma. (*b*) Savannas develop in tropical areas that have seasonal rainfall. They typically have grasses as the dominant vegetation with drought- and fire-resistant trees scattered through the area.

typically tropical, with 50–150 centimeters of rain per year. However, the rain is not distributed evenly throughout the year. There is typically a period with heavy rainfall followed by a prolonged drought. This results in a very seasonally structured ecosystem. The plants and animals in these areas time their reproductive activities to coincide with the rainy period, when limiting factors are least confining. The predominant plants are grasses, but many drought-resistant, flat-topped, thorny trees are common. The trees are also resistant to fire damage. Many of these trees are particularly important because they are involved in nitrogen fixation. They also provide shade and nesting sites for many kinds of animals. As with grasslands, the predominant mammals are the grazers. Kangaroos in Australia, various species of antelope in Africa, and llamas in South America are examples. Many kinds of rodents, birds, insects, and reptiles are associated with this biome. Among the insects, mound-building termites are particularly common.

Tropical Rain Forest

Tropical rain forests are located near the equator in Central and South America, Africa, Southeast Asia, and some islands in the Caribbean Sea and Pacific Ocean. (See figure 5.12.) The temperature is normally warm and relatively constant. There is no frost, and it rains nearly every day. Most areas receive in excess of 200 centimeters of rain per year. Some receive

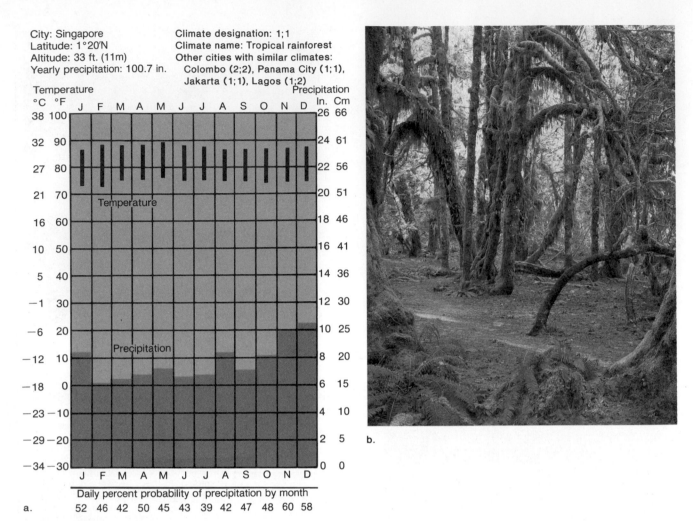

City: Singapore
Latitude: 1°20'N
Altitude: 33 ft. (11m)
Yearly precipitation: 100.7 in.

Climate designation: 1;1
Climate name: Tropical rainforest
Other cities with similar climates:
   Colombo (2;2), Panama City (1;1),
   Jakarta (1;1), Lagos (1;2)

Daily percent probability of precipitation by month
a.   52 46 42 50 45 43 39 42 47 48 60 58

FIGURE 5.12

Tropical Rain Forest. (*a*) Climagraph
for Singapore. (*b*) Tropical rain forests
develop in areas with high rainfall and
warm temperatures. They have an
extremely diverse mixture of plants and
animals.

500 centimeters or more. Because of the warm temperatures and abundant rainfall, most plants grow very rapidly. However, soils are often poor in nutrients because water tends to carry away any nutrients that are not immediately taken up by plants. Many of the trees have extensive root networks, associated with fungi, near the surface of the soil that allow them to capture nutrients from decaying vegetation before they can be carried away. Most of the nutrients in a tropical rain forest are tied up in the biomass, not in the soil, and, therefore, most of these areas do not make good farmland.

These forests have a very large number of species. A small area of a few square kilometers is likely to have hundreds of species of trees. Balsawood, teakwood, and many other ornamental woods are from tropical trees. The trees are usually in two or three layers, with some standing taller than their neighbors. Since most of the sunlight is captured by the trees, many shade-tolerant plants live beneath the trees' canopy. Each tree serves as a surface for the growth of ferns, mosses, orchids, and various kinds of vines. Associated with this variety of plants is an equally large variety of animals. Different kinds of insects, such as ants, termites, moths, butterflies, and beetles, are particularly abundant. Birds also are extremely common, as are many climbing mammals, lizards, and tree frogs. Since flowers and fruits are available throughout the year, there are many kinds of nectar- and fruit-feeding birds and mammals. Because of the low light levels, many of the animals communicate by making noise.

Ecological Principles and Their Application

BOX 5.1
Farming in a Tropical Rain Forest

Because tropical rain forests are so lush and fast-growing, many people have suggested that they would be excellent agricultural areas. In fact, primitive people in such areas have practiced slash-and-burn agriculture for centuries. This method of use involves cutting and burning the vegetation on a small plot of land. The burning releases the nutrients in the vegetation. If planting and tilling are done quickly, the crop covers the soil and prevents its exposure to the hot sun and erosion caused by the frequent rains. During the first year, a good crop can be harvested, but the yield declines each succeeding year unless massive amounts of fertilizer are used. For the primitive inhabitants of these areas, this was no problem because they simply abandoned the garden and cleared a new site. The old garden was quickly repopulated by the seeds of trees in the surrounding forest and succession occurred, resulting in a return to the original forest community. Most attempts to farm large areas using temperate-region agricultural techniques have resulted in failures because of the infertility of the soil and the soil's tendency to become hard.

One kind of soil found in the hot and humid tropical rain forests such as those found in South America is called a laterite soil. Abundant rainfall is common in the areas where laterite soil is found. Since the soil is porous, it is subjected to a great deal of leaching, which results in the removal of soil nutrients as the water flows over and through the exposed soil. Silica is removed, but the soil retains a relatively high concentration of oxides of iron and aluminum. The presence of these compounds contributes to the reddish color of this soil. Laterite soil gets its name from the Latin word for brick, because whenever it is directly exposed to weathering, it hardens into a bricklike mass. In the past, these blocks were used as building materials. A temple in the People's Republic of Kampuchea (Cambodia) is constructed entirely of blocks of laterite soil.

If this is such an infertile soil, why does it support such a lush tropical forest? The high temperature and high amount of moisture encourage the growth of plants; however, there are few nutrients in the soil. The plants rely on a rapid uptake of nutrients from the plants and animals that die. Whenever a plant dies, it decays rapidly, and surrounding plants absorb the nutrients needed for their growth. Tropical forest trees bear mutualistic fungi called mycorrhiza in association with an extensive root network near the surface of the soil. The mycorrhiza quickly penetrate each fallen leaf and take all nutrients back into the roots of their host tree. This activity prevents the loss of valuable nutrients due to the leaching action of high rainfall and porous soil. Nutrients from decaying organisms do not usually go unused in a healthy tropical rain forest.

Other trees perform a different role. It is estimated that 10 to 15 percent of the trees in tropical rain forests are legumes, which have symbiotic nitrogen-fixing bacteria associated with their roots. These trees, which are mixed throughout the forest, provide this much-needed nutrient for other trees that surround them. Thus, the lush vegetation in tropical rain forests is the result of a very finely tuned ecological system that recycles nutrients as fast as they are released. The nutrients are in the plants, not in the soil.

Because of this, many tropical soils are poor in nutrients and are not suitable for long-term agricultural use. The slash-and-burn agriculture practiced by small tribes of people did not disrupt the system. Since the garden plots were small, they were quickly overgrown by the surrounding forest, and the exhausted garden plot reverted to the original forest type. The soil was not exposed to extensive erosion and exposure to the sun, and it was not changed to a red-colored, bricklike texture. The use of this ecosystem for agriculture or forestry must take into account the unique nutrient cycling mechanisms and fragile soil structure, or the tropical rain forest will be permanently destroyed.

More different species of organisms probably are found in the tropical rain forests of the world than in the rest of the world combined. Recently, biologists discovered a whole new community of organisms that live in the canopy (leafy tops) of these forests. The dense vegetation reduces the amount of light on the forest floor. This, coupled with the bewildering array of plants and animals, makes study of this kind of ecosystem difficult.

Temperate Deciduous Forest

Forests in temperate areas of the world that have a winter-summer change of seasons typically have trees that lose their leaves during the winter and

City: Chicago, Illinois
Latitude: 41°52′N
Altitude: 595 ft. (181m)
Yearly precipitation: 33.3 in.

Climate designation: 14;2
Climate name: Humid continental (warm summer)
Other cities with similar climates:
New York (10;2), Berlin (11;2),
Warsaw (15;2)

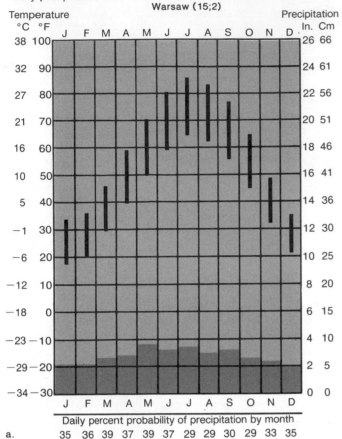

Daily percent probability of precipitation by month
35 36 39 37 39 37 29 29 30 29 33 35

a.

b.

FIGURE 5.13

Temperate Deciduous Forest.
(*a*) Climagraph for Chicago, Illinois.
(*b*) A temperate deciduous forest
develops in areas that have significant
amounts of moisture throughout the
year, but where the temperature falls
below freezing for parts of the year.
During this time the trees lose their
leaves. This kind of forest once
dominated the eastern half of the United
States.

replace them the following spring. This kind of forest is called a **temperate deciduous forest** and is typical of the eastern half of the United States, parts of south central and southeastern Canada, southern Africa, and many areas of Europe and Asia.

These areas generally receive 100 or more centimeters of relatively evenly distributed precipitation per year. Each area of the world has certain species of trees that are the major producers for the biome. In contrast to tropical rain forests, where individuals of a tree species are scattered throughout the forest, temperate deciduous forests generally have many fewer species, and many forests may consist of two or three dominant tree species. In deciduous forests of North America and Europe, common species are maples, aspen, birch, beech, oaks, and hickories. These tall trees shade the forest floor, where many small flowering plants bloom in the spring. These spring wildflowers store food in underground structures. In the spring, before the leaves come out on the trees, the wildflowers can capture sunlight and go through their reproduction before they become shaded. Many smaller shrubs also may be found in many of these forests. (See figure 5.13.)

These forests are homes for a great variety of insects, many of which use the leaves and wood of trees as food. Beetles, moth larvae, wasps, and ants are examples. The birds that live in these forests are primarily migrants that arrive in the spring of the year, raise their young during the summer,

Ecological Principles and Their Application

City: Moscow, USSR
Latitude: 55°46′N
Altitude: 505 ft. (154m)
Yearly precipitation: 21.8 in.

Climate designation: 15;6
Climate name: Humid continental (cool summer)
Other cities with similar climates:
    Montreal (14;6), Winnipeg (15;6),
    Leningrad (15;6)

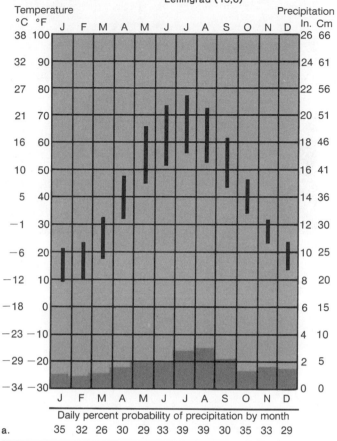

Daily percent probability of precipitation by month

a.   35 32 26 30 29 33 39 39 30 35 33 29

b.

FIGURE 5.14

Taiga, Northern Coniferous Forest, or Boreal Forest. (*a*) Climagraph for Moscow. (*b*) The taiga, northern coniferous forest, or Boreal forest occurs in areas with long winters and heavy snowfall. The trees have adapted to these conditions and provide food and shelter for many of the animals that live there.

and leave in the fall. Many of these birds rely on the large summer insect population for their food. A few kinds of birds are year-round residents, including woodpeckers, grouse, turkeys, and some of the finches. Several kinds of small and large mammals inhabit these areas. Mice, squirrels, deer, moles, and rabbits are common examples. Major predators of these animals are foxes, badgers, weasels, coyotes, and birds of prey.

Taiga, Northern Coniferous Forest, or Boreal Forest

Throughout the southern half of Canada, parts of northern Europe, and much of the USSR, there is a coniferous forest known as the **taiga, northern coniferous forest,** or **boreal forest.** Spruces, firs, and larches are the trees most common in these areas. (See figure 5.14.) The climate is one of short, cool summers and long winters with abundant snowfall. The winters are extremely harsh and can last as long as six months. Typically, the soil freezes during the winter. Precipitation ranges between 25 and 100 centimeters per year, but the climate is humid because of the generally low temperatures during all parts of the year. The landscape is typically dotted with lakes, ponds, and bogs.

The trees are specifically adapted to winter conditions. Winter in this area is relatively dry as far as the trees are concerned because the moisture

City: Fairbanks, Alaska
Latitude: 64°51'N
Altitude: 440 ft. (134m)
Yearly precipitation: 12.4 in.

Climate designation: 16;6
Climate name: Subarctic tundra
Other cities with similar climates:
    Yellowknife (15;6), Yakutsk (16;7)

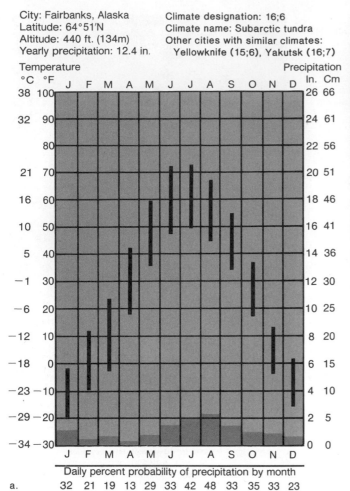

Temperature

Precipitation

Daily percent probability of precipitation by month

a.   32 21 19 13 29 33 42 48 33 35 33 23

b.

FIGURE 5.15

Tundra. (*a*) Climagraph for Fairbanks, Alaska. (*b*) In the northern latitudes and on the tops of some mountains, the growing season is short and plants grow very slowly. Trees are unable to live in these extremely cold areas, in part because there is a permanently frozen layer of soil beneath the surface, known as the permafrost. Because growth is so slow, damage to the tundra can still be seen generations later.

falls as snow and stays above the soil until it melts in the spring. The needle-shaped leaves are adapted to preventing water loss, and the larch loses its needles in the fall of the year. Furthermore, the branches of these trees are flexible, allowing the snow to slide off the pyramid-shaped trees without greatly damaging them. As with the temperate deciduous forest, many of the inhabitants of this biome are temporarily active during the summer. Most birds are migratory and feed on the abundant insect population, which becomes inactive during the long, cold winter. A few woodpeckers, owls, and grouse are permanent residents. Typical mammals are deer, caribou, moose, wolves, weasels, mice, and squirrels. Because of the cold, few reptiles and amphibians live in this biome.

Tundra

North of the taiga is a biome that lacks trees and has a permanently frozen soil layer or **permafrost.** This biome is known as the **tundra.** (See figure 5.15.) Scattered patches of tundralike communities also are found on mountaintops throughout the world. These are known as **alpine tundra.** Because of the permanently frozen soil and extremely cold, windy climate (up to ten months of winter), no trees can live in the area. During the brief summer, the top few centimeters of the soil thaw, and many plants and lichens (reindeer moss) can grow. The plants are short, usually less than 20 centimeters. A few hardy mammals like musk oxen, caribou or reindeer,

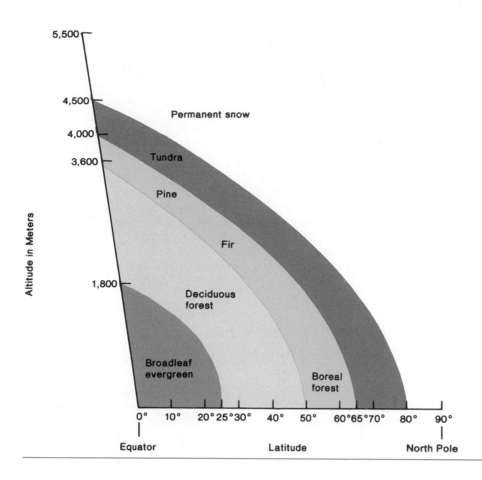

FIGURE 5.16

Relationship between Altitude, Latitude, and Vegetation. As one travels up a mountain, the climate changes. The higher the elevation, the cooler the climate. Even in the tropics tall mountains can have snow on the top. Thus it is possible to experience the same change in vegetation by traveling up a mountain as one would experience traveling from the equator to the North Pole.

arctic hare, and lemmings can survive by feeding on the grasses and other plants that grow during the short, cool summer. Although the amount of precipitation is similar to that in some deserts (25 centimeters per year), the short summer is generally moist because of the melting of the snow that falls during the winter. In addition, the permafrost does not let the water sink into the soil, resulting in waterlogged soils and many shallow ponds and pools. Consequently, many waterfowl like ducks and geese migrate to the tundra for mating purposes. Insects arc common during the summer and serve as food for migratory birds. Permanent resident birds are the ptarmigan and snowy owl. No reptiles or amphibians survive in this extreme climate. Because of the very short growing season, damage to this kind of ecosystem is slow to heal, so it must be handled with a great deal of care.

Altitude and Latitude

The distribution of terrestrial ecosystems is primarily related to precipitation and temperature. The temperature is warmest near the equator and becomes cooler as the poles are approached. Similarly, as the altitude increases, the average temperature decreases. This means that even at the equator it is possible to have cold temperatures on the peaks of tall mountains. As one proceeds from sea level to the tops of mountains, it is possible to pass through a series of biomes that are similar to what would be encountered as one traveled from the equator to the North Pole. (See figure 5.16.)

The Atlantic salmon is a highly prized fish for the table. However, the salmon's life history does not allow for large numbers of them to be caught. They are pelagic fish and spend most of their lives in the open ocean, where they feed on other fish. Although they occur in schools, the amount of time required to locate the schools and catch the fish makes them extremely expensive. However, they typically swim upstream to spawn for a short period of time each year. During this time, they can be caught with relative ease, but most of the fishing is limited to sport fishing.

In Scotland, commercial salmon farming has developed to fill the demand for more salmon. Huge, baglike nets are suspended from floats in sheltered bays and are stocked with young salmon. The "farmer" uses a small boat to visit the floating "farm" on a regular basis to feed the fish a commercial fish food. The fish grow rapidly in this seminatural environment and can be easily harvested for shipment to fish shops and restaurants. This kind of farming makes use of a portion of the natural environment of the fish but restricts their movement to make harvesting economically feasible.

Major Aquatic Ecosystems

The kind of terrestrial biomes that develop in an area is determined by the amount and kind of precipitation and by the temperatures common in the area. Other factors, such as soil type and wind, also play a part. Aquatic ecosystems also have primary determiners that shape the nature of an ecosystem. Four such factors are the ability of the sun's rays to penetrate the water, the nature of the bottom substrate, water temperature, and the amount of dissolved materials.

Marine Ecosystems

An important determiner of the nature of aquatic ecosystems is the amount of salt dissolved in the water. Those that have little dissolved salt are called **freshwater ecosystems** and those that have a high salt content are called **marine ecosystems.**

Pelagic Marine Ecosystems

In the open ocean are many kinds of crustaceans, fish, and whales that swim actively as they pursue food. These kinds of animals are called **pelagic,** and the ecosystem they are a part of is called a **pelagic ecosystem.** As with all ecosystems, the organisms at the bottom of the energy pyramid carry on photosynthesis. A majority of these organisms in the ocean are small, microscopic, floating algae called **phytoplankton.** Since sunlight cannot penetrate to great depths in bodies of water, these phytoplankton are located in the upper regions of bodies of water. This region where the sun's rays penetrate is known as the **euphotic zone.** Small, weakly swimming animals of many kinds, known as **zooplankton,** feed on the phytoplankton. Zooplankton are often located at a greater depth in the ocean than the phytoplankton but migrate upward at night and feed on the large population of phytoplankton. The zooplankton are in turn eaten by larger animals like fish and larger shrimp, which are eaten by larger fish like salmon, tuna, sharks, and mackerel. (See figure 5.17.)

A major factor that influences the nature of a marine community is the kind and amount of material dissolved in the water. Probably more important is the amount of nutrients available to the organisms carrying on photosynthesis. Phosphorus, nitrogen, and carbon are all required for the

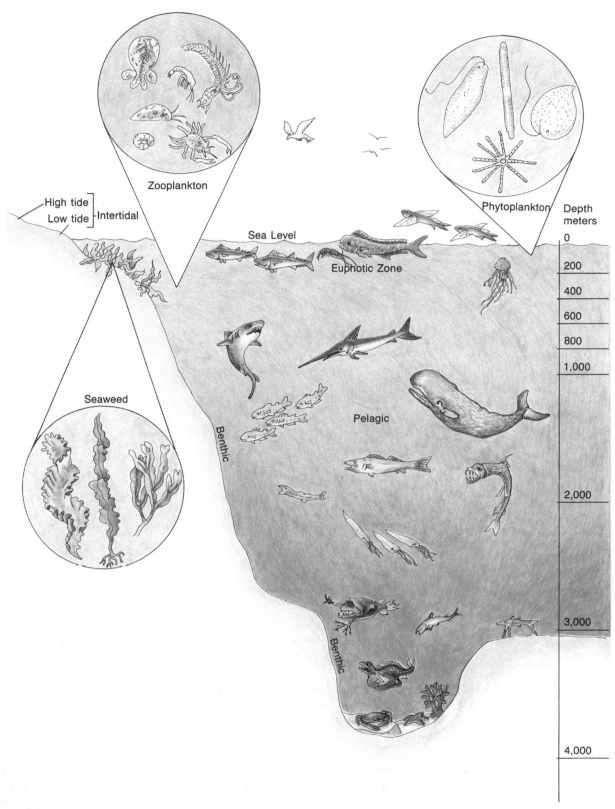

FIGURE 5.17

Marine Ecosystem. All of the photosynthetic activity of the ocean occurs in shallow water called the euphotic zone, either by attached algae near the shore or by minute phytoplankton in the upper levels of the open ocean. Consumers are either free-swimming pelagic organisms or benthic organisms that live on the bottom. Small animals that feed on phytoplankton are known as zooplankton.

construction of new living material. In water, these are often in short supply. Therefore, the most productive aquatic ecosystems are those where these essential nutrients are most common. These areas include places in oceans where currents bring up nutrients that have settled to the bottom and areas where rivers deposit their load of suspended and dissolved materials.

Benthic Marine Ecosystems

Those organisms that live on the ocean bottom, whether attached or not, are known as **benthic** organisms, and the kind of ecosystem that consists of these organisms is called a **benthic ecosystem.** Some fish, clams, oysters, various crustaceans, sponges, sea anemones, and many other kinds of organisms live on the bottom. In shallow water, sunlight can penetrate to the bottom, and a variety of attached photosynthetic organisms like kelp (commonly called seaweeds) are common. Since they are attached and can grow to very large size, many other bottom-dwelling organisms are associated with them.

The kind of substrate is very important in determining the kind of benthic community that develops. Sand tends to shift and move, making it difficult for large plants or algae to become established, although some clams, burrowing worms, and small crustaceans find sand to be a suitable habitat. The clams filter water for plankton and detritus, or burrow through the sand, feeding on the other inhabitants. Muds may provide suitable habitats for some kinds of rooted plants. Muds generally have little oxygen in them but still may be inhabited by a variety of burrowing organisms that feed by filtering the water above them or feed on other animals in the mud. Rocky surfaces in the ocean provide a good substrate for many kinds of large algae to become established. Associated with this profuse growth of algae is a large number of different kinds of animals. (See figure 5.18.)

At great depths in the ocean is an ecosystem that must rely on a continuous rain of organic matter from the euphotic zone. These areas are known as abyssal areas, and the ecosystem is known as an **abyssal ecosystem.** Essentially, all of the organisms in this environment are scavengers that feed on whatever drifts their way.

Temperature also has an impact on the kind of benthic community established. Some communities, such as coral reefs or mangrove swamps, are only found in areas where the water is warm. **Coral reef ecosystems** are the result of large numbers of small animals that build cup-shaped external skeletons around themselves. They are able to protrude from their skeletons to capture food and expose themselves to the sun. This is important because corals contain single-celled algae within their bodies. These algae carry on photosynthesis and provide both themselves and the coral with the nutrients necessary for growth. Because of the requirements for warm water, coral ecosytems are only found near the equator. Coral ecosystems must also be in shallow, clear water since the algae must have ample sunlight to carry on photosynthesis. This mutualistic relationship between algae and coral is the basis for a very productive community of organisms. The skeletons of the corals provide a surface upon which many other kinds of animals live. Some of these animals feed on corals directly, while others feed on small plankton and bits of algae that establish themselves among the coral organisms. Many kinds of fish, crustaceans, sponges, clams, and

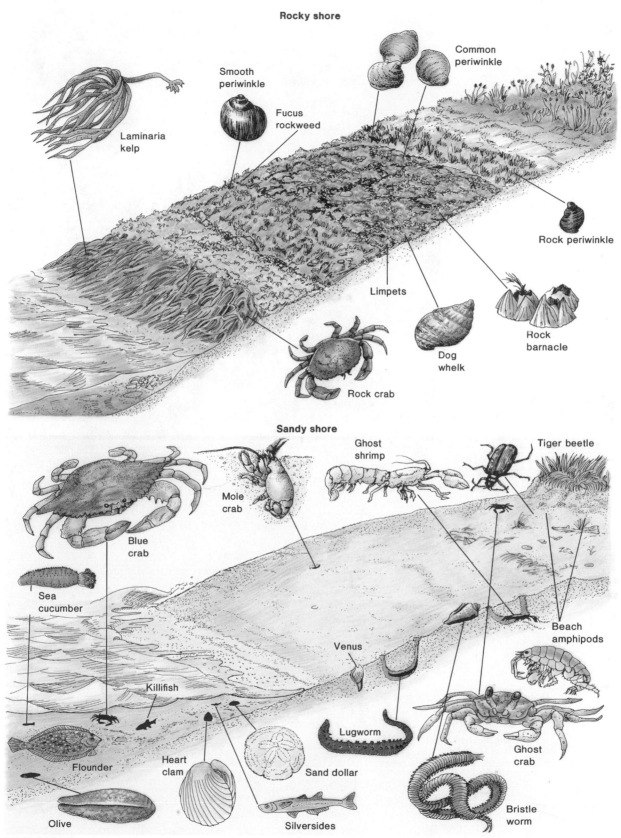

Rocky shore

Laminaria kelp

Smooth periwinkle

Fucus rockweed

Common periwinkle

Rock periwinkle

Limpets

Dog whelk

Rock barnacle

Rock crab

Sandy shore

Ghost shrimp

Tiger beetle

Mole crab

Blue crab

Beach amphipods

Sea cucumber

Venus

Killifish

Lugworm

Ghost crab

Flounder

Heart clam

Sand dollar

Bristle worm

Olive

Silversides

FIGURE 5.18

Types of Shores. The kind of substrate determines the kind of organisms that can live near the shore. Rocks provide areas for attachment that sands do not, since sands are constantly shifting. Muds usually have little oxygen in them; therefore, the organisms that live there must be adapted to those kinds of conditions.

FIGURE 5.19
Coral Reef. Corals are small sea animals that secrete external skeletons. They have a mutualistic relationship with certain algae, which allows both kinds of organisms to be very successful. The skeletal material serves as a substrate for many other kinds of organisms to live.

snails are members of coral reef ecosystems. Coral reefs are considered one of the most productive ecosystems on earth. (See figure 5.19.)

Mangrove swamp ecosystems occupy a region near the shore. The dominant organisms are special kinds of trees that are able to tolerate the high salt content of the ocean. In areas where the water is shallow and wave action is not too great, the trees can become established. They have long seeds that float in the water. When the seed becomes trapped in mud, it takes root. The trees can excrete salt from their leaves. They also have extensively developed roots that are above the water, where they can obtain oxygen and prop up the plant. The trees trap much sediment and provide many places for crabs, jellyfish, sponges, and fish to live. The trapping of sediment and the continual movement of mangroves into shallow areas nearby result in the development of a terrestrial ecosystem in what was once shallow ocean. Mangroves are found in the Caribbean, Southeast Asia, Africa, and other parts of the world where conditions are suitable. (See figure 5.20.)

Estuaries

Estuaries, a special category of marine ecosystem, consist of shallow, partially enclosed areas where fresh water enters the ocean. Because the saltiness of the water in the estuary changes with tides and the flow of water from rivers, the organisms that live here are specially adapted to this set of physical conditions, and the number of species is less than in the ocean or in fresh water. Estuaries are particularly productive ecosystems because of the large amounts of nutrients introduced into the basin from the rivers

FIGURE 5.20
Mangrove Swamp. Mangroves are
tropical trees that are able to live in very
wet, salty muds found along the ocean
shore. Since they are able to trap
additional sediment, they tend to extend
farther seaward as they reproduce.

that run into them. This is further enhanced by the fact that the shallow
water allows light to penetrate to most of the water in the basin. Phytoplankton and attached algae and plants are able to use the sunlight and the
nutrients for rapid growth. This photosynthetic activity supports many kinds
of organisms in the estuary. Estuaries are especially important as nursery
sites for many kinds of fish and crustaceans like flounder and shrimp. The
adults return to these productive, sheltered areas to reproduce before returning to the ocean. The young spend their early life in the estuary and
eventually leave as they get larger and are more able to survive in the ocean.
Estuaries also trap sediment. This activity tends to prevent many kinds of
pollutants from reaching the ocean and also results in the gradual filling
in of the estuary, which may eventually become a salt marsh and then part
of a terrestrial ecosytem.

Freshwater Ecosystems

Freshwater ecosystems differ from marine ecosystems in several ways. The
amount of salt present is much less, the temperature of the water can change
greatly, the water is in the process of moving downhill, oxygen can often
be in short supply, and the organisms that inhabit freshwater systems are
different.

Freshwater ecosystems can be divided into two categories: those in
which the water is relatively stationary, such as lakes, ponds, and reservoirs,
and those in which the water is running downhill, such as streams and rivers.

Large lakes have many of the same characteristics as the ocean. If the
lake is deep, there is a euphotic zone at the lake's top, with many kinds of
phytoplankton, and zooplankton that feed on the phytoplankton. Small fish
feed on the zooplankton, which are in turn eaten by other larger fish. The
species of organisms found in freshwater lakes are different from those found
in the ocean, but the kinds of roles played are similar, so the same termi-
nology is used.

Along the shore and in the shallower parts of lakes, the euphotic zone
contains many kinds of flowering plants. They are rooted in the bottom and
some have leaves that float on the surface or protrude above the water. These
are called **emergent plants.** Cattails, bullrushes, arrowhead plants, and

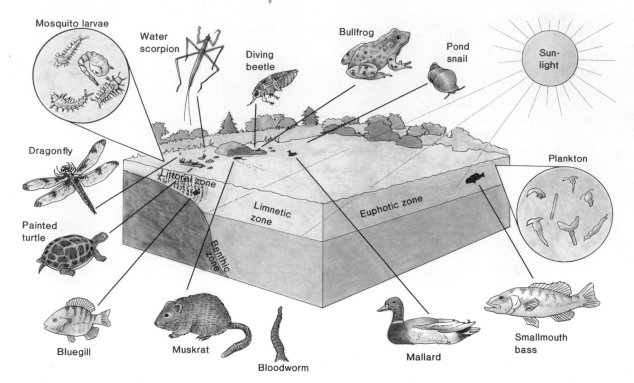

Mosquito larvae

Water scorpion

Diving beetle

Bullfrog

Pond snail

Sun- light

Dragonfly

Plankton

Littoral zone

Limnetic zone

Euphotic zone

Painted turtle

Benthic zone

Bluegill

Muskrat

Bloodworm

Mallard

Smallmouth bass

FIGURE 5.21

Lake Ecosystem. Lakes are similar in structure to oceans except that the species are different because most marine organisms cannot live in fresh water. Insects are common organisms in freshwater lakes, as are many kinds of fish, zooplankton, and phytoplankton.

water lilies are examples. Rooted plants that stay submerged below the surface of the water are called **submerged plants.** *Elodea* and *Chara* are examples.

Many kinds of freshwater algae also grow in the shallow water, where they may appear as mats on the bottom or attached to vegetation and other objects in the water. Associated with the plants and algae are a large number of different kinds of animals. Fish, crayfish, clams, and many kinds of aquatic insects are common inhabitants of this mixture of plants and algae. This region, with rooted vegetation, is known as the **littoral zone,** and the portion of the lake that does not have rooted vegetation is called the **limnetic zone.** (See figure 5.21.)

The productivity of the lake is determined by several factors. Temperature is important, since cold temperatures tend to reduce the amount of photosynthesis. Shallow lakes tend to be more productive because most of the water has sunlight penetrating it and, therefore, photosynthesis can occur to the bottom of the lake. Shallow lakes also tend to be warmer as a result of the warming effects of the sun's rays. A third factor that influences the productivity of lakes is the amount of nutrients present. This is primarily determined by the rivers and streams that carry nutrients to the lake. River systems that run through areas that donate many nutrients will carry the nutrients to the lakes. Exposed soil and farmland tend to release nutrients, as do other human activities like depositing sewage into streams and lakes. Deep, cold, nutrient-poor lakes are low in productivity and are called **oligotrophic lakes.** Shallow, warm, nutrient-rich lakes are called **eutrophic lakes.** Because of the large amount of organic matter that can accumulate in eutrophic lakes, the decomposition of this organic matter by bacteria and chemical processes can require the use of large amounts of oxygen from the water. This is called the **biochemical oxygen demand** or **BOD.**

Although water (H_2O) has oxygen as part of its molecular structure, this oxygen is not available to organisms. The oxygen that they need is molecular oxygen (O_2), which enters water from the air or as a result of photosynthesis by aquatic plants. Wave action as wind blows across the surface tends to mix air and water and allows more oxygen to dissolve in the water.

Dissolved oxygen content in the water is an important factor since the quantity determines the kinds of organisms that can inhabit the lake. Organic materials in the lake result from the metabolic wastes of organisms that live in and around the lake, plants and animals that die, and parts of organisms, such as leaves from trees, that fall into the lake. The breakdown of these organic materials by bacteria and fungi uses oxygen from the water. The amount and kinds of organic matter determine, in part, how much oxygen is left to be used by other organisms, such as fish, crustaceans, and snails. Many lakes may experience periods of time when oxygen is low, resulting in the death of fish and other organisms. These topics are discussed in greater depth in chapter 15.

Streams and rivers represent a second category of freshwater ecosystem. Since the water is moving, planktonic organisms are less important than are attached organisms. Most algae grow attached to rocks and other objects on the bottom. This collection of algae, animals, and fungi is called the **periphyton.** Since the water is shallow (except for large or extremely muddy rivers), light can penetrate to the bottom very easily. However, it is difficult for photosynthetic organisms to accumulate the nutrients necessary for growth, and most clear-water streams are not very productive. As a matter of fact, the major input of nutrients is from organic matter that falls into the stream from terrestrial sources. These are primarily the leaves from trees and other vegetation, as well as the bodies of living and dead insects. Within the stream is a community of organisms that are specifically adapted to use the debris as a source of food. Bacteria and fungi colonize the organic matter, and many kinds of insects shred the material as they eat it and the fungi and bacteria living on it. The feces (intestinal wastes) of these insects and the tiny particles produced during the eating process become food for other insects that build nets to capture the tiny bits of organic matter that drift their way. These insects are in turn eaten by carnivorous insects and fish.

Organisms in larger rivers and muddy streams, which have less light penetration, rely on the accumulation of food that drifts their way from the many streams that empty into the river. These larger rivers tend to be warmer and to have slower-moving water. Consequently, the amount of oxygen is usually less, and the species of plants and animals change. Any additional organic matter added to the river system adds to the BOD, further reducing the oxygen present. Plants may become established along the shore and contribute to the ecosystem by carrying on photosynthesis and providing hiding places for animals.

Just as estuaries are a bridge between freshwater and marine ecosystems, swamps and marshes are a transition between aquatic and terrestrial ecosystems. **Swamps** are areas of trees that are able to live in places that are either permanently flooded or flooded for a major part of the year. **Marshes** are similar areas that are dominated by grasses and reeds. Many swamps and marshes are successional stages that eventually become totally terrestrial communities.

Summary

Ecosystems change as one kind of organism replaces another in a process called succession. Ultimately, a relatively stable stage is reached, called the climax community. Succession may begin with bare rock or water, in which case it is called primary succession, or may occur when the original ecosystem is destroyed, in which case it is called secondary succession. The stages that lead to the climax are called successional stages.

Major regional terrestrial climax communities are called biomes. The primary determiners of the kinds of biomes that develop are the amount and yearly distribution of rainfall and the yearly temperature cycle. Major biomes are desert, grassland, savanna, tropical rain forest, temperate deciduous forest, taiga, and tundra. Each has a particular set of organisms that is adapted to the climatic conditions typical for the area. As one proceeds up a mountainside, it is possible to witness the same kind of change in biomes that occurs if one were to travel from the equator to the North Pole.

Aquatic ecosystems can be divided into marine (saltwater) and freshwater ecosystems. In the ocean, some organisms live in open water and are called pelagic organisms. Light penetrates only the upper few meters of water; therefore, this region is called the euphotic zone. Tiny photosynthetic organisms that float near the surface are called phytoplankton. They are eaten by small animals known as zooplankton, which in turn are eaten by fish and other larger organisms.

The kind of material that makes up the shore determines the mixture of organisms that lives there. Rocky shores provide surfaces for organisms' attachment; sandy shores do not. Muddy shores are often poor in oxygen, but marshes and swamps may develop in these areas. Coral reefs are tropical marine ecosystems dominated by coral animals. Mangrove swamps are tropical marine shoreline ecosystems dominated by trees. Estuaries occur where freshwater streams enter the ocean. They are usually shallow, very productive areas. Many marine organisms use estuaries for reproduction.

Insects are common in fresh water and are absent in marine systems. Lakes show similar structure to the ocean, but the species present are different. Deep, cold-water lakes with poor productivity are called oligotrophic lakes, while shallow, warm-water, highly productive lakes are called eutrophic. Streams differ from lakes in that most of the organic matter present in streams falls into it from the surrounding land. Thus, organisms in streams are highly sensitive to the land uses that occur near the streams.

Review Questions

1. Describe the process of succession. How does primary succession differ from secondary succession?
2. How does a climax community differ from a successional community?
3. List three characteristics typical of each of the following biomes: tropical rain forest, desert, tundra, taiga, savanna, grassland, and temperate deciduous forest.
4. What two primary factors determine the kind of terrestrial biome that will develop in an area?
5. How does altitude affect the kind of biome present?
6. What areas of the ocean are the most productive?
7. How does the nature of the substrate affect the kinds of organisms found at the shore?
8. What is the role of each of the following organisms in a marine ecosystem: phytoplankton, zooplankton, algae, coral organisms, and fish?
9. List three differences between freshwater and marine ecosystems.
10. What is an estuary?

CHAPTER SIX
Population Principles

Objectives

After reading this chapter, you should be able to:

Understand that birthrate and deathrate together determine population growth rate.

Define the following characteristics of a population: natality, mortality, sex ratio, age distribution, reproductive potential, and spatial distribution.

Explain why the biotic potential for most populations is much greater than needed.

Describe how, as it grows, a population goes through lag, exponential growth, and stable equilibrium phases.

Describe how the carrying capacity of a region for a population is determined by environmental resistance.

List the four categories under which limiting factors can be classified.

Describe how some populations enter a death phase after the stable equilibrium phase.

Explain why humans are subject to the same forces of environmental resistance as other organisms.

Understand the implications of overreproduction.

Understand that the human population is still growing rapidly.

Explain how human population growth is influenced by social, theological, philosophical, and political thinking.

Chapter Outline

Key Terms

age distribution
biotic potential
birthrate (natality)
carrying capacity
death phase
deathrate (mortality)
density-dependent
   factors
density-independent
   factors
dispersal

emigration
environmental resistance
exponential growth
   phase
immigration
K-strategists
lag phase
limiting factors
r-strategists
sex ratio
stable equilibrium phase

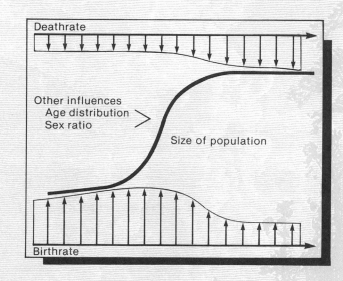

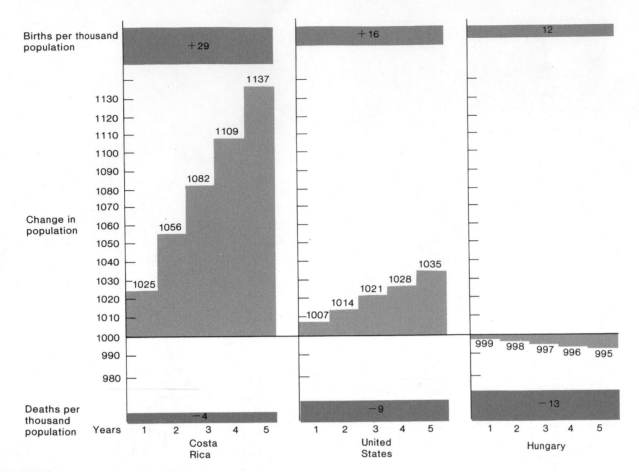

FIGURE 6.1

Effect of Birthrate and Deathrate on Population Size. For a population to grow, the birthrate must exceed the deathrate for a period of time. These three human populations illustrate how the combined effects of births and deaths would change population size if birthrates and deathrates were maintained for a five-year period.

Sources: Data from *World Population Data Sheet,* 1987 and 1990, Population Reference Bureau, Inc., Washington, DC.

Population Characteristics

A population can be defined as a group of individuals of the same species that inhabits an area. Just as individuals within a population are recognizable, populations themselves have specific characteristics that give them identity. Some of these characteristics are birthrate, deathrate, sex ratio, age distribution, and spatial distribution. **Birthrate (natality)** refers to the number of individuals added to the population through reproduction. In human population studies, it is usually expressed as the number of births per thousand individuals per year. For example, if a population of one thousand individuals has ten offspring in one year, the natality is given as ten per thousand per year. This is sometimes called the raw natality of a population. It is not the same as population growth because the growth of a population includes both natality and mortality. **Deathrate (mortality)** is the number of deaths per thousand individuals per year. Natality and mortality together determine whether a population is increasing or decreasing and how rapidly it is changing. For a population to grow, the birthrate must exceed the deathrate for a period. (See figure 6.1.)

A population's birthrate and deathrate depend on its sex ratio and age distribution. The **sex ratio** refers to the number of males relative to the number of females. In humans, about 106 males are born for every 100 females. However, in the United States, by the time people reach their middle twenties, a higher deathrate for males has equalized the sex ratio.

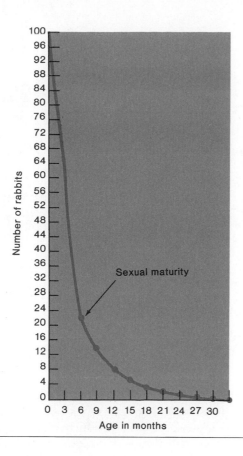

FIGURE 6.2
Survivorship Curve of Cottontail Rabbit. In most natural populations, mortality rates are so high that very few individuals reach sexual maturity, and even fewer reach old age.

The higher deathrate for males continues into old age, when women out-number men. In many insect populations, such as bees, ants, and wasps, the number of females greatly exceeds the number of males at all times, although most of the females are sterile.

Another feature of a population is its **age distribution.** Not all members of a population are the same age. Some are prereproductive juveniles, some are reproducing adults, and some are postreproductive adults. A stable population balances these three age classifications so that the proportion of prereproductive, reproductive, and postreproductive individuals remains constant. Typically, this means that there are more prereproductive individuals than reproductive individuals, and more reproductive individuals than postreproductive individuals.

The age structure of a population is directly related to its mortality rate. Since most species of small animals have a very high mortality rate, they have a large number of prereproductive individuals that declines sharply with increasing age. A good example of this type of population is the cottontail rabbit. Many young are produced, but most of them die, a few reach sexual maturity, and very few live to old age. (See figure 6.2.)

If the majority of a population is postreproductive, the population declines. This age structure develops in many insect populations in the fall after the eggs have been laid. In human populations, selective death due to war or cultural taboos that restrict sexual contact can lead to this kind of age structure in a population.

If the majority of a population is made up of reproducing adults, a "baby boom" can be expected. This population age structure naturally occurs in the spring, when many animal populations begin to mate and to raise offspring.

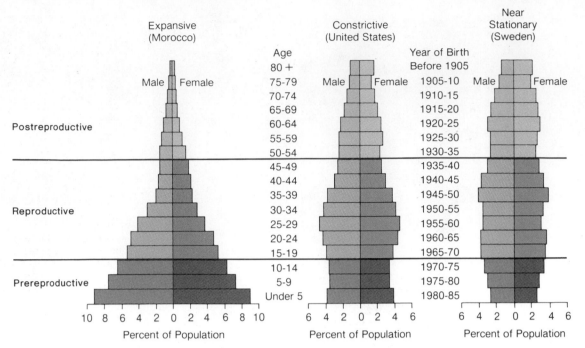

FIGURE 6.3

Age Distribution in Human Populations. The relative numbers of individuals in each of the three categories (prereproductive, reproductive, and postreproductive) are good clues to the future of the population. Morocco has a large number of young individuals that will become reproducing adults. Therefore, this population is likely to grow rapidly. The United States has a declining proportion of prereproductive individuals, and Sweden has a relatively balanced population. Therefore, they will probably have declining populations sometime in the future, with Sweden's population declining first.

Human populations exhibit several types of age distributions. (See figure 6.3.) Morocco has a large prereproductive and reproductive component in its population. This means that Morocco's population will continue to increase rapidly for some time into the future. Sweden has a relatively even age distribution and will remain nearly constant in its population size, although it will probably have a declining population in the future. The United States has a large reproductive component with a declining number of prereproductive individuals. So eventually, the U.S. population will begin to decline.

Just as populations are not usually divided into equal thirds according to their age distribution, organisms are not usually equally distributed spatially. Because of such factors as soil type, quality of habitat, and availability of water, organisms normally are distributed unevenly. When the density of organisms is too great, all organisms within the population are injured because they compete severely with each other for necessary resources. Plant populations may compete for water, soil nutrients, or sunlight. In animal populations, overcrowding might cause exploration and migration into new areas. This **dispersal** of organisms relieves the overcrowded conditions in the home area and, at the same time, leads to the establishment of new populations. Often, juvenile individuals relieve the overcrowding by leaving the home area. The pressure for out-migration **(emigration)** may be a result of seasonal reproduction leading to a rapid increase in the size of the local population, or environmental changes that cause competition to intensify among members of the same species. For example, as water holes dry up, competition for water increases, and many desert birds emigrate to areas where water is still available.

The organisms that leave one population often become members of a different population. This in-migration **(immigration)** may introduce new characteristics that were not in the original population. When Europeans immigrated to North America, they brought genetic and cultural characteristics that had a tremendous impact on the existing Native American

Ecological Principles and Their Application

FIGURE 6.4
Biotic Potential. The ability of a population to reproduce greatly exceeds the number necessary to replace those who die. Here are some examples of the prodigious reproductive abilities of some species.

population. In addition, Europeans brought diseases that were "foreign" to the Native American population. These diseases increased the deathrate and lowered the birthrate of the natives.

Droughts, wars, and political persecution have caused people to emigrate from their native lands to other countries. The United Nations High Commissioner for Refugees in 1987 estimated that there are fifteen million refugees worldwide. Often, the receiving countries are unable to cope with the large influx of new inhabitants. Since 1975, the United States has accepted over one million refugees from Cuba, Haiti, Laos, Vietnam, and many other countries. Immigration to the United States accounts for about one-third of the country's annual population growth. Other countries, such as Thailand, Pakistan, and Sudan, have centers to accept and process refugees, but these countries will not permanently support this refugee population.

A Population Growth Curve

Sex ratios and age distributions within a population directly influence the rate of reproduction within a population. Each species has an inherent reproductive capacity, or **biotic potential,** which is its ability to produce offspring. Generally, this biotic potential is many times greater than the number of offspring needed to replace the parents when the parents die. (See figure 6.4.) Consequently, most of the young die, and only a few survive to become reproductive adults.

However, this high reproductive potential results in a natural tendency for populations to increase. For example, two mice produce four offspring, which, if they live, will also produce offspring while their parents are also reproducing. Therefore, the population will tend to grow in an exponential fashion (2, 4, 8, 16, 32, etc.).

Population growth tends to follow a particular pattern, consisting of a lag phase, an exponential growth phase, and a stable equilibrium phase. Figure 6.5 shows a typical population growth curve. During the first portion of the curve, known as the **lag phase,** the population grows very slowly because the process of reproduction and growth of offspring takes time. Most organisms do not reproduce instantaneously but must first mature into adults. This period is followed by mating and the development of the young into independent organisms. By the time the first batch of young have reached sexual maturity, the parents may be in the process of producing a second set of offspring. Since more total organisms now are reproducing, the population begins to increase at an exponential rate. This stage in the population growth curve is known as the **exponential growth phase.** This

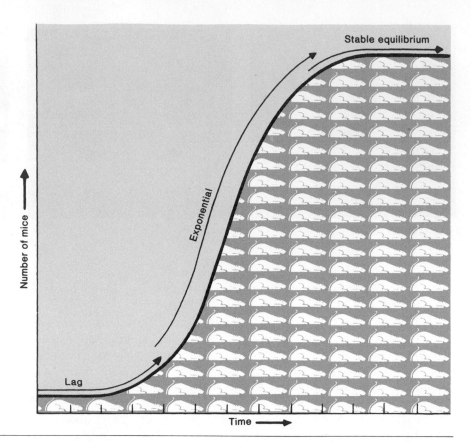

Stable equilibrium

Number of mice

Exponential

Lag

Time

FIGURE 6.5
Typical Population Growth Curve. In
this mouse population, there is little
growth during the lag phase. During the
exponential growth phase, the
population rapidly increases as the
offspring reach reproductive age.
Eventually the population reaches a
stationary growth phase, during which
the birthrate equals the deathrate.

growth will continue for as long as the birthrate exceeds the deathrate.
Eventually, however, the deathrate and the birthrate will come to equal one
another, and the population will stop growing and reach a relatively stable
population size. This stage is known as the **stable equilibrium phase.** For
most organisms, an initial increase in the deathrate brings about the stable
equilibrium phase, but in some situations, the birthrate also decreases. To
understand why population growth tends to level off, it is necessary to discuss the concept of carrying capacity.

Carrying Capacity

The **carrying capacity** of a specific area is the number of individuals of
a species that can survive in that area over time. In most populations, four
broad categories of factors interact to set the carrying capacity for a population. These factors are: (1) the availability of raw materials, (2) the availability of energy, (3) the accumulation of waste products and their means
of disposal, and (4) interactions among organisms. The total of all of these
forces acting together to limit population size is known as **environmental
resistance,** and certain **limiting factors** have a primary role in limiting
the size of a population. In some cases, these limiting factors are easy to
identify and may involve lack of food, lack of oxygen, competition with
other species, or disease. (See figure 6.6.)

For example, in grass plants, nitrogen and magnesium in the soil are
necessary raw materials for the manufacture of chlorophyll. If these minerals are not present in sufficient quantities, the grass population cannot
increase. However, the application of fertilizers containing these minerals

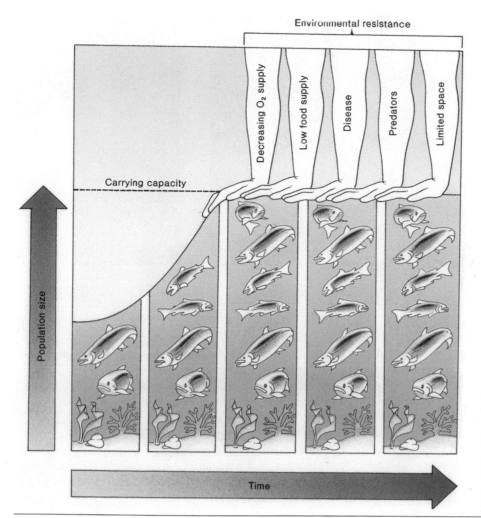

Decreasing O$_2$ supply

Low food supply

Disease

Predators

Limited space

Carrying capacity

Population size

Time

FIGURE 6.6

Carrying Capacity. A number of factors in the environment, such as food, oxygen supply, diseases, predators, and space, determine the number of organisms that can survive in a given area—the carrying capacity of that area. The environmental factors that limit populations are collectively known as environmental resistance.

removes this limiting factor, and the individual grass plants grow and re-produce, resulting in a larger total population. In effect, the carrying capacity has been increased because this specific limiting factor has been removed. The carrying capacity has not been eliminated because now some new limiting factor will predominate. Perhaps it will be the amount of water, the number of insects that feed on the grass, or competition for sunlight.

Because plants require energy in the form of sunlight for photosynthesis, the amount of light can be a limiting factor for many plants. When small plants are in the shade of trees, they often do not grow well and have small populations.

Accumulation of waste products is not normally a limiting factor for plant populations, since plants produce few waste products, but it can be for other kinds of organisms. Bacteria and other tiny organisms, and many kinds of aquatic organisms that live in small ecosystems like puddles, pools, or aquariums, may have waste accumulation as a limiting factor. When a small number of a species of bacterium are placed on a petri plate with nutrient agar (a jellylike material containing food substances), the population growth follows the curve shown in figure 6.7. As expected, the population begins with a lag phase, continues through an exponential growth phase, and eventually levels off in a stable equilibrium phase. However, in

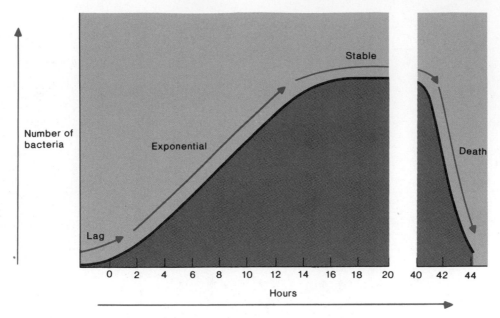

FIGURE 6.7

Bacterial Growth Curve. The change in population size follows a typical population growth curve until waste products become lethal. The buildup of waste products lowers the carrying capacity. When a population begins to decline, it enters the death phase.

this small, enclosed situation, there is no way to get rid of the toxic waste products, and they accumulate, eventually causing the death of the bacteria. This decline in population size is known as the **death phase.** When a population decreases rapidly, it is said to crash.

Interactions among organisms are also very important in controlling population size. Since many birds rely on the seeds of grasses for food, birds have a limiting effect on the size of grass populations. Decomposer organisms may allow for increased populations because they prevent the buildup of toxic wastes. Parasites and predators may cause the premature death of individuals, thus limiting the size of the population. A good example of predator-prey interaction within a population is the interaction between the cat known as the Canada lynx and a member of the rabbit family known as the varying hare. (See figure 6.8.) The varying hare has a high biotic potential that the lynx helps to control by using the hare as food.

Some studies indicate that populations can be controlled by interaction among the individuals within a population. A study of rats shows that crowding causes a breakdown in normal social behavior, which leads to fewer births and increased deaths in laboratory rat populations. The kinds of changes observed included abnormal mating behavior, decreased litter size, fewer litters per year, lack of maternal care, and increased aggression in some rats or withdrawal in others. Thus, limiting factors can reduce birthrates as well as increase deathrates.

Reproductive Strategies and Population Fluctuations

So far, we have talked about population growth as if all organisms reach a stable population when they reach the carrying capacity. That is an appropriate way to begin to understand population changes, but the real world is much more complicated. First of all, species can be divided according to whether their reproductive strategy is a K-strategy or an r-strategy. **K-strategists** are usually large organisms that have relatively long lives,

Ecological Principles and Their Application

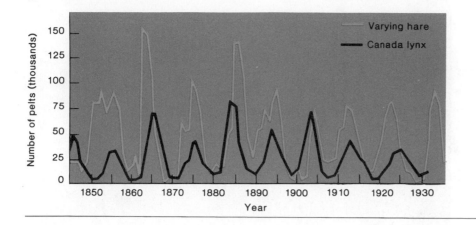

FIGURE 6.8
Predator/Prey Interaction. Interaction between predator and prey species is complex and often difficult to interpret. These data were collected from the records of the number of pelts purchased by the Hudson Bay Company. They show that the two populations fluctuate, with changes in the lynx population usually following changes in the varying hare population.

Data from D. A. MacLulich, *Fluctuations in the Numbers of the Varying Hare* (Lepus americanus). University of Toronto Press, Toronto, 1937, reprinted 1974.

produce few offspring, provide care for their offspring, and typically have populations that stabilize at the carrying capacity. Their reproductive strategy is to invest a great deal of energy in producing a few offspring that have a good chance of living to reproduce. Examples of this kind of organism include deer, lions, swans, and other large animals. Humans generally produce single offspring and, even in countries with high infant mortality, have 80 percent survival of the children. Generally, populations of K-strategists are limited by density-dependent factors. **Density-dependent factors** are those limiting factors that become more severe as the size of the population increases. For example, as a population of hawks increases, the hawks begin to compete for the available food, such as mice, snakes, and small birds. When food is in short supply, many of the young in the nest die, and population growth slows. They have reached the carrying capacity.

The **r-strategist** is typically a small organism that has a short life, produces large numbers of offspring, and does not reach a carrying capacity. Examples of this kind of organism include grasshoppers, gypsy moths, and some mice. The reproductive strategy of r-strategists is to expend large amounts of energy in the production of many offspring, but to provide limited care (often none) to the offspring. Consequently, there is high mortality among the offspring. For example, one female oyster may produce a million eggs, but few of them ever find suitable places to attach themselves and grow. Typically, these populations are limited by **density-independent factors** in which the size of the population has nothing to do with the limiting factor. Typical examples of density-independent factors are changing weather conditions that kill large numbers of organisms, the drying up of a small pond, or the death of entire populations due to the destruction of their food source. The population size of r-strategists is likely to fluctuate wildly. They reproduce rapidly, population size increases until some factor causes the population to crash, and then they begin the whole process all over again.

Even K-strategists, however, have population fluctuations for a variety of reasons. First of all, the environment is not constant from year to year. Flood, drought, fire, extreme cold, and similar events may affect the carrying capacity of an area, thus causing fluctuations in population size. Furthermore, epidemic disease or increased predation may also lead to populations that vary from year to year. Figure 6.8 shows rather substantial changes in the populations of lynx and varying hare. The size of the lynx population seems to be tied to the size of the hare population, which is logical since lynx eat hares. However, the causes of the fluctuations in the

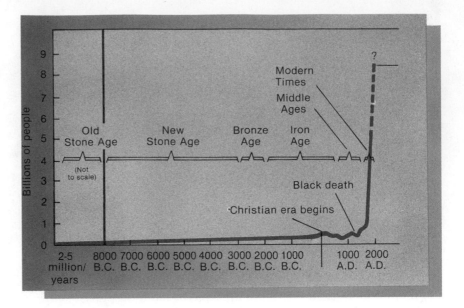

FIGURE 6.9

Human Population Growth. From A.D. 1850 to A.D. 1930, the number of humans doubled (from one billion to two billion) and then doubled again by 1975 (four billion) and could double again (eight billion) by the year 2020. How long can this pattern continue before the earth's ultimate carrying capacity is reached?

varying hare population are unclear. One possibility is that periodic epidemic disease may cause dense populations of organisms like hares to crash, leading to the crash of their predators' populations.

Although local human populations often show fluctuations, the worldwide human population has increased continually for the past several hundred years. Humans have been able to reduce environmental resistance by eliminating competing organisms, increasing food production, and controlling disease organisms.

Prolific Humans

The human population growth curve has a long lag phase followed by a very sharp, rapid exponential growth phase that is still rapidly increasing. (See figure 6.9.) A major reason for the rapid increase is that the human species has reduced its deathrate. When any country has reduced environmental resistance by increasing food production or controlling disease, this technology has been shared throughout the world. Developed countries have sent health-care personnel to all parts of the globe to improve the quality of life for people in the less developed countries. Physicians advise on nutrition, and engineers develop wastewater treatment systems. Improved sanitary facilities in India and Indonesia, for example, decreased mortality caused by cholera. These advancements tend to reduce mortality and directly increase the *quantity* of life because birthrates remain high.

Let us examine this situation from a different perspective. The world population is currently increasing at an annual rate of 1.8 percent. However, even at a 1.8 percent annual increase, the population is growing rapidly. It is often difficult to comprehend the impact of a 1 or 2 percent annual increase on a population. Remember that a growth rate in any population compounds itself, since the additional individuals entering a population will eventually reproduce, thus adding more individuals to the population. Another way to look at population growth is to determine how much time is needed to double the population. This is a valuable way to examine population growth because most of us can appreciate what life would be like

Ecological Principles and Their Application

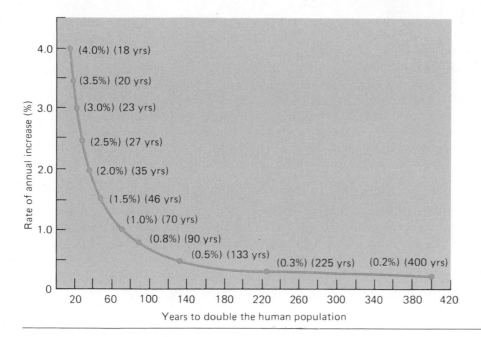

FIGURE 6.10

Doubling Time for the Human Population. This graph shows the relationship between the rate of annual increase in percent and doubling time. A population growth rate of 1 percent per year would result in the doubling of the population in about seventy years. A population growth rate of 3 percent per year would result in a population doubling in about twenty-three years.

if the number of people in our locality were doubled, particularly if the number were to occur within our lifetime.

Figure 6.10 shows the relationship between the rate of annual increase for the human population and the number of years it would take to double the population if that rate were to continue. At a 1 percent rate of annual increase, the population will double in approximately seventy years. At a 2 percent rate of annual increase, the population will double in about thirty-five years. The current worldwide rate of annual increase of about 1.8 percent will double the world population by the year 2030.

What does this very rapid rate of growth mean to the human species? First, as a species, humans are subject to the same limiting factors as all other species. We cannot increase beyond our ability to acquire raw materials, acquire energy, or safely dispose of our waste materials. We also must remember that interactions with other species and with other humans may be involved in determining our carrying capacity.

Let us look at these four factors in more detail. For many of us, raw materials simply mean the amount of food available. However, we have become increasingly dependent on technology, and our life-styles are tied directly to our use of many other kinds of resources. Food production is becoming a limiting factor for some segments of the world's human population. Malnutrition is a major problem in many parts of the world because food is not available in sufficient quantities. Chapter 7 deals with the problems of food production and distribution and their relationship to human population growth.

The second factor, available energy, involves problems similar to those of available raw materials. One important fact is that all species on earth are ultimately dependent on sunlight for their energy. New, less disruptive methods of harnessing this energy must be developed to support an increasing population. Currently, the world population depends on large amounts of fossil fuels to raise food, modify the environment, and allow for movement. When energy prices increase, a large portion of the world's population is placed in jeopardy because their incomes are not sufficient to pay the increased costs for food and other essentials.

Waste disposal is the third factor determining the carrying capacity for humans. Most pollution is, in reality, the waste product of human activity. Some people are convinced that disregard for the quality of our environment will be a major limiting factor. In any case, it just makes good sense to control pollution and to work toward cleaning our environment.

The fourth factor that determines the carrying capacity of a species is interaction with other organisms. We need to become aware that we are not the only species of importance. When we convert land to meet our needs, we displace other species from their habitats. Many of these displaced organisms are not able to compete with us successfully and must migrate or become extinct. Unfortunately, as humans expand their domain, the areas available to these displaced organisms become more rare. Parks and natural areas have become tiny refuges for the plants and animals that once occupied vast expanses of land. If these refuges fall to the developer's bulldozer or are converted to agricultural use, many organisms will become extinct. What today seems like an unimportant organism, one that we could easily do without, may someday be seen as an important link to our very survival.

Humans Are Social Animals

Human survival depends upon interaction and cooperation with other humans. Current technology and medical knowledge are available to control human population growth. (See box 6.1.) and to improve the health of the people of the world. Why then does the population continue to increase, and why does poverty become greater every year? Humans are social animals who have freedom of choice and frequently do not do what is considered to be "best" from an unemotional, uninvolved, biological point of view. People make decisions based on history, social situations, ethical and religious considerations, and personal desires. The biggest obstacle to controlling human population is not biological but falls into the province of philosophers, theologians, politicians, sociologists, and others. People in all fields need to understand the cause of the population problem if they are to successfully deal with every aspect of the problem.

Ultimate Size Limitation

The human population cannot increase indefinitely because, eventually, the weight of the human tissue would equal the weight of the earth. We can say with certainty that the population will ultimately reach its carrying capacity and level off.

Many speculate about the limits of human population growth. In its 1981 report, the United Nations Fund for Population Activities suggests that a moderate estimate of future human population is 10.5 billion people by the year 2110. However, if world human population continues to grow at its current rate of 1.8 percent per year, the population will exceed 10 billion by the year 2030. Some people suggest that a lack of food, a lack of water, or increased waste heat will ultimately control the size of the human population. Still others suggest that, in the future, social controls will limit population growth. Others are concerned that politically related destruction of humanity will occur. No one knows what the ultimate human population size will be or what the most potent limiting factors will be, but most agree that we are approaching the maximum sustainable human population. If the human population continues to increase, the amount of agricultural land available will not be enough to satisfy the demand for food.

BOX 6.1
Control of Births

The use of technology to control disease and famine has greatly reduced the deathrate of the human population. Technological developments can also be used to control the birthrate. A variety of contraceptive methods are available to help people regulate their fertility. Research is continuing to develop new, more effective, more acceptable, and less expensive methods of controlling conception. Because of cultural and religious differences, some forms of contraception may be more acceptable to one segment of the world's population than another.

The most common methods of contraception are oral contraceptive pills, diaphragms and spermicidal jelly, intrauterine contraceptive devices, spermicidal vaginal foam, condoms, vasectomy, and tubal ligation. The range of effectiveness of these methods, shown in the table, is the result of individual fertility differences and the degree of care employed in the use of each method.

In addition to various methods of conception control, abortion can terminate unwanted conceptions early in pregnancy. Most countries with low birthrates, such as the United States, Japan, and many European countries, allow abortions.

Effectiveness of fertility control methods.

Method	Average pregnancy rate per 100 women per year
Abstinence	0
Sterilization (vasectomy or tubal ligation)	0–0.003
Oral contraceptive pill	0.1–1
Intrauterine contraceptive device (IUD)	1–6
Diaphragm with jelly	2–20
Vaginal foam*	2–29
Condom (rubber)*	3–36
Withdrawal (coitus interruptus)	18–23
Rhythm (periodic abstinence)	1–47
Rhythm (temperature method)	1–20
None	60–80

*Use of condom and vaginal foam together is more effective than either used singly.

Oral contraceptive pills

Diaphragm and spermicidal jelly

Intrauterine contraceptive device

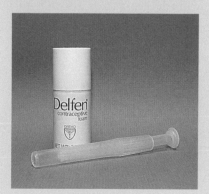

Spermicidal vaginal foam

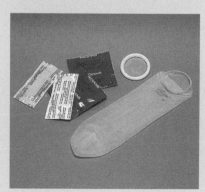

Condom

125

In 1930, the George Reserve was established as a wildlife study area through the generous gift of Detroit industrialist Colonel Edwin S. George. This property, about 464 hectares in southeastern Michigan, was donated to the University of Michigan to further research of natural populations. It is bounded by a game fence 2.9 meters high. The reserve includes some wooded areas, both deciduous and coniferous, some wet and marshy areas, some permanent and temporary ponds, and some open, grassy, meadowlike areas. Originally, six deer were imported into the reserve: four females (thought to be pregnant) and two males. Natural predators were excluded from the reserve, so the deer population was expected to increase rapidly.

One goal of the early research was to keep an accurate record of how the deer population grew. Researchers decided to count the number of deer in various age classes so that they could make a cross-check whenever an animal died.

Each year, a census was conducted in the reserve. During this census, a group of individuals (usually graduate students) would line up and walk from one end of the reserve to the other. The people taking the census would ideally be close enough to each other to maintain contact. Anytime a deer passed through the line of census takers, it was counted. Skeletal remains of deer also were counted. At the end of the census, the total deer population was tabulated.

The graph shows the actual population changes for almost forty years. After only six years, the number of deer had increased from 6 to over 160. Why was there such a rapid increase in just six years? The answer to this question requires some knowledge of deer biology. For example, deer are sometimes sexually mature before they are one year old, and twins or triplets can occur. To answer this question, we also need to know in what phase of population growth the deer population was in 1930 and again in 1940. What do you think would prevent the number of deer from increasing forever? Note that the curve is not a neat, smooth line but has a series of abrupt spikes. What is the reason for this? What is the explanation for the drastic decline in population in the mid-1930s? Why was there an increase in the next few years?

Recently, one of the research activities at the George Reserve has been an attempt to manipulate the population by harvest and to reduce it to a particular level. Contrary to expectations, when the population harvest was increased, there was not an equivalent steplike decrease in population.

Can you think of any reason why there is a gradual decline in the population, rather than the steplike reduction anticipated?

Summary

A population is a group of organisms of the same species that inhabits an area. The birthrate (natality) is the number of individuals entering the population by reproduction during a certain period. The deathrate (mortality) measures the number of individuals that die in a population during a certain period. Population growth is determined by the combined effects of the birthrate and deathrate.

The sex ratio of a population is a way of stating the relative number of males and females in a population. Age distribution within a population and the sex ratio have a profound impact on future population growth. Most organisms have a biotic potential much greater than needed to replace dying organisms.

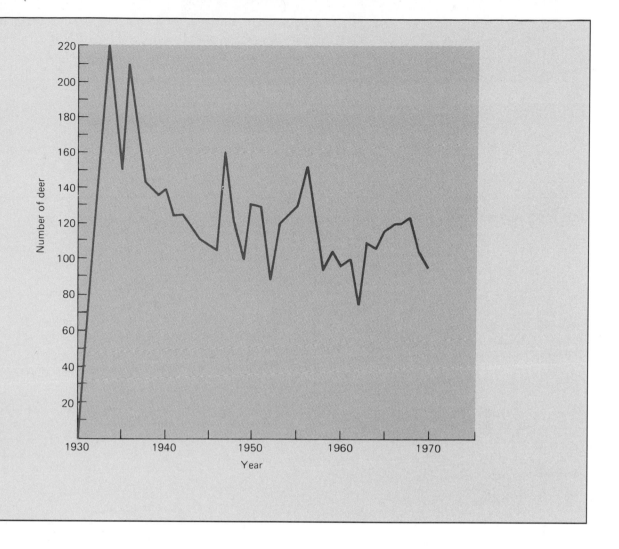

Interactions among individuals in a population, such as competition, predation, and parasitism, are also important in determining population size. Organisms may migrate into (immigrate) or migrate out of (emigrate) an area as a result of competitive pressure.

A typical population growth curve shows a lag phase followed by an exponential growth phase and a stable equilibrium phase at the carrying capacity. The carrying capacity is determined by many limiting factors that are collectively known as environmental resistance. The four major categories of environmental resistance are available raw materials, available energy, disposal of wastes, and interactions among organisms. Some populations experience a death phase following the stable equilibrium phase.

K-strategists typically are large, long-living organisms that reach a stable carrying capacity. Their population size is usually controlled by density-dependent factors. Organisms that are r-strategists are generally small, short-living organisms that reproduce very quickly. Their populations do not generally reach a carrying capacity but are caused to crash by some density-independent factor.

The human population is still increasing at a rapid rate. The earth's ultimate carrying capacity for humans is not known. The causes for the human population growth are not just biological; they are also social, political, philosophical, and theological.

Review Questions

1. How are biotic potential and age distribution interrelated?
2. List three characteristics populations might have.
3. Why do some populations grow? What factors help to determine the rate of this growth?
4. Under what kind of conditions might a death phase occur?
5. List four factors that could determine the carrying capacity of an animal species.
6. How do the concepts of birthrate and population growth differ?
7. How does the population growth curve of humans compare with that of bacteria on a petri dish?
8. How do r-strategists and K-strategists differ?
9. As the human population continues to increase, what might happen to other species?
10. All successful organisms overproduce. What advantage does this provide for the species? What disadvantages may occur?

Ecological Principles and Their Application

CHAPTER SEVEN
Human Population Issues

Objectives

After reading this chapter, you should be able to:

Apply some of the principles discussed in chapter 6 to the human population.

Differentiate between birthrate and population growth rate.

Explain the current population situation in the United States.

Explain how age distribution affects population projections.

Recognize that U.S. society is in the process of adjusting as the average age increases.

Recognize that most of the world still has a rapidly growing population.

Describe the implications of the hypothesis that a demographic transition occurs.

Understand how an increasing world population will alter the worldwide ecosystem.

Understand the implications of the fact that most of the countries of the world are not able to produce enough food to feed themselves.

Explain why less developed nations have high birthrates and why they will continue to have a low standard of living.

Recognize that the developed nations of the world will be under greater pressure to share their abundance.

Chapter Outline

Social Implications of Population Growth
U.S. Population Changes
World Population Changes
The Demographic Transition
Standard of Living: A Comparison
Environmental Costs
Anticipated Changes with Continued Population Growth
    Box 7.1 Governmental Policy and Population Control
Consider This Case Study: Population Growth in Mexico

Key Terms

demographic transition
demography
gross national product
    (GNP)

postwar "baby boom"
replacement fertility
standard of living
zero population growth

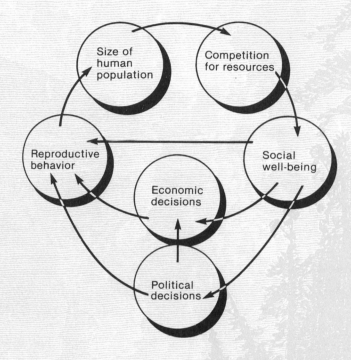

Social Implications of Population Growth

In chapter 6, we examined population characteristics, the way populations grow, and the forces that cause populations to stabilize. These forces also act on human populations. However, unlike other kinds of populations, human populations also are influenced by a variety of social, political, and economic forces. Humans are able to predict the likely course of events and adjust their activities accordingly. They can make conscious decisions based on what they perceive the future political, economic, or social pressures will be. Therefore, people spend considerable time and energy trying to ascertain the future so they can plan their lives.

Decisions made by individuals can affect a whole population and cause changes at the social or political level, which may have international implications. For example, war has been a factor in population pressure. During World War II, both Germany and Japan were partially motivated by a desire to relieve population pressures by acquiring new land.

Unfortunately, many of the news stories and discussions of population trends are confusing. Projections of population growth are logical guesses about what will happen in the future. Experts differ because they use different sets of data to make their projections. Many projections assume major changes in technology that will cause population growth to slow, while others are based on statistics of past population trends and assume that future populations will grow at the same rate.

Another important point is that some experts have a history of being accurate while others do not. When we listen to accounts of what our future will be like, we should evaluate the credibility of the experts. Because humans are highly social organisms, we cannot look at the human population from a strictly biological point of view. How are the biological and social natures of the human species interrelated, and to what extent will they determine future population trends?

U.S. Population Changes

Individuals in a population tend to produce more offspring than the number required to replace the parents when they die. Therefore, populations tend to increase until the environment cannot support any additional individuals. At this point, the number of births equals the number of deaths. Whether this regulation involves either a decrease in births or an increase in deaths, the result is the same. Some countries, such as Ethiopia and Afghanistan, have high birthrates and high deathrates. These countries usually have an extremely high deathrate among children because of disease and malnutrition. Other countries, such as the United States and much of Europe, have low birthrates and deathrates. (See table 7.1.) The children in these countries have much higher life expectancies, since life expectancy is in part related to economic conditions, political stability, and educational opportunity.

The study of populations, their characteristics, and what happens to them is known as **demography.** Demographers have made several predictions about what will happen to the U.S. population by the year 2000. The current U.S. population is about 250 million people. The U.S. Commerce Department estimates that the U.S. population is increasing by about 1 percent per year. Yet, we continue to hear that the United States is reproducing at a rate that will lead to **zero population growth.** A population at zero population growth has reached a stable stage where births equal deaths. However, even after a country reaches a birthrate that just replaces

Table 7.1
Population growth rates in selected countries (1990).

Country	Births per 1,000 individuals	Deaths per 1,000 individuals	Rate of natural increase (annual %)	Time needed to double population (years)
Germany (West)	11	11	0.0	—
Belgium	12	11	0.2	462
Sweden	14	11	0.2	311
United Kingdom	14	12	0.2	301
Japan	10	6	0.4	175
United States	16	9	0.8	92
USSR	19	10	0.9	80
Argentina	21	9	1.3	54
Ethiopia	44	24	2.0	34
Turkey	29	8	2.1	32
Afghanistan	48	22	2.6	27
Paraguay	35	7	2.8	25
Guatemala	40	9	3.1	23
Zimbabwe	42	10	3.2	22
Syria	45	7	3.8	18

Source: Carl Haub, Mary Mederios Kent, and Machiko Yanagishita, *World Population Data Sheet* 1990 (Washington, D.C.: Population Reference Bureau, 1990).

parents, it takes many years for the population growth to stop. This condition exists because the deathrate may continue to fall as people live longer.

Another factor affecting population growth rate is the amount of immigration. In the United States, about one-third of the annual population increase is due to immigrants, which number about 700,000 persons per year. This number is somewhat uncertain, since the number of illegal immigrants has been estimated to be between 100,000 and 500,000 per year. There are 500,000–600,000 legal immigrants per year. Although the U.S. population is growing by about 1 percent per year, only 0.7 percent of the increase is due to the difference between birthrates and deathrates. The remaining 0.3 percent is due to immigration.

A third reason populations continue to grow even after reaching a reproductive rate that leads to zero population growth is the age structure of a population. If a population has a large number of young people who are in the process of raising families or who will be raising families in the near future, the population will increase even if the families limit themselves to two children. Certainly, the U.S. population is not growing as rapidly today as it was twenty years ago, but it is still growing. (See figure 7.1.) In the long run, if the number of children produced in the population is 2.1 children per woman, the population growth rate will become zero. This is known as **replacement fertility.** A fertility rate of 2.1 children per woman is used because some babies die soon after birth and do not contribute to the population for very long.

Depending on the number of young people in a population, it may take twenty years to a century for the population to stabilize so that there is no net growth. Regardless of what the birthrate is, there is no way, short of disaster, that the population of the United States can be prevented from growing between now and the year 2020. Demographers will be watching the U.S. birthrate closely for the next few years to determine if the U.S. population size will become constant or decline.

FIGURE 7.1

Population Growth. The population of the United States has grown continuously until the present. The graph indicates that the size of the population was about 230 million people in 1980. The U.S. Census Bureau has made projections based on a birthrate of 2.1 children per woman (the rate at which the population will eventually become stable) and 1.7 children per woman (the current birthrate). Both rates will eventually result in a leveling off of population growth. The ultimate size of the U.S. population differs considerably, depending on which of the estimates is used. In both cases, the population will continue to increase at least until the year 2020, and since current fertility is about two children per woman, the population will continue to increase past the year 2020.

Data from the U.S. Department of Commerce, Bureau of the Census.

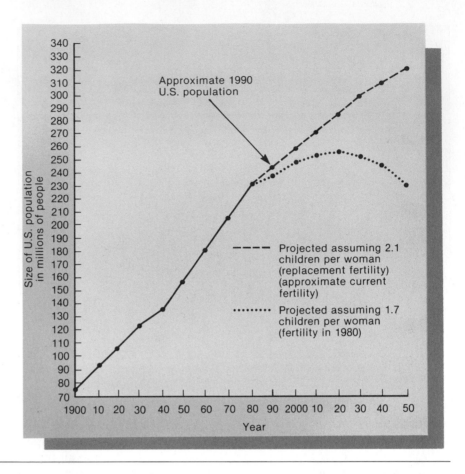

Another item of interest to demographers is the age distribution within the population. Populations may have a majority of individuals in the young age group, or they may have some other age distribution. Figure 7.2 shows the age distribution in the United States in 1988. A decided "bulge" begins at the 40–44 age bracket. These people were born shortly after the end of World War II and marked the beginning of the **postwar "baby boom."** This "baby boom" peaked about fifteen years later, and births have continued at lower levels until recently. The women who were born during this period are now in the middle of their reproductive years. Many of these women have delayed getting married and are having fewer children than their mothers did. Later marriages tend to reduce fertility because there are fewer fertile years per woman in which to produce children. If this trend continues, it will have a decided effect on the U.S. population beyond the year 2020.

Data from the Population Reference Bureau indicate that about 50 percent of sub-Saharan African women in the 15- to 19-year-old age group are married and that the fertility rate is about 6.3 children per woman. In South Korea, only 5 percent of the 15- to 19-year-old women are married, and the fertility rate is 1.6 children per woman. Many people suggest that social changes in North America and Europe, particularly changes in the role of women, are related to declines in birthrate in those countries. As women become better educated and obtain higher-paying jobs, they become financially independent and can afford to marry later and consequently have fewer children. Better-educated women are also more likely to have access

Ecological Principles and Their Application

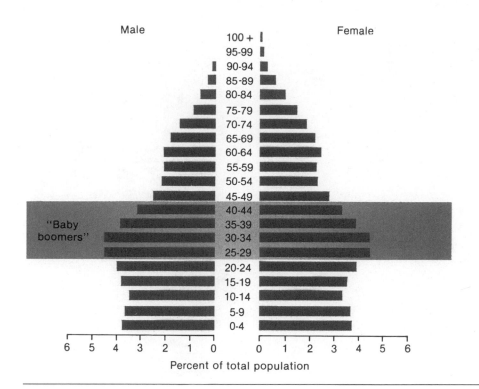

Male   Female

100 +
95-99
90-94
85-89
80-84
75-79
70-74
65-69
60-64
55-59
50-54
45-49
40-44
"Baby   35-39
boomers"   30-34
             25-29
20-24
15-19
10-14
5-9
0-4

6 5 4 3 2 1 0   0 1 2 3 4 5 6

Percent of total population

FIGURE 7.2

Age Distribution of U.S. Population (1988). This graph shows the number of people at each age level. Notice that at age forty to forty-four, a bulge begins that ends at about age twenty-five to twenty-nine. These people represent the "baby boom" that followed World War II. In recent years, births have increased.

Source: Data from the U.S. Department of Commerce, Bureau of the Census.

to and use birth control. In economically advanced countries, 60–80 percent of women typically use contraception. In the less developed countries, contraceptive use is much lower, often less than 30 percent. Changing attitudes toward divorce may also be tied to reduced fertility.

With a decline in the birthrate comes a corresponding change in the businesses that serve specific age groups. The "baby boom" of the late 1940s and the 1950s in the United States resulted in growth in the service industries needed by young families. Maternity wards had to be expanded, schools could not be built fast enough, baby-care companies saw unprecedented sales, and the toy industry flourished. Today, these "babies" are in their thirties and forties and are now buying homes, cars, and appliances. Schools are closing for lack of students. Hospitals are reducing their maternity wards and are beginning to anticipate the need for geriatric services as the average age of the U.S. population increases. What will the social needs be in the year 2020, when many of these "baby boom" individuals retire? Will they be demanding social services similar to those provided to their parents? As population size stabilizes, the average age will increase. A complication is that, in the last five years, U.S. birthrates have increased. Will schools need to be opened again in a few years?

World Population Changes

Currently, the world population is over 5 billion people. By the year 2000, this is expected to increase to about 6.3 billion people. Much of this increase is expected to occur in Africa, Asia, and Latin America, which already have over 79 percent of the world population. (See figure 7.3.)

If world population trends continue, Africa, Asia, and Latin America will increase from their current population size of 4 billion people to over 5 billion people by the year 2000. If projections are correct, these countries would then constitute nearly 82 percent of the world's population. These

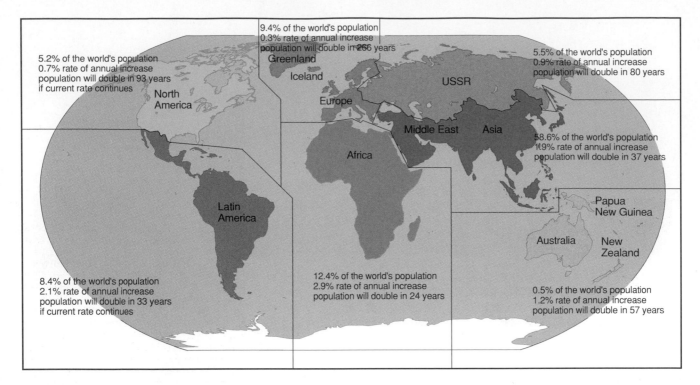

5.2% of the world's population
0.7% rate of annual increase
population will double in 93 years
if current rate continues

North America

9.4% of the world's population
0.3% rate of annual increase
population will double in 266 years

Greenland

Iceland

Europe

USSR

Middle East   Asia

Africa

5.5% of the world's population
0.9% rate of annual increase
population will double in 80 years

58.6% of the world's population
1.9% rate of annual increase
population will double in 37 years

Latin America

Papua New Guinea

Australia   New Zealand

8.4% of the world's population
2.1% rate of annual increase
population will double in 33 years
if current rate continues

12.4% of the world's population
2.9% rate of annual increase
population will double in 24 years

0.5% of the world's population
1.2% rate of annual increase
population will double in 57 years

FIGURE 7.3

Population Growth in the World (1990). The population of the world is not evenly distributed. Currently over 79 percent of the world's population is in the regions indicated in the map. These areas also have the highest rates of increase and are generally considered less developed. Because of the high birthrates, they are likely to remain less developed and will constitute nearly 82 percent of the world's population by the year 2000.

regions not only have the highest population growth rates but also the lowest per capita **gross national product (GNP).** The GNP is an index that measures the total goods and services generated within a country. (See figure 7.4.) Obviously, a wide gap exists between less developed countries with a low GNP and developed countries with a high GNP. Yet, less developed countries aspire to the same **standard of living** exhibited by developed countries of the world.

The Demographic Transition

The relationship between the standard of living and the population growth rate seems to be that those countries with the highest standard of living have the lowest population growth rate, and those with the lowest standard of living have the highest population growth rate. This has led many people to suggest that countries naturally go through a series of stages called a **demographic transition.** This model is based on the historical, social, and economic development of Europe and North America. It suggests that, in a demographic transition, the following four stages occur (see figure 7.5):

1. Initially, countries have a stable population with a high birthrate and a high deathrate. Deathrates are often quite variable because of famine and epidemic disease.
2. Improved economic and social conditions (control of disease and increased food availability) bring about a period of rapid population growth as deathrates fall. Birthrates remain high.
3. As countries become industrialized, the birthrates begin to drop because people make use of contraceptives.
4. Eventually, birthrates and deathrates again become balanced.

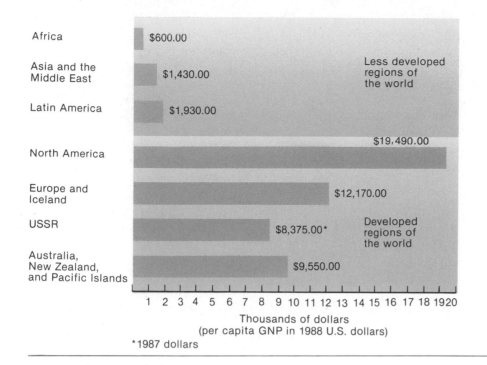

FIGURE 7.4

Per Capita Gross National Product of Nations. The world can be divided into developed and less-developed regions. One measure of the degree to which a country is developed is its per capita gross national product. A less-developed country produces very little for each person in the country, while a developed country has high productivity, and, therefore, also has goods and services available for consumption by its people.

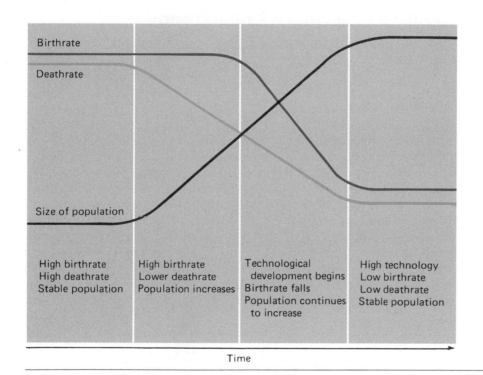

FIGURE 7.5

Demographic Transition. The demographic transition model suggests that, as a country develops technologically, it automatically experiences a drop in the birthrate. This certainly has been the experience of the developed countries of the world. However, the developed countries make up only about 20 percent of the world's population. It is doubtful whether the less-developed countries can achieve the kind of technological advances experienced in the developed world.

This is a very comfortable model because it suggests that, if a country can become industrialized, then social, political, and economic processes will naturally cause its population to stabilize.

However, this leads to some serious questions. Can the historical pattern exhibited by Europe and North America be repeated in the less developed countries? Europe, North America, Japan, and Australia passed through this transition period when world population was lower and when energy and natural resources were still abundant. It is doubtful whether these supplies are adequate to allow for the industrialization of the major portion of the world currently classified as less developed.

A second concern is the time element. With the world population increasing as rapidly as it is, industrialization probably cannot occur fast enough to have a significant impact on population growth. In many less developed countries, the number of children produced is a form of social security: Children take care of their elderly parents. Only people in developed countries can save money for their old age. Thus, they can make decisions about whether to have children, who are expensive to raise, or whether to invest money in some other way to provide for their old age. In other words, in less developed countries, children are considered an economic asset, while in developed countries, they are an economic liability.

When Europe and North America passed through the demographic transition, they had access to large expanses of unexploited lands, either in their native country or as colonies. This provided a safety valve for expanding populations during the early stages of the transition. Without this safety valve, it would have been impossible to deal adequately with the population while simultaneously encouraging economic development. Today, less developed countries may be unable to accumulate the necessary capital to develop economically, since an ever-increasing population is a severe economic drain.

Standard of Living: A Comparison

Standard of living is a difficult concept to quantify since different cultures have different attitudes and feelings about what is good and desirable. Here, we compare averages of several aspects of the cultures in three countries: (1) the United States, which is an example of a highly developed industrialized country; (2) Argentina, which is a moderately developed country; and (3) Zimbabwe, which is less developed. Figure 7.6 shows several statistics that relate to the standard of living in these three countries. The United States produces approximately 7.5 times more goods and services per person than does Argentina and about thirty times more goods and services than Zimbabwe. U.S. citizens consume over five times more energy than the average Argentinean and fourteen times more than the average Zimbabwean. Infant death in Zimbabwe is high, in Argentina it is intermediate, and the U.S. infant deathrate is low. In addition, people live longer in the United States and go to school longer than in the other two countries. Zimbabwe is more densely populated than Argentina. The average Zimbabwean is less well fed, a factor related to high mortality and low life expectancy, whereas the average citizen of the United States or Argentina has more food than needed.

Obviously, tremendous differences exist in the standard of living among these three countries. What the average U.S. citizen would consider a poverty level of existence would be considered a luxurious life for the average person in a poorly developed country. Standard of living seems to be closely tied to energy consumption. In general, the higher the per capita energy consumption, the higher the standard of living of the people. This is related to the degree of industrialization or development of the country. However, with industrialization comes pollution, as huge amounts of fossil fuels are consumed to provide the goods and services that contribute to the standard of living.

Ecological Principles and Their Application

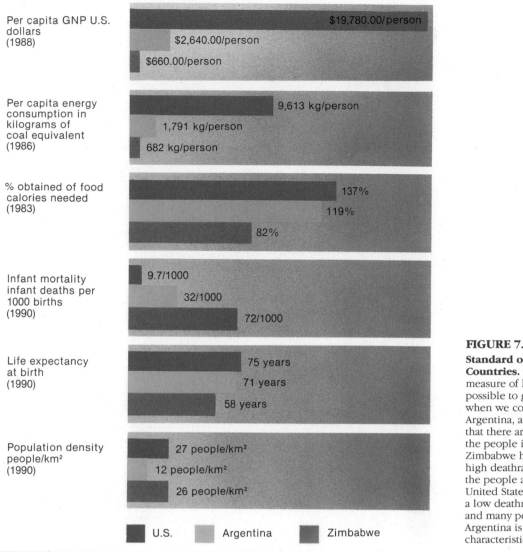

FIGURE 7.6

Standard of Living in Three Countries. Standard of living is a measure of how well one lives. It is not possible to get a precise definition, but when we compare the United States, Argentina, and Zimbabwe, it is obvious that there are great differences in how the people in these countries live. Zimbabwe has a low life expectancy, a high deathrate, and low productivity, and the people are often starving. The United States has a high life expectancy, a low deathrate, and high productivity, and many people eat too much. Argentina is intermediate in all of these characteristics.

Environmental Costs

The human population can increase only at the expense of the populations of other animals and plants. Each ecosystem has a finite carrying capacity and, therefore, has a maximum biomass that can exist within that ecosystem. There can be shifts within ecosystems to allow an increase in the population of one species, but this always adversely affects certain other populations because they are competing for the same basic resources. When the population of farmers increased in the prairie, the population of buffalo declined. (See box 11.3.) One basic need of all people is food. When humans need food, they turn to agricultural practices and convert natural ecosystems to artificially maintained agricultural ecosystems. Mismanaged agricultural resources are often irreversibly destroyed. The Dust Bowl of North America, desertification in Africa, and destruction of tropical rain forests are well-known examples. (See chapter 13 and box 5.1.) Humans may

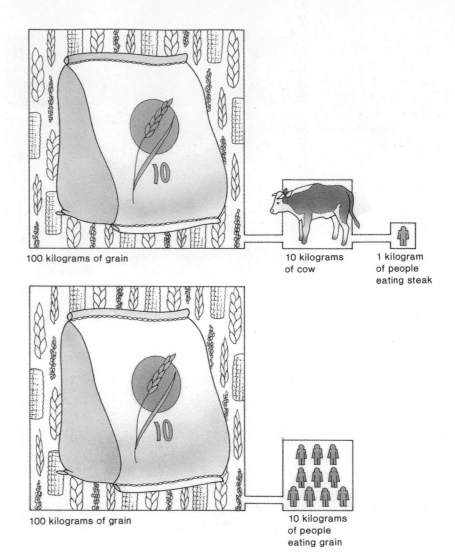

100 kilograms of grain   10 kilograms of cow   1 kilogram of people eating steak

100 kilograms of grain   10 kilograms of people eating grain

FIGURE 7.7

Populations and Trophic Levels. The larger a population, the more energy it takes to sustain the population. Every time one organism is eaten by another organism, approximately 90 percent of the energy is lost. Therefore, when populations are very large, they usually feed at the herbivore trophic level because they cannot afford the 90 percent energy loss that occurs when plants are fed to animals. The same amount of grain can support ten times more people at the herbivore level than at the carnivore level.

eventually degrade the biological productivity of a region rather than increase it. However, to a starving population, the short-term gain is all that is important.

In countries where the population is increasing, the pressure on those countries' agricultural resources is also increasing. A consequence of this basic need for food is that people in less developed countries generally feed at lower trophic levels than do those in the developed world. (See figure 7.7.) Converting low-quality plant material into high-quality animal protein is an expensive process. During this conversion, 90 percent of the energy present in the plants is lost. Therefore, in terms of economics and energy, people in less developed countries must consume the plants themselves rather than feed the plants to animals and then consume the animals. In most cases, if the plants were fed to animals, many people would starve to death. On the other hand, a lack of protein in the diet leads to malnutrition.

In countries where food is in short supply, agricultural land is already being exploited to its limit, and there is still a need for more food. This leaves the United States, Canada, Australia, Argentina, New Zealand, and the European Economic Community as net food exporters. Many countries, like India and China, are able to grow enough food for their people but do not have any left for export. Many others, including the Soviet Union, are not able to grow enough to meet their own needs and, therefore, must import food.

A country that is a net food importer is not necessarily destitute. The Soviet Union, Japan, and some European countries are net food importers but have enough economic assets to purchase what they need. Hunger occurs when countries do not produce enough food to feed their people and cannot obtain food through purchase or humanitarian aid.

The current situation with respect to world food production and hunger is very complicated. It involves the resources needed to produce food, such as arable land, labor and machines, appropriate crop selection, and economic incentives. It also involves the maldistribution of food within countries. This is often an economic problem, since the poorest in most countries have difficulty finding the basic necessities of life, while the richer have an excess of food and other resources. In addition, political activities are often very important in determining food availability. War, payment of foreign debt, and poor management are often contributing factors to hunger and malnutrition.

Improved plant varieties, irrigation, and improved agricultural methods have dramatically increased food production in some parts of the world. In recent years, India, China and much of southern Asia have moved from being food importers to being self-sufficient, and in some cases, food exporters.

The areas of greatest need are in sub-Saharan Africa. Africa is the only major region of the world where per capita grain production has decreased over the past few decades. These regions are trying to use marginal lands for food production, as forests, scrubland, and grasslands are converted to agriculture. Often, this land is not able to support continued agricultural production, which leads to erosion and desertification.

Anticipated Changes with Continued Population Growth

If the human population continues to increase (and it appears that it will), the pressure for the necessities of life will become greater. Differences in standard of living between developed and less developed countries will remain great because most population increases will occur in less developed countries. The supply of fuel and other resources is dwindling. The pressure for these resources will intensify as the industrialized countries seek to maintain their current standard of living. People in less developed countries will continue to seek more land to raise the crops needed to feed themselves unless major increases in food production per hectare occur. Since most of these people live in tropical areas, tropical forests will be cleared for farmland. Use of tropical forest as farmland often causes erosion or alteration of the soil, which can no longer support either forest or crops.

BOX 7.1
Governmental Policy and Population Control

The actions of government can have a significant impact on the population growth pattern within a nation. Some policies are aimed at either stimulating a population increase or attempting to control population. The Canadian government pays a bonus to the parents upon the birth of a child. Other countries offer incentives to couples to encourage them to have no children or to limit their family to one child. Many countries make contraceptives available at no cost and have programs to educate the people about effective conception control.

Some governments are more subtle in their policies. Most countries of industrialized Europe have a low birthrate but need a large labor force. They rely on importing labor from less industrialized nations, and therefore, do not need to encourage births in their own countries. They can simply import labor as needed and deport workers when needs decrease. This also reduces the cost of production, since they do not need to pay many fringe benefits to alien workers.

A government's immigration policies can have a significant effect on the country's population growth. Some countries, such as Australia, encourage people (particularly Europeans) to immigrate. Their passage is paid if they remain in the country. However, most countries have strict limits on the number of people who can enter the country each year.

Some governments have a profound effect on population growth within their country by taking no action at all. For example, many governments in South and Central America take no action to control their populations because of the anti-birth-control position of the Catholic Church. This assures a continued high birthrate in these countries.

For over twenty years, Nicolae Ceausescu's government actively sought to increase the birthrate of the Romanian population. Ceausescu's announced goal was a birthrate of twenty per thousand of the population. (The current average for Europe is thirteen per thousand.) To accomplish this goal, abortions were prohibited, access to contraceptives was denied without a doctor's prescription, importation of contraceptives was banned, women twenty to thirty years of age were required to undergo regular pregnancy tests, and many other pressures were applied to force women to become pregnant. This policy produced an increase in the fertility rate but also resulted in increased maternal and infant mortality and child abandonment. With Ceausescu's fall from power in 1989 and his subsequent execution, access to contraceptives is no longer prohibited, and Romania's citizens are no longer "required" to have children.

China has long been the most populated nation; it now contains about one-fourth of the world's population. When the People's Republic of China was established in 1949, the population was 540 million. Because of a high birthrate, the population increased to 614 million by 1955. This rapid increase resulted in some changing attitudes. Abortions became legal in 1953, and the first family-planning campaign began in 1955, followed by a second family-planning program in 1962. (See the box figure.)

The forerunner of the present family-planning policy began in 1971 with the launching of the *wan xi shao* campaign. Translated, this means "later" (marriages), "longer" (intervals between births), "fewer" (children). As part of this program, the legal ages for marriages were raised. For women and men in rural areas, the ages were raised to twenty-three and twenty-five, respectively; for women and men in urban areas, the ages were raised to twenty-five and twenty-eight, respectively. These steps reduced the birthrate. (See the box figure.)

Now the People's Republic of China has the one-child campaign. (See the box figure.) Initial surveys indicated that only 20 percent of married couples favored this program. To secure more participation, a series of penalties and rewards has been established for those who pledge to have no more than one child. These couples who comply are awarded the "only-child glory certificate." Recipients of these certificates receive free medical care, cash bonuses for their work, special housing treatment, and extra old-age pensions. The child receives free school tuition and preferential treatment when entering the job market.

Those who break their pledge and have additional children forfeit all benefits after the birth of their second child. If they have a third child, their wages are reduced by 10 percent, they must live in a housing area designed for a smaller family, and they are required to pay for the food for the third child.

Since the introduction of rewards and punishments, the one-child program has been favorably received. Several large urban areas report a 95 percent adherence to the policy; however, in rural areas, only 25 percent favor the program. How effective this policy will be remains to be seen, but as the population of China ages, it may be necessary to encourage births to supply the needed work force.

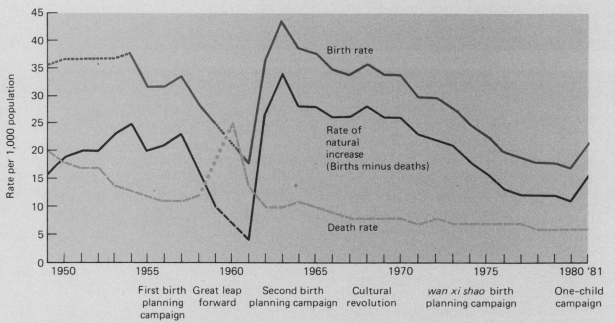

a.

b.

(*a*) Birth, death, and natural increase rates: China, 1919–1982. (*b*) One-child program.

(*a*) From *Population Bulletin,* "China Demographic Billionaire," by H. Yuan Tien, 1983, Population Reference Bureau, Inc.

This could cause profound changes in the world ecosystem as natural ecosystems are converted to agricultural ecosystems. Developed countries may have to choose between helping the less developed countries and maintaining their friendship, or isolating themselves from the problems of the less developed nations. Neither of these governmental policies will be able to prevent a change in life-style as world population increases. The resources of the world are finite. Even if the industrialized countries continue to get a disproportionate share of the world's resources, the amount of resource per person will decline as population rises. It seems that, as world population increases, the less developed areas will maintain their low standard of living. It is difficult to see how their standard of living could get much lower since some of the people are already starving to death. Developed nations will probably see their life-style become less consumption-oriented. What some view as current necessities (meals in restaurants, vacations to remote sites, and two cars per family) will probably become luxuries. Many people who enjoy the freedom of mobility associated with the automobile will have their travel limited as public transportation replaces private transportation. Recreation may not involve expensive, energy-demanding machines (powerboats, motorcycles, and electric-powered toys) but will emphasize such activities as hiking, bicycling, and reading. These changes will not come quickly, unless some catastrophic political or economic force causes major worldwide adjustments. Most likely, many changes will occur only as economic pressures affect families. As a result, we may be healthier, under less pressure, and certainly more in balance with the rest of the world in the future.

Demography is the study of human populations and the things that affect them. Demographers study the sex ratio and age distribution within a population to predict population growth.

Many social changes, such as later marriages and the role of women, affect population growth. The current U.S. birthrate will eventually result in zero population growth, a stable stage in which births equal deaths. If current trends continue, Africa, Asia, and Latin America will be the three areas of most rapid future population growth. These are also the areas with the lowest GNP and lower standards of living than in the industrialized nations.

The demographic transition model suggests that, as a country becomes industrialized, its population begins to stabilize, but there is little hope that the earth can support the entire world in the style of industrialized nations. It is doubtful whether there are enough energy and natural resources to develop these areas or enough time to change the trends of population growth. Highly developed countries should anticipate increased pressure in the future to share their wealth with less developed countries.

Summary

Review Questions

1. What is demography?
2. What is demographic transition? What is it based upon?
3. What is a "baby boom"?
4. What does age distribution of a population mean?
5. List ten differences between your standard of living and that of someone in a less developed country.
6. Why do people who live in overpopulated countries use plants as their main source of food?
7. Although predicting the future is difficult, what do you think your life will be like in the year 2000? Why?
8. List five changes you might anticipate if world population were to double in the next fifty years.
9. Which three areas of the world have the highest population growth rate? Which three areas of the world have the lowest standard of living?
10. How many children per woman would lead to a stable U.S. population?

PART THREE
Energy

All living systems can be described by the flow of energy through them: Energy enables simple forms of matter to be changed into more complex forms; energy is needed to maintain this complexity; and energy expenditure is also necessary to sustain the complex technical and social units typical of human populations.

All living things (including humans) rely on the sun as a source of energy. Coal, oil, and natural gas are energy sources available today because organisms in the past captured sunlight energy and stored it in the complex organic molecules that made up their bodies, which were then compressed and concentrated.

Technological development and fossil-fuel exploitation are directly related to one another and have allowed us an increasingly higher standard of living. Chapter 8 traces the development of energy consumption, its interrelationship with economic development and life-styles, and current uses and demands. Chapter 9 discusses the sources of energy currently being used, as well as the significance of each source and its impact. Chapter 10 discusses nuclear energy and the concerns that its use generates.

CHAPTER EIGHT
Energy and Civilization: Patterns of Consumption

Objectives

After reading this chapter, you should be able to:

Explain why all organisms require a constant input of energy.

Describe how per capita energy consumption increased as civilization developed from hunting and gathering through primitive agriculture to advanced ancient cultures.

Describe the development of advanced modern civilizations as new fuels were used to run machines.

Recognize that coal deposits are not uniformly distributed throughout the world.

Correlate the Industrial Revolution with various social and economic changes.

Explain why cheap oil and natural gas led to the development of a consumption-oriented society.

Explain how the automobile changed people's life-styles.

List the four uses of energy.

Explain why overall energy use in the United States declined during the 1970s and 1980s.

Chapter Outline

Key Term

Industrial Revolution

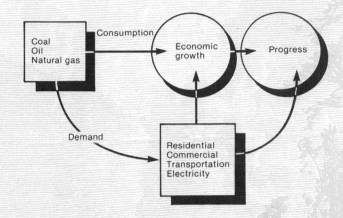

FIGURE 8.1
Hunter-Gatherer Society. In this type of society, people obtain nearly all of their energy from the collection of wild plants and the hunting of animals. These societies do not make large demands on fossil fuels.

History of Energy Consumption

Every form of life and all societies require a constant input of energy. If the flow of energy through organisms or societies ceases, they stop functioning and begin to disintegrate. Some organisms and societies are more energy efficient than others. In general, history shows that more complex industrial societies have the greatest energy needs. If societies are to survive, they must continue to expend energy. However, the pattern of energy consumption may need to change as traditional sources become limited.

Biological Energy Requirements

An energy input is essential to maintain life. In any ecosystem, the sun provides the energy to sustain all forms of life. (See chapter 4.) The first transfer of energy occurs during the process of photosynthesis, when plants convert light energy into chemical energy in the production of food. Herbivorous animals utilize the food energy in the plants. The herbivores, in turn, provide a source of energy for the carnivores. Because nearly all of their energy requirements were supplied by food, primitive humans were no different from other animals in an ecosystem. In such hunter-gatherer cultures, only biological energy demands are made. (See figure 8.1.)

Early in human history, people began to use additional sources of energy to make their lives more comfortable. The development of domesticated plants and animals provided a more dependable supply of food. People no longer needed to depend solely upon the gathering of wild plants and the hunting of wild animals for sustenance. In addition, domesticated animals furnished a source of energy for transportation, farming, and other tasks. (See figure 8.2.)

Early civilizations, such as the Aztecs, Greeks, Egyptians, Romans, and Chinese, were based on human muscle, animal muscle, and fire as sources of energy. Although these civilizations were culturally advanced, they were still directly dependent upon plants for their energy.

FIGURE 8.2
Animal Power. This bas-relief panel from an Egyptian tomb depicts an important accomplishment in the development of human civilization. With the use of domesticated animals, people had a source of power other than their own muscles.

Wood

Except for limited use of some wind-powered and water-powered machines, wood furnished most of the energy and materials needed for developing civilizations. Home construction and shipbuilding utilized wood as lumber. The controlled use of fire was the first use of energy in a form other than food. Wood provided the fuel to meet this demand. The energy provided by wood enabled humans to cook their food, heat their dwellings, and develop a primitive form of metallurgy. Such advances separated humans from other animals. The heavy use of wood eventually resulted in a shortage, and people were forced to seek alternative forms of fuel.

Because of a long history of high population density, India and some other parts of the world experienced a wood shortage hundreds of years before Europe and North America did. In many of these areas, animal dung replaced wood as a fuel source and is still used today in some parts of the world.

Western Europe and North America were able to use wood as a fuel for a longer period of time. The forests of Europe supplied sufficient fuel until the thirteenth century. In North America, the vast expanses of virgin forests supplied adequate fuel until the late nineteenth century. Fortunately, when local supplies of wood declined in Europe and North America, coal, formed from fossilized plant remains, was available as an alternative energy source. (See figure 8.3.)

Fossil Fuels

Fossil fuels are the remains of plants, animals, and microorganisms that lived millions of years ago. During the Carboniferous period, 275 to 350 million years ago, conditions in the world were conducive to the formation of large deposits of fossil fuels. (See figure 8.4.) Ever since machines replaced muscle power, the major energy sources for the world have been fossil remains from the distant past.

Historically, the first fossil fuel to be used extensively was coal. Those regions of the world that had readily available coal deposits were able to

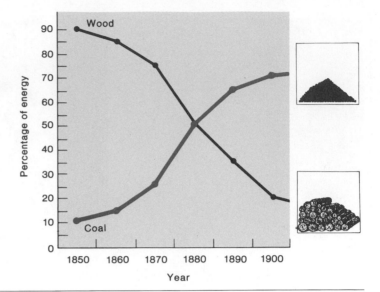

FIGURE 8.3
Coal Replaces Wood. In 1850, wood furnished 90 percent of U.S. energy, and coal most of the remaining 10 percent. Fifty years later, wood supplied only 20 percent of the energy and coal supplied 70 percent. The remainder was furnished by oil and natural gas.

Source: Data from the U.S. Bureau of Mines.

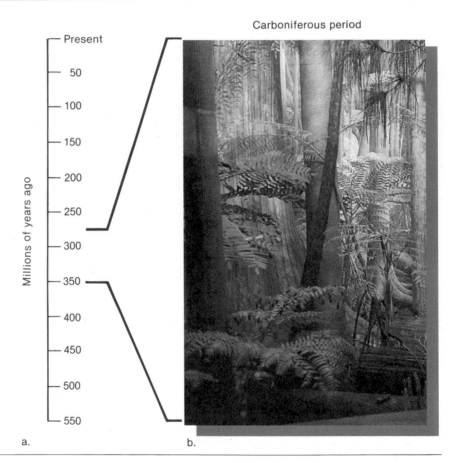

FIGURE 8.4

Carboniferous Period. Approximately three hundred million years ago, this kind of ecosystem was common throughout the world. Plant material accumulated in these swamps and was ultimately converted to coal.

switch to this new fuel and participated in a period of history known as the **Industrial Revolution** in the early eighteenth century. During the Industrial Revolution, the burning of coal provided the energy to run machines that replaced human and animal labor in the manufacturing and transporting of goods. Nations without a source of coal, or those possessing coal reserves that were not easily exploited, did not participate in the Industrial Revolution.

Figure 8.5 shows the current distribution of the world's coal reserves. The location of these coal deposits is extremely important in today's world economy. Currently, China has plans to move toward a more industrially

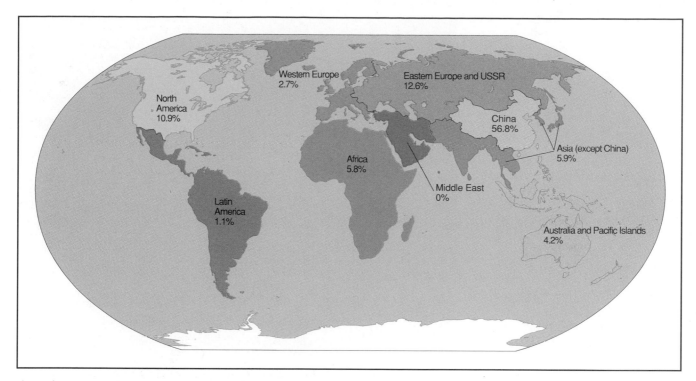

based economy by exploiting its massive coal reserves. (See the case study at the end of the chapter.) Many people are concerned that this will contribute to further greenhouse warming by increasing the amount of carbon dioxide in the atmosphere. See chapter 17 for a discussion of the greenhouse effect and the consequences of global warming.

Prior to the Industrial Revolution, Europe and North America were predominately rural. Goods were manufactured on a small scale in the home. As machines and the coal to power them became increasingly available, the factory system of manufacturing products replaced the small home-based operation. Because growing factories required a constantly increasing labor supply, people left the farms and congregated in the areas surrounding the factories. Villages became towns, and towns became cities. Widespread use of coal in cities resulted in increased air pollution. In spite of these changes, the Industrial Revolution was viewed as progress. Energy consumption increased, economic growth continued, and people prospered. Within a span of two hundred years, the daily per capita energy consumption of industrialized nations increased eightfold. This energy was furnished primarily by coal, but a new source of energy was about to be discovered: oil.

Even though the Chinese used some gas and oil as early as 1000 B.C., these resources remained virtually untapped until fairly recently. The oil well that Edwin L. Drake, an early oil prospector, drilled in Pennsylvania in 1859 was not the world's first oil well, but it was the beginning of the modern petroleum era. By 1870, oil production in the United States had reached over four million barrels a year and supplied 1 percent of the nation's energy requirements. This grew to nearly 50 percent by 1970 and currently contributes just over 40 percent. (See figure 8.6.)

For the first sixty years of production, the principal use of oil was to make kerosene, a fuel for lamps, while the gasoline produced was discarded as a waste product. During this time, the supply of oil exceeded the demand. However, the development of the automobile caused a dramatic increase in demand. In 1900, the United States had only eight thousand automobiles. By 1920, this had increased to eight million cars, and by 1987 to 139 million. Oil products were in great demand as automobile fuel and lubricants.

FIGURE 8.5

Recoverable Coal Reserves of the World. The percentages indicate the coal reserves in different parts of the world. This coal can be recovered under present local economic conditions using available technology.

Source: Data from World Energy Conference.

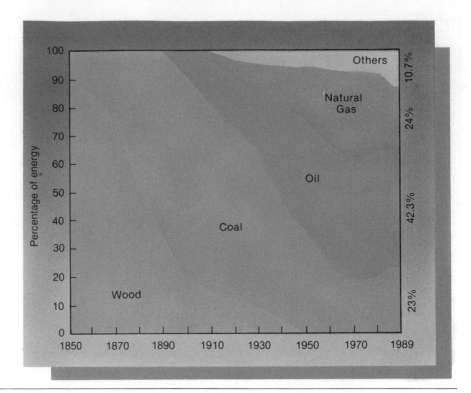

FIGURE 8.6

Oil Replaces Coal. Just as wood was replaced by coal, coal was later replaced by oil. This graph represents the production of energy by various sources in the United States. We can see that oil has remained a dominant energy source for the last thirty years, coal has decreased slightly until recently, natural gas increased and then declined, hydroelectric power has remained about the same, and nuclear power has been increasing for the last fifteen years.

Source: Annual Energy Review, 1986, Energy Information Administration.

The use of natural gas did not increase as rapidly as the use of oil. Its primary use was for home heating. In the early 1900s, 90 percent of the natural gas was "flared," burned as a waste product at the wells.

Major oil and natural gas fields were located in the warm southwestern regions of the United States, but the demand for natural gas was mainly in the cold midwestern and northern regions. Since demand was largely seasonal (occurring during only the colder periods of the year) and at such a great distance from the supply, transportation and storage were problems. In 1920, natural gas supplied only 5 percent of U.S. energy requirements.

A series of events involving the federal government and the need for more oil was ultimately responsible for increased use of natural gas in the United States. World War II brought new energy requirements for manufacturing and transportation. The war machinery needed energy to operate. In 1943, a federally financed pipeline was constructed to transport oil within the United States. This 2,000-kilometer pipeline allowed oil to be transported more efficiently from the wells in Texas, Louisiana, and Oklahoma to the refineries and factories in the eastern section of the country. In 1944, a longer (2,400 kilometers), federally financed pipeline was built to increase the flow of oil to the country's eastern and midwestern regions.

After the war, the federal government sold these pipelines to private corporations. The corporations converted the pipelines to transport natural gas. Thus, a direct link was established between the natural gas fields in the Southwest and the markets in the Midwest and East. By 1971, there were 400,000 kilometers of long-range transmission pipelines and 986,000 kilometers of distribution pipelines in the United States. Approximately 1,613,000 kilometers of natural gas distribution pipelines are used. These improvements in transporting the natural gas were coupled with the development of large, underground storage sites. As a result, natural gas now supplies 24 percent of U.S. energy.

Energy and Economics

Another factor that provided the United States with an abundant source of cheap energy was government regulation and intervention. In addition to financing the early pipelines that enabled the transport of natural gas and oil, the government regulated the price of these products at artificially low levels. The government also reinforced a policy of maintaining low prices for oil and natural gas by continuing to grant tax advantages to oil and natural gas companies, which helped companies to be profitable without raising prices. Overall, however, the artificially low prices for oil and natural gas encouraged a high consumption of fuel.

Economic Growth and Energy Consumption

The policy of encouraging fossil-fuel consumption led to the development of a technology-oriented society. The replacement of human and animal energy with fossil fuels began with the Industrial Revolution and was greatly accelerated by the supply of cheap, easy-to-handle, and highly efficient fuels. The result was unprecedented economic growth in North America and the rest of the industrialized world.

World War II was a prime factor in ending the depression of the 1930s. The demands of the military created millions of defense jobs. Almost everyone was employed, but there was a scarcity of consumer goods. After World War II, consumer goods that had been unavailable during the war years were in great demand. Industry made the transition from the production of military goods to the production of consumer goods. A high employment level, a rapidly expanding population, and a supply of inexpensive energy encouraged a period of rapid economic growth. Today, the per capita energy consumption in the United States is about four times greater than in 1945, the year World War II ended.

The Role of the Automobile

The cheap, abundant energy that fueled industries produced an ever-increasing amount and array of consumer goods. One such product was the automobile. At the beginning of this century, people began to drive cars. The growth of the automobile industry led to the construction of improved roadways, which required energy for their construction. Therefore, the energy costs of driving a car were greater than just the fuel consumed. As roads improved, higher speeds were possible. The demand for faster cars grew, and automobile companies were quick to meet the demands. Bigger cars required even more fuel and better roads. So roads were continually being improved, and more cars were being produced. A cycle of *more chasing more* had begun.

The rise of the two-car family increased demand for cars, which was coupled with demand for more energy. It requires energy to mine the ore, process the ore into metals, form the metals into automobile components, and transport all the materials. As the economy grew, so did energy requirements.

More cars meant more jobs in the automobile industry, the steel industry, the glass industry, and hundreds of other related industries. The construction of thousands of kilometers of roads created additional jobs. The oil companies grew from being the suppliers of lamp oil to being the largest industry in North America. The automobile industry played a major role in the economic development of the industrialized world. All this

The price of a liter of gasoline is determined by two major factors: (1) the cost of purchasing and processing crude oil into gasoline and (2) various taxes. Most of the differences in gasoline prices between countries are a result of the differences in taxes and reflect differences in government policy toward motor vehicle transportation.

A major objective of governments is to collect money to build and repair roads, and governments often charge the user by taxing the fuel used by the car or truck. Governments can also discourage the use of automobiles by increasing the cost of fuel. An increase in fuel costs also creates a demand for increased fuel efficiency in all forms of motor transport.

Many European countries raise more money from fuel taxes than they spend on building and repairing roads, while the United States raises approximately 60 percent of the monies needed for roads from fuel

taxes. The relatively low cost of fuel in the United States encourages more travel and increases road repair costs. The cost of taxes to the U.S. consumer is about 20 percent of the cost of the fuel, while in Japan and many European countries, the percentage is 60 to 75 percent. (See box table.)

Cost of liter of gasoline in January 1991 (in U.S. dollars).	
Italy	$1.25
France	$1.03
Japan	$0.97
United Kingdom	$0.84
Germany	$0.81
United States	$0.22

FIGURE 8.7
Energy-Demanding Life-Style.
Building private homes on large individual lots some distance from shopping areas and places of employment is directly related to the heavy use of the automobile as a mode of transportation. Heating and cooling a large enclosed shopping mall, along with the gasoline consumed in driving to the shopping center, increase the demand for energy.

wealth gave people more money for cars and other necessities of life. The car, originally a luxury, was now considered a necessity.

The car not only created new jobs in the automobile and related industries, but altered people's life-styles as well. It helped people to travel greater distances during vacations. New resorts and chains of motels, restaurants, and other service industries developed to serve the motoring public. Thousands of new jobs were created. Because people could live farther from work, they began to move to the suburbs. (See figure 8.7.) The growth of large shopping centers in the suburban areas hastened the decline of some of the central business districts in the United States. Today, fewer than 50 percent of retail sales are made in central business districts of U.S. cities, which has resulted in a loss of jobs in these areas. In Philadelphia, 79 percent of all retail jobs were within the city in 1930; by 1970, this had declined to 43 percent.

While people were moving to the suburbs, they were also changing their buying habits. Labor-saving, energy-consuming devices became essential in the home. The vacuum cleaner, dishwasher, garbage disposal, and automatic garage door opener are only a few of the ways human power is replaced with electrical power. Eleven percent of the electrical energy in the United States is used to operate home appliances. In addition, other aspects of the life-style point out our energy dependence. The small, horse-powered farm of yesterday has grown into the huge, diesel-powered farm of today. Regardless of where we live, we expect Central American bananas, Florida oranges, California lettuce, Texas beef, Hawaiian pineapples, and Nova Scotian lobsters to be readily available at all times. What we often fail to consider is the amount of energy required to process, refrigerate, and transport these items. The car, the modern home, the farm, and the variety of items on our grocery shelves are only a few indications of how our life-style is based on cheap, abundant energy.

Table 8.1	
Energy consumption, 1989.	
Region	**Energy consumption per capita (tonnes of oil equivalent)**
Africa	0.3
Latin America	1.0
Japan	3.4
France	3.6
West Germany	4.3
United States	7.9
Canada	9.3

How Energy Is Used

If the world's population growth continues at the current rate and energy conservation does not improve, the worldwide demand for energy will more than double by the year 2020. Not all countries use energy in the same way. In industrialized nations, energy is used about equally for four purposes: (1) residential and commercial, (2) industrial, (3) transportation, and (4) electrical utility. In less developed nations, most of the energy used is for residential purposes. In developing countries, most of the energy is for industrial purposes.

Different nations vary in the amount of energy they use as well as in what uses they make of the energy. (See table 8.1.) To maintain their style of living, individuals in the United States use over twice as much energy as someone in France or Japan and over twenty-five times as much energy as a person in Africa.

Residential and Commercial Energy Demand

The amount of energy required for residential and commercial use varies greatly throughout the world. A country with a high GNP uses a lower percentage of its energy per capita for residential and commercial needs than a less developed country. For example, about 30 percent of the energy used in the United States is for residential and commercial energy, while in India, 90 percent of the energy is for residential uses. The types of uses that different nations make of residential and commercial energy also vary widely. Of the residential and commercial energy usage in the United States, 75 percent is for air conditioning, refrigeration, water heating, and space heating. In India, almost all of the energy used in the home is for cooking since the scarcity and high cost of fuel precludes uses for other purposes.

Therefore, when residential and commercial energy conservation is considered, the current pattern of energy use in the region of the world determines the type of conservation methods that could be effective. In Canada, which has a cold climate, 40 percent of the residential energy is used for heating. Proper conservation practices could reduce this by 50 percent. In Africa, almost half of the energy used in the home is for cooking. (See figure 8.8.) Using more efficient stoves instead of open fires could reduce these energy requirements by 50 percent. North Americans and people in other developed countries also could reduce energy consumption in many ways.

FIGURE 8.8

Open Fire Cooking. About half of the energy demand in Africa is for cooking. Using a stove instead of an open fire could reduce this energy need by nearly 50 percent.

Industrial Energy Demand

The amount of energy used for industrial processes varies considerably. Nearly 60 percent of all the commercial energy used in the Soviet Union is used by industry. In addition, many of the industrial processes used in the Soviet Union are not energy efficient. In the United States, about 30 percent of the energy is used by industry. Therefore, replacing obsolete machinery in the Soviet Union would lessen energy demand more in that country than similar measures in the United States.

The types of industrial processes in various countries are major factors in the energy demand of those countries. In the Soviet Union, which has large coal, oil, and natural gas reserves, steel is processed from ore because energy is readily available. Spain and Italy, on the other hand, lack large deposits of fossil fuel and produce steel from scrap steel because it requires less energy. Producing 1 metric ton of steel from iron ore in the Soviet Union uses twice as much energy as producing a metric ton of steel from scrap in Spain or Italy.

Large capital investment is necessary to upgrade processes and reduce industrial energy consumption. Many countries cannot afford to convert to a more efficient method. For example, India, a nation with few coal deposits, still uses the outdated open-hearth furnace to produce steel. This requires nearly double the worldwide energy average for producing a metric ton of steel. Yet, the high cost of converting to modern methods forces India to continue to use this energy-expensive method.

Transportation Energy Demand

As with residential, commercial, and industrial uses of energy, the amount of energy used for transportation varies widely throughout the world. In some of the less developed nations, transportation uses are very small. Per capita energy use for transportation is larger in developing countries, and highly developed countries have the highest consumption. (See table 8.2.)

Once a country's state of development has been taken into account, the mix of bus, rail, water, and private automobiles is the main factor in determining the country's energy use for transportation. In Europe, Latin America, and many other parts of the world, rail and bus transport are widely used because they are more efficient than private automobile travel. (See figure 8.9.) In these countries, automobiles require about four times more energy per passenger kilometer than bus or rail transport. In addition, most

TABLE 8.2
Per capita energy use for transportation (1986).

Country	Energy use in gigajoules/capita
India	1
Zimbabwe	2
Mexico	12
Argentina	17
USSR	25
Japan	25
Netherlands	38
Denmark	40
Australia	75
United States	100

FIGURE 8.9
Public Transportation. In regions of the world where energy is expensive, people make maximum use of cheaper public transportation.

of these countries have high taxes on fuel, which raise the cost to the consumer and encourage the use of public transport. In the United States, the situation is somewhat different. The government policy has been to keep the cost of energy low. (See box 8.1.) Consequently, the automobile plays a dominant role, and public transport is primarily used only in metropolitan areas. In the United States, rail and bus transport are about twice as energy efficient as private automobiles. Private automobiles in North America consume over 15 percent of the world's oil production, while the rest of the automobiles in the world consume 7 percent. Air travel is relatively expensive in terms of energy, although it is slightly more efficient than private automobiles. Passengers, however, are paying for the convenience of rapid travel over long distances.

Electrical Energy Demand

As with other forms of energy use, electrical consumption in different regions of the world varies widely. The amount of electricity used by all of the less developed nations, with over 75 percent of the world's population, is only 66 percent of that used by the United States alone. The per capita use of electricity in the United States is twenty-five times greater than average per capita use in the less developed countries. In Nepal, the per capita

In 1973, headlines such as "OPEC Increases Oil Prices" appeared in newspapers throughout the world. Most readers asked: "What is OPEC?" In fact, many people today still are not certain what OPEC is or how OPEC influences their lives.

OPEC, the Organization of the Petroleum Exporting Countries, had its beginning in September 1960 when the governments of five of the world's leading oil exporting countries agreed to form an oil cartel. Three of the original members—Saudi Arabia, Iraq, and Kuwait—were Arab countries, while Venezuela and Iran were non-Arab members. Today, thirteen countries belong to OPEC. These include seven Arab states: Saudi Arabia, Kuwait, Libya, Algeria, Iraq, Qatar, and United Arab Emirates; and six non-Arab members: Iran, Indonesia, Nigeria, Gabon, Ecuador, and Venezuela. (See the map.) OPEC nations control over 60 percent of the world's estimated oil reserves of 890 billion barrels, which makes OPEC an important world influence.

The foundation for OPEC had its origin in the 1940s. During this time, the western nations experienced a drastic increase in the demand for oil, first to meet the military demands of World War II and then the increased civilian demand following the war. Domestic oil production could not meet the increased demand, and new sources were needed.

To a large extent, British and American oil companies began to develop the vast reserves in the Middle East and South America. These companies entered into an agreement with local governments whereby the oil companies supplied the capital and developed the oil fields, and a "royalty" based on production was paid to the local government. In essence, the oil belonged to the companies. Originally, the countries did not have the money or the technology to produce the oil, and they welcomed outside investors.

However, from the outset, many of the producer countries began to pressure the oil companies for more direct control of the oil fields. As a result of continuous renegotiation of the agreements with the oil companies, the oil-producing countries gradually gained control of the oil fields. Eventually, these oil-producing industries were nationalized as governments assumed complete ownership and management. This nationalization was completed by the time of the Arab-Israeli War of 1973. In 1973, when OPEC was supplying 54.6 percent of the world's oil, seven Arab members of OPEC reduced their production. This resulted in a worldwide oil shortage, which caused all oil companies, in OPEC and non-OPEC countries, to increase their prices. In the United States, the price of oil went from $3.39 per barrel to $13.93 per barrel.

Throughout the world, local oil shortages developed. There were long lines of cars at the gas stations; people were urged to lower their thermostats to conserve oil; tax advantages were enacted to encourage various forms of energy conservation; research activity into other forms of energy, such as solar, wind, and biomass conversion, increased; and cars were downsized. With the large increase in oil prices also came an increase in oil exploration and production.

The conservation of energy and the increased domestic production resulted in oil prices falling to eight dollars a barrel by 1975. However a second round of OPEC price increases in 1979–1980 caused the

use of electricity is twenty-three kilowatt hours per year, which is enough to light a 100-watt light bulb for one week. By contrast, the per capita consumption of electricity in the United States is 270 times greater than in Nepal. There is also a wide variation in electrical consumption within developed countries. For example, the per capita use in Europe is about half of that in the United States.

The difference in the per capita electrical energy consumption for different countries is only part of the picture. In more developed nations, about a quarter of the electricity is used by industry. In less developed nations, over half of the electricity is used by industry: for example, 55 percent in Mexico and 70 percent in South Korea.

Current Energy Trends

From a historical point of view, it is possible to plot changes in energy consumption. The role of economics, political changes, public attitudes, and many other factors must be incorporated into an analysis of changing energy use trends.

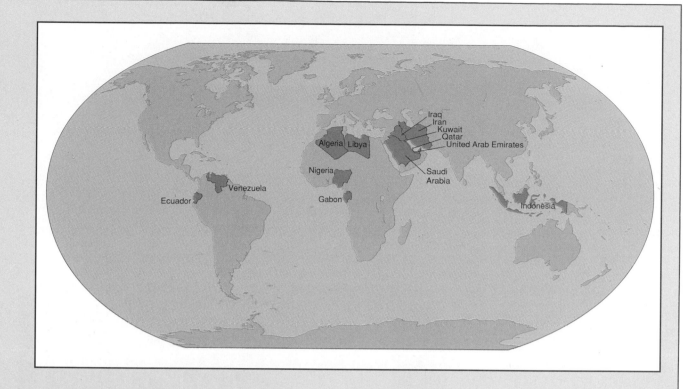

price to increase to twenty dollars a barrel. As with the first round of increases in 1973–1974, there was a renewed demand for conservation, alternate forms of energy, and increased oil production. The combination of conservation and increased domestic production again erased the oil shortages and actually lead to an oil surplus in the 1980s. This resulted in a decline in oil prices and convinced many people that there was no oil shortage.

OPEC's share of the world's oil market has declined to about 30 percent. This has important economic implications because many of the OPEC nations are less developed, and during the period of high oil prices, they borrowed large sums of money from the industrial nations. Now they are unable to repay this money, and many banks in the western nations could experience large financial losses.

During the 1980s, there were important differences among OPEC members concerning pricing of oil and production rules. This weakened OPEC. The 1990 invasion of Kuwait by Iraq deeply divided many of the OPEC countries.

Since 1950, U.S. energy consumption has steadily increased. However, there were two periods of decline, one beginning in 1973 and the other in 1979. Both of these episodes were the result of political turmoil in the Middle East and the increasing influence of OPEC. (See box 8.2.) (See figure 8.10.) The same trend occurred in Western Europe, Japan, and Australia. (See figure 8.11.) Several factors contributed to these declines in energy consumption. Increased prices for oil and all other forms of energy forced businesses and individuals to become more energy conscious and to expand efforts to conserve energy. From 1970 to 1983, the amount of energy used for heat per dwelling in the United States declined by 20 percent. Comparable reductions were made in Denmark, West Germany, and Sweden. In the United States, the government insisted that energy consumption data be made available to the consumer. The government forced the design of automobiles and appliances to become more energy efficient. Consumers could make purchases based on energy efficiency. In addition to financial considerations, many people altered their energy consumption for philosophical reasons. They found that they could do so without significantly altering their standard of living.

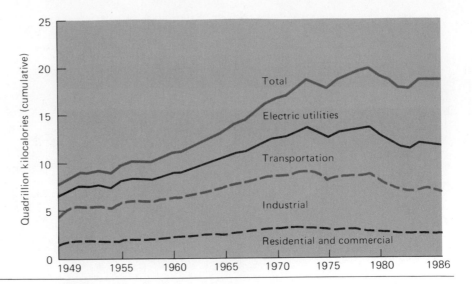

FIGURE 8.10

Changes in U.S. Energy Consumption. Energy consumption in the United States experienced a gradual annual increase until 1979. Since that time, there has been a general decline in energy use.

Annual Energy Review 1986.

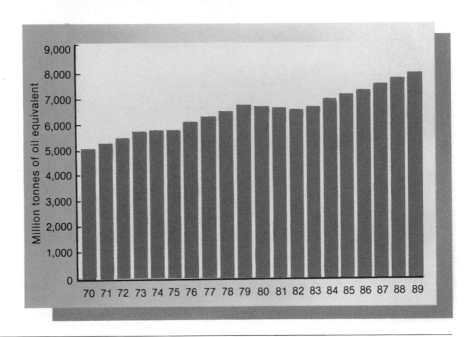

FIGURE 8.11

World Energy Consumption. Except for periods of decline in North America, Europe, Japan, and Australia in 1973 and 1979, energy consumption has increased since 1965.

Source: *British Petroleum Statistical Review of World Energy,* June 1990.

However, since 1980, energy consumption has increased in the United States and Europe. (See figure 8.11.) Like the decrease in energy use in 1973 and 1979, the increase since 1980 is due to the price of oil. In 1979, oil was selling for about forty dollars a barrel. Beginning in 1980, the price began to drop, and in 1990, oil was selling for less than twenty dollars a barrel. When Iraq invaded Kuwait in August 1990, prices rose dramatically. (See box 8.3.)

During the 1980s, with energy costs declining, people in the United States and Europe became less concerned about their energy consumption. More energy was used to heat or cool homes and buildings, the use of home appliances increased, and the sales of big cars increased. From 1970 to 1990, worldwide use of energy increased 45 percent. Most of this increase was in countries that increased their industrial production. Japan registered a 32-percent increase, Mexico a 144-percent increase, and South Korea a 226-percent increase, while developed countries showed modest increases.

CONSIDER THIS CASE STUDY
Energy Development in China

China has approximately 20 percent of the world's population yet uses less than 10 percent of the energy consumed in the world. About 80 percent of China's population live in rural areas, where most of the energy is produced from biomass, including fuelwood and animal wastes that generate methane gas. China has significant oil reserves and huge supplies of coal, but much of the energy technology is out-of-date. Many industries are fueled by the burning of coal, which contributes to very poor air quality, particularly in the winter months when coal is also used for space heating.

China has plans to quadruple its industrial production by exploiting its vast coal reserves. Without improved air-pollution technology, however, the quality of the air would suffer. China needs industrialized countries to provide the technology to develop industry and control pollution.

Should western industrialized nations supply the technology and thereby encourage China's industrialization?

The burning of coal contributes to the amount of carbon dioxide in the air and to greenhouse warming. Should China be discouraged from using its coal resources?

Do other countries have an obligation to supply advanced technology so that China will use its energy efficiently and reduce its contribution of carbon dioxide?

Energy is a prerequisite for all life and any type of civilization. A direct correlation exists between the amount of energy used and the complexity of the civilization.

Wood furnished most of the energy and construction materials for early civilizations. Heavy use of wood in densely populated areas eventually resulted in a wood shortage. Fossil fuels replaced wood as a prime source of energy. Fossil fuels were formed from the remains of plants, animals, and microorganisms that lived about 300 million years ago. Fossil-fuel consumption led to the development of a technology-oriented society.

Summary

Throughout the world, residential and commercial, industrial, transportation, and electrical utilities require energy. Because of financial, political, and other factors, different nations vary in the amount of energy they use as well as in what uses they make of the energy. Analysts expect the worldwide demand for energy to increase steadily.

Review Questions

1. Why was the sun able to provide all the energy requirements for human needs before the Industrial Revolution?
2. In addition to food, what other energy requirements does a civilization have?
3. Why were some countries unable to use the technology that began to develop with the Industrial Revolution?
4. What factors caused a shift from wood to coal as a source of energy?
5. How were the needs for energy in World War II responsible for the subsequent increased consumption of natural gas?
6. What part does government regulation play in changing the consumption of natural gas and oil?
7. Why was much of the natural gas that was first produced wasted?
8. What was the initial use of oil? What single factor was responsible for a rapid increase in oil consumption?
9. List the four purposes for which a civilization uses energy.
10. Why is OPEC important in the world's economy?

CHAPTER NINE
Energy Sources

Objectives

After reading this chapter, you should be able to:

Differentiate between resources and reserves.

Identify peat, lignite, bituminous coal, and anthracite coal as steps in the process of coal formation.

Recognize that natural gas and oil are formed from ancient marine deposits.

Explain the different methods of coal mining.

Explain how some methods of coal mining can have negative environmental impacts.

Explain why surface mining of coal is used in some areas and underground mining in other areas.

Explain why more expensive processes and exploration are necessary for locating oil.

Explain why secondary recovery methods are used to obtain oil and natural gas.

Explain why transport of natural gas is still a problem in some areas.

Explain why the amount of energy supplied by water power has increased.

Explain the limitations of water power.

Describe how geothermal and tidal energy produce electricity.

Recognize that geothermal and tidal energy can only be used in areas with the proper geologic features.

Describe how wind-generated electricity is produced.

Explain the use of solar energy in a passive heating system, in an active heating system, and in the generation of electricity.

Recognize the limitations of solar energy and wind power.

Explain why use of fuelwood has increased, and describe fuelwood's limitations.

Describe the potential and limitations of biomass conversion as a source of energy.

Recognize that wastes represent a source of energy.

Recognize that energy conservation can significantly reduce our need for additional energy sources.

Chapter Outline

Key Terms

acid mine drainage
active solar system
biomass
black lung disease
geothermal energy
liquefied natural gas
nonrenewable energy
  sources
overburden
passive solar system

peat
photovoltaic cell
renewable energy
  sources
reserves
resources
secondary recovery
surface mining
underground mining

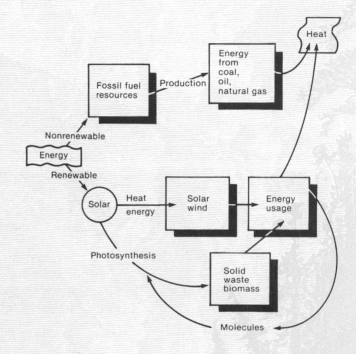

Energy Sources

Chapter 8 outlined the history of energy consumption and how civilization's advances were closely linked to the availability of energy. As new manufacturing processes developed, dependable sources of energy were required. The development of a technologically based civilization accelerated in the twentieth century. For example, in 1990, the world's population was over three times larger than in 1900, but the quantity of products manufactured increased by about thirtyfold in the same amount of time. To produce these goods, world energy consumption in 1990 was twelve times greater than in 1900. (See table 9.1.)

The most commonly used energy sources for the industrialized nations are fossil fuels: coal, oil, and natural gas, which supply 80 percent of the world's energy. Fossil fuels were formed hundreds of millions of years ago, and they represent the accumulation of millions of years worth of energy-rich organic molecules. The rate of formation of fossil fuels is so slow as to be ineffective over the course of human history. Since we are using these resources much faster than they can be produced and the amount of these materials is finite, they are known as **nonrenewable energy sources.**

Eventually, human demands will exhaust the supplies of coal, oil, and natural gas. At the present rate of consumption, there are sufficient coal reserves to last over one thousand years. The world's oil reserves are estimated to be 890 billion barrels. With current world usage of about 20 billion barrels per year, this is a forty-five-year supply of oil. The 113 trillion cubic meters of natural gas reserve will probably be exhausted in fifty-five years. Therefore, alternative forms of energy, in addition to coal, oil, and natural gas, should be utilized to allow the supply of fossil fuels to last longer.

Many alternative energy sources are **renewable energy sources,** which are resources that are replaced in a reasonable length of time or that are usually readily available. In plants, photosynthesis converts light energy into chemical energy. Some of this energy is stored in the organic molecules of wood. Any form of biomass, either plant or animal, can be traced to the energy of the sun. Since wood and the various forms of biomass are constantly being produced, they are forms of renewable energy. Solid wastes, such as paper products or plastics, are also renewable forms of energy since they are constantly being produced. Solar, geothermal, and tidal energy are renewable energy sources because they are continuously available. Anyone who has ever laid in the sun, seen a geyser or hot springs, or been swimming in the surf has experienced these forms of energy. However, many technical problems must be solved before these renewable energy forms can contribute significantly to meeting humans' energy demands.

Resources and Reserves

When discussing deposits of nonrenewable resources, such as fossil fuels, we must differentiate between deposits that can be exploited and those that cannot. From a technical point of view, a **resource** is a naturally occurring substance of use to humans that has the potential for being feasibly extracted under prevailing conditions. **Reserves** are the known deposits from which materials can be extracted profitably with existing technology under present economic conditions. Therefore, reserves are smaller quantities than resources. (See figure 9.1.) Both terms are used when discussing the amount of a mineral or the fossil-fuel deposits a country has at its disposal. This can cause considerable confusion if the difference between these two concepts is not understood. By definition, the reserves of a mineral or fossil

TABLE 9.1
World population economic output, and fossil-fuel consumption.

	Population (billions)	Gross world product (trillion 1980 dollars)	Fossil-fuel consumption (billion tons coal equivalent)
1900	1.6	0.6	1
1950	2.5	2.9	3
1990	5.3	18.4	12

Source: Population statistics from United Nations; gross world product in 1900, author's estimate, and in 1950 from Herbert R. Block, *The Planetary Product in 1980: A Creative Pause?* (Washington, D.C.: U.S. Department of State, 1981), with updates from International Monetary Fund; fossil fuel consumption in 1900 from M. King Hubbert, "Energy Resources" in *Resources and Man* (Washington, D.C.: National Academy of Science, 1969); for remaining years, Worldwatch estimates based on data from American Petroleum Institute and U.S. Department of Energy.

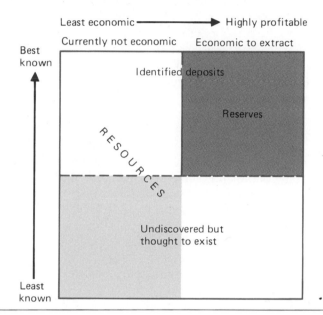

FIGURE 9.1

Resources and Reserves. Each term describes the amount of a natural resource present. Reserves are those known deposits that can be profitably obtained using current technology under current economic conditions. Reserves are shown in the box in the upper left-hand corner in this diagram. The darker the color, the more valuable the reserve. Resources are a much larger quantity that includes undiscovered deposits and deposits that currently cannot be profitably used, although it might be feasible to do so if technology or market conditions change.

Source: Adapted from the United States Bureau of Mines.

fuel will change as technology changes, as new deposits are discovered, and as economic conditions change. Therefore, there can be changes in the amount of reserves, but the total amount of the material will remain constant. It has simply been shifted from the resources category to the reserves category.

When we read about the availability of certain resources, we must remember that energy is needed to extract the wanted material from the earth. The more dispersed the material is, the more expensive it is to collect. So when people talk about how much fossil fuel is available, we must determine what they mean. Much of the world's fossil-fuel resources is located in deposits that cannot be used. This may be due to economic reasons. If the cost of removing and processing a fuel is greater than the fuel's market value, no one is going to produce that fuel. Also, if the amount of energy used to produce, refine, and transport a fuel is greater than the potential energy of that fuel, the fuel will not be produced. A net useful energy yield is necessary to exploit the resource. However, in the future, new technology or changing prices may permit the profitable removal of some resources that currently are not profitable. If so, those resources will be reclassified as reserves.

The ancient Chinese are said to have been the first to use oil as a fuel. The only oil available to them was the small amount that naturally seeped out of the ground. These seepages represented the known oil reserves as well as the known oil resources at that time.

It was nearly two thousand years before there was any significant increase in oil reserves. When the first oil well in North America was drilled in Pennsylvania in 1859, it greatly expanded the estimate of the amount of oil present. There was a sudden increase in the known oil reserves. In the years that followed, new deposits were discovered. Advanced drilling techniques led to the discovery of deeper oil deposits, and offshore drilling established the location of oil under the ocean floor. At the time of their initial discovery, these deep deposits and the offshore deposits added to the estimated size of the world's oil resources. But they did not necessarily add to the reserves because it was not always profitable to extract the oil. However, with advances in drilling and pumping methods and increases in oil prices, it eventually became profitable to obtain oil from many of these deposits. As it became economical to extract these deposits, they were reclassified from the category of resources to the category of reserves.

When a new oil field is discovered, it increases the known oil resource. However, its size can only be estimated. Therefore, the figures given for any resource or reserve are estimates of the amount of material, not precise measurements.

Thus, the relationship between resources and reserves is in a constant state of flux. For example, prior to 1973, many oil wells were capped because it was not profitable to remove the oil. The oil still in the ground at these wells was a resource, not a reserve. During the oil embargo of 1973–1974, when some of the OPEC countries reduced oil production, the price of oil increased. With the increase in oil prices in 1973–1974 and again in 1979–1980, these wells became profitable, and the oil in these fields became reserves. However, in the 1980s, an oil surplus developed because the high oil prices resulting from the embargo caused increased production worldwide. As a result of this oil glut, prices fell. Thus, many of these same wells were capped again, and the oil was no longer a reserve. As the 1990s begin, the price of oil is very unstable but is expected to increase. Similarly, the statuses of resources and reserves for coal and natural gas, as well as for most mineral ores, are constantly changing as economic conditions and technology change.

Fossil Fuels—Nonrenewable Sources of Energy

Of the world's commercial energy, 88 percent is furnished by the three nonrenewable fossil fuel resources: coal, oil, and natural gas. Coal supplies about 28 percent, oil supplies about 39 percent, and natural gas supplies about 21 percent. Each fuel has specific advantages and disadvantages and requires special techniques for its production and use.

Fossil-Fuel Formation

Tropical freshwater swamps covered many regions of the earth 300 million years ago. Conditions in these swamps favored extremely rapid plant growth, resulting in large accumulations of plant material. Because this plant material collected under water, decay was inhibited, and a spongy mass of organic material formed, called **peat.** Estimates indicate that it required three hundred years to accumulate 9 meters of peat. Peat is 90 percent water, 5 percent carbon, and 5 percent volatile materials. (See figure 9.2.) In parts

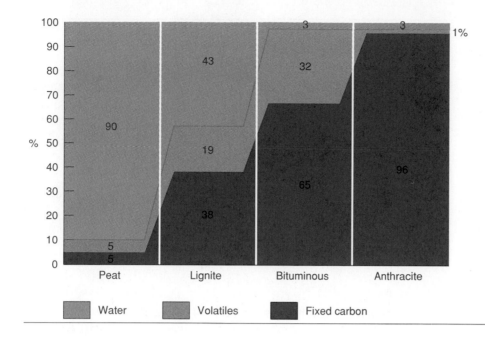

FIGURE 9.2

Composition of Coal. The high percentage of water in peat and lignite make them low-energy-yielding fuels. Increased pressure and heat decreases the moisture content and increases the percentage of carbon during the formation of bituminous or anthracite coal. Bituminous and anthracite coal, therefore, are better sources of energy.

of the world, peat is cut, dried, and used as a fuel source. However, because of its high water content, peat is regarded as a low grade of fuel.

Due to geological changes in the earth, some of the swamps containing peat became seas. The plant material that had collected in the swamps was now at the bottom of a sea, where it was covered by sediment. The weight of the plant material plus the weight of the accumulated sediment compressed the material into a harder form: a low-grade type of coal known as lignite.

A combination of several factors upgraded some of the lignite into a better fuel. If the weight of the sediment was great enough, the heat from the earth high enough, and the length of time long enough, the lignite was changed into bituminous (soft) coal. The major change from lignite to bituminous involved a reduction in the water content.

If the heat and pressure continued over time, some of the bituminous coal was changed to anthracite (hard) coal. Through this combination of events, which occurred over hundreds of millions of years, present-day coal deposits were created.

Oil, like coal, is a product from the past and probably originated from microscopic marine organisms. When these organisms died and accumulated on the ocean bottom, their decay released oil droplets. Gradually, the muddy sediment formed rock called shale, which contained dispersed oil droplets. Although shale is very common and contains a great deal of oil, extraction from shale is very difficult because the oil is not concentrated. When a layer of sandstone formed on top of the oil-containing rock and an impermeable layer of rock formed on top of the sandstone, conditions were suitable for the formation of oil pools. Usually, the trapped oil does not exist as a liquid mass, but rather as a concentration of oil within the pores of sandstone, where it accumulates because water and gas pressure force it out of the shale. (See figure 9.3.) This is particularly true if the rock layers were folded by geological forces.

Natural gas, like coal and oil, was formed from fossil remains. In fact, the geological conditions conducive for oil formation are the same as those for natural gas, and the two fuels are often found together. However, in the

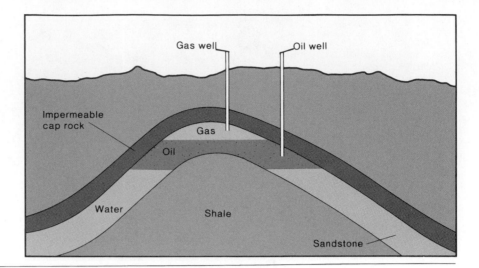

FIGURE 9.3
Crude Oil and Natural Gas Pool.
Water and gas pressure force oil and gas out of the shale and into sandstone beneath the impermeable rock.

formation of natural gas, the organic material changed to lighter, more volatile (easily evaporated) hydrocarbons than those found in oil. The most common hydrocarbon in natural gas is the gas methane (CH_4). Water, liquid hydrocarbons, and other gases may be present in natural gas as it is pumped from a well.

The conditions that led to the formation of oil and gas deposits were not evenly distributed throughout the world. Figure 9.4 illustrates this uneven distribution for oil. Furthermore, some of these deposits are easy to exploit, while others are not.

Mining and Use of Coal

Coal is the world's most abundant fossil fuel, but it varies in quality and is generally classified into three categories: lignite, bituminous, and anthracite. Lignite has a high moisture content and is crumbly in nature, which makes it the least desirable form of coal. Bituminous coal is the most widely used coal because it is the easiest to mine and the most abundant. It supplies about 20 percent of the world's energy requirements. For most uses, anthracite coal is the most desirable because it furnishes more energy than the other grades of coal and is the cleanest-burning coal. But anthracite is not as common and is usually more expensive because it is found at great depth and is difficult to obtain.

Because coal was formed as a result of plant material being buried under layers of sediment, mining operations are needed to extract the coal. There are two methods of mining coal: surface mining and underground mining. **Surface mining** (strip mining) involves removing the material on top of a vein of coal, called **overburden,** to get at the coal beneath. (See figure 9.5.) Coal is usually surface mined when the overburden is less than 100 meters thick. This type of mining operation is very efficient because it removes 100 percent of the coal in a vein and can be profitably used for a seam of coal as thin as half a meter. For these reasons, surface mining results in the best utilization of coal reserves. Advances in the methods of surface mining and the development of better equipment have increased the amount of surface mining in the United States from 30 percent of the coal production in 1970 to 60 percent today. The trend of more coal from surface mining has also occurred in other countries. Surface mining is the most commonly used method in Canada, Australia, and the Soviet Union.

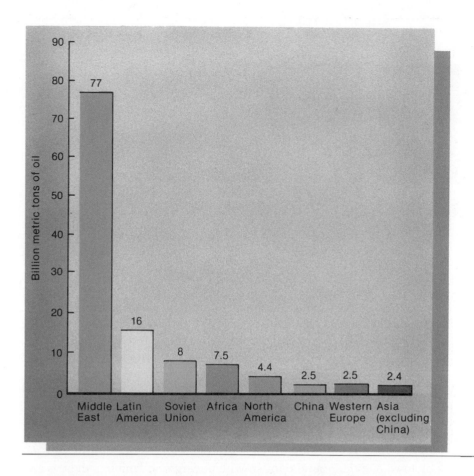

FIGURE 9.4

World's Oil Reserves. The world's supply of oil is not distributed equally. Certain areas of the world enjoy an economic advantage because they control the vast amounts of oil. Reserves are given in billions of metric tons of oil.

Source: Data from World Energy Conference.

FIGURE 9.5

Surface Mining. Large power draglines are used to remove the overburden, which is piled to the side. The coal can then be loaded into trucks. When the coal has been removed, the overburden is placed back in the trench.

If the overburden is thick, surface mining becomes too expensive, and the coal is extracted through **underground mining.** The deeply buried coal seam can be reached in two ways: In flat country, where the vein of coal lies buried beneath a thick overburden, the coal is reached by a vertical shaft. (See figure 9.6a.) In hilly areas, where the coal seam often comes to the surface along the side of a hill, the coal is reached from a drift-mine opening. (See figure 9.6b.)

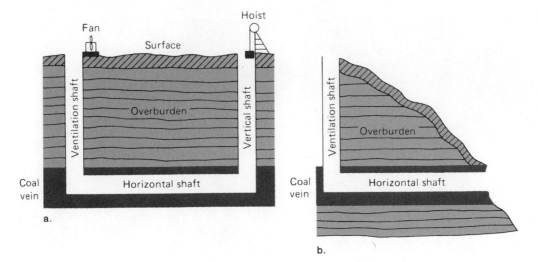

Hoist
Fan
Surface
Ventilation shaft
Overburden
Vertical shaft
Coal vein
Horizontal shaft
a.

Ventilation shaft
Overburden
Coal vein
Horizontal shaft
b.

FIGURE 9.6

Underground Mining. If the overburden is too thick to allow surface mining, underground mining must be used. (*a*) If the coal vein is not exposed, a vertical shaft is sunk to reach the coal. (*b*) In hilly areas, if the vein is exposed, a drift mine is used in which miners enter from the side of the hill.

Coal has problems associated with its mining, transportation, and use. Surface mining disrupts the landscape. However, it is possible to minimize this disturbance by reclaiming the area after mining operations are completed. (See figure 9.7.) The cost of such reclamation is passed on to the consumer in the form of higher coal prices. Reclamation rarely, if ever, returns the land to its previous level of productivity. Although underground mining methods do not disrupt the surface environment as much as surface mining, subsidence occurs if the mine collapses, and large waste heaps are produced.

Miner safety is a concern with all types of mining, but it is a much greater problem with underground mining. With underground mining, there is also the problem of **black lung disease,** which is a respiratory condition that results from the accumulation of large amounts of fine coal-dust particles in the miner's lungs, which inhibits the exchange of gases between the lungs and the blood. The health-care costs and death benefits related to black lung disease are actually an indirect cost of coal mining. Since these are partially paid by the federal government, their full price is not reflected in the price of coal but is paid by the consumer in the form of federal taxes.

Because coal is bulky, rail shipment is the most economical way of transporting it from the mine. Rail shipment costs include the expense of constructing and maintaining the tracks, as well as the cost of the energy required by the engines to move the long strings of railroad cars, called unit trains. In some areas, the coal is transferred from trains to ships. The large amount of coal dust released into the atmosphere at the loading and unloading sites can cause local air-pollution problems. Also, if a boat or railroad car is used to transport coal, there is the expense of cleaning it before other types of goods can be shipped. In some cases, the coal can be ground and mixed with water to form a slurry that can be pumped through pipelines. This helps to alleviate some of the air-pollution problems without causing significant water-pollution problems.

Each year in the United States, the burning of coal releases millions of metric tons of material into the atmosphere and is responsible for millions of dollars of damage to the environment. Coal is a major source of air pollution, and the burning of coal for electric generation is the prime source of this type of pollution.

Since coal is a fossil fuel formed from plant remains, it contains sulfur, because proteins, which are a part of all organisms, contain sulfur. This

a.

b.

sulfur is associated with the problem of **acid mine drainage** and air pollution. Acid mine drainage results from the combined action of oxygen, water, and certain bacteria causing the sulfur in coal to form sulfuric acid. Sulfuric acid can seep out of a vein of coal even before the coal is mined. However, the problem becomes most acute when the coal is mined and the overburden is disturbed, allowing rains to wash the sulfuric acid from the mines into streams. Streams may become so acidic that they can only support certain species of bacteria and algae. Today, regulations limit the amount of runoff from mines, but underground mines and surface mines abandoned before these regulations were enacted continue to pollute the water.

Currently, a form of acid pollution called acid deposition is becoming serious. Acid deposition occurs because of the accumulation of potential acid-forming particles on a surface when coal is burned and sulfur oxides are released into the atmosphere. Each year, over 150 million metric tons of sulfur dioxide are released into the atmosphere worldwide. This problem is discussed in greater detail in chapter 17.

Because coal is bulky and dusty and often has a high sulfur content resulting in air pollution, alternative sources of fuel are sought. The most common alternative fuels for coal are oil and natural gas.

Production and Use of Oil

In the 1920s, as the automobile became common, oil use was encouraged because it was more convenient and often less expensive than coal. This stimulated exploration for oil. Today, geologists use a series of tests to locate underground formations that may contain oil. When a likely area is identified, a test well is drilled to determine if oil is actually present. The many easy-to-reach oil fields have already been tapped. Drilling now focuses on smaller amounts of oil in less accessible sites, which results in more expensive oil. As the oil deposits on land have become more difficult to locate, geologists have widened the search for possible sites to include the ocean floor. The cost of building an offshore drilling platform can be millions of dollars. To reduce the cost of drilling, as many as seventy wells may be sunk from a single offshore platform. (See figure 9.8.)

If the water or gas pressure associated with a pool of oil is great enough, the oil is forced to the surface when a well is drilled. When the natural pressure is not great enough, the oil must be pumped to the surface. Present technology allows only about one-third of the oil in the ground to be removed. This means that two barrels of oil are left in the ground for every

FIGURE 9.7

Surface-Mine Reclamation. (*a*) This photograph shows a large area that has been surface mined with little effort to reclaim the land. The windrows created by past mining activity are clearly evident, and little effort has been made to reforest the land. By contrast, (*b*) is an example of proper surface-mine reclamation. The sides of the cut have been graded and planted with trees. The topsoil has been returned, and the level land is now productive farmland.

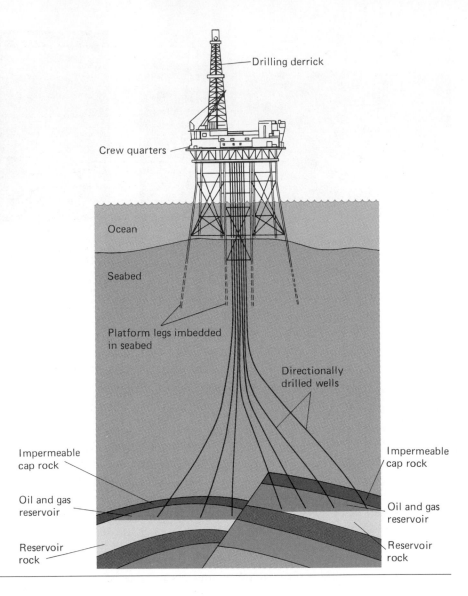

Drilling derrick

Crew quarters

Ocean

Seabed

Platform legs imbedded
in seabed

Directionally
drilled wells

Impermeable
cap rock

Oil and gas
reservoir

Reservoir
rock

Impermeable
cap rock

Oil and gas
reservoir

Reservoir
rock

FIGURE 9.8

Offshore Drilling. Once the drilling
platform is secured to the ocean floor, a
number of wells can be sunk to obtain
the gas or oil.

Source: American Petroleum Institute.

barrel produced. In most oil fields, **secondary recovery** is used to recover
a greater amount of the oil. Secondary recovery methods can include
pumping water or gas into the well to drive the oil out or even starting a
fire in the oil-soaked rock to liquefy thick oil. As oil prices increase, more
expensive secondary recovery methods will be used. Oil obtained from
more expensive deep wells and offshore wells also will increase.

Oil as it comes from the ground is not in a form suitable for use. It
must be refined to get the desired products. The various components of the
crude oil can be separated and collected, however, by heating the oil in a
distillation tower. (See figure 9.9.) After distillation, the products may be
further refined by the "cracking" process. In this process, heat, pressure,
and catalysts are used to produce a higher percentage of volatile chemicals,
such as gasoline, from less volatile liquids, such as diesel fuel and furnace
oils. Therefore, it is possible, within limits, to obtain different products from
a barrel of oil. In addition to fuel uses, petrochemicals from oil serve as
raw materials for a variety of synthetic compounds. (See figure 9.10.)

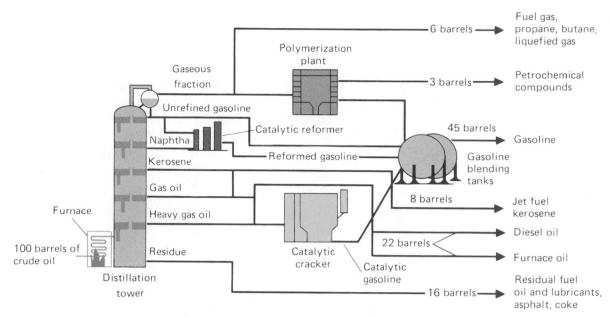

		6 barrels → Fuel gas, propane, butane, liquefied gas
Gaseous fraction	Polymerization plant	3 barrels → Petrochemical compounds
Unrefined gasoline		
	Catalytic reformer	45 barrels → Gasoline
Naphtha		
Kerosene	Reformed gasoline	Gasoline blending tanks
Gas oil		
Heavy gas oil		8 barrels → Jet fuel kerosene
Furnace	Catalytic cracker	Diesel oil
100 barrels of crude oil		22 barrels → Furnace oil
Residue	Catalytic gasoline	
Distillation tower		16 barrels → Residual fuel oil and lubricants, asphalt, coke

FIGURE 9.9

Uses of Crude Oil. A great variety of products can be obtained from the distilling and refining of crude oil. A barrel of crude oil produces slightly less than half a barrel of gasoline. This figure shows the many steps in the refining process and the variety of products that can be obtained from crude oil.

FIGURE 9.10

Oil-Based Synthetic Materials. These common household items are produced from chemicals derived from oil. Although petrochemicals represent only about 3 percent of each barrel of oil, they are extremely profitable for the oil companies.

The environmental impacts of producing, transporting, and using oil are somewhat different from those encountered with coal. Oil spills in the oceans have been widely reported by the news media. (See figure 9.11.) However, these accidental spills are only responsible for about a third of the oil pollution resulting from shipping. Almost 70 percent of the oil pollution in the oceans is the result of normal, routine shipping operations. Oil spills on land can contaminate soil and underground water, while the evaporation of oil products and the incomplete burning of oil fuels result in air-pollution problems. These are discussed in chapter 17.

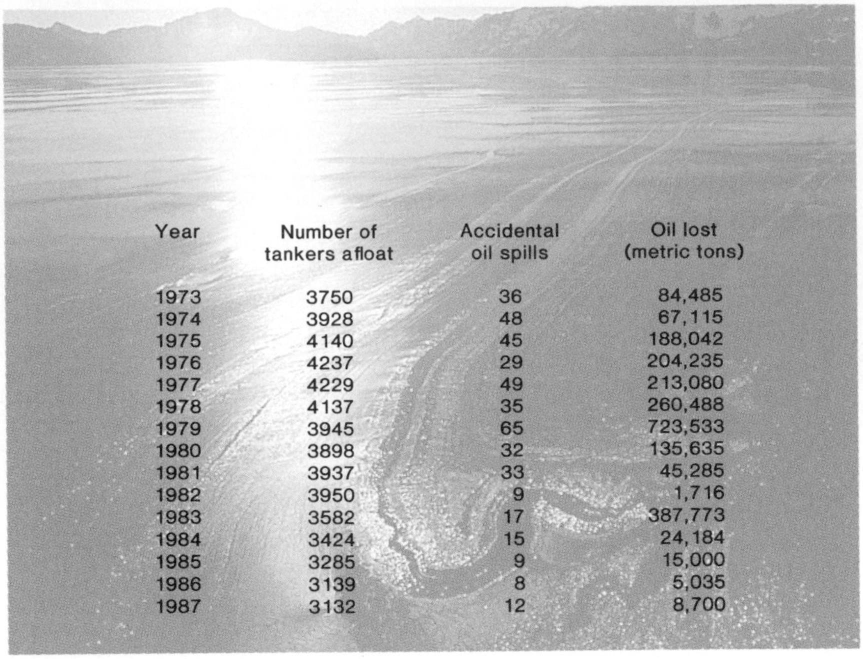

Year	Number of tankers afloat	Accidental oil spills	Oil lost (metric tons)
1973	3750	36	84,485
1974	3928	48	67,115
1975	4140	45	188,042
1976	4237	29	204,235
1977	4229	49	213,080
1978	4137	35	260,488
1979	3945	65	723,533
1980	3898	32	135,635
1981	3937	33	45,285
1982	3950	9	1,716
1983	3582	17	387,773
1984	3424	15	24,184
1985	3285	9	15,000
1986	3139	8	5,035
1987	3132	12	8,700

FIGURE 9.11

Oil Spills. Accidents involving oil tankers are sources of water pollution.

Production and Use of Natural Gas

Natural gas, the third major source of fossil-fuel energy, supplies 17 percent of the world's energy. The drilling operations for obtaining natural gas are similar to those used for oil. In fact, a well may yield both oil and natural gas. As with oil, various secondary recovery methods that pump air or water into a well are used to obtain the maximum amount of natural gas from a deposit. After processing, the gas is piped to the consumer for use.

Transport of natural gas still presents a problem in some parts of the world. Because wells are too far from consumers to make pipelines practical, much of the natural gas is burned as a waste product at the wells in the Middle East, Mexico, Venezuela, and Nigeria. However, several new methods of transporting natural gas and converting it into other products are being explored. At −162° C, natural gas becomes a liquid and has only 1/600 of the volume of its gaseous form. Tankers have been designed to transport **liquefied natural gas** from the area of production to the area of demand. Many people, however, are concerned about possible accidents that would cause the tankers to explode. Another process converts natural gas to methanol, a liquid alcohol, and transports it in that form. As the demand for natural gas increases, the amount of it wasted will decrease and new methods of transportation will be employed. Higher prices will make it profitable to transport natural gas greater distances from the wells to the consumers.

Of the three fossil fuels, natural gas is the least disruptive to the environment. A natural gas well does not produce any unsightly waste, although there may be local odor problems. Except for the danger of an explosion or fire, it poses no danger to the environment while being transported. Since it is clean burning, almost no air pollution results from its use. The products of its combustion are carbon dioxide and water.

Although natural gas is used primarily for heat energy, it does have other uses, such as for the manufacture of petrochemicals and fertilizer. Methane contains hydrogen atoms that are combined with nitrogen from the air to form ammonia, which can be used as fertilizer.

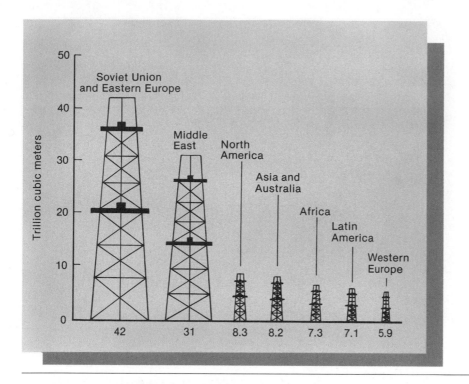

FIGURE 9.12
Natural Gas Reserves (1990). Natural gas reserves, like oil and coal, are concentrated in certain regions of the world. Figures are trillion cubic meters.
Source: Data from World Energy Conference.

From 1970 to 1986, world production of natural gas rose by more than 60 percent. Even though there was a general increase, the production of natural gas declined in the United States by 30 percent during this period. Meanwhile, in the Soviet Union, which has the largest natural gas reserves (figure 9.12), production increased by 225 percent.

Because of the development of oil from beneath the North Sea, several European countries (the United Kingdom, Norway, and the Netherlands) have sizeable reserves of natural gas. During oil shortages in the 1970s, they increased their use of natural gas to compensate for the increased cost of oil. In fact, Western Europe increased its consumption of natural gas by 46 percent from 1979 to 1984. But this did not free these countries from imported energy since most of the increase in natural gas usage was the result of increased imports from the Soviet Union.

Natural gas use in less developed nations has also steadily increased. Fifty less developed countries have usable reserves of natural gas, yet thirty of these countries import oil. The World Bank estimates that, in these fifty countries, current production of natural gas is only 16 percent of its potential. The biggest obstacle to developing these reserves is the lack of a distribution system. However, projections call for production in less developed countries to increase to over 60 percent of the potential by 1995.

Renewable Sources of Energy Currently Being Used

The three nonrenewable fossil-fuel sources of energy—coal, oil, and natural gas—furnish about 88 percent of the world's commercial energy. Nuclear power provides 5 percent of the world's energy. The remainder of the world's energy is supplied by renewable energy sources, including water power, tidal power, geothermal power, wind power, solar energy, fuelwood, biomass conversion, and solid waste.

FIGURE 9.13
Hydroelectric Power Plant. The water impounded in this reservoir is used to produce electricity. In addition, this reservoir serves as a means of flood control and provides an area for recreation.

Water Power

People have used water to power a variety of machines for a long time. Some early uses of water power were to mill grain, saw wood, and power machinery for the textile industry. However, today, water power is used almost exclusively to generate electricity. Flowing water supplies the energy to turn a generator and produce electricity. Hydroelectric power plants are commonly located on artificial reservoirs. (See figure 9.13.) The impounded water represents a potential energy source. In some areas of the world, where the streams have steep gradients and a constant flow of water, hydroelectricity may be generated without the construction of a reservoir. However, such sites are usually found in mountainous regions, and only small power-generating stations can be operated. At present, hydroelectricity produces 7 percent of the world's energy.

The construction of a reservoir for a hydroelectric plant presents environmental and social problems, including loss of fertile farmland, destruction of the natural aquatic ecosystem, relocation of entire communities (including the people and the buildings of that community), and a reduction in the amount of nutrient-rich silt deposited on downriver agricultural lands. The building of the Tellico Dam in Tennessee was delayed for several years because it might have caused the extinction of the snail darter, a fish restricted to the streams that would be flooded. The construction of the Aswan Dam in Egypt resulted in the displacement of eighty thousand people and provided an environment that increased the amount of schistosomiasis. Schistosomiasis is a waterborne disease caused by flatworm parasites that spend part of their life cycle in snails that live in slowly moving water. Because the irrigation canals built to distribute the water from the Aswan Dam provide ideal conditions for the snails, many of the people who use the water in the canals for cooking, drinking, bathing, and as a sewer are infected with the flatworm parasites.

Even though problems are associated with hydroelectric projects, new sites continue to be developed. From 1973 to 1984, the energy furnished by hydroelectricity for world use increased by 46 percent.

Slightly over 17 percent of the potential hydroelectric sites of the world have been developed. (See the box table.) The World Energy Conference estimates that the electricity produced by hydropower will increase six times by the year 2020. The less developed countries, which have developed about 10 percent of their hydropower, will experience most of this growth.

The projected increase is based mainly on the development of hydroelectric plants on large reservoirs. However, construction of "mini-hydro" (less than 10 megawatts) and "micro-hydro" (less than 1 megawatt) hydroelectric plants also is increasing. Such plants can be built in remote areas and supply electricity to local areas. China has built over eighty thousand such small stations, and the United States has nearly fifteen hundred.

About 50 percent of the U.S. hydroelectric capacity has been developed. However, this does not reveal the entire picture. For example, the Wild and Scenic Rivers Act (1968) prevents the construction of dams on certain designated streams. Presently, thirty-seven potential hydroelectric sites are on streams protected by this act. The hydroelectric generating potential often quoted for the United States includes these areas even though, at present, construction of generating plants at these sites is not possible. The U.S. political climate that favors the protection of certain rivers might be modified if the demand for more energy becomes acute.

Developed hydroelectric sites, 1987.

Region	Percent of hydropower developed
Asia	12
South America	14
Africa	7
North America	48
USSR	8
Europe	88
Oceania	32

Tidal Power

Another source of energy related to local geologic conditions involves tidal flow. The gravitational pull of the sun and the moon, along with the earth's rotation, cause tides. The tidal movement of water represents a great deal of energy. For years, engineers have suggested that this moving water could be used to produce electricity. The principle is the same as that employed in a hydroelectric plant. However, tidal changes are greatest near the poles and are accentuated in narrow bays and estuaries.

In the 1930s, the United States explored the possibility of constructing a tidal electrical generating facility at Passamaquoddy, Maine, on the Bay of Fundy, but after spending considerable time and money on a feasibility study, the idea was abandoned.

In 1966, France constructed a commercial tidal generating station. (See figure 9.14.) Located on the Rance Estuary on the Brittany coast of France, this is the world's only large tidal generating station. The station was built to generate 240 megawatts but, because of the tides, it usually produces only 62 megawatts. At present, the British government has undertaken a feasibility study regarding the construction of a 7,200 megawatt tidal station in the Severn Estuary in southwestern England. This would have thirty times the capacity of the French station. Early reports are that the facility can be built for a price that is comparable to a coal-fired plant of similar generating capacity. A negative aspect is that the plant would disrupt the normal estuary flow and would concentrate the pollutants in the area.

FIGURE 9.14

Tidal Generating Station. The Rance River Estuary Power Plant in France is the world's largest tidal electrical generating station.

FIGURE 9.15
Geothermal Power Plant. In this plant, the steam obtained from geothermal wells is used in the production of electricity.

Geothermal Power

The earth's core is a molten mass of material possessing vast amounts of energy. In some regions of the earth, this material sometimes breaks through the earth's crust and produces volcanoes. In other regions, the hot material is close enough to the surface to heat underground water and form steam. Geysers and hot springs are natural areas where this steam and hot water come to the surface. In areas where the steam is trapped underground, **geothermal energy** is tapped by drilling wells to obtain the steam. This steam is then used to power electrical generators. At present, geothermal energy is only practical in areas where this hot mass is near the surface. (See figure 9.15.)

In the United States, the Pacific Gas and Electric Company (PG&E) has been producing electricity from geothermal energy since 1960. PG&E's complex of generating units, which constitute the world's largest geothermal plant, is located in northern California. The power produced by these generators equals the hydroelectric generating capacity of Hoover Dam and produces the electrical power for the city of San Francisco.

The United States has over half of the world's geothermal electrical generating capacity, but over 130 other generating plants are operating in twelve other countries. In addition to the Philippines, which have half the generating power that the United States does, Italy, Mexico, Japan, New Zealand, and Iceland produce a sizeable amount of electricity by geothermal methods.

The state of Hawaii has plans to build two 25-megawatt geothermal plants on the island of Hawaii. Since most of the electricity currently generated in Hawaii comes from oil-fired power plants, geothermal-generated electricity would be less costly and reduce the risk of accidental oil spills. However, the plants would be built in a rain-forest area that environmentalists feel would be destroyed by the construction.

In addition to producing electricity, geothermal energy is used directly for space heating. In Iceland, half of the geothermal energy is used to produce electricity, and half is used for heating. In the capital, Reykjavík, all of the buildings are heated with geothermal energy at a cost that is less than 25 percent of what it would be if oil were used.

FIGURE 9.16
Wind Energy. Fields of wind-powered generators such as these can produce large amounts of electricity.

Although geothermal energy does not seem to be an environmental concern, it does present some problems. The steam contains hydrogen sulfide gas, which has the odor of rotten eggs and is an unpleasant form of air pollution. The minerals in the steam are corrosive to pipes and equipment, resulting in constant maintenance problems. In addition, the minerals in the water are toxic to fish.

Wind Power

As the sun's radiant energy strikes the earth, it is converted into heat, which warms the atmosphere. The earth is unequally heated because various portions receive different amounts of sunlight. Since warm air is less dense and rises, cooler, denser air flows in to take its place. This flow of air is wind. For centuries, wind has been used to move ships, grind grains, pump water, and do other forms of work. In more recent times, wind has been used to generate electricity. (See figure 9.16.)

In 1987, California was using sixteen thousand wind turbines to produce 1.5 gigawatts of electricity at a cost competitive with newly constructed coal or nuclear plants. The California Energy Commission plans to have 8 percent of the state's electricity produced by wind by the year 2000. Denmark, Sweden, the Netherlands, and India also are planning to develop more facilities for wind-generated electricity. The Netherlands plans to use wind to generate 7 percent of its electricity by the year 2000, while India plans to generate 5 gigawatts by the same year.

In Inner Mongolia, nomadic herdsmen have two thousand small, portable, wind-driven generators that travel with them and provide electricity for electric light, television, and movies in their tents as well as for electric fences to contain their animals.

A steady and dependable source of wind makes the use of wind power more productive in some regions than others. Wide, open areas, such as the Great Plains in the United States, are better suited for wind power than

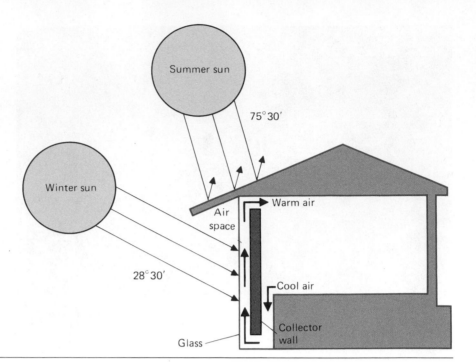

FIGURE 9.17

Passive Solar Heating. The length of overhang in this home is designed for solar heating at the latitude of St. Louis, Missouri (38°N). In this design, a wall 30–40 centimeters thick is used to collect and store heat. The collector wall is located behind a glass wall and faces south. During a midwinter day, when the sun's angle is 28 degrees, light energy is collected by the wall and stored as heat. At night, the heat stored in the wall is used to warm the house. Natural convection causes the air to circulate past the wall, and the house is heated. During a midsummer day, when the sun's angle is 75 degrees, the overhang shades the collector wall from the sun.

are heavily wooded areas. In addition, some people consider the sight of a large number of wind generators, such as are found in California, to be visual pollution.

Solar Energy

The sun is often mentioned as the ultimate answer to the world's energy problems. It provides a continuous supply of energy that far exceeds the world's demands. In fact, the amount of energy received from the sun each day is six hundred times greater than the amount of energy produced each day by all other energy sources combined.

Solar energy is utilized in three ways:

1. In a passive heating system, the sun's energy is converted directly into heat for use at the site where it is collected.
2. In an active heating system, the sun's energy is converted into heat, but the heat must be transferred from the collection area to the place of use.
3. The sun's energy also can be used to generate electricity, which may be used to operate solar batteries or may be transmitted along normal transmission lines.

Passive Solar Systems

Anyone who has walked barefoot on a sidewalk or a blacktopped surface on a sunny day has experienced the effects of passive solar heating. In a **passive solar system,** the light energy is transformed to heat energy when it is absorbed by a surface. Some of the earliest uses of passive solar energy were to dry food and clothes and to evaporate seawater to produce salt. In fact, solar energy is still used for these purposes. Homes and buildings can also be designed to use passive solar energy for heating. (See figure 9.17.)

Since there are no moving parts, a passive solar system is maintenance free. Therefore, none of the energy is used to transfer heat within the system, and there are no operating costs. However, passive solar design is usually only practical in new construction.

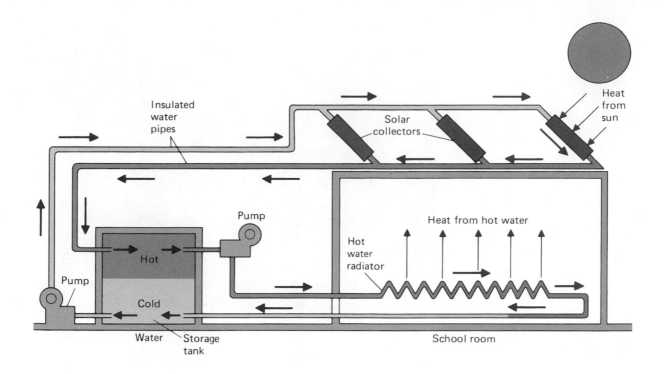

FIGURE 9.18
Active Solar System. Collectors
provide 65 percent of the hot water in
this dental school building in California.

Active Solar Systems

Another method of using solar energy is an active solar system. In addition
to a solar collector, an **active solar system** requires a pump and a system
of pipes to transfer the heat from the site of its production to the area to
be heated.

Active solar systems are most easily installed in new buildings. (See
figure 9.18.) However, in some cases, they can be installed in existing structures. The major obstacle to the use of active systems is the cost. An active
system requires a more complex collector, consisting of a series of liquid-
filled tubes and a system of pipes to transfer the warm liquid, than a passive

system. Added to the cost of the collector is the expense of the pumps and pipes needed to transfer the heat. Because an active system has moving parts, it also has operation and maintenance costs.

Since the sun is only available part of the time, heat energy must be stored for later use. Rock or water is often used to store heat. A Saskatchewan-based company has developed a storage system using sodium sulfate. Polyethylene trays filled with sodium sulfate have a storage capacity that is ten to fifteen times the amount of an equivalent volume of rock or water. The sodium sulfate crystals absorb the heat and change into a liquid. When the liquid sodium sulfate recrystallizes, most of this stored heat is released. The efficiency of this storage system makes it possible and economically sound to retrofit older homes with solar energy. Savings in heating cost are estimated at more than $300 per year. At this rate, construction costs could be recouped in three to ten years.

Solar-Generated Electricity

When the first **photovoltaic cell,** a bimetallic unit that allows the direct conversion of sunlight to electricity, was developed by Bell Laboratories in 1954, it was regarded as an expensive novelty. However, as more efficient batteries were developed and production costs were reduced, practical uses were found for photovoltaic cells. By the mid 1980s, over sixty million solar calculators were being produced annually. These calculators used over 10 percent of the photovoltaic cells manufactured.

Photovoltaic cells appear to be emerging as sources of small amounts of electricity for special uses like calculators and for running equipment in remote regions. Since the normal system of generating electricity in large, centrally located plants and distributing it by high tension lines is costly, it is only practical for use in highly populated areas. Photovoltaic cells are a more practical method for producing energy in the remote regions of the world. Over ten thousand solar-electric homes are found in parts of Alaska and the Australian outback. (See figure 9.19.) The French government has subsidized the installation of over two thousand solar electric units on eighteen islands in the Pacific. These units provide electricity for a thousand homes and five hospitals.

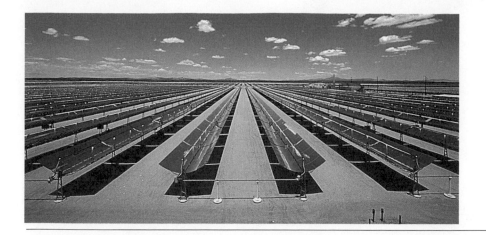

FIGURE 9.20
Solar Generation of Electricity. This solar-powered electricity generating plant is capable of generating electricity at a cost that is competitive with other methods of generating electricity.

The price of photovoltaic cells has been falling as better technology is developed. Eventually, the cells may become competitive with other energy sources, particularly as the cost of fossil fuels rises.

Solar energy is also being used to generate electricity in a more conventional way. A company named Luz International in California has built solar collector troughs that can heat oil in pipes to 390° C. (See figure 9.20.) This heat can be transferred to water, which is turned to steam that is used to run conventional electricity-generating turbines. As with photovoltaic cells, the cost of producing electricity in this manner is falling and is becoming competitive with conventional sources.

Limitations of Solar Energy

Solar energy provides less than 1 percent of the world's energy for several reasons. The most obvious is that it only works during the day, which means that some type of heat or electrical storage mechanism is needed for night use, which adds to the expense of relying on solar energy. The fact that solar heating is most practical in new construction also limits its use. In colder climates, solar heat is inadequate as the sole source of heat, and some type of a conventional heating system is required for backup. Climate is also a problem since many areas have extensive cloudy periods, which reduce the amount of energy that can be collected. Although the price of collectors and related equipment has decreased in recent years, many collector systems are still expensive. For example, photovoltaic cells are three times more expensive to construct than other types of electrical generating equipment.

Fuelwood

The oil shortages in the 1970s resulted in a renewed interest in the developed countries in using fuelwood as a source of energy. Because of its bulk and low level of energy when compared to equal amounts of coal or oil, it is not practical to transport fuelwood over a long distance, and most fuelwood is used locally. In the United States, Norway, and Sweden, fuelwood furnishes 10 percent of the energy for home heating. Canada obtains 3 percent of its total energy, not just home-heating energy, from fuelwood. Most of this energy is used in forest product industries, such as lumbering and paper mills.

In less developed countries, fuelwood has been the major source of fuel for centuries. In fact, fuelwood is the primary source of energy for nearly

FIGURE 9.21
Desertification. The demand for
fuelwood in many regions has resulted
in the destruction of forests. This is a
major cause of desertification.

half of the world's population. In ninety-five countries, mainly in less developed countries such as Kenya, Sri Lanka, and Tanzania, fuelwood supplies over half of the energy, and in twenty-one of these countries, fuelwood supplies over 75 percent of the energy. In those regions where fuelwood is the main source of energy, the primary use is for cooking.

The use of fuelwood as a prime energy source, a rapid population increase, and the high cost of other types of fuel have combined to create some serious environmental problems in many areas. In the world, 1.3 billion people are not able to obtain enough fuelwood or must harvest the fuelwood at a rate that exceeds its growth. This has resulted in the destruction of much forestland in Asia and Africa and has hastened the rate of desertification in these regions. (See figure 9.21.)

Burning wood is also a source of air pollution. Environmental Protection Agency studies indicate that over seventy-five organic compounds are released when wood is burned, twenty-two of which are hydrocarbons known or suspected to be carcinogens. However, these are usually released in small amounts, and no studies have been conducted to determine if wood burning produces them in large enough quantities to be harmful. Higher amounts of carbon monoxide are released from burning wood than from burning oil or natural gas. In addition, many deaths have been caused by carbon-monoxide emissions from poorly operating wood-burning stoves.

In areas with a high population density, the heavy use of fuelwood releases large amounts of fly ash into the air. In Missoula, Montana, 55 percent of the particles in the air during the summer were from burning wood. In the winter, wood was responsible for 75 percent of the particles. A number of steps have been taken to reduce air pollution resulting from burning wood. London has a total ban on burning wood, Oregon requires all stoves installed after 1986 to be equipped with pollution controls, and Vail, Colorado, only permits one wood-burning stove per dwelling.

Biomass Conversion

Biomass is an accumulation of living material. Biomass conversion is the process of obtaining energy or fuel from the chemical energy stored in biomass. It is not a new idea; burning wood is a form of biomass conversion that has been used for thousands of years. In the 1930s, studies were conducted to determine whether alcohol from the fermentation of plant material could be substituted for gasoline. During World War II, charcoal-fueled engines were used in the trucks and cars in Europe.

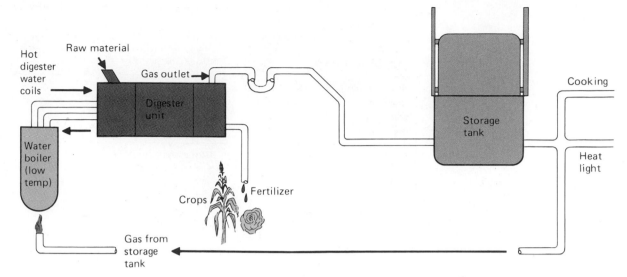

Hot digester water coils

Raw material

Gas outlet

Digester unit

Water boiler (low temp)

Crops

Fertilizer

Gas from storage tank

Storage tank

Cooking

Heat light

FIGURE 9.22

Methane Digester. In the digester unit, anaerobic bacteria convert animal waste into methane gas. This gas is then used as a source of fuel. The sludge from this process serves as a fertilizer. In many less-developed countries, this type of digester has the advantages of providing a source of energy, providing a supply of fertilizer, and managing animal wastes, which helps reduce disease.

Biomass can be burned directly as a source of heat for cooking, burned to produce electricity, converted to alcohol, or used to generate methane. (See figure 9.22.) The People's Republic of China has five hundred thousand small methane digesters in homes and on farms, India has one hundred thousand, and Korea has fifty thousand. Brazil is the largest producer of alcohol from biomass. The low price for sugar coupled with the high price for oil has prompted Brazil to use its large crop of sugar cane as a source of energy. Alcohol provides 50 percent of Brazil's automobile fuel.

In October 1985, twenty less developed nations formed the Biomass Users Network (BUN). BUN sponsors workshops, publishes new biomass methods, and maintains a list of experts in various areas of biomass conversion. The first workshops featured a wood-burning electrical generating method developed in the Philippines, methane generation from Guatemala, and alcohol production from Brazil. Today, over forty nations are members of BUN.

Some environmental and economic concerns are associated with biomass conversion. Those countries that use large amounts of biomass for energy are usually those countries that have food shortages. Biomass conversion means that fewer of the nutrients are being returned to the soil, and this compounds the food shortage. If the price of food rises or the price of oil falls, there could be less biomass conversion.

In addition, the energy required to produce usable energy stocks from biomass must be taken into account. The growing of corn to produce alcohol in the United States requires large energy inputs. The amount of energy present in the alcohol produced from the growing of corn is actually less than the amount of energy that went into the production of the alcohol. Obviously, this makes no sense from an energy point of view. However, the convenience of a liquid fuel may be worth paying for in economic terms, even though it does not make sense from an energy balance point of view.

Solid Waste

Residents of New York City discard 24,000 metric tons of waste each day, or 1.8 kilograms per person. In fact, New Yorkers lead the world in the

Table 9.2
Amount of solid waste produced per capita.

High-income cities	Kilograms per day
New York, United States	2.0
Singapore	1.6
Tokyo, Japan	0.94
Rome, Italy	0.72

Low-income cities	Kilograms per day
Tunis, Tunisia	0.56
Medellín, Colombia	0.54
Calcutta, India	0.51
Kano, Nigeria	0.46

FIGURE 9.23
Waste to Energy. Municipal trash can be burned to produce heat and electricity. This refuse pit is used to feed hoppers of high-temperature furnaces.

production of municipal trash. High-income cities such as New York usually produce more waste than low-income cities. (See table 9.2.) About 80 percent of this waste is combustible and, therefore, represents a potential energy source. (See figure 9.23.)

"Trash power," the use of municipal waste as a source of energy, requires several steps. First, the material must be sorted so that the burnable organic material is separated from the inorganic material. The sorting is accomplished most economically by the person who produces the waste, which requires the producer to separate trash before putting it out for pickup. The trash must be separated into garbage, burnable materials, glass, and metals, and compartmentalized collecting trucks are necessary. Second, the feasibility of using waste as a source of fuel requires a large volume and a dependable supply of waste to be most efficient. If a community constructs a facility to burn 200 metric tons of waste a day, the community must generate and collect 200 metric tons of waste each day.

Communities have been burning trash as a means of reducing the volume of waste for a number of years. The first consolidated incineration of waste was done in Nottingham, England, in 1874. Burning the municipal waste of Munich, Germany, not only reduces the volume of the waste but supplies the energy for 12 percent of the city's electricity. Rotterdam, the Netherlands, operates a 55-megawatt power plant from its garbage. In the United States, only 3 percent of household waste is burned. This is low in comparison to the 26 percent burned in Japan, 51 percent in Sweden, and 75 percent in Switzerland.

Burning trash is not a way of making a profit. It is a way of decreasing the cost of trash disposal because it reduces the need for landfill sites. Baltimore produces gas from 1,000 metric tons of waste per day. This gas generates steam for the Baltimore Gas and Electric Company. The daily waste also yields 80 metric tons of a charcoal-like material (char), 70 tons of ferrous metals, and 170 metric tons of glass. These products all have potential uses. The char can be burned as an additional source of energy. The ferrous metals can be sold, reducing the cost of operating the plant, as well as conserving mineral resources. The glass can be sold for recycling, further reducing the economic costs of the plant.

Although the burning of trash reduces the trash volume and furnishes energy, there are environmental concerns, one of which is air pollution. Many of the older incinerating plants no longer comply with today's air-quality standards. Also, much of the waste material, such as bleached paper and plastics, contains chlorine-containing organic compounds. When burned, these chlorine-containing compounds can form dioxins, which are highly toxic and suspected carcinogens.

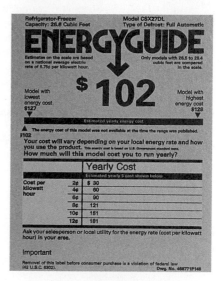

Energy Conservation

Although conservation is not a way of generating energy, it is a way of reducing the need for additional energy consumption. Some conservation technologies are sophisticated, while others are quite simple. For example, if a small, inexpensive wood-burning stove was developed and put into use to replace open fires in the less developed world, energy consumption in these regions could be reduced by 50 percent.

Since the first oil crisis of 1973, U.S. economic activity has increased by about 2.5 percent per year with almost no increase in energy consumption. Significant attention was paid to conservation efforts during this time. Automobiles were redesigned, houses were insulated, and people were encouraged to save energy. During the Reagan years, these efforts stopped, and energy funding was shifted from conservation to new ways of exploiting fossil fuels and nuclear fuels. Fortunately, these trends seem to be reversing, and conservation is again receiving considerable attention. The events in the Middle East and the 1991 war with Iraq will encourage conservation.

Several technologies are now available that reduce energy consumption. (See figure 9.24.) Highly efficient fluorescent light bulbs that can be used in regular incandescent fixtures give the same amount of light for 25 percent of the energy. Since lighting and air conditioning (which removes the heat from inefficient incandescent lighting) account for 25 percent of U.S. electricity consumption, widespread use of these lights could significantly reduce energy consumption. Low-emissive glass for windows can reduce the amount of heat entering a building while allowing light to enter. The use of this glass in new construction and replacement windows

FIGURE 9.24
Energy Conservation. The use of fluorescent light bulbs, energy-efficient appliances, and low-emissive glass could reduce energy consumption significantly.

BOX 9.2
The James Bay Power Project

The James Bay power project is a development that would harness the energy of almost every drop of water in the rivers flowing through 350,000 square kilometers of northeastern Quebec—more than one-fifth of Canada's largest province. The water would be collected in vast reservoirs behind powerhouses on the main rivers. While some water would be released year-round to spin turbines and generate electricity, the system would be geared more to winter months, when demand for power is at its peak, primarily in the heavily populated northeastern United States. The reservoir levels would drop as much as 20 meters as water is released and generating stations are pushed to capacity. Cascading rivers would be dammed and diverted to create the reservoirs, flooding a combined area larger than the surface of Lake Ontario. Some rivers would be reduced to a trickle; others would simply be submerged.

The scene of all this activity would be a wilderness of lakes, rivers, lichens, and peat bogs along the east side of James Bay and the southeast coast of Hudson Bay. The region is home to many animal species, including moose, caribou, beaver, and lynx. The cold lakes and fast-flowing rivers are filled with fish, and the coastline is a rich habitat for fish, birds, whales, and seals. These resources are a crucial source of food and income for the Cree and Inuit natives. The coastal waters are internationally renowned resting and breeding grounds for millions of migratory birds.

According to Hydro-Quebec, the company undertaking the development, the $40 billion project of dikes, powerhouses, roads, and transmission lines could be inserted into an unspoiled northern wilderness with minimal environmental harm. Supporters of the project extol the jobs and income that would result and the subsequent exports of electricity it would bring for Quebec.

Hydro-Quebec has awarded billions of dollars worth of engineering and supply contracts to Quebec firms, enabling the firms to develop high-technology products and become international competitors. In addition, Hydro-Quebec says that every kilowatt of power from James Bay would cut the amount that power plants fueled by coal, oil, or nuclear energy would have to generate at greater risk to the environment.

Critics have argued that the project would create few long-term jobs while taking a devastating toll on the environment. They also worry that the nine thousand local Cree and Inuit would lose their source of food, their livelihood, and their identity. Native people have battled the project from the outset. In 1975, after winning an injunction in Quebec Superior Court and then losing on appeal, the Grand Council of the Cree agreed to let the project proceed in return for some control over use of the land and in the environmental review process. In 1990, the Cree asked the Federal Environmental Assessment Review Office for a public review of the project.

Environmental concerns of the project include erosion, siltation, destruction of critical habitat, nutrient loading and algae blooms from decaying vegetation, water temperature increase, and the mixing of salt and fresh water. In addition, a small clam that is the main source of food for millions of birds would be severely impacted. The loss of this clam from changes in salinity and temperature could threaten many species of migratory birds.

While many questions remain unanswered, the overriding question is: Can humans limit their appetite for power so that megaprojects such as the one at James Bay do not need to be considered? To date, the answer appears to be no.

could have a major impact on the U.S. energy picture. Many other technologies, such as automatic dimming devices or automatic light-shutoff devices, are being used in new construction.

The economics of conservation need encouragement. Often, poorly designed, energy-expensive buildings and machines can be produced inexpensively. The short-term cost is low, but the long-term cost is high. The public needs to be educated to look at the long-term economic and energy costs of purchasing poorly designed buildings and appliances.

Electric utilities have recently become part of the energy conservation picture. They have been allowed to make money on conservation efforts in some states, where previously they could only make money by building more power plants. This encourages them to become involved in energy conservation education efforts, which allow them to serve more people

(*a*) Location of the James Bay power project. (*b*) Native Cree fishing for trout on a free-flowing river in the area scheduled for development. (*c*) Dams like these have created vast, often sediment-laden reservoirs in the Northern Quebec wilderness.

without building new power plants, since their new customers are more efficient.

Finally, changes in government policies could lead to funding of conservation efforts. Tax increases on energy-expensive activities would encourage business, industry, and the public to look for less expensive, more energy-efficient ways of doing things. In June 1990, the people of California passed a proposal to double the state tax on gasoline over the next five years. Although the tax is supposed to support the construction and maintenance of roads, it will also raise the price of gasoline and cause people to look for more energy-efficient methods of travel. In October 1991, a five-cent-per-gallon U.S. tax was imposed. Many individuals believe that there should be larger federal taxes on gasoline. It is argued that such taxes would promote conservation.

"In the conflict between nature and resources development, nature invariably loses." These words by an environmentalist have been applied to the Alaskan pipeline. The $7.7 billion pipeline was the largest construction job ever undertaken by private developers and the biggest project since the digging of the Panama Canal. People first began thinking about a pipeline in 1968, when oil was discovered on the Alaskan north slope. For nearly seven years, construction of the pipeline to transport this oil was delayed. First, the Alaskan natives (Eskimos, Aleuts, and Native Americans) delayed construction by demanding a share of the oil's wealth. In addition, environmentalists demanded design changes. As a result, the plans were changed from a buried, heated line, which would have destroyed sections of the permafrost, to an elevated pipeline.

Construction began in 1974 and was completed three years later. A 1,285-kilometer pipeline now extends from Prudhoe Bay in the north to the port of Valdez in the south. (See the map.) This pipeline can carry 1.6 million barrels of oil each day. This is 7 percent of the daily oil consumption of the lower forty-eight states and represents an annual savings in excess of $6 billion over the cost of importing the same amount of foreign oil.

To safeguard against any disruption in animal migration, four hundred animal crossings were constructed. These involved burying the pipe or elevating it 4.5 meters above the surface. Thus far, the wildlife seem to have adjusted to the pipeline, and there has been no disruption in the migration of caribou or other wildlife. In fact, the caribou and moose cross under the pipeline anywhere it is 1.5 meters above the ground. In addition, much of the pipe was built aboveground so that the 35° C oil would not thaw the permafrost. Where the pipe is buried, it is insulated, and refrigerated brine prevents thawing of the permafrost. As a protection against earthquakes, the pipeline has joints at three locations that allow it to move 6 meters vertically and 6 meters horizontally. Alyeska, the corporation that constructed the pipeline, states that all is well with the environment, but not everyone agrees.

Some pipeline workers have reported that the supports elevating the pipeline are sinking. Alyeska denies this. The agreement that, during the construction process, bulldozers were only to operate in the winter, when they would do less damage, was not honored. In addition, 2.1 million liters of oil were spilled on the right-of-way during construction. This is approximately 1 liter per 60 centimeters of pipeline. There are also concerns about the surveillance of the pipeline during its construction and also now that it is in operation. The surveillance is still being done largely by engineers. Ecologists, who can identify damage to the environment, are only in advisory roles.

The biggest problem seems to be summed up by Guy Martin, Assistant Secretary of Interior for Land and Water Resources at the time the pipeline was being built: "The major environmental effect of this pipeline and the road that parallels it is that they're there."

Summary

A resource is a naturally occurring substance that is potentially feasible to extract under prevailing conditions. Reserves are known deposits from which materials can be extracted profitably with existing technology under present economic conditions.

Coal is the world's most abundant fossil fuel. Coal is obtained either by surface mining or underground mining. Problems associated with coal extraction are disruption in the landscape due to surface mining and subsidence due to underground mining. Black lung disease, waste heaps, water and air pollution, and acid mine drainage are additional problems. Oil was originally chosen as an alternative to coal because it was more convenient and less expensive. However, the supply of oil is limited. As oil becomes less readily available, multiple offshore wells, secondary recovery methods, and increased oil exploration will become more common. Natural gas is another major source of fossil-fuel energy. The primary problem associated with natural gas is transport of the gas to consumers.

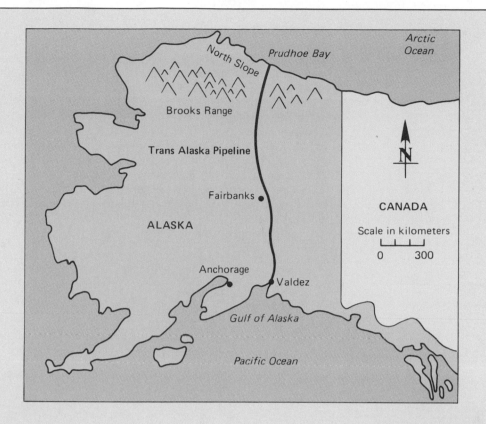

The delays in construction supposedly tripled the cost of the pipeline. Were the delays worth the cost?

What additional environmental dangers are present in the port of Valdez as a result of the pipeline?

Should Alyeska be held responsible for the material spilled during the construction?

Who should be responsible for surveillance of the pipeline? Why?

Is the oil more important than the protection of a fragile Arctic ecosystem?

Fossil fuels are nonrenewable: The amounts of these fuels are finite. When the fossil fuels are exhausted, they will have to be replaced with other forms of energy, probably renewable forms. Water power, which generates 7 percent of the world's electrical energy, is only developed to 17 percent of its potential. The use of geothermal and tidal energy is limited by geographic locations. Wind power may be used to generate electricity but may require wide, open areas and a large number of wind generators. Solar energy can be collected and used in either passive or active systems and can also be used to generate electricity. Lack of a constant supply of sunlight is solar energy's primary limitation. Fuelwood is a minor source of energy in industrialized countries but is the major source of fuel in many less developed nations. Biomass can be burned to provide heat for cooking or to produce electricity, or it can be converted to alcohol or used to generate methane. In some communities, solid waste is burned to reduce the volume of the waste and also to supply energy.

Energy conservation can reduce energy demands without noticeably changing standard of living.

Review Questions

1. Why are fossil fuels important?
2. Distinguish between reserves and resources.
3. What are the advantages of surface mining of coal as compared to underground mining? What are the disadvantages associated with surface mining?
4. Compare the environmental impact of the use of coal with the use of oil.
5. What are some limiting factors in the development of new hydroelectric generating sites?
6. What factors limit the development of tidal power as a source of electricity?
7. In what parts of the world is geothermal energy available?
8. Why can wind be considered a form of solar energy?
9. Compare a passive solar heating system with an active solar heating system.
10. What problems are associated with the use of solid waste as a source of energy?
11. List three energy conservation techniques.

CHAPTER TEN
Nuclear Power: Solution or Problem?

Objectives

After reading this chapter, you should be able to:

Explain how nuclear fission has the potential to provide large amounts of energy.

Describe how a nuclear reactor produces electricity.

Describe the basic types of nuclear reactors.

Explain the steps involved in the nuclear fuel cycle.

List concerns regarding the use of nuclear power.

Explain the problem of decommissioning a nuclear plant.

Describe how high-level radiation waste is stored.

Describe the incident at Chernobyl.

Explain how a breeder reactor differs from other nuclear reactors.

List the technical problems associated with the design and operation of a liquid metal fast-breeder reactor.

Explain the process of fusion.

Chapter Outline

Nuclear Fission

Nuclear Reactors

The Nuclear Fuel Cycle

Nuclear Power Concerns
    Exposure to Radiation
    Thermal Pollution
    Decommissioning Costs
    Radioactive Waste Disposal
    Box 10.1 Three Mile Island

The Accident at Chernobyl

Breeder Reactors
    Consider This Case Study: Converting Nuclear
        Power Plants to Fossil-Fuel Plants

Nuclear Fusion

Key Terms

alpha radiation
atomic fission
beta radiation
boiling-water reactor
chain reaction
decommissioning
gamma radiation
gas-cooled reactor
heavy-water reactor
high-temperature, gas-
  cooled reactor
light-water reactor
liquid metal fast-breeder
  reactor

moderator
nuclear breeder reactor
nuclear fusion
nuclear reactor
plutonium-239 (Pu-239)
pressurized-water reactor
radiation
radioactive
radioactive half-life
rem
thermal pollution
uranium-235 (U-235)

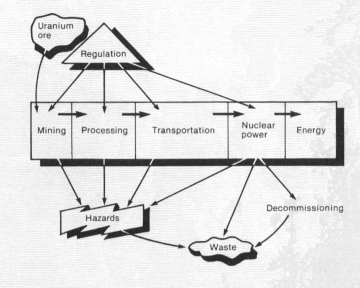

Nuclear Fission

The first controlled fission of an atom occurred in Germany in 1938. However, the United States was the first to produce an instrument of destruction, the atomic bomb. Except for some medical uses of radiation, the main thrust for developing atomic energy for the first dozen years was to produce better bombs. After World War II, however, there were attempts to use atomic energy for peaceful purposes, such as nuclear power generation and treatment of disease. The world's first electricity-generating reactor was constructed in the United States in 1951. The Soviet Union built its first reactor in 1954.

In December 1953, President Dwight D. Eisenhower, in his "Atoms for Peace" speech, made the following prediction:

> Nuclear reactors will produce electricity so cheaply that it will not be necessary to meter it. The users will pay an annual fee and use as much electricity as they want. Atoms will provide a safe, clean, and dependable source of electricity.

Nearly forty years have passed since the promises of "Atoms for Peace" were made, and nuclear power has been found to be generally more expensive than coal-generated electricity. How is atomic energy used to generate electricity? Why is it expensive? This chapter attempts to answer these and other questions concerning atomic energy.

Atomic structure was presented in chapter 3. All atoms are composed of a central region called the nucleus, containing positively charged protons and neutrons that have no charge. Moving around the nucleus are smaller, negatively charged electrons.

The neutrons and protons are held together in the nucleus by energy. Since the positively charged particles in the nucleus repel one another, energy is needed to hold the nucleus together. However, some isotopes of atoms are **radioactive;** that is, the nuclei of these atoms are unstable and spontaneously decompose. Neutrons, electrons, and protons are released during nuclear disintegration, and a great deal of energy is released as well.

The rate of decomposition is consistent for any given isotope. It is measured and expressed as **radioactive half-life,** which is the time it takes for one-half of the radioactive material to spontaneously decompose. Table 10.1 lists the half-lives of several isotopes.

Nuclear disintegration releases **radiation,** a form of energy traveling from one place to another. There are three major types of radiation: **Alpha radiation** consists of a moving particle composed of two neutrons and two protons. Alpha radiation usually only travels through air for about a meter and can be stopped by a sheet of paper. **Beta radiation** consists of electrons released from nuclei. The electrons also only travel a short distance through air and are stopped by a layer of clothing. **Gamma radiation** is a type of electromagnetic radiation, like X rays, light, and radio waves. It can pass through several centimeters of concrete. Beta and gamma radiation cause equal amounts of biological damage. Alpha radiation can cause up to twenty times more damage than beta or gamma radiation. In addition to alpha, beta, and gamma radiation, neutrons are released from the nuclei of unstable atoms. When a neutron hits another nucleus, it causes the nucleus to split. This process is known as **atomic fission.**

Nuclear Reactors

A **nuclear reactor** is a device that permits a controlled nuclear fission chain reaction. In the reactor, neutrons are used to cause a controlled fission of

TABLE 10.1

The half-life of some radioactive isotopes.

Radioactive isotope	Half-life
Iodine 132	2.4 hours
Technetium 99	6.0 hours
Rhodium 105	36.0 hours
Xenon 133	5.3 days
Barium 140	12.8 days
Cerium 144	284.0 days
Cesium 137	30 years
Carbon 14	5,730 years
Uranium 234	250,000 years
Chlorine 36	300,000 years
Beryllium 10	4.5 million years
Potassium 40	1.3 billion years
Helium 4	12.5 billion years

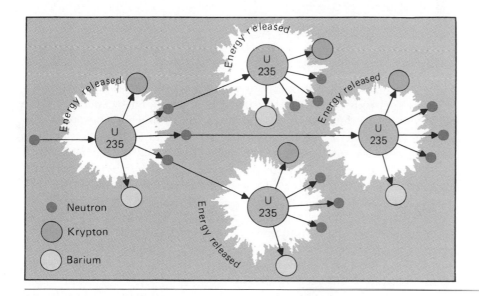

FIGURE 10.1

Nuclear Fission. When a neutron strikes a nucleus of U-235, energy is released, krypton and barium are produced, and several neutrons are released. These new neutrons may strike other atoms of U-235, causing their nuclei to split. This series of events is called a chain reaction.

heavy atoms, such as uranium. **Uranium-235 (U-235)** is a radioactive isotope of uranium used to fuel nuclear fission reactors. When the nucleus of a U-235 atom is struck by a slowly moving neutron from another atom, the nucleus of the atom of U-235 is split. When this occurs, two to three rapidly moving neutrons are released, along with large amounts of energy. This energy is an important product of nuclear fission reactions. The released neutrons strike the nuclei of other atoms of U-235 and also cause them to undergo fission, which, in turn, releases more energy and more neutrons, thus resulting in a **chain reaction.** Once begun, this chain reaction continues to release energy until the fuel is spent or the neutrons are prevented from striking other nuclei. (See figure 10.1.)

In addition to fuel rods containing uranium, reactors contain control rods of cadmium, boron, or other nonfissionable materials used to control the rate of fission by absorbing neutrons. When control rods are lowered into a reactor, they absorb the neutrons produced by fissioning uranium. Therefore, there are fewer neutrons to continue a chain reaction, and the rate of fission decreases. If the control rods are withdrawn, more fission occurs, and more particles, radiation, and heat are produced.

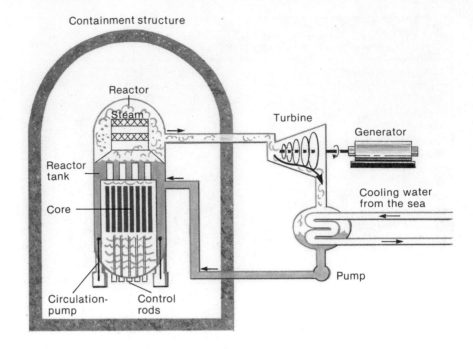

Containment structure

Reactor

Steam

Turbine

Generator

Reactor tank

Cooling water from the sea

Core

Circulation-pump

Control rods

Pump

FIGURE 10.2
Boiling-Water Reactor (BWR). A boiling-water reactor is a type of light water reactor that produces steam to use directly to power the turbine and produce electricity. Water is used as a moderator and as a reactor core coolant.

The fuel rods housed in a reactor are surrounded by water or some other type of moderator. The water is called a **moderator** because it absorbs energy, which slows neutrons, enabling them to split the nuclei of other atoms. Fast-moving neutrons are less effective at splitting atoms than slow-moving neutrons. As U-235 undergoes fission, the energy of the fast-moving neutrons is transferred to water; the neutrons slow down, and the water is heated.

In the production of electricity, a nuclear-powered reactor serves the same function as any fossil-fueled boiler. It produces heat, which converts water to steam to operate a turbine that generates electricity. After passing through the turbine, the steam must be cooled, and the water is returned to the reactor to be heated again.

Various types of reactors have been constructed to furnish heat for the production of steam. They differ in the moderator used, in how the reactor core is cooled, and in how the heat from the core is used to generate steam. Water is the most commonly used reactor-core coolant and also serves as a neutron moderator. **Light-water reactors (LWR)** use ordinary water, which contains the light, most common isotope of hydrogen, which has an atomic mass of one. The two types of LWR are **boiling-water reactors (BWR)** and **pressurized-water reactors (PWR).**

In the BWR (the type of construction used in 50 percent of the nuclear reactors in the world), the water functions as both a moderator and reactor-core coolant. (See figure 10.2.) Steam is formed within the reactor and transferred directly to the turbine, which generates electricity. A disadvantage of the BWR is that the steam passing to the turbine must be treated to remove any radiation. However, some radioactive material is left in the steam even after treatment; therefore, the generating building must be shielded.

In the PWR (the type of construction used in 22 percent of the nuclear reactors in the world), the water is kept under high pressure so that steam is not allowed to form in the reactor . (See figure 10.3.) A secondary loop transfers the heat from the pressurized water in the reactor to a steam generator. The steam is used to turn the turbine and generate electricity. Such

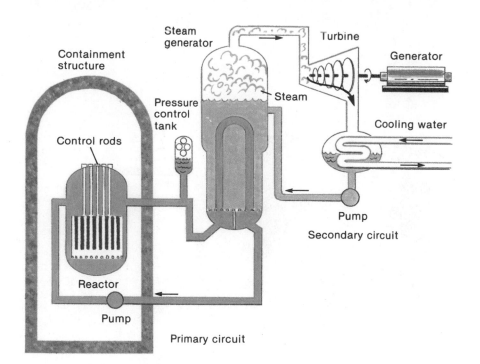

Steam generator

Steam

Turbine

Generator

Containment structure

Pressure control tank

Cooling water

Control rods

Reactor

Pump

Primary circuit

Pump

Secondary circuit

FIGURE 10.3
Pressurized-Water Reactor (PWR).
A pressurized-water reactor is a type of light-water reactor that uses a steam generator to form steam and a secondary loop to transfer this steam to the turbine.

an arrangement reduces the risk of radiation in the steam but adds to the cost of the construction by requiring a secondary loop for the steam generator.

A second type of reactor using water as a coolant is a **heavy-water reactor (HWR).** This type of reactor was developed by the Canadians and uses water containing the hydrogen isotope deuterium in its molecular structure as the reactor-core coolant and moderator. Since the deuterium atom is twice as heavy as the more common hydrogen isotope, the water that contains deuterium weighs slightly more than ordinary water. A HWR also uses a steam generator to convert regular water to steam in a secondary loop. Thus, it is similar in structure to a PWR. The major advantage of a HWR is that naturally occurring uranium isotopic mixtures serve as a suitable fuel. This is possible because heavy water is a better neutron moderator than regular water, while other reactors require that the amount of U-235 be enriched to get a suitable fuel. Since it does not require enriched fuel, the operating costs of a HWR are less than that of a LWR.

The **gas-cooled reactor (GCR)** was developed by atomic scientists in the United Kingdom. Originally, CO_2 served as a coolant for a graphite-moderated core. Like the HWR, natural isotopic mixtures of uranium are used as a fuel. Presently, the GCR has been modified into the **high-temperature, gas-cooled reactor (HTGCR)** (see figure 10.4). In this reactor, helium serves as the coolant for an enriched fuel in a graphite-moderated core. The HTGCR is similar to a PWR in the use of a steam generator and a secondary loop.

The various types of reactors represent differing approaches to building safe, economical plants. Each method has its advantages and disadvantages, its supporters and critics, and modifications will continue to be made to these basic types.

At both Three Mile Island in the United States and Chernobyl in the Soviet Union, mechanical processes were necessary to shut the reactor down. The failure of these mechanisms or their overriding by technicians

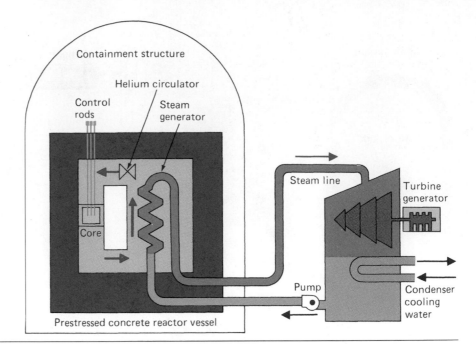

FIGURE 10.4
High-Temperature, Gas-Cooled Reactor (HTGCR). This type of reactor uses graphite as a moderator and the inert gas helium as the reactor-core coolant. A steam generator forms steam and a secondary loop is used to transfer steam to the turbine.

contributed to the two accidents at these sites. Many of the new generation of nuclear reactors being considered are to be designed in such a way that mechanical processes are not necessary to shut down the reactor. Passive mechanisms, such as gravity, would cause control rods to shut down the reactor, or water would flood the reactor.

When people think of nuclear energy, they usually think of the reactor because it is the most visible and most costly step in the production of nuclear power, and they worry about reactor safety. However, the production and handling of the fuel is also important.

The Nuclear Fuel Cycle

The nuclear fuel cycle begins with the mining operation. (See figure 10.5.) Low-grade uranium ore is obtained by underground or surface mining. The ore contains about 0.2 percent uranium by weight. After it is mined, the ore goes through a milling process. The ore is crushed and treated with a solvent to concentrate the uranium. Milling produces *yellowcake,* a material containing 70 to 90 percent uranium oxide.

After milling, the uranium in the yellowcake is a mixture of 0.7 percent U-235 and 93.3 percent U-238. This is not a high enough concentration of U-235 for most types of reactors, so the amount of U-235 must be increased by enrichment. Since the masses of the isotopes U-235 and U-238 vary only slightly, and there is no chemical difference, enrichment is a difficult and expensive process. As a result of enrichment, the U-235 content is increased from 0.7 percent to 3 percent.

Fuel fabrication converts the enriched material into a powder, which is then compacted into pellets about the size of a pencil eraser. These pellets are sealed in metal fuel rods about 4 meters in length, which are then loaded into the reactor.

As fission occurs, the concentration of U-235 atoms decreases. After about three years, a fuel rod does not have enough radioactive material to sustain a chain reaction, and the spent fuel rods must be replaced by new fuel rods. However, the spent rods are still very radioactive, containing about

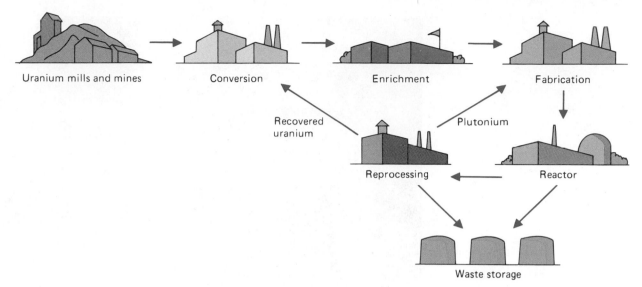

Uranium mills and mines → Conversion → Enrichment → Fabrication

Recovered uranium

Plutonium

Reprocessing ← Reactor

Waste storage

FIGURE 10.5

Steps in the Nuclear Fuel Cycle. In the United States, the material goes from the reactor to waste storage. In England and the Soviet Union, the material may be reprocessed.

U.S. Department of Energy.

1 percent U-235 and 1 percent plutonium. These spent rods are a major source of the radioactive waste material produced in a nuclear reactor.

When nuclear reactors were first being built, it was proposed that spent fuel rods could be reprocessed: The remaining U-235 could be enriched and used to manufacture new fuel rods. Since plutonium is radioactive, it could also be fabricated into fuel rods. Besides providing new fuel, reprocessing would reduce the amount of nuclear waste. However, the cost of producing fuel rods by reprocessing was found to be more expensive than producing fuel rods from ore, and the United States closed its reprocessing facilities. At present, only the Soviet Union, France, and the United Kingdom reprocess spent fuel rods as an alternative to storing them as nuclear waste.

Each step in the nuclear fuel cycle involves the transport of radioactive materials. The uranium mine is some distance from the processing plant. The fuel rods must be transported to the power plant, and the spent rods must be moved to a reprocessing plant or to a storage area. Each of these links in the fuel cycle represents a possible accident or mishandling that could result in the release of radioactive material. Therefore, the methods of transport are extremely carefully designed and tested before they are used. Many people are convinced that the transport of radioactive materials is hazardous, while others are satisfied that the utmost care is being taken and that the risks are extremely small.

Ultimately, both high-level and low-level radioactive wastes must be stored. Thus, each step in the nuclear fuel cycle, from the mining of uranium to the storage of nuclear waste, has health and environmental concerns associated with it.

Nuclear Power Concerns

The public has many concerns about the dangers of radiation that can be released from nuclear power plants and other points in the nuclear fuel cycle.

Exposure to Radiation

Radiation is converted to other forms of energy when it is absorbed by matter. This energy conversion results in damage at a cellular, tissue, organ, or organism level when organisms are irradiated. The degree and kind of damage

FIGURE 10.6
Protective Equipment. Persons working in an area subjected to radiation must take steps to protect themselves. These workers are wearing protective clothing and filtering the air they breathe.

vary with the kind of radiation, the amount of radiation, the duration of the exposure, and the particular type of cells irradiated.

The radiation produced in a nuclear blast or in the immediate vicinity of a nuclear reactor can result in the immediate death of cells and organisms. (See figure 10.6.) Lesser doses may produce changes within certain cell structures. For example, moderate doses destroy membranous structures in cells, which will eventually result in the death of the cells.

Radiation can also cause mutations, which are changes in the genetic messages within cells. Mutations can cause two quite different kinds of problems. Mutations that occur in the ovaries or testes can form mutated eggs or sperm, which can lead to abnormal children. Therefore, care is usually taken to shield these organs from unnecessary radiation. Mutations that occur in other tissues of the body may manifest themselves as abnormal tissue growths known as cancer. Two common cancers that are strongly linked to increased radiation exposure are leukemia and breast cancer. Because mutations are essentially permanent, they may accumulate over an extended period of time. Therefore, the accumulated effects of radiation over a person's lifetime may result in the development of cancer later in life.

Human exposure to radiation is usually measured in **rems** (*R*oentgen *E*quivalent *M*an), a measure of the biological damage to tissue. The effects of large doses (1,000 to 1,000,000 rems) are easily seen and can be quantified, because there is a high incidence of death at these levels, but demonstrating known harmful biological effects from smaller doses is much more difficult. (See table 10.2.) Moderate doses (10 to 1,000 rems) are known to increase the likelihood of cancer and birth defects. The higher the dose, the higher the incidence of abnormality. Lower doses may cause temporary cellular changes, but it is difficult to demonstrate long-term effects. Thus, the effects of low-level chronic radiation generate much controversy. Some people feel that all radiation is harmful, that there is no safe level. Others feel that the increased risk of low-level radiation is extremely small and that current radiation standards are adequate to protect the public, especially in light of the benefits that radiation can give in terms of medical diagnoses and electrical energy. Current research is trying to assess the risks associated with repeated exposure to low-level radiation.

TABLE 10.2
Radiation effects.

Source	Dose	Biological effect
Nuclear bomb blast or exposure in a nuclear facility	100,000 rems/incident	Immediate death
X rays for cancer patients	10,000 rems/incident	Coma, death within one to two days
	1,000 rems/incident	Nausea, lining of intestine damaged, death in one to two weeks
	100 rems/incident	Increased probability of leukemia
	10 rems/incident	Early embryos may show abnormalities
Upper limit for occupationally exposed people	5 rems/year	Effects difficult to demonstrate
X ray of the intestine	1 rem/procedure	Effects difficult to demonstrate
Upper limit for release from nuclear installations (except nuclear power plants)	0.5 rem/year	Effects difficult to demonstrate
Natural background radiation	0.1–0.2 rem/year	Effects difficult to demonstrate
Upper limit for release by nuclear power plants	0.005 rem/year	Effects difficult to demonstrate

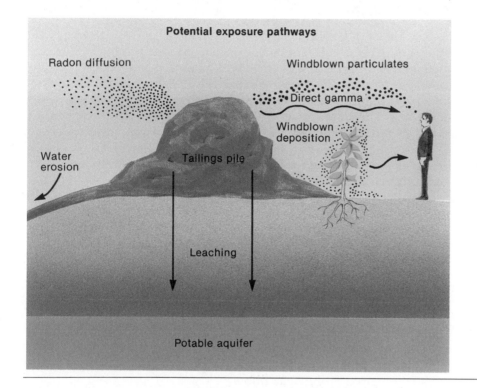

Potential exposure pathways

Radon diffusion

Windblown particulates

Direct gamma

Windblown deposition

Water erosion

Tailings pile

Leaching

Potable aquifer

FIGURE 10.7

Uranium Mine Tailings. Even though the amount of radiation in the tailings is low, the radiation still represents a threat. The radioactivity is dispersed through the environment in several ways, including seepage into the water supply, dispersal of radioactive particles by wind, and diffusion of radioactive gases. In addition, the mine tailings are a source of gamma rays, which can cause negative health effects.

Each step in the nuclear fuel cycle poses a radiation exposure problem. Mining of uranium presents a radiation hazard. Although the radiation level in the ore is low, prolonged exposure of miners to low-level radiation increases the rates of certain cancers like lung cancer. Uranium miners who smoke have an even higher lung cancer rate.

Several hundred thousand metric tons of ore are required to produce the uranium needed to fuel one reactor. The ore must be crushed in a milling process, which releases radioactive dust into the atmosphere, and workers are chronically exposed to low levels of radioactivity. The crushed rock is left on the surface as mine tailings. There are over 150 million metric tons of low-level radioactive mine tailings in the United States and at least as much in the rest of the world. These tailings constitute a hazard because they can be dispersed into the environment. (See figure 10.7.)

FIGURE 10.8
Cooling Towers. Cooling towers draw air over wet surfaces. The evaporation of the water cools the surface, removing heat from the nuclear generator and releasing the heat into the air.

In the enrichment and fabrication processes of the nuclear fuel cycle, the main dangers again are exposure to radiation and accidental release of radioactive material into the environment. Transport of radioactive material also involves risk of exposure. Most radioactive material is transported by highways or railroads. If a transporting vehicle were involved in an accident, radioactive material could be released into the environment.

Even after fuel rods have been loaded into the reactor, people who work in the area of the reactor risk radiation exposure. Probably the single greatest danger in the nuclear fuel cycle is a core meltdown.

Countries that reprocess spent fuel rods must also be concerned about possible theft of rods or reprocessed material by terrorist organizations that would use the radioactive material to construct an atomic bomb.

In addition to the problems resulting from radioactive material, nuclear power plants have three other problems: **thermal pollution, decommissioning,** and radioactive waste disposal.

Thermal Pollution

The problem of thermal pollution is not unique to nuclear power plants. Both fossil-fuel and nuclear plants generate steam to produce electricity, which results in a great deal of waste heat. In a fossil-fuel plant, half of the heat energy produces electricity, and half is lost as waste heat. In a nuclear power plant, only one-third of the heat generates electricity, and two-thirds is waste heat. Therefore, the less efficient nuclear power plant increases the amount of thermal pollution in the environment. To reduce this, costly cooling facilities are constructed and operated. Cooling processes usually involve water, so nuclear power plants often are constructed next to a water source. In some cases, water is drawn directly from lakes, rivers, or the oceans and returned. In other cases, giant cooling towers are constructed. (See figure 10.8.)

Decommissioning Costs

Most industrial facilities have a life expectancy: the number of years they can be profitably operated. The life expectancy for an electrical generating plant, whether fossil fuel or nuclear, is about thirty years, after which time the plant is demolished. With a fossil-fuel plant, the demolition is relatively simple and quick. A wrecking ball and bulldozers reduce the plant to rubble, which is trucked off to a landfill. The only harm to the environment is usually the dust raised by the demolition.

Demolition of a nuclear plant is not so simple. In fact, nuclear plants are not demolished, they are decommissioned. Today, several nuclear plants in the world are shut down and waiting to be decommissioned, and the number of plants scheduled for decommissioning will continue to grow. Two problems are associated with decommissioning a nuclear power plant. The first is that no one is exactly certain how it can be safely done. The second is that no one is certain how much it will cost.

When a nuclear plant is constructed, none of it is radioactive. However, once the plant is in operation, some areas become contaminated. By the time a plant is decommissioned, much of it is radioactive. This increased radioactivity is the result of (1) contamination and (2) activated material.

Radioactivity resulting from contamination could involve the pipes, containment building, control building, and certain auxiliary buildings. The

Energy

amount of such contamination is dependent upon the amount of fuel leakage that occurred when the plant was in operation. Washing decontaminates the surface of these buildings. But even after surface decontamination, most of the equipment and building material is still radioactive. Such material, along with all the solutions used in the decontamination, must be disposed of safely.

Activated material results from the conversion of nonradioactive atoms in the reactor and the shield into radioactive isotopes as a result of being bombarded by neutrons. Those materials that have been converted to radioactive forms are said to have been activated. The half-life of the activated materials may vary from five years for cobalt to eighty thousand years for nickel. The radioactivity cannot be removed from the activated materials because it is within the structures themselves, not just sitting on the surface.

When a plant must be decommissioned, the plant is shut down; all the fuel rods are removed; all of the water used as a reactor-core coolant, as a moderator, or to produce steam must be drained; and the reactor and generator pipes must be cleaned and flushed. The spent fuel rods, the drained water, and the material used to clean the pipes are all radioactive and require safe storage or disposal.

After these initial procedures, there are three decommissioning options: (1) decontaminate and dismantle the plant as soon as it is shut down; (2) put the plant in mothballs for twenty to one hundred years to allow for disintegration of radioactive materials having a short half-life and then dismantle the plant; (3) entomb the plant by covering the reactor with reinforced concrete and placing a barrier around the plant.

Originally, entombment was thought to be the best method. But because of the long half-life of some of the radioactive material and a danger of groundwater contamination, this method now appears to be the least favorable. Two-thirds of U.S. utilities with nuclear plants favor immediate decontamination and dismantling. Japan plans to mothball the plants for ten years before dismantling. France and Canada plan to mothball for a longer period and then dismantle.

Costs associated with decommissioning include dismantling, followed by packaging, transporting, and burial of wastes. Because of contamination and activated material, ordinary dismantling methods cannot be used to demolish the buildings. Special remote-control machinery must be developed. Techniques that do not release dust into the atmosphere must be used (one suggestion is to do the work under water), and workers must wear protective clothing. To further add to the problems and the cost is the fact that much of the protective equipment and the expensive remote-control machinery may become so contaminated that they will have to be disposed of safely. Waste disposal is estimated to be 40 percent of the decommissioning costs. Table 10.3 shows the volume of material that would need to be handled from a 1,100-megawatt power plant.

Cost projections for decommissioning a single plant range from a low estimate of $50 million to a high of $3 billion. If this last figure is accurate, it may cost more to decommission some plants than it cost to construct them. The first nuclear power plant to be decommissioned was a small, 22-megawatt reactor at Elk River, Minnesota, that had only operated for four years. The Department of Energy was responsible for decommissioning the Elk River plant. The process was completed in 1974, three years after the decommissioning was begun, at a cost of $6.15 million. Part of the decommissioning was the burial of 3,360 cubic meters of decontaminated material from the plant.

TABLE 10.3

Low-level radioactive contaminants from decommissioning a 1,100-megawatt pressurized-water reactor.

Material	Volume (cubic meters)
Radioactive	618
Activated	
Metal	484
Concrete	707
Contaminated	
Metal	5,465
Concrete	10,613
Total	17,887

Source: *Worldwatch Paper 69,* "Decommissioning: Nuclear Power's Missing Link" and U.S. Nuclear Regulatory Commission.

It has been proposed that, when a nuclear plant is constructed, a decommissioning fund be established. This has not been favorably received, since utilities already have difficulty raising the money to build the plants in the first place. They are reluctant to add decommissioning costs, which would raise electrical rates even more. However, nine states in the United States require that money be collected and placed into an external account to finance the costs of decommissioning. On the other hand, the state of Michigan prohibits the collection of decommissioning costs in advance.

In Sweden, the federal government collects an annual tax for decommissioning. A separate account is set up for each individual reactor. A similar arrangement is used in Switzerland and Germany. However, most utilities worldwide seem to ignore the cost of decommissioning, which is why most of the reactors that have been shut down have not been decommissioned. The problem of decommissioning may become a legacy for future generations, which means that the costs of decommissioning nuclear plants will be borne by those who had no voice in their construction and who did not even benefit from the use of the electricity these plants produced.

Radioactive Waste Disposal

When the world entered the atomic age, the problem of the disposal of nuclear waste was not fully appreciated. Low-level radioactive waste is generated by nuclear power plants, military facilities, hospitals, and research institutions. High-level radioactive waste results from spent fuel rods and obsolete nuclear weapons.

High-Level Radioactive Waste

High-level radioactive waste has been accumulating. In the United States, 280 million liters of liquid, highly radioactive military waste are temporarily stored in Idaho, South Carolina, and Washington. Two million cubic meters of low-level radioactive military and commercial waste are buried in various sites. In addition, about 21,000 metric tons of high-level radioactive waste from spent fuel rods are being stored in cooling ponds at reactor sites.

In other countries, spent fuel rods are either stored in cooling ponds or sent to reprocessing plants. Even though reprocessing is more expensive than manufacturing fuel rods from ore, some countries use reprocessing as

BOX 10.1
Three Mile Island

On 14 March 1979, the valves in three auxiliary pumps at unit two reactor at the Three Mile Island nuclear plant closed. At 4:00 A.M. on 28 March 1979, there was a breakdown of the main pump at unit two. The auxiliary pumps failed to operate because of the closed valves. When the pumps failed, the electrical-generating turbine stopped. At this time, an emergency coolant should have flooded the reactor and stabilized the temperature. The coolant did start to flood the reactor, but a faulty gauge indicated that the reactor was flooded. As a result of this faulty reading, an operator overrode the automatic emergency cooling system and manually closed it down.

Without the emergency coolant, the reactor temperature rose rapidly. Two hours after the main pump failed, the temperature and pressure were still rising, and an alarm was sounded. An hour later, the control rods stopped fission, but a partial core meltdown had occurred. Later that day, as pressure in the primary loop increased, radioactive steam was vented into the atmosphere. Also, part of the plant's ventilating system was contaminated with radioactive gases.

Problems continued to develop, and on 29 March, a bubble of hydrogen formed at the top of the reactor. The next day, pregnant women and children were advised to evacuate the area. Five days after the accident, it was announced that the immediate danger

had passed. But by this time, unit two had been severely damaged. It was over a year later before anyone entered the plant.

More than ten years have passed since the accident at Three Mile Island. During that time, Metropolitan Edison, the operators of the plant, and the Nuclear Regulatory Commission have been criticized for their action at the time of the accident and their failure to clean up afterward. The crippled reactor was eventually defueled in 1990 at a cost of about $1 billion. It will be placed in monitored storage until 2010.

an alternative to waste storage. The reprocessing plants in France and the United Kingdom accept domestic and foreign fuel rods, and any fuel rods fabricated by the Soviet Union can be returned for reprocessing.

One proposal for permanent storage of high-level radioactive waste is to bury it in a stable geologic formation. France, United States, Spain, and Germany are investigating storage in salt deposits. The waste would be placed in borosilicate glass containers. Each container would then be sealed in a thick-walled, stainless steel canister. Due to the high temperature of the radioactive waste, the containers would be stored aboveground for ten years. At the end of this time, the temperature of the container would be reduced, and the material could be buried in a salt deposit 600 meters below the surface. (See figure 10.9.)

Sweden intends to store the waste 500 meters underground in granite. Site construction is not planned until 2010, when Sweden's twelve reactors will be decommissioned. France plans to use its reprocessing plants and "temporary" storage indefinitely.

The politics concerning the disposal of high-level radioactive waste are probably as critical as developing a suitable method. No one wants a radioactive disposal site in their area. In December 1982, the U.S. Congress passed legislation calling for a high-level radioactive disposal site to be

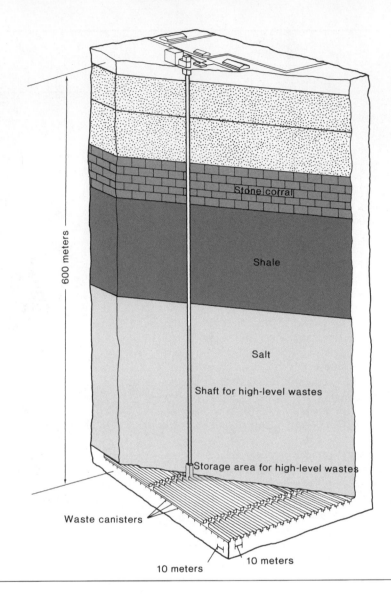

FIGURE 10.9

High-Level Nuclear Waste Disposal.
After being sealed in glass containers
and placed in stainless steel canisters,
the waste from nuclear power plants
could be deposited in salt domes 600
meters underground.

selected by March 1987 and to be completed by 1998. In 1984, the Department of Energy stated that it was already three years behind schedule.
Final site selection occured in 1989, three years later than the original date.
The location is Yucca Mountain, Nevada. However, local protests are continuing.

Low-Level Radioactive Waste

Low-level radioactive waste includes the cooling water from nuclear reactors, material from decommissioned reactors, radioactive materials used
in the medical field, and protective clothing worn by persons working with
radioactive materials, and also results from many other modern uses of radioactive isotopes. Disposal of this type of radioactive material is very difficult to control. Estimates indicate that much low-level radioactive waste
is not disposed of properly.

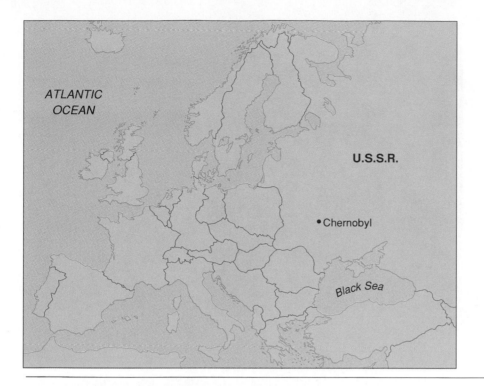

FIGURE 10.10
Chernobyl. The town of Chernobyl became infamous as the location of the world's worst nuclear power plant accident.

In 1970, a moratorium halted the dumping of U.S. radioactive waste in the oceans. But, prior to this, some ninety thousand barrels of radioactive waste had been placed on the ocean floor. Until 1983, European countries had been dumping both high-level and low-level radioactive material into the Atlantic Ocean. Before 1983, when ocean dumping was halted, 90,000 metric tons of radioactive waste had been disposed of in the ocean.

The United States produces a half million cubic meters of low-level radioactive waste per year. This is presently being buried in disposal sites in Nevada, South Carolina, and Washington. However, these states balked at accepting all the country's waste. In 1980, Congress set a deadline of 1986 (later extended to 1993) for each state to provide for its own low-level radioactive site. Later, a change allowed several states to cooperate and form regional coalitions, called compacts. In this arrangement, one state would provide a disposal site for the entire compact. Only California and Texas did not enter into a compact arrangement. The original deadline of 1986 has passed. However, most states still do not have a place to deposit low-level radioactive wastes, and most will not meet the 1993 deadline established by Congress.

The Accident at Chernobyl

Chernobyl is a small city in the Soviet Ukraine and, like many small cities in the world, most people had never heard of it. (See figure 10.10.) However, in the spring of 1986, the world's largest nuclear accident catapulted Chernobyl into the news.

At 1:00 A.M. on 25 April 1986, a test was begun on unit four at the nuclear reactor site in Chernobyl. The test was designed to measure the amount

FIGURE 10.11
Chernobyl. An uncontrolled chain of reactions in the reactor of unit four resulted in a series of explosions and fires. See the circled area in the photograph.

of electricity that the still-spinning turbine would produce if the steam were shut off. To reduce the output of steam, the control rods were lowered into the core. However, due to a demand for power, the test was delayed. At 2:00 A.M., to prevent further delay in the testing, the emergency core-cooling system was manually overridden. This was a serious safety violation.

Insertion of the control rods reduced the operating level of the reactor from its usual 3,200 megawatts of thermal power to 1,600 megawatts. At 11:10 P.M. on April 25th, the monitoring units were set for low-level operation. At this time, the operators failed to program the computer to maintain power at 700 megawatts, and output dropped to 30 megawatts. To increase power, many of the control rods were withdrawn. However, an inert gas (xenon) that had accumulated on the fuel rods absorbed the neutrons, thus slowing down the rate of power increase. In an attempt to obtain more power, all the control rods were withdrawn. This was a second serious safety violation.

At 1:00 A.M. on April 26, the operators shut off most emergency warning signals and turned on all eight pumps to provide adequate cooling after the testing was completed. Just as final stages of the test were beginning, a signal indicated excessive reaction in the reactor. The operators manually blocked the automatic reactor shutdown and began the test.

As the tests continued, the power output of the reactor rose to its normal level and continued to rise. The operators activated the emergency system designed to put the control rods back into the reactor and stop the fission. But the core had already been deformed, and the rods would not fit properly. Then there were two explosions.

In 4.5 seconds, the energy level of the reactor increased some two thousand times. The reactor was now operating at one hundred times its rated capacity. The fuel rods ruptured, and the cooling water turned into steam. The explosion that followed ripped off the 1,000-metric-ton concrete roof above the reactor. In less than ten seconds, a reactor with one of the best safety records in the Soviet Union became the scene of the world's worst nuclear accident. (See figure 10.11.) It required ten days to bring the runaway reaction under control.

TABLE 10.4
Estimated financial losses from the Chernobyl accident.

Lost	Estimated cost (million dollars)
Plant replacement	1,040–1,250
Agricultural losses	1,000–1,900
Site cleanup	350– 690
Health costs	280– 560
Export losses	220– 660
Relocation	70
Total	2,960–5,130

Source: "Economic Consequences of the Accident at Chernobyl Nuclear Plant," *PlanEcon Reports,* and Worldwatch Paper 75, *Reassessing Nuclear Power: The Fallout from Chernobyl, 1987.*

The immediate consequences were 31 fatalities, 500 persons hospitalized, including 203 with acute radiation sickness, and 116,000 people evacuated. Of the evacuees, 24,000 received high amounts of radiation, above 45 rems. This is a level expected to cause increased instances of leukemia. Chernobyl put 300 to 400 million people in fifteen countries at risk. Estimates concerning the number of deaths that will eventually be caused by the fallout from Chernobyl in the Soviet Union and Europe range from five thousand to seventy-five thousand. Some scientists project as many as half a million deaths.

Over a year after the disaster at Chernobyl, Soviet officials halted the decontamination of twenty-seven cities and villages within 40 kilometers of Chernobyl. The largest city to be affected was Pripyat, which had a population of fifty thousand and was only 4 kilometers from the reactor. A new town was built to accommodate those displaced by the accident, and Pripyat remains a ghost town. Sixteen other communities within the 40-kilometer radius have been cleaned up, and the original inhabitants have returned, although there is still controversy about the safety.

It is as difficult to put a figure on the financial damage as it was to put a figure on the amount of human suffering. Direct financial losses to the Soviet Union are placed between $2.96 and $5.13 billion (see table 10.4).

Financial losses outside the Soviet Union are even more difficult to calculate. In the spring of 1986, some twenty nations measured the level of radiation in all types of food. (See figure 10.12.) No one knows the losses suffered by farmers, wholesalers, shippers, and retailers. Sheep farmers in the United Kingdom estimated their losses at $15 million. Swedish farm losses are estimated at $145 million, and German losses are given as $240 million. Australia refused a shipment of Italian wheat because of the high level of radioactivity.

In addition to the human suffering and the financial losses, Chernobyl had another effect. It deepened the concern of the public about the safety of nuclear reactors. Even before Chernobyl, between 1980 and 1986, the governments of Australia, Denmark, Greece, Luxembourg, and New Zealand had officially adopted a "no nuclear" policy. In addition, since 1980, ten countries have canceled nuclear plant orders or mothballed plants under construction. Argentina canceled four plants; Brazil, eight; China, eight; Mexico, eighteen; and the United States, fifty-four. There have been no orders for new plants in the United States since 1974. In addition, Sweden, Austria, and the Philippines have decided to phase out and dismantle their nuclear power plants.

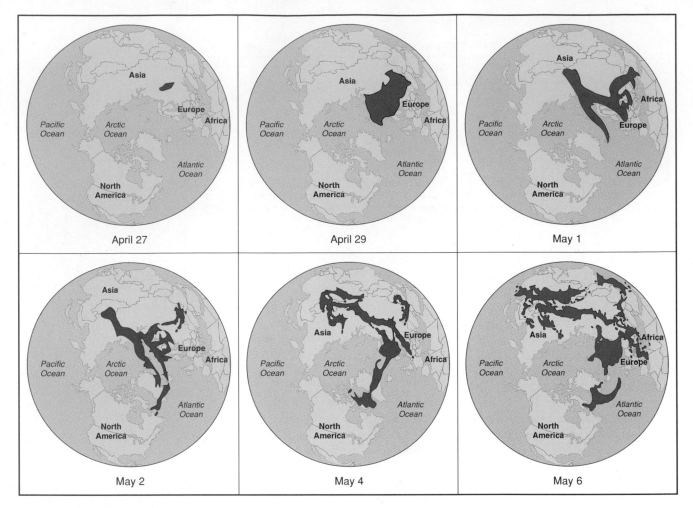

FIGURE 10.12

Radioactive Fallout from Chernobyl.
Many European countries lost millions of
dollars because the crops and livestock
exposed to the fallout from Chernobyl
were considered unsafe to consume.

Before Chernobyl, 65 percent of the people in the United Kingdom were opposed to nuclear plants; after Chernobyl, 83 percent were against them. In Germany, opposition increased from 46 percent to 83 percent. Opposition in the United States rose from 67 to 78 percent. (See figure 10.13.) Even the French, who have the greatest commitment to nuclear power, were against it by 52 percent. Nuclear fission has not lived up to the promises put forth in the "Atoms for Peace" program. In fact, many are predicting that the amount of energy furnished by nuclear fission reactors will actually decline in the future. With the present political and economic conditions regarding nuclear fission, the future of two newer technologies—breeder reactors and nuclear fusion—also appear bleak.

Breeder Reactors

During the early stages of the development of nuclear power plants, breeder reactor construction was seen as the logical step after nuclear fission development. A regular fission reactor produces heat to generate electricity but does not form radioactive products useful as a fuel. A **nuclear breeder reactor** is a nuclear fission reactor that produces heat to be converted to steam to generate electricity and also forms a new supply of radioactive isotopes. If a fast-moving neutron hits a nonradioactive uranium-238 (U-238) nucleus and is absorbed, an atom of radioactive **plutonium-239**

FIGURE 10.14

Formation of Pu-239 in a Breeder Reactor. When a fast-moving neutron (N) is absorbed by the nucleus of a U-238 atom, a series of reactions results in the formation of Pu-239 from U-238. Two intermediate atoms are U-239 and Np-239, which release beta particles (electrons) from their nuclei. (A neutron in the nucleus can release a beta particle and become a proton.) This is an important reaction because, while the U-238 does not disintegrate readily and, therefore, is not a nuclear fuel, the Pu-239 is fissionable and can serve as a nuclear fuel.

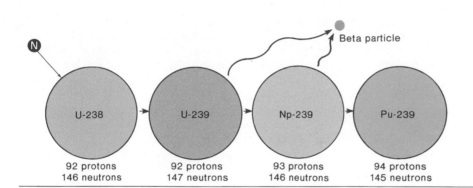

(Pu-239) is produced. (See figure 10.14.) In a breeder reactor, water is not used as a moderator because water slows the neutrons needed to produce the Pu-239 too much. A moderator that allows the neutrons to move more rapidly and that also has good heat transfer properties is required.

The **liquid metal fast-breeder reactor (LMFBR)** appears to be the most promising model. In this type of reactor, the fuel rods in the core are surrounded by rods of U-238 and liquid sodium. The energy of the neutrons released from U-235 in the fuel rods heats the sodium to a temperature of 620°C. In addition to furnishing heat, these fast-moving neutrons are absorbed by the rods containing U-238, and some of these atoms are converted into Pu-239. After approximately ten years of operation during which electricity is produced, the LMFBR will have also produced enough radioactive material to operate a second reactor.

There are some serious drawbacks to the LMFBR. Sodium reacts explosively if allowed to come into contact with water or air. Therefore, costly, highly specialized equipment is required to contain and pump the sodium. If these systems fail, the sodium boils, which would allow the chain reaction to proceed at a faster rate and could damage the reactor, leading to a nuclear accident.

There are also problems in the startup of a LMFBR. When a breeder reactor is starting up or going on-line after a shutdown, the solid sodium cannot be moved by the pumps until it becomes a liquid. This presents a technical problem in the development of a LMFBR.

Another problem is that reaction rates are extremely rapid and very difficult to regulate. The instrumentation needed to monitor a LMFBR must be of extremely high quality because control of the reaction involves precise adjustments over very short periods of time.

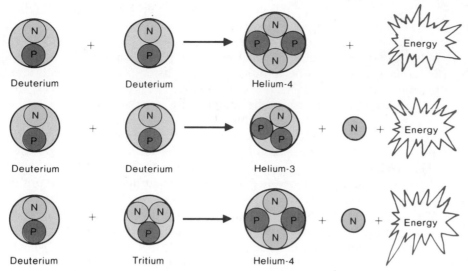

FIGURE 10.15

Nuclear Fusion. In nuclear fusion, small atomic nuclei are combined to form heavier nuclei. Large amounts of energy are released when this occurs. Different isotopes of hydrogen can be used in the process of fusion. There are three possible types of fusion: (1) Two deuterium isotopes can combine to form helium-4 and energy; (2) two deuterium isotopes can combine to form helium-3, a free neutron, and energy; and (3) a deuterium and a tritium isotope can combine to form helium-4, a free neutron, and energy.

Finally, the product of a breeder reactor, plutonium-239, is extremely hazardous to humans who come in contact with it. Additionally, because plutonium-239 can be made into nuclear weapons, very close security must be practiced whenever plutonium-239 is transported, processed, or produced. The larger the number of breeder reactors in use, the more difficult the security problems become and the more likely that the small amount needed to manufacture a bomb could be stolen.

Because of these problems, no breeder reactors are scheduled for commercial use in the United States. In fact, the only experimental breeder reactor in the United States is scheduled to be shut down. The Clinch River Fast-Breeder Reactor Plant near Knoxville, Tennessee, which was to be operational in the 1970s, never was completed.

In Europe, as in the United States, there has been a slowdown in the development of breeder reactors. In 1981, after running smoothly for eight years, a 250-megawatt French prototype plant developed a leak in the cooling system that resulted in a sodium fire. This raised some serious questions regarding the safety of breeder reactors. In 1982, the French government announced that it was scaling down the planned construction of breeder reactors from five to one. Today, it is very doubtful if even that one will be built.

Also, as early as 1982, the governments of the Soviet Union, the United Kingdom, Belgium, and the Netherlands cut back on their commitment to breeder reactors. At that time, the reasons were mainly economic. After the accident at Chernobyl, however, political pressure against all types of nuclear plants, including breeder reactors, will be considerable.

Nuclear Fusion

When two lightweight atomic nuclei combine to form a heavier nucleus, a large amount of energy is released. This process is known as **nuclear fusion.** The energy produced by the sun is the result of fusion. Most studies of fusion have involved small atoms like hydrogen. Most hydrogen atoms have one proton and no neutrons in the nucleus. The hydrogen isotope deuterium (H^2) has a neutron and a proton in the nucleus. Tritium (H^3) has a proton and two neutrons. When deuterium and tritium isotopes combine to form heavier atoms, large amounts of energy are released. (See figure 10.15.) The energy that could be released by combining the deuterium in

1 cubic kilometer of ocean water would be greater than that contained in the world's entire supply of fossil fuels.

Although fusion could solve the world's energy problems, technology must answer several questions before fusion power can become a reality. Three factors must be met simultaneously if fusion is to occur: high temperature, adequate density, and confinement. If heat is used to provide the energy necessary for fusion, the temperature must approach that of the center of the sun. At the same time, the walls of the vessel confining the atoms must be protected from the heat, or they will vaporize. However, the main problem is containment of the nuclei. Because they have a positive electric charge, the nuclei repel one another.

Even though, in theory, fusion promises to furnish a large amount of energy, technical difficulties appear to prevent its commercial use anywhere in the near future. Even the governments of nuclear nations are only budgeting modest amounts for fusion research. And, like nuclear fission and the breeder reactor, economic costs and the reaction to Chernobyl may continue to delay the development of fusion reactors.

Summary

Splitting the nucleus of an atom of uranium-235 is a type of nuclear fission. The heat energy released from a nuclear fission chain reaction in a nuclear reactor is used to heat water to produce steam that generates electricity. Various types of nuclear reactors have been constructed, including light-water reactors, heavy-water reactors, and gas-cooled reactors.

The nuclear fuel cycle includes mining uranium, uranium enrichment, nuclear reactors, and the storage of nuclear waste. Potential environmental and safety problems are associated with each step in the nuclear fuel cycle. Safe storage of low-level and high-level radioactive waste remains a problem, as does decommissioning of nuclear power plants.

Public approval of nuclear power plants began to decline in the 1960s. The accident at Chernobyl accelerated mounting concerns regarding nuclear power plants.

The increased cost of constructing nuclear power plants and questions relating to nuclear safety have caused a reduction in the number of nuclear power plants being constructed. These factors have also limited the development of nuclear breeder reactors and nuclear fusion reasearch.

Review Questions

1. How does a nuclear power plant generate electricity?
2. Name the steps in the nuclear fuel cycle.
3. What is a rem?
4. Why is there no specific amount of allowable radiation to which we may be exposed without endangering our health?
5. How will nuclear fuel supplies, the cost of decommissioning facilities, and storage and ultimate disposal of nuclear wastes influence the nuclear power industry?
6. What happened at Chernobyl, and why did it happen?
7. Describe a liquid metal fast-breeder reactor, and explain how it works.
8. How is plutonium-239 produced in a breeder reactor?
9. Why is plutonium-239 considered dangerous?
10. Why is fusion not being used as a source of energy?

PART FOUR
Human Influences on Ecosystems

The success of the human species is a result of our ability to change our surroundings so that we have tools, food, water, and shelter. Our use of mineral resources is important to many aspects of modern society. Chapter 11 discusses how humans exploit mineral resources and use natural ecosystems. Often, this means that existing natural ecosystems are altered significantly or replaced by agricultural ecosystems. In some cases, organisms are driven to extinction by human activities.

Chapter 12 emphasizes the use of the land surface, while chapter 13 discusses the nature and wise use of soil. Land and soil are not the same. Land is the part of the world that is not covered by water, while soil is a thin layer on the land's surface that supports plant growth. Land-use planning must take the nature of the soil into account when land-use decisions are made.

Agriculture allows us to support over five billion people on the earth. The use of the soil, discussed in chapter 13, and the use of pesticides, discussed in chapter 14, are both important to the process of food production. Misuse of the soil or pesticides can lead to the loss of usable farmland or to the poisoning of people.

Water is a necessity for human life. It is important as a nutrient, but also for sustaining agriculture, allowing industrial processes, facilitating transportation, and many other functions. Chapter 15 deals with a wide range of water-use issues and conservation practices.

CHAPTER ELEVEN
Human Impact on Resources and Ecosystems

Objectives

After reading this chapter, you should be able to:

Recognize that humans have an increasing impact on natural ecosystems.

Define pollution.

Differentiate between renewable and nonrenewable resources.

Recognize that mineral resources are unevenly distributed, which creates international trade in these commodities.

List three types of costs associated with mineral exploitation.

Understand that some wilderness areas still have minimal human influence.

Appreciate the ways humans modify forests.

Identify causes of desertification.

Recognize that aquatic systems are modified by terrestrial changes.

Identify changes that occur to aquatic systems as a result of human activity.

Recognize that wildlife management focuses on specific species.

Appreciate that waterfowl management is an international problem.

Recognize that extinction is a natural process.

Recognize that humans increase the rate of extinction.

Identify ways that humans cause extinctions.

Recognize that many extinctions can be prevented if societies are willing to preserve critical habitats and prevent the hunting of endangered species.

Chapter Outline

Key Terms

biodegradable
clear-cutting
cover
desertification
economic costs
endangered species
energy costs
environmental costs
extinction
habitat management

migratory birds
natural resources
nonrenewable resources
patchwork clear-cutting
pollution
reforestation
renewable resources
selective harvesting
threatened species
wilderness

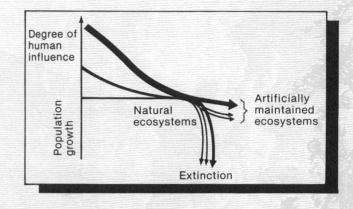

The Changing Role of Human Impact

At one time, a human was just another consumer somewhere in the food chain. Humans fell prey to predators and died as a result of disease and accident just like other animals. The tools available to exploit their surroundings were primitive, so these people did not have a long-term effect on their surroundings. Mineral and energy resources were only minimally exploited. Chert, obsidian, and raw copper were mined for the manufacture of tools, and certain other mineral materials, such as salt, clay, and ocher, were used as nutrients, for making pots, or as pigments. Aside from occasional use of surface seams of coal or oil, most of the energy needs of early humans were filled by biomass.

As human populations grew, and as their tools and systems of use became more advanced, the impact that a single human could have on his or her surroundings increased tremendously. The use of fire was one of the first events that marked the capability of humans to change ecosystems. Although fires were also natural events in many ecosystems, the more frequent use of fire to capture game and to clear land for gardens returned climax communities to earlier successional stages more frequently than normal.

As technology advanced, wood was needed for fuel and building materials, land was cleared for farming, streams were dammed to provide water power, and various mineral resources were exploited to provide energy and build machines. These modifications allowed larger human populations to survive, but always at the expense of previously existing ecosystems.

Today, with over five billion people on the earth, nearly all of the surface of the earth has been affected in some way by human activity. Even the polar ice caps show the effects of human activity. Lead residues from the burning of leaded gasoline and various kinds of organic pollutants can be identified in the layers of ice that build up from the continuous accumulation of snow in these areas. (The amount of lead is currently decreasing because some countries have been making the transition from burning leaded fuel to using unleaded fuel in automobiles.)

Historical Basis of Pollution

Humans have always been concerned about potential threats to their welfare. Ancient people attempted to describe these forces in their folklore and mythology. "The Four Horsemen of the Apocalypse" represents such an attempt: Four riders—famine, pestilence, disease, and war—were pictured as riding across the earth destroying human life. Three of the riders (famine, pestilence, and disease) are natural phenomena. The fourth (war) is a product of human activity.

Throughout history, there have been numerous attempts to eliminate the misery caused by hunger and disease. In general, we rely on science and technology to improve the quality of life. However, technological progress often offers a short-term solution to a specific problem but in the process can create an additional problem: pollution. While scientific advancement and technology attempt to alleviate human misery, they might actually degrade the quality of life for future generations. For example, improving sanitary conditions in Thailand does decrease the spread of waterborne diseases but, in turn, results in an increase in the population, which must then be fed, housed, and clothed. This attempt to improve life results, ultimately, in increased crowding and hunger.

Human Influences on Ecosystems

Not everyone defines pollution in the same way. **Pollution** is sometimes defined as something that people produce in large enough quantities that it interferes with our health or well-being. In many cases, it is difficult to prove whether a substance is causing harm or whether the amount produced is large enough to cause damage.

In 1981, a small city in the U.S. Midwest discovered traces of arsenic in its groundwater supply. It still has not been determined whether trace amounts can be considered hazardous to the health of the local population or what the specific source of the arsenic is. In this case, whether arsenic should be considered a pollutant is unclear.

If we accept the definition of pollution as materials that can be hazardous to health, heavy metals such as arsenic certainly should be considered toxic, but we do not know how much arsenic contamination is allowable before harm occurs. Are the small quantities of arsenic in groundwater great enough that, over a period of time, they accumulate and damage the population? Is one part of the population more susceptible than another part? Will children with smaller body masses be more affected than adults? How did the arsenic get into the groundwater? Was it the result of intense spraying of apple trees in past generations, waste disposal from an industry no longer in operation, or the dissolution of the parent rock material? The answers to these questions are necessary before arsenic can be labelled a pollutant. Once a material is labelled a pollutant, potential problems connected to the pollutant and ways of dealing with those problems still need to be identified. A first step in this determination is to decide whether the contaminant is merely an annoyance or if it is truly a health hazard. (See figure 11.1.)

Early Roman history cites examples of air and water pollution. However, widespread pollution began more recently. Increases in the human population and the advance of technology are the main factors that contribute to modern pollution. When the human population was small and people lived in a primitive manner, the wastes produced were biological and so dilute that they often did not constitute a pollution problem. People used what was naturally available and did not manufacture many products. Humans, like any other animal, fit into their natural ecosystems. Their waste products were **biodegradable** materials. Biodegradable materials are a source of food for decomposers, the organisms that break the material down into simpler chemicals, such as water and carbon dioxide.

Pollution began when human populations became so concentrated that their waste materials could not be broken down as fast as they were produced. As the human population increased, people began to congregate and establish cities. The release of large amounts of smoke and other forms of waste into the air caused an unhealthy condition because the pollutants were released faster than they could be absorbed by the atmosphere.

Renewable and Nonrenewable Resources

Natural resources are those structures and processes that can be used by humans for their own purposes but cannot be created by them. If the supply of a resource is very large and the demand for it is low, the resource is often thought of as free. Sunlight, land, and air are often not even thought of as natural resources because their supply is so large. If a resource has been consumed or if it was rare initially, it is expensive. Pearls and gold and other precious metals fall into this category. In the past, land was considered a limitless natural resource, but as the population grew and the demand for

Fish kill—an indication of water contamination

Health hazard ↑

Nuclear power plant cooling towers—possible thermal pollution and radiation hazards

Smoke from stack—contains particulate material, which could cause lung problems

Smog—an indication that thermal inversion has kept the air contamination in the valley

Traffic—fumes (HC, NO_x, PAN, etc.) from internal combustion engines cause eye irritation

Feed lot—odor pollution as well as a source of water contamination from surface runoff

Strip—visual pollution, which is an annoyance but not a health hazard

Annoyance ↓

Litter—an indication that we need to become more aware of how we dispose of materials

FIGURE 11.1

Forms of Pollution. Pollution is produced in many forms. Some are major health concerns, whereas others merely annoy.

food, lodging, and transportation increased, we began to realize that land is a finite, nonrenewable resource. (See figure 11.2.) The landscape is also a natural resource, as is readily seen in countries with the proper combination of mountainous terrain and high rainfall, which can be used to generate hydroelectric power. Rivers, forests, scenery, climates, and wildlife populations are additional examples of natural resources. Resources, especially those like land, air, and water, become more valuable as we begin to exploit them more intensively.

Natural resources are usually categorized as either renewable or nonrenewable. **Renewable resources** are those that can be formed or regenerated by natural processes. Soil, vegetation, animal life, air, and water are renewable primarily because they naturally repair and cleanse themselves.

Nonrenewable resources are those that are not replaced by natural processes or whose rate of replacement is so slow as to be ineffective.

FIGURE 11.2
**Mismanagement of a Renewable
Resource.** Although soil is a renewable
resource, extensive use can permanently
damage it. Many of the world's deserts
were formed or extended by unwise use
of farmland. This photograph shows a
once-productive farm, now abandoned to
the wind and sand because the soil was
mistreated and allowed to erode.

Therefore, when nonrenewable resources are used up, they are gone, and a substitute must be found or we must do without. In this chapter and chapter 12, we deal with two aspects of land as a nonrenewable resource: (1) minerals that are not only finite in amount but that are also no longer able to be formed or regenerated; and (2) land surface, which, although nonrenewable, has to some extent multiple and concurrent uses. Minerals and land are resources that require thoughtful decisions about how and where and for what purposes they will be used.

Mineral Resources

Nature does not respect political boundaries. The geological and biological processes that caused the formation of a mineral resource in specific places occurred many millions of years ago. Subsequently, landmasses have been divided into various political entities. Some have a greater wealth of mineral resources than others. (See figure 11.3.) Because no country has within its territorial limits all of the mineral resources it needs, an international exchange has developed. In particular, the industrially developed countries import many of the things they need from countries with mineral resources but without economic resources to develop them. Table 11.1 shows several minerals and the countries that supply them. The United States lacks a large number of materials and must import them from other countries, often in less developed areas of the world.

Not only are mineral resources unevenly distributed, but those that are easiest to use and are the least costly to extract have been exploited. Therefore, if we continue to use mineral resources, they will be harder to find and more costly to develop. As with energy, the United States is one of the primary consumers of the world's mineral resources. Reasonable estimates are that the United States consumes about 30 percent of the minerals produced in the world each year, which is a disproportionate share, given that U.S. population is less than 5 percent of the world's population.

Costs Associated with Mineral Exploitation

Costs are always associated with the exploitation of any natural resource. These costs fall into three different categories. First, the **economic costs**

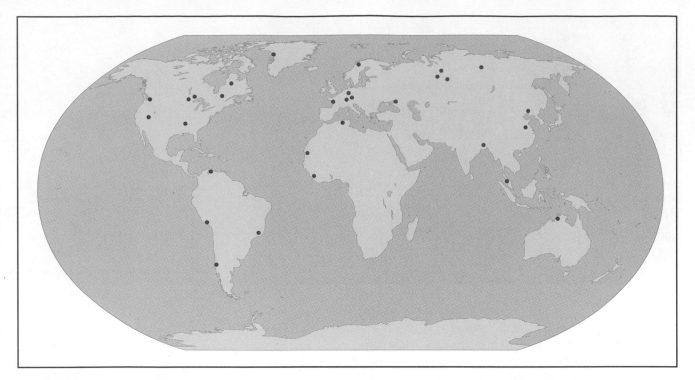

FIGURE 11.3

Distribution of Iron Ore Mining Areas. Obviously, most countries in the world have no active iron mines. Although mines do not exactly correspond with the distribution of ore bodies, the map does give an indication of the scattered distribution of iron ore deposits throughout the world.

TABLE 11.1
Distribution of some mineral resources.

Mineral	% consumed in U.S. that was imported	Major supplier countries (1982–1985)
Columbium	100	Brazil, Canada, Thailand
Mica (sheet)	100	India, Belgium, France
Strontium	100	Mexico, Spain
Manganese	100	South Africa, Gabon, Brazil
Tantalum	91	Thailand, Brazil, Australia
Bauxite	96	Jamaica, Guinea, Brazil
Cobalt	85	Zaire, Zambia, Canada
Chromium	79	South Africa, Zimbabwe, Turkey
Platinum group	92	South Africa, USSR, United Kingdom
Tin	74	Malaysia, Thailand, Indonesia, Bolivia
Nickel	75	Canada, Australia, Botswana
Potassium	77	Canada, Israel
Cadmium	66	Canada, Australia, Mexico
Zinc	73	Canada, Mexico, Honduras
Mercury	50	Algeria, Spain, Japan

Source: Data from *Statistical Abstract of the United States,* 1989.

are those monetary costs necessary to exploit the resource. Money is needed to lease or buy land, build equipment, pay for labor, and buy the energy necessary to run the equipment.

A second category is the **energy cost** of exploiting the resource. As the cost of energy rises, some currently profitable processes may become uneconomic. For example, today it may make sense economically to talk about extracting oil from shale. But if more energy must be invested than is actually obtained, the extraction is not a practical endeavor. As fossil fuels become rarer, economic costs will rise, and current thinking may need to be reevaluated.

A third way to look at costs is in terms of environmental effects. Air pollution, water pollution, animal extinction, and loss of scenic quality are all **environmental costs** of resource exploitation. Environmental costs are often deferred costs. They may not even be recognized as costs at first but become important after several years. Environmental costs are also often represented by lost opportunities or lost values because the resource could not be used for another purpose. Environmental costs are being converted to economic costs as more strict controls on the pollution of the environment are enacted and enforced. It takes money to clean up polluted water and air, or to reclaim land that has been removed from biological production by mining.

These three categories of costs (economic, energy, and environmental) are associated with the several steps that lead from the mineral source in its undisturbed state to the manufacture of a finished product. These steps are exploration, mining, refining, transportation, and manufacturing.

Steps in Mineral Exploitation

The costs associated with locating new sources of minerals (exploration) are primarily economic because exploration takes time and new technology. There are also some energy costs and some very small environmental costs. As the better sources of mineral resources are used up, it will be necessary to look for minerals in areas that are more difficult to explore, such as under the oceans. Therefore, both economic and energy costs will increase. In addition, some areas, such as national parks and preserves, have been off-limits to mineral exploration. As current reserves of mineral resources are used up, pressures will build to explore in these protected areas, resulting in increased environmental costs.

Once a mineral resource has been located and the decision made to exploit it, the resource must be taken from the earth. Mining involves large expenditures of money to pay for labor and the construction of machines and equipment. Energy must be purchased for operation.

In addition to the economic and energy costs are significant environmental costs. Mining affects the environment in several ways. All mining operations involve the separation of the valuable mineral from the surrounding rock. The surrounding rock must then be disposed of in some way. These mine tailings are usually piled on the surface of the earth, where they present an eyesore. It is also difficult to get vegetation to grow on these deposits. Some mine tailings contain materials (such as asbestos, arsenic, lead, and radioactive materials) that can be harmful to humans and other living things.

Many types of mining operations require vast quantities of water for the extraction process. The quality of this water is degraded, so it is unsuitable for drinking, irrigation, or recreation. Since mining disturbs the natural vegetation in an area, water may carry soil particles into streams and cause erosion and siltation. Some mining operations, such as strip mining, rearrange the top layers of the soil, which lessens or eliminates its productivity for a long period of time. (See figure 11.4.) Strip mining has disturbed approximately 75,000 square kilometers of U.S. land, an area equivalent to the state of Maine.

After ore is mined, it must be processed to remove the desired materials from the rocks in which they are embedded. Table 11.2 shows the

FIGURE 11.4

A Strip-Mining Operation. It is easy to see the important impact a mine of this type has on the local environment. Unfortunately, many mining operations are associated with areas that are also known for their scenic beauty.

TABLE 11.2
Metals in ores.

Metal	% metal needed in the ore for profitable extraction	Price range per kilogram
Iron	30.0	
Chromium	30.0	Less than a dollar
Aluminium	20.0	
Nickel	1.5	
Tin	1.0	Several dollars
Copper	0.5	
Uranium	0.1	Tens of dollars

percentage of desired material present in several ores. Ore processing requires economic costs: physical facilities and equipment must be built, people must be employed, and fuel must be purchased for the operation. The energy costs are also significant because refining is basically the process of concentrating the desired material. Natural physical processes tend to disperse materials; therefore, energy must be expended to concentrate them. In many cases, this energy expenditure is the major cost associated with obtaining the mineral. As energy costs increase and ores become more diluted, the economics of producing some materials may change so drastically that substitute materials that require less energy expenditure may need to be sought.

Extracting materials from ores also involves environmental costs in the form of air and water pollution. These environmental costs are being converted to economic costs as regulations on industry require less environmental damage than had been previously tolerated.

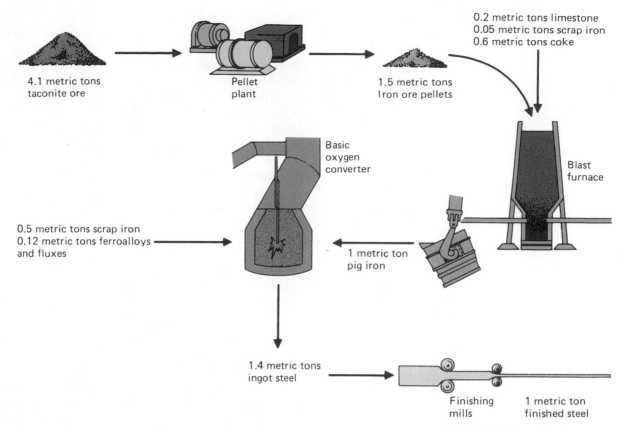

0.2 metric tons limestone
0.05 metric tons scrap iron
0.6 metric tons coke

4.1 metric tons
taconite ore

Pellet
plant

1.5 metric tons
Iron ore pellets

Blast
furnace

Basic
oxygen
converter

0.5 metric tons scrap iron
0.12 metric tons ferroalloys
and fluxes

1 metric ton
pig iron

1.4 metric tons
ingot steel

Finishing
mills

1 metric ton
finished steel

FIGURE 11.5

Steps in Steel Manufacture. Each step involves the production of pollutants, labor and capital costs, and the expenditure of energy to concentrate the ore.

Transportation is another component of the overall cost of extracting minerals from the earth. Transportation is involved in the actual mining process, in getting the ore to the refinery, and in moving the concentrated mineral to the site where it will be made into a finished product. Transportation costs are primarily in the form of money and energy.

Finally, there is the cost of making the finished product from the concentrated ore. Figure 11.5 reviews this entire process by looking at what happens from the time iron atoms are mined until finished steel is produced.

Recycling As an Alternative

A full appreciation of the mineral resources situation requires consideration of the concept of recycling. Many minerals extracted from the earth are not actually consumed or used up; they are just temporarily within a structure or a process. When the mineral is reclaimed and used over for another structure or process, it is recycled.

Empty aluminum beverage cans are no longer useful. However, the aluminum atom can be reprocessed into new cans or other aluminum products. The obsolete items must be recaptured before they have dispersed into the environment. For example, waste oil is relatively easy to recapture at the gasoline station where engine oil is replaced, but difficult to recapture if the oil is dumped on the ground or into a sewer.

Costs from all categories (economic, energy, and environmental) are involved in reprocessing, but the costs are often less because mining or refining is unnecessary. In many industries, however, the economic cost is still lowest if new materials are mined and refined rather than if used materials are reprocessed. For example, disassembling a building and reusing its parts is more costly than using new construction materials. However, when the energy cost and environmental costs are taken into account, recycling is generally a less expensive alternative.

Government subsidies can reduce the profitability of recycling. Government subsidies occur in a variety of forms. Some are direct subsidies, such as special lower prices for the transportation of new materials and price penalties for the transportation of used materials. Most metal ores are shipped at special preferred low rates. An indirect subsidy is apparent in government purchases of huge quantities of goods that must meet very exact specifications. These specifications often discourage the use of recycled materials. Indirect subsidies have as great an effect as many direct subsidies.

Another reason recycling is not more common is that, historically, the monetary costs for energy have been extremely low. As the cost of energy continues to rise, more attention will be given to recycling minerals rather than mining new ones. Also, we all pay for the environmental cost of altering our surroundings. This is reflected in the price tag of anything we buy. Manufacturing products in an environmentally harmonious way will become economically advantageous.

Areas with Minimal Human Impact—Wilderness and Remote Areas

There are still many areas of the world that have had minimal human impact. Some of these are remote areas with harsh environmental conditions, such as the continent of Antarctica, northern arctic areas, the tops of some tall mountains, or extremely arid areas, such as desert areas of Africa, central Australia, and central Asia. Most of these areas have been explored and found to be too harsh to sustain human habitation because growing food would be impossible.

Many other areas of the world, such as tropical rain forests, currently support ecosystems that are little affected by humans but are under threat because they are capable of supporting agriculture or other human uses. The primary factor that will determine the survival of these natural areas is population pressure. As the human population grows, it will need more food and more space. This ultimately leads to the destruction of natural ecosystems and their replacement by human modified ecosystems. Many countries have established parks and other special designations of land use to protect specific areas of natural beauty or communities of organisms thought worthy of protection. This has been most noticeable in Africa, Central and South America, North America, and Australia. Until recently, these continents had relatively large amounts of land that were relatively untouched. (See figure 11.6.)

Europe and the British Isles have almost none of the original vegetation remaining. The forests were cut to provide for shipbuilding, fuel, and agriculture. The forests that remain are small remnants of the original communities and have been modified by the introduction of many exotic species of plants and animals that have replaced the original inhabitants. The pressures to modify the environment were greatest in this area because of the large population and the development of the Industrial Revolution.

Parks and protected lands allow a variety of uses, depending on the country and its laws. Some are used primarily as tourist attractions, which generate funds for the use of the country. Others are set up primarily to protect a particular species or community of organisms, while others simply

Human Influences on Ecosystems

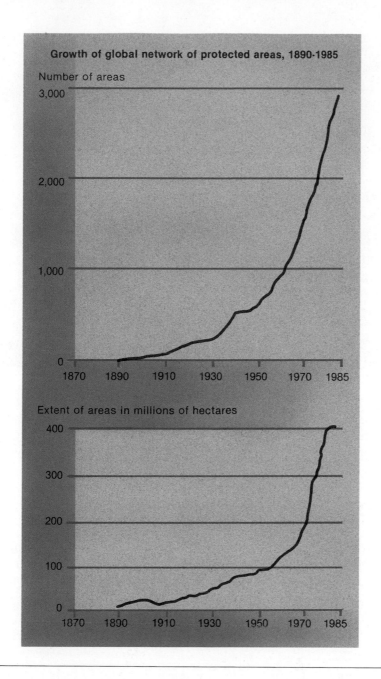

Growth of global network of protected areas, 1890-1985

Number of areas

(Graph showing number of areas from 0 to 3,000 on the y-axis, years 1870 to 1985 on the x-axis)

Extent of areas in millions of hectares

(Graph showing extent from 0 to 400 on the y-axis, years 1870 to 1985 on the x-axis)

FIGURE 11.6

Growth in Major Park and Wilderness Areas of the World. Although there has been a steady increase in the number of protected areas and the number of hectares protected, many of the protected areas are in the more developed countries that can afford to set aside land for nonproductive uses.

seek to restrict the use of the area so that certain standards are met. A particularly sensitive issue is the designation of certain areas as wilderness areas. Although definitions of wilderness may vary from country to country, the U.S. Congress, in the Wilderness Act of 1964, defined **wilderness** as "an area where the earth and its community of life are untrampled by man, where man himself is a visitor who does not remain."

Recently, there has been intense pressure from within the U.S. federal government to allow oil exploration activities to be conducted in areas currently designated as wilderness, such as the Alaskan Arctic National Wildlife Range. As population increases and resources become more scarce, this pressure will become greater throughout the world.

FIGURE 11.7

A Specialized Marsupial—the Numbat. This small marsupial mammal requires termite- and ant-infested trees for its survival. The insects serve as food and the hollow limbs and logs provide hiding places. Loss of old-growth forests with diseased trees will lead to the numbat's extinction.

The oceans of the world represent a major area that has been modified very little by human activities. Some areas receive intense use, particularly areas near the shore, but many areas of the open ocean are poorly understood and little affected by humans.

Ecosystems Modified by Human Use

Humans modify natural ecosystems in several fundamental ways. The removal of some individuals or species from the ecosystem is typical. Management to encourage the reproduction of certain species and the introduction of exotic species are other common activities. Typically, the most productive natural ecosystems are the first to be modified by human use and are the most intensely managed.

Forests

Most forest ecosystems have been extensively modified by human activity. Originally, almost half of the United States, three-fourths of Canada, almost all of Europe, and significant portions of the rest of the world were forested. The forests were removed for fuel, building materials, to clear land for farming, and just because they were in the way. This activity returned the forest to an earlier successional stage, which resulted in the loss of certain animal and plant species that required mature forests as part of their niche.

Today in Australia, a small squirrel-sized marsupial known as a numbat is totally dependent on old forests that have termite- and ant-infested trees. The insects are numbats' major source of food, and the hollow trees and limbs caused by the insects' activities provide numbats with places to hide. Clearly, tree harvesting will have a negative impact on this species, which is already in danger of extinction. (See figure 11.7.)

Human Influences on Ecosystems

TABLE 11.3
Distribution of world forest resources, 1985.

Region of the world	Millions of hectares of closed forest	% of the world total
Latin America	692	24.2
USSR	792	27.6
North America	469	16.4
Asia	469	16.4
Africa	218	7.6
Europe	153	5.3
Australia, and Pacific area	72	2.5
World total	2865	100.0

Source: Data from the Food and Agricultural Organization of the UN.

TABLE 11.4
World wood production.

Region of the world	Production in million cubic meters of round wood—1986	% of world total
Latin America	314.7	9.7
USSR	377.6	11.6
North America	723.2	22.2
United States	484.5	14.9
Canada	180.5	5.5
Other	58.2	1.8
Asia	998.1	30.7
Africa	449.5	13.8
Europe	351.3	10.8
Pacific area	38	1.2
World total	3,252.4	100.0

Source: Data from the *United Nations Statistical Yearbook*, 1986.

Because of increasing world population, forested areas are being placed under pressure to provide the needed firewood, lumber, and agricultural land resources. These same forests are important refuges for many species of plants and animals, and are often essential as protectors of watersheds. If the trees are removed, flooding is more common, and much valuable water is lost as runoff.

In general, all continents contain significant amounts of forested land. Table 11.3 shows that North America, Latin America, and the USSR contain nearly 70 percent of the densely forested land in the world. However, these three regions are responsible for less than 45 percent of the wood products produced. (See table 11.4.) The discrepancy between the distribution of forest resources and the production of wood products exists because Latin America produces little wood relative to its capacity. Asia, on the other hand, produces over 30 percent of the world's wood products with only 16 percent of the world's growing timber. The USSR produces 12 percent of the world's wood with 28 percent of the world's growing timber.

BOX 11.1
The Northern Spotted Owl

In June 1990, the northern spotted owl was listed as a threatened species. Threatened species are those that are likely to become endangered if current conditions do not change. The northern spotted owl is found in old-growth coniferous forests in the Pacific Northwest. As with most carnivores at the top of the food chain, the northern spotted owl requires large areas of forest for hunting. The listing of the owl as threatened requires that the U.S. Forest Service develop and implement a plan to protect the owl. Since the owl is only found in mature old-growth forests, and since it requires large areas for hunting, the plan must include the setting aside of large tracts of forested land and protecting it from logging.

Logging in the Pacific Northwest has already used up most of the trees on private land. Most of the remaining old-growth forest, which is about 10 percent of its original area, is on land managed by the U.S. Forest Service. Many of the trees in these forests are several centuries old, and forest replacement following logging takes an extremely long time. This time scale would not allow for the continued existence of the northern spotted owl if major sections of the remaining old-growth forest were logged.

The timber industry is an important element in the economy of the Pacific Northwest. The U.S. Forest Service had been selling about 300 square kilometers

of forested land per year. If the old-growth forest is not available for harvest, many people will lose their jobs, and the federal government will lose over $200 million per year. Those who make their living from the forest argue that there are already adequate wilderness areas and areas unfit for logging to supply the northern spotted owl with all the space it needs. They also argue that their livelihood is endangered in a way similar to the owl, and ask who is more important.

The areas of the world that could increase timber production are the tropical forests and the northern forests of the USSR and Canada. Both of these areas present some problems. The far north of the USSR and Canada is largely forested, but these remote areas would require extensive transportation expansion to allow for their exploitation. In addition, these are slow-growing coniferous forests, so there are questions about how intensively these forests can be exploited before the point is reached when they are not growing as fast as they are being harvested. Good forest management seeks a sustained yield from the forest. This requires that the rate of harvest be approximately the same as the rate of regrowth.

The tropical forests of Asia, South America, and Africa represent another underutilized source of timber. However, tropical forests have a diverse mixture of tree species that require different kinds of harvesting techniques than have traditionally been used in the northern temperate forests of the world. Also, tropical forests are not as likely to regenerate after logging as are many of the temperate forests. If they do not regenerate following logging, they must be considered nonrenewable resources. If these forests are to be used, a new set of forestry principles will need to be developed to allow for the establishment of a renewable tropical forest industry. Currently, few of these forests are being managed for long-term productivity; they are being harvested on a short-term economic basis only,

Human Influences on Ecosystems

as if they were nonrenewable resources. Tropical forests in Central America have been reduced by nearly 50 percent in the last forty years, while those in Africa have declined over 20 percent.

The deforestation of large tracts of land has significant climate impacts. Forested land is very effective in trapping rainfall and preventing its rapid runoff from the surface. Furthermore, the large amount of water transpired from the leaves of trees tends to increase the humidity of the air in forested areas. The shade provided by trees and the evaporation of water also tend to moderate the temperature extremes experienced in the local area. Destruction of huge areas of forest can result in regional climate change.

More recently, people have become concerned about preserving the carbon dioxide trapping potential of forests. Trees trap large amounts of carbon dioxide as a result of photosynthesis and may help to prevent increased carbon dioxide levels that contribute to global warming. See chapter 17 for a discussion of global warming.

Another complicating factor is that human population pressures are the greatest in tropical regions of the world. More people need more food, which means that more forestland will be taken for agriculture, thus reducing the stocks of timber available and also reducing the land available for forest production.

Forest Management Practices

Whenever a resource is exploited, several different interests are put into conflict. Two major ones are economic interests and environmental interests. Economic factors are easy to measure. The cost of exploitation and the financial return for this expenditure are the primary issues. The environmental viewpoint is often difficult to put into monetary terms and must often rely on ethical or biological arguments to temper the economic arguments. Modern forest management practices in many parts of the world involve a compromise between these two points of view.

The forests of the world are known quantities. The economic worth of the standing timber can be assessed, and the value of forests for wildlife and watershed protection can be given a value. However, beyond that, the harvesting of forests changes the ecosystem from its original character. Logging removes the trees and, therefore, the habitat for many kinds of animals that require mature stands of timber. The pine martin, grizzly bear, and wolf all require forested habitat that is relatively untouched by human activity. Human activity also destroys the "wilderness" character of forested areas. An area that has recently been logged is not very scenic, and the roads and other changes necessary to allow the logging process to continue often irreversibly alter the area's wilderness character. Obviously, wilderness and logging cannot coexist. Therefore, it becomes necessary to consciously decide whether a forest is going to be used for the production of timber or if it is to be designated a wilderness area.

Several areas of concern must be addressed when it is decided that a forested area will be logged. Removal of trees from an area exposes the soil to increased erosion. Because the soil's water-holding ability is related to the amount of organic material and roots in the soil, denuded land allows water to run off rather than sink into it. Soil particles are carried by this water and can wash into streams, where they cause siltation. The loss of soil particles reduces the soil's fertility. The particles that enter streams may cover spawning sites and eliminate fish populations. If the trees along the stream are removed, the water will be warmed by the increased sunlight. This may also have a negative effect on the fish population.

FIGURE 11.8

Clear-Cutting. Clear-cutting is the forest harvesting method favored by lumbering interests because it is the most economic way to remove trees. However, clear-cutting exposes the soil to erosion and should only be used on sites that have little slope and where regrowth is rapid.

The roads necessary for moving equipment and removing the logs constitute another problem. Constant travel over these roads removes any vegetation and exposes the bare soil to more rapid erosion. When the roads are not properly located and constructed, they eventually become gulleys and serve as channels for the flow of water. Most of these environmental problems can be minimized by using properly engineered roads and appropriate harvesting methods.

One of the most controversial forest-logging practices is **clear-cutting.** (See figure 11.8.) As the name implies, all of the trees in a large area are removed, which is a very economical method of harvesting trees. However, clear-cutting exposes the soil to significant erosive forces. If done in large blocks, it may slow reestablishment of forest and have significant effects on wildlife.

On some sites, clear-cutting is still a reasonable method of harvesting trees. Sites with gentle slopes, where regrowth is rapid, have little erosion, and environmental damage is limited. This is especially true if any streams in the area are left with a border of mature trees. The roots of the trees help to retain the stream banks and retard siltation. The shade provided by the trees also helps to prevent warming of the water, which might be detrimental to some fish species.

Clear-cutting can be very destructive on sites with steep slopes or where regrowth is slow. Under these circumstances, it may be possible to use **patchwork clear-cutting.** With this method of harvesting timber, smaller areas are clear-cut among patches of untouched forest. This reduces many of the problems associated with clear-cutting and can also improve conditions for many species of game animals that flourish in successional forests, but not in mature forests. Deer, grouse, and rabbits are examples of animals that benefit from a mixture of mature forest and early-stage successional forest.

Sites where natural reseeding or regrowth is slow may need to be replanted with trees, a process called **reforestation.** Reforestation is especially important in many of the conifer species, which often require bare soil to become established. Many of the deciduous trees will resprout from stumps and grow quickly from the seeds that litter the forest floor. Therefore, reforestation is not as important in deciduous forests.

Selective harvesting of some species of trees is also possible but is not as efficient as other methods of harvesting from the point of view of the timber harvesters. It allows, however, for individual, mature, high-value trees to be harvested without causing much change to the forest ecosystem.

The previous discussion shows that forestlands can be managed to meet a variety of needs. The forest products industry would prefer even-aged stands of a single species of tree to enable the most efficient harvest. Wildlife managers would prefer to see forests with a variety of tree species at different stages of maturity to encourage a variety of species of wildlife. A single-species, even-aged forest does not support as wide a variety of wildlife as a mixed-age forest. For example, in the southern pine forests of the United States, the red-cockaded woodpecker requires old, diseased pine trees in which to build its nests. (See figure 11.9.) A well-managed forest plantation does not provide these sites. Proper planning and effort allow integration of both of these uses of the forest. Tree harvesting can be done in small enough patches to encourage wildlife, but in patches large enough to be economical. Some older trees, or some patches of old-stand timber, can be left as refuges for species that require them as a part of their niche.

Many lumber companies maintain forests as crops and seek to manage forests in the same way farmers manage crops. Single species, even-aged forests are often planted because this makes harvesting more efficient. Often, the forests are planted with fast-growing hybrid trees that have been developed in the same way that high-yielding agricultural crops have been. Competing species are controlled by the use of fire in some forests, and insects are controlled by aerial spraying. In these intensively managed forests, some single-species forest plantations mature to harvestable size in twenty years, rather than in the approximately one hundred years that is typical for naturally reproducing mixed forests. However, the quality of the lumber products is reduced for many purposes. Such single-species forests also have a decreased variety of organisms associated with them since different species of wildlife have differing food and cover requirements.

The trees planted in many managed forests may be exotic species. *Eucalyptus* forests have been planted in South America, and most of the forests in northern England and Scotland have a mixture of native pines and imported species from the European mainland and North America.

Road construction, fire control, and control of fungus and insect pests modify forests used for timber purposes. For this reason, many people are attempting to protect forests being exploited for the first time. Of particular concern are the tropical rain forests that are under intense pressure to provide both wood products for the economies of the nations that possess these forests and land that can temporarily be used for agriculture. (See box 5.1.)

FIGURE 11.9

Red-Cockaded Woodpecker. The red-cockaded woodpecker requires old, living pines that have a disease known as red heart in which to build its nest. Diseased, old trees are rare in intensely managed forests. Therefore, the birds are endangered.

Rangelands

Many of the arid and semiarid lands of the world have been converted to raising domesticated or semidomesticated animals in permanent open ranges or nomadic herding. (See figure 11.10.) Most of these areas will not support agriculture but can support sparse populations of native or introduced species of grazing animals. These areas have been modified because the grazing animals have selective eating habits, which tend to reduce certain species of plants and encourage others.

In most of these areas, the animals that graze are exotic, nonnative species. Sheep, cattle, and goats that are native to Europe and Asia have been introduced into the Americas, Australia, New Zealand, and many areas of Africa. The introduction of exotic species has reduced the numbers of certain native species. For example, sheep and cattle have replaced the bison in North America, and many native antelope species have been negatively

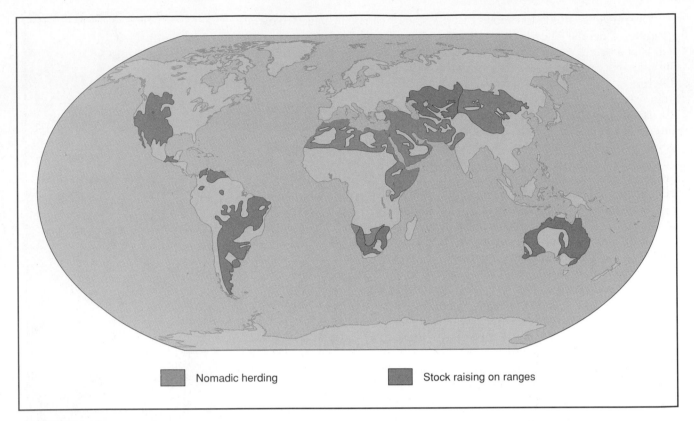

Nomadic herding

Stock raising on ranges

FIGURE 11.10

Nomadic Herding. The arid and semiarid regions of the world will not support farming without irrigation. In many of these areas livestock can be raised. Permanent ranges occur where rainfall is low but regular. Nomadic herders can utilize areas that have irregular, sparse rainfall.

affected by cattle in Africa. In an effort to increase the productivity of rangelands, management techniques may specifically eliminate certain species of plants not useful as food for the grazing animals, or specific grasses may be planted that are not native to the area.

In many parts of the world, where human population pressures are great, overgrazing is a severe problem. As populations increase, people who become desperate for land to farm attempt to graze too many animals on the land and cut down the trees for firewood. If overgrazed, many plants die, and the loss of plant cover allows the soil to begin blowing, resulting in a loss of fertility, which further reduces the land's ability to support vegetation. The cutting of trees for firewood has a similar effect, but it is also significant since many of these trees are legumes, which are important in nitrogen fixation. Therefore, their removal further reduces soil fertility. All of this results in degradation of the land to a more desertlike ecosystem. This process of converting arid and semiarid land to desert because of the improper use of the land by humans is called **desertification.**

Desertification can be found throughout the world but is particularly prevalent in northern Africa, where rainfall is irregular and unpredictable, and where many of the people are subsistence farmers or nomadic herders who are under considerable pressure to provide food for their families. (See figure 11.11.)

Modified Aquatic Ecosystems

For several reasons, the most productive areas of the ocean are those close to land. In shallow water, the entire depth (water column) is exposed to sunlight so that biological productivity is high. The nutrients from the land also tend to make these waters more fertile, the oceans near land tend to

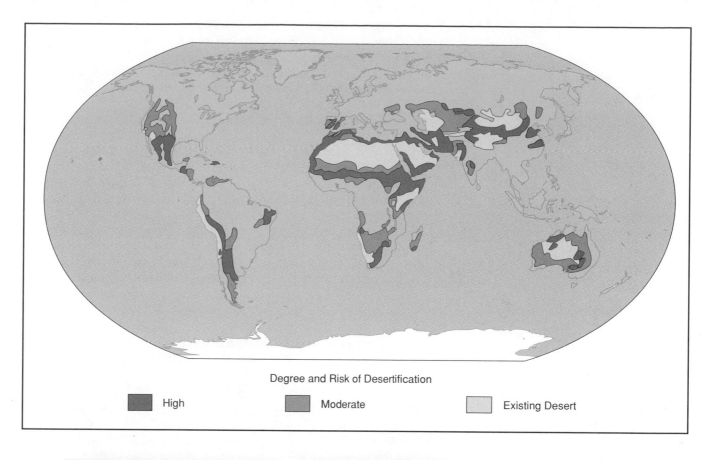

Degree and Risk of Desertification

■ High ■ Moderate □ Existing Desert

FIGURE 11.11

Desertification. Arid and semiarid areas can be converted to deserts by overgrazing or unsuccessful farming practices. The loss of vegetation increases erosion by wind and water, increases the evaporation rate, and reduces the amount of water that infiltrates the soil. All of these conditions encourage the development of desert-like areas.

(Top) Sources: From the *United Nations Map of World Desertification,* United Nations Food and Agriculture Organization. United Nations Educational, Scientific, and Cultural Organization, and the World Meteorological Organization for the United Nations Conference on Desertification, 1977, Nairobi, Kenya.

be shallow, and land masses modify currents that bring nutrients up from the ocean bottom. Many of the commercially important fish species are bottom-dwellers, but fishing for them at great depths is not practical. Therefore, fishing pressure is concentrated on areas where the water is shallow and relatively nutrient rich.

Since commercial fishing removes certain species, it can change species composition in heavily fished areas. The commercial fishing industry has been increasingly attempting to market fish species that previously were regarded as unacceptable to the consumer. These activities are the result of reduced catches of previously desired species. Examples of "newly discovered" fish in this category are monkfish and orange roughy.

Twenty years ago, anchovy fishing off the coast of Peru was a major industry. From 1971 to 1972, the catch dropped dramatically. Overfishing was believed to be one of the major contributing factors. This was aggravated by an increase in the area's water temperature, which prevented nutrient-rich layers from rising to the euphotic zone. Thus, productivity at all trophic levels decreased. Today, productivity in this Peruvian industry is beginning to increase slightly.

Because freshwater ecosystems are small and more intimately associated with human activities, few have not been considerably altered. Modification of water quality and the introduction of exotic species are two primary alterations to freshwater ecosystems. The warming of water because of thermal pollution and the addition of nutrients to the water have made it impossible for some native species to survive. The effects of water pollution are discussed in chapter 15.

The management of fish populations is similar to that of other wild animals. Fish require cover, such as logs, stumps, rocks, and weed beds, so they can escape from predators. They also need special areas for spawning and raising young. This might be a gravel bed in a stream for salmon, a sandy area in a lake for bluegills or bass, or an estuary for many species of marine fish. A fisheries biologist tries to manipulate some of the features of the habitat to enhance them for the desired species of game fish. This might take the form of providing artificial spawning areas or cover for species. Regulation of the fishing season so that the fish have an opportunity to breed is also important.

In addition to these basic concerns, the fisheries biologist gives special attention to water quality. Whenever people use water or disturb the land near the water, water quality is affected. For example, toxic substances kill fish directly, and organic matter in the water may reduce the oxygen present. The use of water by industry or the removal of trees lining a stream warms the water and makes it unsuitable for certain species. Poor watershed management results in siltation, which covers spawning areas, clogs the gills of young fish, and changes the bottom so food organisms cannot live there. As a result, the fisheries biologist is probably as concerned about what happens outside the lake or stream as what happens within it.

The introduction of exotic fish species has greatly affected the naturally occurring freshwater ecosystems. As pointed out in chapter 1, the Great Lakes have been altered considerably by the accidental and purposeful introduction of fish species. The sea lamprey, smelt, carp, alewife, brown trout, and several species of salmon are all new to this ecosystem. (See figure 11.12.) The sea lamprey is parasitic on lake trout and other species and nearly eliminated the native lake-trout population. The lamprey problem was brought under control by the use of a very specific larvicide that kills the immature lamprey in the streams. This technique works because mature lamprey migrate up streams to spawn, and the larvae spend a year or two in the stream before migrating downstream into the lake. Since the lamprey have been partially controlled, lake-trout populations have been increasing. (See figure 11.13.) Recovery of the lake-trout population is particularly desirable, since at one time it was an important commercial species.

Native	Introduced
Brook lamprey	Sea lamprey
Lake trout	Brown trout
Brook trout	Rainbow trout
Whitefish	Pink salmon
Herring	Coho salmon
Ciscoes	Chinook salmon
	Atlantic salmon
Suckers	
Chubs	Carp
Shiners	
Catfish	
Bullheads	
	Smelt
	Alewife
White bass	White perch
Smallmouth bass	Crappies
Rock bass	Sunfish
Yellow perch	
Walleye	

FIGURE 11.12

Native and Introduced Fish Species in the Great Lakes. The Great Lakes have been altered considerably by the introduction of many nonnative fish species. Some were introduced accidentally (lamprey, alewife, and carp), while others were introduced on purpose (salmon, brown trout, rainbow trout).

FIGURE 11.13

The Impact of the Lamprey on Commercial Fishing. The lamprey entered the Great Lakes in 1932. Because it is an external parasite on lake trout, it had a drastic effect on the population of lake trout in the Great Lakes. As a result of programs to prevent the lamprey from reproducing, the number of lamprey has been reduced somewhat, and the lake-trout population is recovering with the aid of stocking programs.

Another accidental introduction to the Great Lakes, the alewife, a small fish of little commercial or sport value, became a problem during the 1960s, when alewife populations were so great that they died in large numbers and littered beaches. Various species of salmon were introduced at about this time in an attempt to control alewife populations and replace the lake-trout population, which had been depressed by the lamprey. While this

BOX 11.2
Native American Fishing Rights

Throughout many parts of the United States, particularly in the Pacific Northwest and the Great Lakes states, a controversy over fishing rights is building. This conflict is unique because it involves treaties that were made 100 to 150 years ago between the U.S. government and Native American nations. It has become a major political, economic, social, and legal issue in some states, such as Washington and Michigan, where it has involved the entire court system, from local courts to the U.S. Supreme Court. The precise question revolves around the interpretation of treaty language. This controversy has, on several occasions, turned to violence and has divided the populations of many communities.

According to the wording of many treaties entered into in the 1800s, the rights of Native Americans to engage in fishing would not be infringed upon by the states. Native Americans claim the treaties give them the legal authority to engage in commercial fishing enterprises even when such fishing may be restricted or banned altogether for the general public.

On the other side are many state officials and sport fishing organizations who believe that Native Americans are seriously endangering populations of such fish as salmon and trout by their uncontrolled netting of fish for commercial purposes. They further argue that many species of fish being netted by Native Americans belong to the entire state and not only to a certain group because the fish are stocked or planted by the state. Another concern involves the fishing techniques used by Native Americans. In the 1800s, when the treaties were signed, commercial fishing technology was limited. Today, however, Native Americans and other commercial fishers use nylon nets, power boats, depth finders, and other technological aids that enable them to catch much larger quantities of fish than they could with traditional fishing practices.

In the early 1970s, when sport fishers complained that stocks of fish in Lake Michigan were being depleted because of Native Americans' gill-net fishing, the state of Michigan tried to regulate Native American fishing. The issue ended up in court, and in 1978, a U.S. district court judge in Grand Rapids, Michigan, upheld fishing rights granted under 1836 and 1855 treaties between two Chippewa tribes and the U.S. government. The federal judge ruled that the state had no authority in the matter because the issue in question involved a federal treaty, and the state did not have the right to make regulations contrary to the treaty. Only Congress had such power. Subsequent to the decision, the tribes and the state negotiated changes in the areas in which the tribes could fish in an effort to reduce the potential for conflict between Native American fishers and sport or other commercial fishers.

A similar case was decided by a U.S. district judge in Tacoma, Washington, who interpreted treaties signed in 1854 and 1855 and ruled that Native Americans could catch up to 50 percent of the salmon that passed through their tribal land on their way to other parts of the state. In the face of protests by the nonnative commercial fishing industry, the federal judge took over the regulation of salmon fishing in the state. Commercial fishers, who outnumbered Native American fishers by a ratio of approximately eight to one, feared for their livelihood. The case eventually went to the U.S. Supreme Court. In 1979, the U.S. Supreme Court ruled, in a six-to-three decision, to uphold the federal treaties that entitled Native Americans to half the salmon caught in the area of Puget Sound of the state of Washington. It further held that the 50 percent figure had to include the salmon that Native Americans caught on their lands for home consumption and religious ceremonies as well as the fish they caught for commercial purposes. The high court also voted to uphold the right of the district court to continue supervising the fishing industry because of the state's resistance to the interpretation of the treaties.

Clearly, the issues surrounding Native American fishing rights are complex and broad in scope. There are purely biological questions involving a resource and its wise use, economic issues involving families' livelihoods, cultural and religious concerns pertaining to Native Americans' use of a resource, legal questions involving federal and state conflicts, and moral questions relating to the unfair treatment of Native Americans in the past. In such conflicts, there is seldom one right answer. While the courts have ruled on the cases in Michigan and Washington, and efforts are being made by the states and the Native Americans to allow for wise use of the fishing resource, the problem still exists and is likely to continue for years to come.

salmon introduction has been an economic success and has generated millions of dollars because of its great contribution to the sport fishing industry, it may have had a negative impact on the populations of some native fish like the lake trout, with which the salmon probably compete. Salmon also migrate up streams, where they disrupt the spawning of native fish. In addition, most species of salmon die after spawning, which causes a local odor problem.

Fish management in much of the world includes management for both recreational and commercial food purposes. Therefore, the fisheries resource manager must try to satisfy two different interest groups. Both the person who fishes for fun and the one who harvests for commercial purposes must adhere to regulations. However, the regulations are often different because the commercial fisher is allowed to use different fishing methods, such as nets.

Streams and rivers are often modified for navigation, irrigation, flood control, or power production purposes, all of which may alter the natural ecosystem. These topics are discussed in greater detail in chapter 15.

Ecosystem Modification for Wildlife Management Purposes

Many kinds of terrestrial wild animals are desired for hunting or other purposes, and efforts are made to improve conditions for these species. Several kinds of techniques, some of which result in ecosystem modifications, are used to enhance certain wildlife populations. These techniques include game and habitat analysis, population census methods, stocking of areas with game species, predator control, establishment of refuges, and habitat management.

Each kind of animal has specific requirements, and management of a particular species requires an understanding of the habitat needs of that species. An animal's habitat must provide the following five requirements: food and water, cover for escaping from enemies, cover for protection against the elements, cover for resting and sleeping, and cover for mating and the rearing of young. **Cover** is a term used to refer to any set of physical features that conceals or protects animals from the elements or enemies.

Animals are highly specific in their habitat requirements; not every area satisfies all demands. For example, bobwhite quail must have a winter supply of food that protrudes above the snow. In addition, quail require a supply of small rocks called grit, which the birds use in their gizzards to help in the grinding of food. During most of the year, they need a field of tall grass and weeds as cover from natural predators. However, such protection is not available during the winter, and a thicket of brush is necessary. The thicket serves as cover from predators and shelter from the cold winter weather. Most areas provide suitable cover for resting, but for sleeping, quail prefer an open, elevated location, which allows the birds to quickly take flight if attacked at night. When raising young, they require grassy areas with some patches of bare ground where the young can sun themselves and dry out if they get wet. All of these requirements must be available within a radius of 400 meters, because this is the extent of a quail's normal daily travels.

Once habitat requirements are known, the desirable population size must be ascertained. The population must be checked regularly to see if it is within acceptable limits. Population censuses are used to provide this data. If the population is below the desired number, organisms may be artificially introduced. This is referred to as stocking. Many of the species that are hunted are actually introduced species, and areas are stocked with these introduced species. Many kinds of game birds are imported species. The

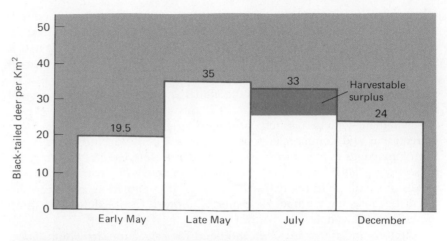

FIGURE 11.14

Managing a Wildlife Population. The seasonal changes in this population of black-tailed deer are typical of many game species. The hunting season is usually timed to occur in the fall so that surplus animals will be harvested prior to winter, when the carrying capacity is lower.

Data from Taber, R. D., and Dasmann, R. F., "The Dynamics of Three Natural Populations of the Deer *Odocoileus hemionus columbianus," Ecology* 38, no. 2 (1957): 233–46.

ringnecked pheasant was originally from Asia but has been introduced into Europe, Great Britain, and North America. All of the large game animals of New Zealand are introduced species, since there were none present in the original biota of these islands. Several species of deer have been introduced into Europe from Asia. Many of these are raised in deer parks, where the animals are very similar to free-ranging domestic cattle. These animals may be hunted for sport or may be slaughtered to provide food.

Since wildlife management often involves the harvesting of animals by hunting for sport and for meat, seasons are usually regulated to assure adequate reproduction and provide the largest possible healthy population during the designated hunting season. Hunting seasons usually occur in the fall because many animals normally die during the winter. Winter taxes the animal's ability to stay warm and is also a time of low food supplies in most temperate regions. A well-managed wildlife resource allows for a large number of animals to be harvested in the fall and still leaves a healthy population to survive the winter and reproduce during the following spring. (See figure 11.14.)

Often, predator or competitor populations have been reduced to allow optimal survival of the game species. In intensely managed European systems, gamekeepers are often charged with the responsibility of killing predators or unwanted competitors. With more wild and less intensely managed species, as is typical in North America, very little predator or competitor control is exercised. At one time, it was thought that populations of game species could be increased substantially if predators were controlled. In Alaska, for example, the salmon-canning industry claimed that bald eagles

were reducing the salmon population, and a bounty was placed on the bald eagle. From 1917 to 1952, 128,000 eagles were killed for bounty money in Alaska. This theory of predator control to increase populations of game species has not proven to be valid in most cases, however, since the predators do not normally take the prime animals anyway. They are more likely to capture sick or injured individuals not suitable for game hunting.

For some species, such as ground nesting birds, it still may make some sense to control predators that eat eggs or capture the young, but in most cases, humans have a greater impact by habitat modification and hunting than do the natural predators. Although at one time they were thought to be helpful, bounties and other forms of predator management have largely been eliminated in North America. Since in many parts of the world, the major predators on game species are humans, regulation of hunting is a form of predator control.

Many species may require the establishment of refuges where they are protected from hunting or have the habitat manipulated to assure their reproduction. Elephants in Africa are often hunted illegally for their ivory. The establishment of protected refuges that are patrolled by armed game wardens is necessary to prevent their extinction in much of Africa. Many native prairie species benefit from the exclusion of introduced grazing animals like cattle and sheep. (See figure 11.15.)

The Kirtland warbler is an endangered species that nests in the central part of Michigan, although evidence now indicates that some of these birds may be mating in the neighboring states of Wisconsin and Minnesota. The nesting cover required for this species is young, even-aged stands of jack pine. The areas where these birds nest have been protected, and, in addition, the habitat has been modified to provide for optimal nesting. Portions of their nesting area are periodically burned to assure future nesting sites. This is important because the jack-pine cone will only open if subjected to fire. So fire is a very important part of the ecological niche of the Kirtland warbler.

The burning of old stands of jack pine to allow for regrowth is an example of **habitat management.** Once the food habits, predators, and cover requirements of a game species are well understood, the habitat can be altered to optimize it for the specific desired species. Habitat management

Quail

	Adults	Young	Total
1st year	2	14	16
2nd year	16	112	128
3rd year	128	896	1024

Deer

	Adults	Yearlings	Fawns	Total
1st year	2	0	2	4
2nd year	2	2	2	6
3rd year	4	2	4	10
4th year	6	4	6	16
5th year	10	6	10	26

FIGURE 11.16

Reproductive Potential. If we assume no mortality, animals have a reproductive capacity far above what is required to just keep the population stable. In the real world, mortality generally keeps populations from growing beyond the ability of their habitat to sustain them.

may take the form of encouraging some species of plants that are the preferred food of the game species. For example, habitat management for deer may involve encouraging the growth of many young trees, saplings, and low-growing shrubs, since deer use these plants for food and cover, by cutting the timber in the area and allowing the natural regrowth to supply the food and cover the deer need. Both forest management and deer management may need to be integrated in this case because some other species of animals, like squirrels, will be excluded if mature trees are cut, since they rely on the seeds of trees as a major food source.

Given suitable habitats and protection, most wild animals can maintain a sizeable population. In general, organisms produce more offspring than can survive. Figure 11.16 illustrates the reproductive potential of both quail and white-tailed deer. High reproductive potential, protection from hunting, and the management and restoration of suitable habitats have reversed the drastic population declines of deer in some parts of the world. In Pennsylvania, where deer were once extinct, the number is now in excess of six hundred thousand. Even in predominately agricultural states, the deer have made a comeback. In Indiana, a herd of nearly one hundred thousand deer has grown from an initial stocking of thirty-five individuals.

Other animals have shown comparable changes in population size. In Zimbabwe, contrary to most African countries, the number of elephants has increased from about two hundred in 1900 to forty thousand in 1987. This is probably not a sustainable population, and the government would like to reduce the number to about thirty thousand. The wild turkey population in the United States has increased from about twenty thousand birds in 1890 to two million birds today.

Waterfowl (ducks, geese, swans, rails, etc.) present some special management problems because they are migratory. **Migratory birds** can fly thousands of kilometers and, therefore, can travel north in the spring to reproduce during the summer months and return to the south when cold weather freezes the ponds, lakes, and streams that serve as their summer homes. (See figure 11.17.) Because many waterfowl nest in Canada and winter in parts of the United States and Central America, an international agreement among the involved countries is necessary to manage and prevent the destruction of this wildlife resource. Habitat management has taken several different forms. In Canada, where much of the breeding occurs,

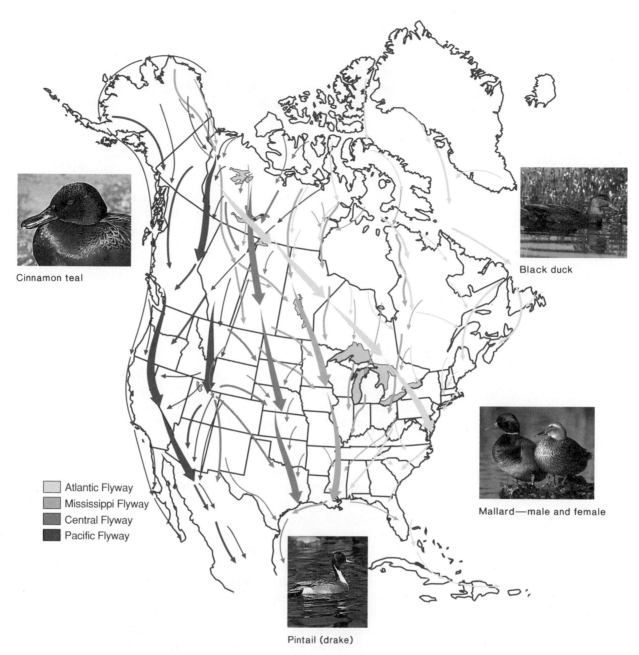

Cinnamon teal

Black duck

Atlantic Flyway
Mississippi Flyway
Central Flyway
Pacific Flyway

Mallard—male and female

Pintail (drake)

FIGURE 11.17
Migration Routes for North American Waterfowl. Migratory waterfowl follow traditional routes when they migrate. These have become known as the Atlantic, Mississippi, Central, and Pacific flyways. Many of these waterfowl are hatched in Canada, migrate through the United States, and winter in the southern United States or Mexico.

government and private organizations have worked to prevent the draining of small ponds and lakes that provide nesting areas for the birds. In addition, new impoundments have been created where it is practical. Because birds migrate southward during the fall hunting season, a series of wildlife refuges provide havens from hunting pressure. In addition, these refuges may be used to raise local populations of birds. During the winter, many of these birds congregate in the southern United States. Refuges in these areas are important overwintering areas where the waterfowl can find food and shelter.

Replacement of Natural Ecosystems with Human-Managed Ecosystems

Over large parts of the world, the natural ecosystems that once existed have been completely destroyed and replaced with intensely managed agricultural ecosystems. This is a natural consequence of human population growth. As the population increased, more and more land was converted to agriculture. Today, very little land can be converted to agriculture. All of the best land has already been converted, leaving only marginal areas that have relatively low productivity. The major ecosystems affected by this conversion to agriculture were temperate forests and grasslands. What today supports wheat, rice, and corn was originally prairie or forest. Chapters 13 and 14 provide a more in-depth look at agricultural ecosystems and how they differ from naturally occurring ecosystems.

Extinction

Extinction is the death of a species, the elimination of all the individuals of a particular kind. It is a natural and common event in the long history of biological evolution. Of the estimated five hundred million species of organisms that are believed to have ever existed on earth since life began, perhaps five to ten million are currently active. This represents an extinction rate of 98 to 99 percent. Obviously, these numbers are estimates, but the fact remains that extinction has been the fate of most species of organisms. However, as some species were going extinct, others were entering the biological arena for the first time.

Each individual organism is a member of a species and shows characteristics that are general for the species and others that are peculiar to the individual. Since a species is made up of individuals with different combinations of characteristics, some individuals are better adapted to fit the prevailing environmental conditions than are others. These more successful individuals are more likely to reproduce and, as a result, their genes are likely to become even more common in future generations. This shift in gene frequencies as a result of different rates of reproductive success can lead to the development of new and separate interbreeding units, which may eventually become new species if they are isolated from the main population for a long enough period of time. (See figure 11.18.) Because the environment is continually changing, a species must adapt or become extinct. If this adaptation occurs over a long period, the original species may evolve into a new species. In this situation, the original species eventually becomes extinct, but before it does, many of its genes (particularly the favorable ones) are passed on to the newly evolving species. An example of this involves the extinction of the dinosaurs. As earth's environment changed, it became less favorable for dinosaurs. However, before their extinction, many of the favorable genes in this reptile line had formed the basis for the newly evolving mammals and birds. This form of gradual extinction and replacement by other, better-adapted groups is common in evolutionary history. The fossil record shows many groups of organisms that flourished for millions of years before more advanced groups replaced them.

Studies of modern local extinctions suggest that certain kinds of species are more likely than others to become extinct. (See table 11.5.) Species populations that already are low are prone to extinction because successful breeding is more difficult than in more numerous species populations. Many organisms at higher trophic levels, such as carnivores, are usually low in numbers compared to their prey species and often have a lower reproductive rate than their prey. Thus, these species find responding to environmental change more difficult.

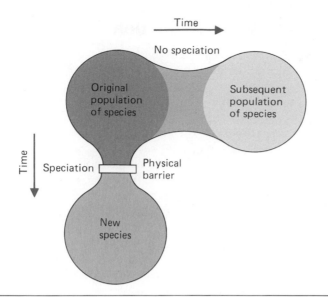

FIGURE 11.18

Speciation. Populations of a species may become isolated from one another for a long time as a result of a barrier. As they encounter new environmental conditions, selection occurs, and the isolated population becomes adapted to local environmental conditions. If isolation is total and lasts for a long time, a new species may develop. If no barrier prevents gene exchange between populations, a new species is not likely to develop, but the population will still adapt to new environmental conditions.

Table 11.5
Probability of becoming extinct.

Most likely to become extinct	Least likely to become extinct
1. Low population density	1. High population density
2. Found in small area	2. Found over large area
3. Specialized niche	3. Generalized niche
4. Low reproductive rates	4. High reproductive rates

Organisms in small, restricted areas are also prone to extinction because an environmental change in their locale can eliminate the entire species at one time. Organisms scattered over large areas are much less likely to have their entire range negatively affected at the same time.

Specialized organisms are also more likely to become extinct than are generalized ones. Since specialized organisms rely on a few key factors in the environment, anything that negatively affects these factors could result in their extinction.

Rabbits and rats are good examples of animals that are not likely to become extinct soon. They have high population density and a wide geographic distribution. In addition, they have high reproductive rates and are generalists that can live under a variety of conditions and use a variety of items as food. Conversely, the cheetah is much more likely to become extinct because it has a low population density, is restricted to certain parts of Africa, has low reproductive rates, and is very specialized in its food habits. It must run down small antelope, in the open, during daylight, by itself. Similarly, the entire wild whooping crane species consists of about 130 individuals who are restricted to small winter and summer ranges that must have isolated marshes. (Captive and experimental populations bring the total number to about 200 individuals.) In addition, the rate of reproduction is low.

Human-Accelerated Extinction

Humans are among the most successful organisms on the face of the earth. We are adaptable, intelligent animals with high reproductive rates and few enemies. As our population increases, we displace other kinds of organisms to support the additional people. This has resulted in an accelerated rate of extinction. Wherever humans become the dominant organism, extinctions occur. Sometimes, we use other animals directly as food. In doing so, we reduce the population of our prey species. Since our population is so large and because we have an advanced technology, catching or killing other animals for food is relatively easy. In some cases, this has led to extinctions. The passenger pigeon in North America, the moas (giant birds) of New Zealand, and the bison and wild cattle of Europe were certainly helped on their way to extinction by people who hunted them for food.

We use organisms for a variety of other purposes in addition to food: Many plants and animals are used as ornaments. Flowers are picked, animals skins are worn, and animal parts are used for their purported aphrodisiac qualities. In the United States, many species of cactus are being severely reduced because people like to have them in their front yards. In other parts of the world, rhinoceros horn is used to make dagger handles or is powdered and sold as an aphrodisiac. Because some people are willing to pay huge amounts of money for these products, unscrupulous people are willing to take the chance of poaching these animals for the quick profit they can realize. These activities have already resulted in local extinctions of some plants and animals and may be a contributing factor to the future extinction of some species.

Some organisms are extinct because they were regarded as pests. Many large predators have been locally exterminated because they prey on the domestic animals that humans use for food. Mountain lions and grizzly bears in North America have been reduced to small, isolated populations, in part because they were hunted to reduce livestock loss. Tigers in Asia and the lion and wolf in Europe were also reduced or eliminated for similar reasons. Even though commercial hunters killed thousands of passenger pigeons, their ultimate extinction was caused primarily by the increased agricultural use of the forests. Passenger pigeons ate the acorns of oaks and the beechnuts from beech, and also relied on the forests for communal nesting sites. When the forests were cleared, the pigeons became pests to farmers, who shot them to protect their crops from being eaten by the birds.

Some extinctions of pest species are considered desirable. Most people would not be disturbed by the extinction of such animals as black widow spiders, mosquitoes, rats, or fleas. In fact, people work hard to drive some species to extinction. For example, in the *Morbidity and Mortality Weekly Report* (26 October 1979), the U.S. Centers for Disease Control triumphantly announced that the virus that causes smallpox was extinct after many years of continuous effort to eliminate it.

The most important cause of extinctions related to human activity involves habitat alteration. (See figure 11.19.) Whenever humans populate an area, they change it by converting the original ecosystem into something that supports the human population. Forests and grasslands have been converted to agricultural and grazing lands. Coupled with this has been the introduction of new species that are useful in agriculture or grazing. These new plants and animals compete with the native organisms for nutrients and living space, and often, the native organisms lose. (See box 11.3.) The African elephant and various species of rhinoceros are in danger because

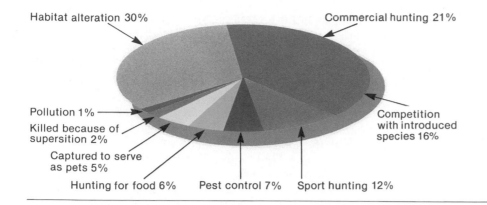

Habitat alteration 30%

Commercial hunting 21%

Pollution 1%

Killed because of supersition 2%

Captured to serve as pets 5%

Hunting for food 6%   Pest control 7%   Sport hunting 12%

Competition with introduced species 16%

FIGURE 11.19

Causes of Extinction. Many organisms have become extinct as a result of human activities. The most important of these is the indirect destruction of habitat by humans modifying the environment for their own purposes.

BOX 11.3
The History of the Bison

The original ecosystem in the central portions of North America was a prairie dominated by a few species of grasses. The eastern prairie, where the moisture was greater, had grasses up to 2 meters tall, while the dryer western grasslands were populated with shorter grasses. Many different kinds of animals lived in this area, including prairie dogs, grasshoppers, many kinds of birds, and bison. The bison was the dominant organism in the area. Millions of these animals roamed the prairie regions of North America with few predators other than the Native Americans, who used the bison for food, their hides for shelter, and their horns for tools and ornaments. The relationship between the bison and Native Americans was a predator-prey relationship in which the humans did not significantly reduce bison numbers.

When Europeans came to North America, they changed this relationship drastically. European-born Americans sought to convert the prairie to agriculture and ranching. However, two things stood in their way: the Native Americans, who resented the intrusion of the "white man" into their territory, and the bison. Since many of the Native American tribes had horses and a history of warlike encounters with other tribes, they attempted to protect their land from this intrusion. The bison was a competitor in the eyes of the settlers, since it was impossible to use the land for agriculture or ranching with the millions of bison occupying so much area. The U.S. government established a policy of controlling the bison and Native Americans: Since bison were the primary food source of Native Americans in many areas, elimination of the bison would result in the starvation of many Native Americans, which would eliminate them as a problem for the frontier settler. In 1874, the secretary of the interior stated that ". . . the civilization of the Indian was impossible while the

buffalo remained on the plains." Another example of this kind of thinking was expressed by a Colonel Dodge, who was quoted as saying, "Kill every buffalo you can, every buffalo dead is an Indian gone." Bison were killed by the millions. Often, only their hides and tongues were taken; the rest of the animal was left to rot. Years later, the bones from these animals were collected and ground up to provide fertilizer. By 1888, the bison was virtually eliminated.

A few bison were left in the Canadian wild and remote mountain areas of the United States, while others were present in small captive herds. Eventually, in the early 1900s, the U.S. government established the national Bison Range near Missoula, Montana. The Canadian government also established a bison reserve in Alberta. These animals have proven to be useful since many people now desire meat with a lower fat content. Crossbreeding cattle with bison has led to a new breed of cattle with reduced fat content known as a beefalo. The animal that was barely saved from extinction by a few thoughtful individuals may contribute to better health for all.

they are hunted for ivory or horn, but it was the original fragmentation of their habitat for agricultural purposes that first placed the animals in conflict with humans.

In much of the tropical world today, rain forests are being cleared to provide grazing land or agricultural land for an expanding human population. This activity results in the destruction of existing forests and their fragmentation into small islands. Scientists studying the effects of this activity have noticed that, as a forest is reduced to small patches, many species of birds disappear from the area. The same kinds of activities certainly happened in Europe, where little of the original forest is left. Similarly, in North America, the eastern deciduous forests were reduced, which probably resulted in the extinctions of some animals and plants that previously existed there. Almost all of the original prairie in the United States has been replaced with agricultural land, resulting in the loss of some species.

Humans also cause extinction in less direct ways. The building of dams changes the character of rivers, making them less suitable for some species. Air and water pollution may kill all life in certain areas. For example, acid precipitation has lowered the pH of some lakes so much that all life has been eliminated. In other cases, indirect human activity may selectively eliminate species that are less tolerant to the pollutants being released, or the introduction of exotic species may eliminate existing species. The accidental introduction of the chestnut blight fungus into the United States resulted in the loss of the chestnut trees over much of the Appalachian region. Some species that were dependent on these trees probably were also eliminated.

Why Worry about Extinction?

If extinction is a natural event, and if those species that become extinct are those that cannot effectively cope with the activities of humans, why should we worry about them? There are several different kinds of answers to that question. First of all, strictly from a selfish point of view, many species we know little about may be useful to us. Many plants have chemicals in them that are used as medicines. If we drive a plant into extinction, we may be eliminating a potentially useful product. Also, as the human population continues to increase, the need for different kinds of food plants and animals will increase. Once they are eliminated, however, we have lost the opportunity to use them for our own ends. Most of the wild ancestors of our most important food grains, like maize (corn), wheat, and rice, are thought to be extinct. What would our world be like if these plants had not been domesticated before they went extinct as wild populations? Because of this concern, many agricultural institutes and universities maintain "gene banks" of wild and primitive stocks of crop plants.

Another interesting idea is that certain organisms in some ecosystems appear to play key or pivotal roles. Accidental extinction of one of these species could be devastating to the ecosystem and the humans that use it. For example, the sardine fishery off the coast of southern California and the anchoveta fishery off the coast of Peru appear to have been fundamentally altered by overfishing. Although the details are not known, it is thought that, once the population was significantly reduced, other organisms filled their niche, making it impossible for them to return to their original populations.

Many people also feel that all species have a fundamental right to exist without being needlessly eliminated by unthinking activity of the human species. This is a philosophical statement that has nothing to do with economic or social value but is an ethical position.

What Is Being Done to Prevent Extinction?

Efforts to prevent human-caused extinctions are difficult to assess. Some countries have enacted legislation to protect species that are in danger of becoming extinct. Those species that receive special protection usually are given some sort of designation, such as endangered or threatened. **Endangered species** are those that are presently in such small numbers that they are in immediate jeopardy of becoming extinct. **Threatened species** could become extinct if a critical factor in their environment were changed.

Most of the active interest in preventing human-caused extinctions comes from developed countries. In developed countries, however, the problem is less acute because vulnerable species have already been eliminated. Extinction is a greater potential problem in tropical, less developed countries, where many biologists estimate that there may be as many species of organisms in the tropical rain forests of the world as in the rest of the world combined. Unfortunately, extinction prevention is not a major issue in many less developed countries. This difference in level of interest is understandable since the developed world has surplus food, higher disposable income, and higher education levels, while people in many of the less developed countries, where population growth is high, are most concerned with immediate needs for food and shelter, not with long-term issues like extinction.

The forests of Central America are already two-thirds gone. In the Amazon basin, about 100 square kilometers of tropical forestland are cleared each year for farms or by logging companies. Because many tropical countries are less developed, many of the logging companies are owned by foreign companies that are only interested in the short-term economic gain from harvesting the forests, not in the long-term ecological health of the forest and the country.

Nevertheless, many of the governments of less developed countries have responded to pressures and suggestions from outside sources and have established preserves, parks, and other protected areas to help protect species in danger of extinction. This does not solve the problem, however, since the areas must be protected from poachers and unauthorized agricultural activity of people who sneak into the preserves. These people are responding to basic biological and economic pressures to provide food for their families. The hiring of security forces to patrol such areas is expensive for many of these countries.

Even in countries where interest levels in extinction prevention are relatively high, there is a cultural bias in favor of the selection of certain kinds of organisms for special endangered species protection. Most endangered or threatened species are birds, mammals, some insects (particularly butterflies), a few mollusks and fish, and certain categories of plants. Bacteria, fungi, reptiles, most insects, and many other inconspicuous organisms rarely show up on endangered species lists, even though they play vital roles in the nitrogen cycle, in the carbon cycle, and as decomposers.

Several international organizations work to prevent the extinction of organisms. The International Union for the Conservation of Nature and Natural Resources (IUCN) estimates that, by the year 2000, at least five hundred thousand species of plants and animals may be biologically exterminated. The IUCN classifies species in danger of extinction into four categories: endangered, vulnerable, rare, and indeterminate. Endangered species are those whose survival is unlikely if the causal factors continue. These organisms need action by people to preserve them, or they will become extinct. Vulnerable species are those that have decreasing populations and will become endangered unless causal factors, such as habitat destruction,

Some animals thrive wherever humans are common. Rats and mice have migrated throughout the world, where they feed on the crops that humans raise. In some areas, as much as 50 percent of the crop may be lost to rodents and insects. In some parts of the world, bounties are given for every rat killed. Skunks, foxes, coyotes, rabbits, and many birds thrive where human activity provides a mixture of woodland, farmland, and land devoted to housing.

Pigeons are common animals in many cities, where predation is reduced and they have access to bits of food left by human activity. Storks in Europe build their large nests of sticks in chimneys, and they are protected by the local people. Many other birds, such as barn swallows, chimney swifts, and the European house sparrow, use human structures (as evidenced by the birds' common names) as places to build their nests.

Peregrine falcons have been introduced into cities, where they nest on window ledges and rooftops and feed on pigeons, which are often considered pests. Many people place birdhouses and bird-feeding stations near their homes. Bird feeders have extended the range of many birds that would not normally be able to survive the winter months.

are stopped. Rare species are primarily those that have small worldwide populations and that could be at risk in the future. Indeterminate species are those that are thought to be extinct, vulnerable, or rare, but so little is known about them that they are impossible to classify.

Although the IUCN is a highly visible international conservation organization, it has very little power to cause change. It generally seeks to protect species in danger by encouraging countries to complete inventories of plants and animals within their borders. They also encourage the training of plant and animal biologists within the countries involved. (There is currently a critical shortage of plant and animal biologists who are familiar with the organisms of the tropics.) The IUCN also encourages the establishment of preserves to protect species in danger of extinction.

The U.S. Endangered Species Act was passed in 1973. This legislation gave the federal government jurisdiction over any species that were designated as endangered. About three hundred U.S. species and subspecies have been so designated by the Office of Endangered Species of the Department of Interior. (See figure 11.20.) The Endangered Species Act directs that no activity by a governmental agency should lead to the extinction of an endangered species and that all governmental agencies must use whatever measures are necessary to preserve these species.

The key to preventing extinctions is preservation of the habitat required by the endangered species. Consequently, many U.S. governmental agencies and private organizations have purchased such sensitive habitats or have managed areas to preserve suitable habitats for endangered species. Setting aside certain land areas or bodies of water forces government and private enterprise to confront the issue of endangered species. The question eventually becomes one of assigning a value to the endangered species. This is not an easy task; often, politics become involved, and the endangered species does not always win.

A case in point is the controversy that surrounded Tennessee's Tellico Dam project in 1978. The U.S. Supreme Court declared that completion of the $116 million federal project would result in a violation of the Endangered Species Act because the dam would threaten the survival of an endangered species called the snail darter, a tiny fish about 8 centimeters long that lived in the stream that would be altered by the dam's construction.

Blackfooted ferret

Whooping crane

Mission blue butterfly

Bald eagle

Manatee

Black rhinoceros

FIGURE 11.20

Endangered Species. Many species have been placed on the endangered species list. These are plants and animals that are present in such low numbers that they are in immediate danger of becoming extinct.

Developers, however, were not deterred. They lobbied in Congress to have all federally funded projects exempted from the act. Conservationists lobbied for the preservation of the act as it was originally written. Eventually, in 1978, Congress amended the Endangered Species Act so that exemptions to the act could be granted for federally declared major disaster areas or for national defense, or by a seven-member Endangered Species Review Committee. If the committee found that the economic benefits of a project outweighed the harmful ecological effects, it could exempt a project from the Endangered Species Act. At their first meeting, the review board denied the request to exempt the Tellico Dam project on the grounds that the project was economically unsound. This should have stopped the Tellico Dam project. However, nine months later, as a result of several political maneuvers, Congress appropriated money to complete the dam. It is now complete and full of water. The snail darters that once dwelled on the site were transplanted to nearby rivers and by 1981 appeared to be reproducing.

The new amendments to the Endangered Species Act also weakened the ability of the U.S. government to add new species to the endangered and threatened lists. Before a new species can be listed, it is now necessary to determine the boundaries of the species' critical habitat, prepare an economic impact study, and hold public hearings—all within two years of the proposal of the listing.

BOX 11.5
Extinction of the Dusky Seaside Sparrow

On Tuesday, 16 June 1987, the last dusky seaside sparrow (*Ammodramus maratimus nigrescens*) died in captivity at Walt Disney World's Discovery Island Zoological Park in Orlando, Florida. The bird was a male that was probably about twelve years old. Originally, this subspecies and several other subspecies were found in the coastal salt marshes on the Atlantic coast of Florida. (A subspecies is a distinct population of a species that has several characteristics that distinguish it from other populations.) One other subspecies, the Smyrna seaside sparrow (*Ammodramus maratimus pelonata*), is believed to have become extinct several years ago, and a third subspecies, the Cape Sable seaside sparrow (*Ammodramus maratimus mirabilis*), was listed as an endangered

species in 1967. Before the deaths of the last remaining dusky seaside sparrows, a few males were crossed with another subspecies, Scott's seaside sparrow (*Ammodramus maratimus peninsulae*). Thus, the hybrid offspring between these two subspecies contain some of the genes that made the dusky seaside sparrow unique.

The endangerment and extinction of these different birds was a direct result of the land development and drainage that destroyed the salt-marsh habitat to which they were adapted. The development of Cape Canaveral as a major center for the U.S. space program also resulted in the modification of much of the birds' original habitat and was a partial cause of their extinction.

CONSIDER THIS CASE STUDY
Serengeti National Park

Serengeti National Park, in the east-central African country of Tanzania, is predominately grassland. Large herds of migratory grazing animals, such as zebra, wildebeest, and Thomson's gazelle, use the park and surrounding land. Together, these three species number about eight hundred thousand animals. They use an area of about 25,000 square kilometers (about the size of Hawaii or Wales), but only half of this area is included within the park. The animals are only protected in the park—not outside the park. Several kinds of predators use this mixed group of grazers

as a source of food. Lions, wild dogs, hyenas, and jackals are the most common. Humans are also significant predators of all the animals in the park, even though park rangers try to prevent poaching.

Tanzania is a developing country with mineral resources and agricultural potential, as well as one of the fastest-growing populations in the world. Currently, food production is just barely keeping up with demand.

What do you think will happen to the animals of Serengeti National Park? Why?

Human Influences on Ecosystems

Pollution is the result of technological advancements and increased population density. It is defined as anything produced by humans in a quantity that interferes with the health or well-being of an organism.

Natural resources are those structures and processes that can be used by people but cannot be created by them. Renewable resources can be regenerated or repaired, while nonrenewable resources are consumed.

Mineral resources must be extracted from ores, a process that requires expenditures of energy and money, and also changes the environment. Mineral resources are not evenly distributed, so most countries must purchase some of their minerals from foreign sources. The major steps of mineral exploitation are exploration, mining, processing the ore, transportation, and manufacturing the finished product. Ultimately, all costs of mineral exploitation are reduced to monetary costs. Recycling reduces the demand for new sources of mineral depsoits but does not necessarily save money.

As the human population increases, more and more ecosystems are altered by us as we attempt to feed and house ourselves. However, some remote areas have been changed very little by humans. Tropical rain forests represent some of the last large wilderness areas, but they are being rapidly converted to other uses by the constant pressure of growing populations. Many of these wilderness areas should be protected because of the rich diversity of species they possess.

We change natural ecosystems by replacing them with agricultural ecosystems, we alter species mixtures by introducing plants and animals, and we reduce populations by harvesting trees and animals for our use. The cutting of forests must be done in a manner that allows for regrowth so that the soil is not exposed to the erosional effects of wind and water. Cutting small areas and reforestation help to prevent these problems.

Grazing of arid and semiarid lands can be a valuable way to provide food for people in these areas. However, the land is often overgrazed and then may be degraded to a desert that is not capable of supporting the animals needed to feed growing populations.

Aquatic ecosystems are modified by pollution and activities that occur on the land adjacent to the water. Land that is devoid of vegetation erodes and fills streams and lakes with sediment. Warming of the water is also likely to occur if trees are removed from the stream side. Many exotic species of fish have been introduced into the freshwater ecosystems of the world. This alters normal food chains and reduces the populations of native species.

Often, areas are managed for specific species of wildlife. This involves careful planning and habitat manipulation to provide the best possible population for hunting purposes. Some areas are intensely managed, as in many European game parks, while in other parts of the world, the game animals lead a more normal wild life. Waterfowl present a unique problem because they migrate across international boundaries.

Extinction is a normal consequence of not being able to adapt to changes in the environment. However, since humans have such a great influence on nearly every ecosystem in the world, they have been the cause of increased rates of extinction. Many people recognize the value of species, both as possible helpers of humans and for their own intrinsic worth, and are trying to preserve sensitive habitats so that species will not be driven to extinction because of the appropriation of their habitats for other uses.

Review Questions

1. Name three ways that humans directly alter ecosystems.
2. Why is the impact of humans greater today than at any time in the past?
3. Define pollution.
4. What are three kinds of costs associated with resource exploitation?
5. Why is recycling usually more energy efficient than mining new raw materials?
6. List three problems associated with forest exploitation.
7. What is desertification? What causes it?
8. What effects do increased temperature and increased organic matter have on aquatic ecosystems?
9. List six techniques utilized by wildlife managers.
10. What special problems are associated with waterfowl management?
11. What is extinction, and why does it occur?
12. Why should humans worry about extinction?
13. List three actions that can be taken to prevent extinctions.

Human Influences on Ecosystems

CHAPTER TWELVE
Land-Use Planning

Objectives

After reading this chapter, you should be able to:

List the relative amounts of land used for crops, livestock, urban development, and other uses in the United States.

Explain the impact that water has on the location and development of cities.

Explain why farmland surrounding cities was used for housing.

Explain how taxation may influence land use.

Explain why floodplains and wetlands are often mismanaged.

List several exclusionary land uses and several multiple land uses.

Describe the economic impact of recreation.

Recognize that people desire outdoor recreation.

Explain why recreational areas are needed in urban locations.

Explain why some land must be designated for particular recreational uses, such as wilderness areas, and why that decision sometimes invites conflict from those who do not desire to use the land in the designated way.

List the steps in the development and implementation of a land-use plan.

Describe methods of enforcing compliance with land-use plans.

List the problems associated with local land-use planning.

Explain the advantage of regional planning and the problems associated with it.

Chapter Outline

Key Terms

floodplain
floodplain zoning
    ordinances
land-use planning
megalopolis
multiple land use
nature centers

outdoor recreation
ribbon sprawl
tract development
urban sprawl
wetlands
zoning

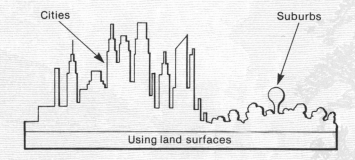

Cities   Suburbs

Using land surfaces

Conflicting desires
• Industry
• Housing
• Transportation
• Commercial development
• Recreation
} Planning resolves conflicts

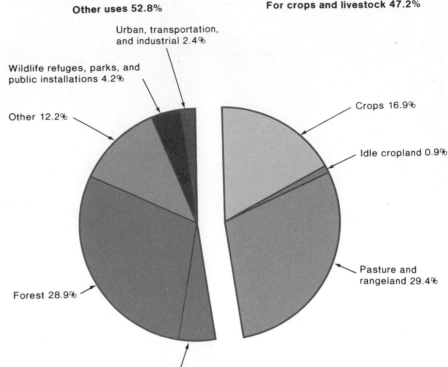

Other uses 52.8%   For crops and livestock 47.2%

Urban, transportation,
and industrial 2.4%

Wildlife refuges, parks, and
public installations 4.2%

Other 12.2%

Crops 16.9%

Idle cropland 0.9%

Pasture and
rangeland 29.4%

Forest 28.9%

Recreation 5.1%

FIGURE 12.1

Land Use in the United States. Forty-seven percent of the land in the United States is used for crops and livestock, while less than 3 percent is used for urban centers and transportation.

Source: Data from the *Statistical Abstract of the United States,* 1989.

Historical Land Use and the Development of Cities

The land surface is a nonrenewable resource because it is not able to be newly formed by natural processes, and once used for certain activities, is unable to be used for other purposes. (See figure 12.1.) Of the land in the United States, 47 percent is used for crops and livestock, about 50 percent is forests and natural areas, and the remaining 2–3 percent is used intensively by people in urban centers. This urban use includes transportation corridors between the urban centers.

This pattern of land use differs greatly from conditions a century ago. Early settlers often immigrated to the New World from crowded European cities. The promise of open space and available land was a primary incentive for those who left their homes. As the North American population increased, the original rural setting developed villages and cities.

Because the developing countries in North America had no system of railroads or highways, the primary method of transportation was by water. Without a reliable transportation network, cities could not grow and industry could not prosper. Thus, the early towns were usually built near water. Typically, cities developed as far inland as rivers were navigable. Abrupt changes in elevation required transshipment of goods, and cities often grew at these points (Buffalo, New York; Sault Sainte Marie, Ontario). Water met many of the needs of villages and small towns, such as drinking water, transportation, power, and waste disposal. Without access to water, St. Louis, Montreal, Chicago, Detroit, Vancouver, and other cities would not have developed. (See figure 12.2.)

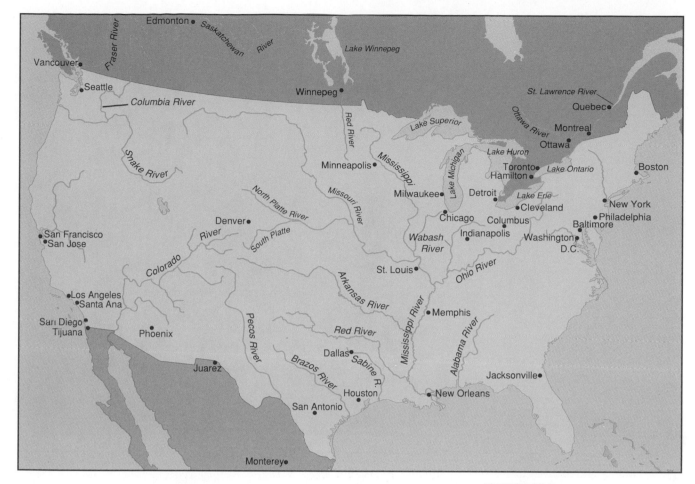

FIGURE 12.2

Water and Urban Centers. Note that most of the large urban centers are located on water. Water is an important means of transportation and was a major determining factor in the growth of cities. The cities shown have populations of 500,000 or more.

Industrial use of water also became important as cities grew, and most industrial development took place on the waterfront. Large industrial developments replaced small gristmills, sawmills, and blacksmith shops. Industry eventually took over most of the waterfront. There was little control of industry activities, so the waterfront became polluted, unhealthy, and an undesirable place to live. Anyone who could afford to do so moved away from the original city center.

Land use and water are intimately interrelated. Development of a water resource for industrial use or transportation affects the land bordering the water. The type of land use similarly affects the watercourse. In urban centers, the "decision" to use waterfront land for industrial development was forced by a need for transportation and power. Once this pattern had begun, a series of events eventually resulted in the development of suburban metropolitan regions.

Cities, which are usually located next to water, are often surrounded by rich farmlands. River valleys provide flat, rich land for the growth of crops. Until transportation systems became well developed, farms needed to be close to the city so that produce could be marketed. This rich farmland adjacent to the city was the most easily developed space for city expansion. As the population of the city grew, demand for land within the city increased. As the city's land prices rose, people and businesses began to look for cheaper land on the outskirts of the city. Developers and real estate

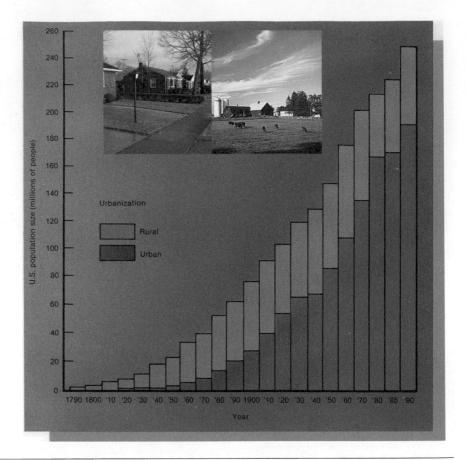

FIGURE 12.3

Rural to Urban Population Shift. In 1800, the United States was essentially a rural country. Industrialization in the late 1800s began the shift to an increasing urban population. In 1990, 74 percent of the U.S. population was urban.

Source: Data from the *Statistical Abstract of the United States,* 1987, and 1990 data from Population Reference Bureau, Inc., Washington, DC.

agents were quick to respond and to assist in the acquisition and conversion of agricultural land to residential or commercial uses. They viewed land as a commodity to be bought and sold for a profit, rather than as a finite resource to be managed. As long as money could be made by converting agricultural land to other purposes, it was impossible to prevent such conversion.

North America remained essentially rural until industrial growth began in the last third of the 1800s, when the population began a trend toward greater urbanization. (See figure 12.3.) This large-scale migration to the cities had several causes: Improvement in agriculture required less farm labor; therefore, people moved from the farm to the city. Second, job opportunities were available in the city because industry was developing. The average person was no longer a farmer but, rather, a factory worker, shopkeeper, or clerk living in a tenement or tiny apartment. This pattern of rural to urban migration occurred throughout North America, as it would in any developing nation. Urbanization was also enhanced by new immigrants. These new citizens tended to settle in towns and cities, where employment was more likely. Many immigrants congregated in parts of cities. As a result, today most large cities have ethnic sections, with populations of Irish, Italian, Chinese, German, and Greek descent, that give North American cities a unique character. A third reason for the growth of cities was that cities offered a greater variety of cultural, social, and artistic opportunities than rural communities. Thus, cities were attractive for cultural as well as economic reasons.

Human Influences on Ecosystems

FIGURE 12.4
Transportation and Suburbia. This traffic congestion is a common experience for people who work in cities but live in the suburbs. The popularity of automobiles and the desire to live in the suburbs are tied closely to one another.

During the last two decades, land development in North America has destroyed many natural areas that people had long enjoyed. The Sunday drive in the countryside has become a battle to escape the evergrowing suburbs. The unique character of neighborhoods and communities has been changed by apartment complexes. Developments have sprung up along beaches, and recreational communities are common sights. Even the deserts and marshes have not been spared; they sprout the little red flags of subdividers. These changes have occurred in a vacuum without concern for other changes that may or may not be desirable. Let us look at the various uses of land to help determine the overall result of this development.

The Rise of Suburbia

As cities continued to grow, certain sections within each city began to deteriorate. Industrial activity continued to be concentrated near water in the city's center. Industrial pollution and urban crowding turned the core of many cities into undesirable living areas. In the early 1900s, people who could afford to leave began to move to the outskirts of the city. This trend continued after World War II, in the 1940s, 1950s, and 1960s, as more people were able to buy homes. Most of the homes were in the attractive suburbs, away from the pollution and congestion of the central city. These houses were built on large lots in decentralized patterns, which increased the cost of supplying services, increased energy needs, and made it very difficult to establish efficient public transportation networks. Rising automobile ownership and improved highway systems also encouraged suburban growth. The convenience of a personal automobile escalated decentralized housing patterns, which, in turn, required better highways, which led to further decentralization. This spiral continued in the United States until economic factors of increased energy costs and high unemployment in the 1970s slowed it somewhat. (See figure 12.4.)

In some areas throughout North America, the growth of suburbs has been slowed due to the increased cost of housing and transportation. People have migrated back to some cities on a limited scale because of the lower cost of urban houses and the fact that public transportation is generally more efficient in the city than in the suburbs, thus freeing urban residents

a.

b.

c.

FIGURE 12.5

Urban Sprawl. Note the three different types of growth depicted in these photos. (*a*) The wealthy suburbs with large lots are adjacent to the city. (*b*) Ribbon sprawl develops as a commercial strip along highways. (*c*) Tract development results in neighborhoods consisting of large numbers of similar houses on small lots.

from the daily cost of commuting. This reverse migration, however, is still greatly offset by the continual growth of the suburban communities.

By 1960, unplanned suburban growth had become known as **urban sprawl.** Large housing tracts surrounded cities, which made it difficult for people to find open space. A city dweller could no longer take a bus to the city limits and enjoy the open space of the countryside.

Urban sprawl occurs in three different ways. (See figure 12.5.) One type of growth is associated with the wealthier urbanites adjacent to the city. These subdivisions of homes are on large individual lots in the more pleasing geographic areas. A second form of urban sprawl involves development along transportation routes. This is referred to as **ribbon sprawl** and usually consists of commercial and industrial buildings. Ribbon sprawl results in high costs for the extension of utilities and other public services. It also makes the extent of urbanization seem much larger than it actually is. A third development pattern is tract development. **Tract development** consists of the construction of similar residential units over large areas. Initially, these tracts are often separated from each other by farmland.

As suburbs continued to grow, cities (once separated by farmland) began to merge, and it became difficult to tell where one city ended and another began. This type of growth led to the development of regional cities. Although these cities maintain their individual names, they are really just part of one large urban area called a **megalopolis.** (See figure 12.6.) The

Human Influences on Ecosystems

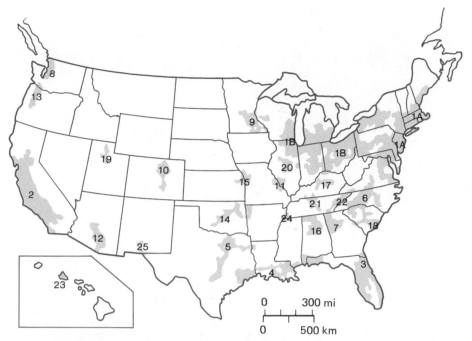

1. Metropolitan belt
   A. Atlantic Seaboard (BoWash)
   B. Lower Great Lakes (ChiPitts)
2. California Region (SanSan)
3. Florida Peninsula (JaMi)
4. Gulf Coast
5. East central Texas — Red River
6. Southern Piedmont
7. North Georgia — South Tennessee
8. Puget Sound
9. Twin Cities Region
10. Colorado Piedmont
11. Saint Louis
12. Metropolitan Arizona
13. Willamette Valley
14. Central Oklahoma—Arkansas Valley
15. Missouri—Kaw Valley
16. North Alabama
17. Blue Grass
18. Southern Coastal Plain
19. Salt Lake Valley
20. Central Illinois
21. Nashville Region
22. East Tennessee
23. Oahu Island
24. Memphis
25. El Paso—Ciudad Juarez

FIGURE 12.6

Regional Cities in the United States.
By the year 2000, twenty-five major urban regional cities are expected to have developed. Each will have a population of at least one million people. Four super cities will each have a great deal more than one million people.

Source: J. P. Pickard, "U.S. Metropolitan Growth and Expansion, 1970–2000, with Population Projections" in *Population Growth and the American Future*. Washington, DC: U.S. Governmental Printing Office, 1972.

eastern seaboard of the United States, from Boston, Massachusetts, to Washington, D.C., is an example of a continuous city. Other examples are London to Dover in England or the Toronto-Mississauga region of Canada or the southern Florida coast from Miami northward.

Problems Associated with Unplanned Urban Growth

Loss of Farmland

One U.S. Department of Agriculture study found that three-fourths of the land recently urbanized was previously used for high-value crop production. An area that is flat, well drained, accessible to transportation, and close to cities is ideal farmland. However, it is also prime development land. Land adjacent to cities that once supported crops now supports new housing developments, shopping centers, and parking lots. (See figure 12.7.)

Urban development of farmland is proceeding at a rapid pace in the United States. Approximately 1 million hectares of land is developed each year. This is equivalent to an area half the size of New Jersey. One reason for this conversion is the way the land is taxed. The Fourteenth Amendment

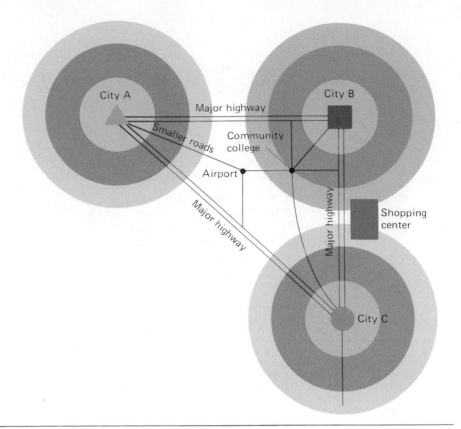

FIGURE 12.7

Loss of Farmland. As cities grow outward, they eventually grow together to form a regional city. The land between them, once used for farming, becomes developed for residential and commercial purposes. Improved transportation routes and joint facilities (such as airports, shopping centers, and community colleges) hasten this loss of farmland.

to the U.S. Constitution specifies that all real property shall be assessed and taxed at its highest and best use. This means that property will be taxed on what can be done with it, not necessarily what is being done with it. For example, if land can be used for both farming and residential development, it is taxed as if it were being used for residential purposes. If a farmer sells a portion of the farm to a developer who builds five houses, local taxing authorities would consider these houses to be the "highest and best use" of the land. All of the farmer's land would then probably be reassessed and the taxes increased substantially. Farmers faced with this situation are often forced to sell their farmland because they are taxed on the commercial value of the land, rather than on its value as farmland. This policy encourages development and forces people out of farming. A number of states have developed different taxation practices to relieve this inequity. These new policies assess taxes based on *current* use of the land and not on the land's highest *potential* use.

Floodplain Problems

Even though certain cities may try to control their growth and maintain some open space for recreation and sight and sound breaks, the city may be in a location that is altogether unsuitable. These are the cities built in areas that are ripe for a natural disaster, such as an earthquake or a flood.

Many cities are located near rivers in areas called **floodplains,** the lowland area on either side of a river. This floodplain is generally flat, and so is an inviting area for residential development. But the floodplain is so named because it is periodically covered with water. Some may flood annually, while others flood less regularly. A better use of these areas is for open space or for recreation, yet developers continue to try to use the space for construction of houses and light industry.

When a floodplain is developed for residential or commercial use, a retaining wall usually is built to prevent the periodic flooding natural to the area. This then increases the cost of the development, increases the cost of insurance protection, and creates high-water problems downstream. Frequently, tax monies are used to repair the damage that results from the unwise use of floodplains. Floodplain development is one of the insidious problems associated with a rapidly expanding population. As long as the population continues to increase, these less desirable areas are likely to be used for housing, whether they are subject to annual flooding or less frequent damage.

Floods are natural phenomena. Contrary to popular impressions, no evidence supports the premise that floods are worse today than they were one hundred or two hundred years ago, except perhaps on small, isolated watersheds. What has increased is the economic loss from the flooding. This loss reflects the fact that more cities, industries, railroads, and highways have been constructed on floodplains. Sometimes, there are no alternatives to floodplain development, but too often, risks are simply ignored. Because flooding causes loss of life and property, floodplains should no longer be developed for uses other than agriculture and recreation.

Many communities have enacted **floodplain zoning ordinances** to restrict future building in floodplains. Although such ordinances may prevent further economic losses from flooding, what happens to those individuals who already live in floodplains? Floodplain building ordinances usually allow current residents to remain. However, these residents often find it extremely difficult to obtain property insurance. Relocation, usually at a financial loss, is the only alternative. Such situations are unfortunate; perhaps, proper planning in the future will prevent these problems.

Wetlands Misuse

Since access to water was and continues to be important to industrial development, many cities are located in areas with extensive wetlands. The **wetlands** include swamps, tidal marshes, coastal wetlands, and estuaries. Because wetlands breed mosquitoes and are sometimes barriers to the free movement of the population, they have often been thought of as useless or harmful. Therefore, these wetlands have often been drained, filled, or used as dumps. Many modern cities have completely covered over extensive areas of wetlands and may even have small streams running under streets, completely enclosed in concrete.

Each different kind of wetland, however, has unique qualities and serves as a home to many different kinds of plants and animals. Frequently, wetlands are associated with the reproductive portion of the life cycle of the organisms inhabiting the area. They are likely to be nesting sites and spawning sites, and commercial and sport fisheries are dependent upon the organisms that use these habitats for the production and protection of their young. Because most wetlands receive constant inputs of nutrients from the water that drains from the surrounding land, they are highly productive and excellent places for the rapid growth of aquatic species. Human impact on wetlands has severely degraded or eliminated these spawning and nursery habitats.

The state of Louisiana has extensive coastal wetlands along the Gulf of Mexico. Approximately 10,000 hectares (100 square kilometers) are lost per year for a variety of reasons, most of them the result of human activity.

Much of the coastline consists of poorly compacted muds that are easily eroded if left exposed to wind and wave action. Control of the Mississippi River through upstream dams and reservoirs has reduced the amount of sediment delivered to the region, and, therefore, replacement of lost sediments is impossible. Furthermore, as the muds compact due to settling, the land subsides and erosion is more likely.

Another major contributing factor to the loss of these coastal wetlands is the activity of oil and gas exploration and production companies that are developing resources from beneath the Gulf of Mexico. Typically, these companies cut canals through the marshes to allow for easy access of barges and other equipment to the drilling platforms and production facilities. When oil or gas is found, it must be transported to land through pipelines. The laying of pipelines often results in the cutting of additional canals through the wetlands. These canals break the wall of vegetation that protects the soft muds and accelerates the rate of erosion.

The nutria, a rodent introduced from South America, is also contributing to the problem. These animals reproduce rapidly and can have such large populations that they eat all the vegetation in a local area. This exposes the muds to erosion and contributes to the loss of wetlands. Trapping these animals for their fur was at one time profitable for the local people and helped to control the animals. However, a shift in public opinion has reduced the acceptability of fur for clothing and depressed the fur market, making the trapping of nutria uneconomical.

Besides providing a necessary habitat for fish and other organisms, wetlands also provide natural filters for sediments and runoff. This natural filtration process reduces sedimentation in other waters and allows time for the biological cleaning of water before it enters a larger body of water. In addition, wetlands often protect the shoreline from erosion. When destroyed, wetlands must be replaced by costly artificial measures, such as breakwalls built to protect the shoreline.

Other Land-Use Problems

The geologic status of an area must also be considered in land-use decisions. The building of cities on the sides of volcanos or on major earthquake-prone faults has led to much loss of life and property. Similarly, the use of unstable hillsides or areas subject to periodic fires as sites for homes and villages is unwise.

Another problem in a few locations is lack of water. Southern California and metropolitan areas in Arizona must import water to sustain the communities. Wise planning would limit growth to whatever could be sustained by available resources. As these areas in California and Arizona continue to grow, the strain on regional water resources will increase.

Multiple Land Use

Once a land-use decision is made, it may be irreversible. For example, when a highway is constructed, it is virtually impossible to use that land for any other purpose. If land is designated as a wilderness area, no human development is allowed. Other single exclusionary uses of land are agriculture, recreation, waste storage, housing, and industrial development.

Some land uses, however, do not have to be exclusionary. Two or more uses of land occurring at the same time is called **multiple land use.** Per-

FIGURE 12.8
Urban Open Space. This is an aerial view of Central Park in New York City. If the land had not been set aside in the late 1800s, it would have been developed like the rest of Manhattan.

haps the best example of multiple land use occurs in U.S. national forests. In 1960, the Multiple Use Sustained Yield Act divided use of national forests into four categories: wildlife habitat preservation, recreation, lumbering, and watershed protection. This act encouraged both economic and recreational use of the forests.

Another example of multiple land use involves the development of parks on floodplains. These parks provide open space, recreational land, and storage reservoirs for storm sewer runoff. Storage reservoirs can even be blended into the park landscape as ponds or lakes.

Once land is recognized as a nonrenewable resource, it may become more fashionable to engage in meaningful land-use planning. Such plans must consider the long-term needs of the region as well as the short-term goals of the population.

Recreational Land Use

Nearly three-fourths of the U.S. population live in urban areas. These urban dwellers value open space because open space breaks up the sights and sounds of the city and provides a place for many kinds of recreation. Inadequate land-use planning in the past has rapidly converted open spaces in or near urban areas to other uses. Until recently, the creation of a new park within a city was considered an uneconomical use of the land, but people are now beginning to realize the need for parks and open spaces.

Some cities recognized the need for open space a long time ago and allocated land for parks. London, Toronto, and Perth, Australia, have centrally located and well-used parks. New York City set aside approximately 200 hectares for Central Park in the late 1800s. (See figure 12.8.) Boston has developed a park system that provides a variety of urban open spaces. However, other cities have not dealt with this need for open space because they lacked either the foresight or the funding.

Urban Recreation

Recreation seems to be a basic human need. The most primitive tribes and cultures all had games or recreational activities. New forms of recreation

FIGURE 12.9

Urban Recreation. In urban areas, recreation often takes the form of sports programs, playgrounds, and walking. Most cities recognize the need for such activities and develop extensive recreation programs for their citizens.

are continually being developed. In the congested urban center, special areas where recreation can take place often must be constructed.

A major problem with urban recreation is the development of recreational activities and facilities conveniently located near the residential areas. Areas that are not conveniently located may be only infrequently used. For example, the hundreds of thousands of square kilometers of national parks in Alaska will be visited by only a very small proportion of the U.S. population. Large urban centers are discovering that they must provide adequate, low-cost recreational opportunities within their jurisdiction. Some of these opportunities are provided in the form of commercial establishments, such as bowling centers, amusement parks, and theaters. Others must be subsidized by the community. (See figure 12.9.) Playgrounds, organized recreational activities, and open space have usually been combined into an arm of the municipal government known as the Parks and Recreation Department. Cities spend millions of dollars to develop and maintain recreation programs. Often, there is conflict over the allocation of financial and land resources. These are very closely tied because open land is scarce in urban areas, and it is expensive. Riverfront property is ideal for park and recreational use, but it is also prime land for the development of industrial, commercial, or high-rise residential buildings. Although conflict is inevitable, many metropolitan areas are beginning to see that recreational resources may be as important as economic growth for maintaining a healthy community.

An outgrowth of the trend toward urbanization is the development of **nature centers.** In many urban areas, there is so little natural area left that the people who live there need to be given an opportunity to learn about nature. Nature centers are basically teaching institutions that provide a variety of methods for people to learn about and appreciate the natural world. Zoos, botanical gardens, and some urban parks, combined with interpretative centers, also provide recreational experiences. Nature centers are usually located near urban centers, in places where some appreciation of the natural processes and phenomena can be developed. In some cases, they are operated by municipal governments. In other cases, they may be run by school systems or other nonprofit organizations.

Human Influences on Ecosystems

TABLE 12.1

Number of people who participated in selected outdoor recreational activities in 1987.

Activity	Number of participants
Bicycling	24.5 million
Fishing	24.3 million
Camping	20.4 million
Running	11.4 million
Hunting	9.5 million
Golf	9.3 million
Hiking	8.0 million
Tennis	7.8 million
Skiing	7.0 million

Source: Data from the *Statistical Abstract of the United States*, 1989.

Outdoor Recreation

Not all people enjoy the same recreation. Some people enjoy reading or watching television. Others prefer commercial recreational activities, such as golf, tennis, bowling, amusement areas, racetracks, and skiing. Another major area of recreation, usually classified as **outdoor recreation,** involves using the natural out-of-doors for hiking, camping, canoeing, and so forth. Millions of people want to use public lands for these activities.

Most recreational activities require the consumption of natural resources, such as minerals, fuels, and timber. Some activities require more resources than others, but even the backpacker, who has traditionally been considered an ecologically frugal individual, requires considerable equipment. Other activities, such as the use of off-road vehicles, require even more resources in the form of equipment and fuel. Table 12.1 lists various outdoor activities and the number of people who participate in each activity in the United States. As energy and other resources become less available, we may be forced to abandon some of our more extravagant recreational activities for those that are more ecologically conservative.

Conflicts over Recreational Land Use

Many people desire to use the natural world for recreational purposes because nature can provide challenges that may be lacking in their day-to-day lives. Whether the challenge is hiking in the wilderness, underwater exploration, climbing mountains, or driving a vehicle through an area that has no roads, people experience a sense of adventure from these activities. Look at table 12.1 again. It lists various outdoor activities and the number of people who participate in each kind of activity. All of these activities use the out-of-doors, but not in the same way. Conflicts develop because some of these activities cannot occur in the same place at the same time. For example, wilderness camping and backpacking often conflict with off-road vehicles.

There is a basic conflict between those who prefer to use motorized vehicles and those who prefer to use muscle power in their recreational pursuits. (See figure 12.10.) This conflict is particularly strong because both groups would like to use the same publicly owned land for their activities. Both have paid taxes, both "own" the land, and both feel that it should be available for them to use as they wish. When everyone "owns" something, there is often little desire to regulate activities.

FIGURE 12.10
Conflict over Recreational Use of Land. This land may be used for both hiking and jeep trails. The people who participate in these two kinds of recreation are often antagonists over the allocation of land for recreational use.

For example, much of the rangeland in the western United States is publicly owned. Based on "Animal Use Months" established by the Bureau of Land Management or the Forest Service, ranchers are allowed to graze cattle, but only on certain public lands. Technically, failure to comply can mean a loss of grazing rights. However, since the bureaucracy of setting and enforcing the regulation is politically motivated and the regulatory agency is understaffed, there is no incentive for ranchers to limit the number of cattle or sheep on the land unless all ranchers agree to do so. The usual result is that a few individuals exploit the land in hopes of a short-term profit. This exploitation could cause permanent damage to the rangeland. Many people want to use the public lands for outdoor recreation and resent the control of the use of this land by one category of user. (See box 16.2.)

An obvious solution to this type of problem is to allocate land to specific uses and to regulate the use once allocations have been made. Several U.S. governmental agencies, such as the National Park Service, the Bureau of Land Management, the U.S. Forest Service, and the U.S. Fish and Wildlife Service, are engaged in allocating and regulating the use of the lands they control. However, all of these agencies have conflicting roles. The U.S. Forest Service has a mandate to manage the forested public lands for timber production. This is often in conflict with recreational uses that people might want to make of the land. Similarly, the Bureau of Land Management has huge tracts of land that can be used for recreation, but it traditionally has been associated with grazing interests.

A particularly sensitive issue is the designation of certain areas as wilderness areas. Obviously, if an area is to be wilderness, human activity must be severely restricted. This means that the vast majority of Americans will never see or make use of wilderness areas. Many people argue that this is unfair because they are paying taxes to provide for the recreation of a select few. Others argue that if everyone were to use these areas, their charm and unique character would be destroyed and that, therefore, the cost of preserving wilderness is justifiable.

Areas designated as wilderness make up a very small proportion of the total public land available for recreation. (See figure 12.11.) The fact that there are relatively few wilderness areas has resulted in a further problem: The areas are being loved to death. People pressure on this resource has

Human Influences on Ecosystems

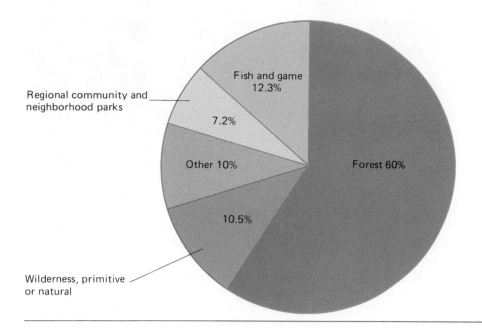

Fish and game
12.3%

Regional community and
neighborhood parks

7.2%

Other 10%

Forest 60%

10.5%

Wilderness, primitive
or natural

FIGURE 12.11

U.S. Federal Recreational Lands in 1983. Of the approximately 108 million hectares of federal recreational lands in the United States, approximately 10 percent is designated as wilderness, primitive, or natural.

Source: Data from *Statistical Abstract of the United States*, 1984.

become so great that, in some cases, the wilderness quality is being tarnished. The designation of additional wilderness would relieve some of this pressure.

Land-Use Planning Principles

As we look back at the location and development of the cities and metropolitan areas, we can see the mistakes that have been made. If we were in a position to start over, we might locate our population centers differently; we might regulate building and development for the least damaging but highest use of the land. But how does one go about determining this *best use* and *least damaging use* of an area? What goes into the process of land-use planning?

. **Land-use planning** is the construction of an orderly list of priorities for the use of available land. The development of any plan involves a data-gathering process in which current use, geological data, biological data, and sociological information are surveyed. From these data, projections are made about what human needs will be. All of the data collected are integrated with human needs projections, and each parcel of land is evaluated and assigned a best use under the circumstances. Finally, mechanisms for implementing the plan are divided into two categories: purchase of the land or regulation of its use.

Probably the simplest way to ensure protection of certain desirable lands is to purchase them. This method has often been used by government or by conservation-minded organizations or individuals. The major problem with purchasing land is the cost. Many communities are not in a financial position to purchase lands; therefore, they attempt to regulate land use by zoning laws.

Zoning is a common type of land-use regulation. When land is zoned, it is designated for specific potential uses. Several common designations are agricultural, commercial, residential, recreational, and industrial. (See figure 12.12.)

Often, local governments lack the funds to hire professional planners. As a result, zoning regulations are frequently made by people who see only

FIGURE 12.12

Zoning. Most communities have a zoning authority that designates areas for particular use. These signs are indications that decisions have been made about the "best" use for these lands.

the short-term gain and not the possible long-term loss. The land is simply zoned for its current use. Even when well-thought-out land-use plans exist, they are usually modified to encourage local short-term growth rather than to provide for the long-range needs of the community. The community needs to be alert to variances from established land-use plans because, once the plan is compromised, it becomes easier to accept future deviations that may not be in the best interests of the community. Many times, individuals who make zoning decisions are realtors, developers, or local businesspeople. These individuals wield significant local political power and are not always unbiased in their decisions. Concerned citizens must try to combat special interests by attending zoning commission meetings and by participating in the planning process.

Regional Planning

Regional planning is more effective than local land-use planning because political boundaries seldom reflect the geological and biological data base used in planning. Larger planning units can afford to hire professional planners, while local units cannot. The concept behind regional planning is coordination because problems do not respect political boundaries. For example, airport locations should be based on a regional plan that incorporates the several local jurisdictions in the region. Three cities only 30 kilometers apart should not build three separate airports when one regional airport could serve their needs better and at a lower cost to the taxpayers.

Although regional planning is increasing in North America, the majority of regional governmental bodies are presently voluntary and lack any power to implement programs. Their only role is to advise the member governments on policy. Unfortunately, members of regional organizations still seem unwilling to give up power and may put their own interests above the goals of the region and view policy from a narrow perspective. An elected, multipurpose, regional government, on the other hand, would be ideal for implementing land-use policy. Such governments exist in only a handful of places and show few signs of spreading.

One way to encourage regional planning is to develop planning policies at the state or provincial level. The first state to develop a comprehensive statewide land-use program was Hawaii. During the early 1960s, much of Hawaii's natural beauty was being destroyed to build houses and apartments for the increasing population. The same land that attracted tourists was being destroyed to provide hotels and supermarkets for them. Local governments had failed to establish and enforce land-use controls. Consequently, the Hawaii State Land-Use Commission was founded in 1961.

Historically, towns and cities grew as a natural by-product of people choosing to live in certain areas for agricultural, business, or recreational reasons. Beginning in the 1920s, private and governmental planners began to think about how an ideal town would be planned. These communities would be completely built before houses were offered for sale. This concept of preplanning, designing, and building an ideal town was not fully developed until the 1960s. By 1967, about forty-three towns could be classified as planned "new towns."

One example of a new town is Reston, Virginia, located about 40 kilometers west of Washington, D.C. Reston began to accept residents in 1964 and has a projected population of eighty thousand. Because developers tried to preserve the great natural beauty of the area and the high quality of architectural design of its buildings, Reston has attracted much attention. Reston also has innovative programs in education, government, transportation, and recreation. For example, the stores in Reston are within easy walking distance of the residential parts of the community, and there are many open spaces for family activities. Because Reston is not dependent upon the automobile, noise and air pollution have been greatly reduced.

Recent research indicates that the residents of Reston have rated their community much higher than residents of less well-planned suburbs.

Although the concept of properly planned new towns is appealing, such communities are still rare for several reasons: To attract citizens, new towns must be located near large urban centers, where land availability is limited. In addition, the cost of creating new towns is extremely high, and many people still prefer the charm of a traditionally developed community.

FIGURE 12.13
Hawaii Land-Use Plan. Hawaii was the first state to develop a comprehensive land-use plan. This development in an agricultural area was stopped when the plan was implemented in the 1960s.

This commission designated all land as urban, agricultural, or conservational. Each parcel of land could be used only for its designated purpose. Other uses were allowed only by special permit. To date, the record for Hawaii's action shows that it has been successful in controlling urban growth and preserving the islands' natural beauty, even though the population continues to grow. (See figure 12.13.)

Several states are attempting to follow Hawaii's lead in state land-use regulation. Some states have passed legislation dealing with special types of land use. Examples include wetland preservation, floodplain protection, and scenic and historic site preservation. Although direct state involvement in land-use regulation is relatively new, it is expected to grow. Only large, well-financed levels of government can afford to pay for the growing cost of adequate land-use planning. State and regional governments are also more likely to have the power to counter the political and economic influences of land developers, lobbyists, and other special-interest groups when conflicts over specific land-use policies arise.

Urban Transportation Planning

A growing concern of urban governments is to develop comprehensive urban transportation plans. While the specifics of such plans might vary from region to region, urban transportation planning usually involves four major goals:

1. Conserve energy and land resources.
2. Provide efficient and inexpensive transportation within the city for the elderly, young, poor, and handicapped.
3. Provide suburban people with the opportunity to commute efficiently.
4. Help to reduce urban pollution.

Any successful urban transportation plan should integrate all of these goals, but funding and intergovernmental cooperation are needed to achieve this. To date, both of these factors have been limited in the United States. The problems associated with current urban transportation will certainly not disappear overnight, but comprehensive planning is the first step to help solve problems.

A primary method of transportation in most urban centers is the automobile. The automobile's advantages include convenience and freedom of movement. However, many urban areas are beginning to recognize that the automobile's disadvantages may outweigh its advantages. The disadvantages of urban automobile transportation include increased air pollution, substantial land use for highways and parking areas, and encouragement of urban sprawl. Recognizing these disadvantages, some cities, such as Toronto, London, San Francisco, and New York, have attempted to dissuade automobile use by developing mass transit systems and by allowing automobile parking costs to increase substantially.

The major types of urban mass transit systems are railroads, subways, trolleys, and buses. Such systems, however, are not without their problems. The major problems associated with mass transit are that it is:

1. Economically feasible only along heavily populated routes
2. Less convenient than the automobile
3. Extremely expensive to build and operate
4. Often crowded and uncomfortable

Although mass transit provides for a substantial portion of the urban transportation in some parts of the world, such as Western Europe and the Soviet

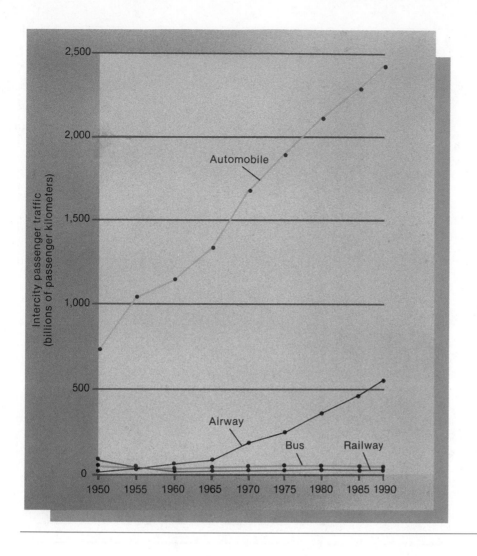

FIGURE 12.14
Decline of Mass Transportation. The automobile has consistently replaced rail and bus transit over the past forty years.

Union, mass transit use in the United States has continued to decline since the 1940s. (See figure 12.14.) A variety of forces has caused this decline. As people became more affluent, they could afford to own automobiles, which are a convenient, individualized method of transportation. The government encouraged automobile use by financing highways and expressways, by maintaining a cheap energy policy, and by withdrawing support for most forms of mass transportation. The government still encourages automobile transportation with hidden subsidies but maintains that rail and bus transportation should pay for themselves.

Rising gas prices and increasing competition for parking will prompt more U.S. citizens to seek alternatives to private automobile use. Car and van pools and dial-a-ride systems have been accepted by many individuals. Other types of transportation, such as the bicycle, should not be overlooked in urban transportation planning.

CONSIDER THIS CASE STUDY
Decision Making in Land-Use Planning

The following situation has happened thousands of times during the last twenty years:

A developer has just announced plans to build a large shopping mall on the outskirts of your city in what is now prime farmland. Many jobs will be created by the construction and operation of the proposed shopping mall. In addition to the stores, three new theaters will be built. Presently, your city has some unemployment, only one theater in the downtown area, and little variety in its retail businesses. On the surface, the proposed mall seems to offer only good news. Is this the case? Before answering, look at the entire situation.

What will happen to the downtown area when the mall is opened?
If you owned a downtown business, would you favor building the mall?
What will happen to the taxes on the farms near the new mall?
If you were a farmer, would you favor the project?
What effect will paving prime farmland have?
How will storm water runoff be affected?
How will future housing development be influenced?
What other problems can be associated with building the complex?
If you were the mayor or city manager, would you favor construction of the mall? Why?
Is the proposed project all good news after all?

Summary

Land and landscape are considered natural resources. Of U.S. land, 47 percent is used for crops and livestock, about 50 percent is forests and natural areas, and the remaining 2–3 percent is used for urban centers and transportation corridors. Historically, our large urban centers began as small towns located near water. The water served the needs of the town in many ways, especially transportation. As towns became larger, the farmland surrounding the town became suburbs surrounding industrial centers.

Many problems have resulted from unplanned urban growth. Current taxation policies encourage residential development of farmland, which results in a loss of valuable agricultural land. Floodplains and wetlands are two areas that are often mismanaged. Loss of property and life results when people build on floodplains. Wetlands protect our shorelines and provide a natural habitat for fish and wildlife.

Some examples of exclusionary land use are housing and highways. Multiple land use is possible where various uses can be integrated.

People in North America spend large amounts of money on leisure-time activities and services. Recreation is important socially and psychologically. People seem to need recreation as a change from their normal working life. Conflicts sometimes arise over the use of resources for recreational purposes because some recreational activities cannot occur at the same time and place. These conflicts are usually resolved by the allocation and regulation of resources.

Land-use planning involves gathering data, projecting needs, and developing mechanisms for implementing the plan. Purchasing and zoning the land are two ways to enforce land-use planning. Local planning is often not on a large enough scale to be effective because problems may not be confined to local political boundaries. Regional planning units can afford professional planners and are better able to withstand political and economic pressures. A growing concern of urban governments is to develop comprehensive urban transportation plans that seek to conserve energy and land resources, provide efficient and inexpensive transportation, provide efficient commuting, and help to reduce urban pollution.

Review Questions

1. Why did urban centers develop near waterways?
2. Describe the typical changes that have occurred in cities from the time they were first founded until now.
3. Why do people move to the suburbs?
4. Why do some farmers near urban areas sell their land for residential or commercial development?
5. What is a megalopolis?
6. What land uses are suitable on floodplains?
7. What is multiple land use?
8. Why is it important to provide recreational space in urban planning?
9. How can recreational activities damage the environment?
10. What is the monetary impact of recreational activities?
11. What are some strictly urban-related recreational activities?
12. List some conflicts that arise when an area is designated strictly as recreational.
13. Describe the steps necessary to develop a land-use plan.
14. What are the advantages of regional or state planning?
15. List three benefits of land-use planning.

CHAPTER THIRTEEN
Soil and Its Use

Objectives

After reading this chapter, you should be able to:

List the physical, chemical, and biological factors responsible for soil formation.

Explain the importance of humus to soil fertility.

Differentiate between soil texture and soil structure.

Explain how texture and structure influence soil atmosphere and soil water.

Explain the role of living things in soil formation and fertility.

Describe the various layers in a soil profile.

Describe the processes of soil erosion by water and wind.

Explain how contour farming, strip farming, terracing, waterways, windbreaks, and conservation tillage reduce soil erosion.

Understand that the misuse of soil reduces soil fertility, pollutes streams, and requires expensive remedial measures.

Explain how land not suited for cultivation may still be productively used for other purposes.

Chapter Outline

Soil and Land
Soil Formation
Soil Properties
Soil Profile
Soil Erosion
Soil Conservation Practices
    Box 13.1 Land Capability Classes
    Contour Farming
    Strip Farming
    Terracing
    Waterways
    Windbreaks
    Conservation Tillage
Uses of Nonfarm Land
Consider This Case Study: Soil Erosion in Virginia

Key Terms

chemical weathering
contour farming
erosion
friability
horizon
humus
land
leaching
loam
mechanical weathering

parent material
soil
soil profile
soil structure
soil texture
strip farming
terrace
waterways
weathering
windbreak

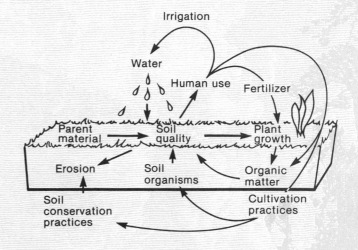

Soil and Land

Soil and land are often thought of as being the same. However, **land** is the part of the world not covered by the oceans, while **soil** is an organized mixture of minerals, organic material, living organisms, air, and water that together support the growth of plant life. Soil is a thin covering over the land. Farmers are particularly concerned with soil because the nature of the soil determines what kinds of crops can be grown and which farming methods must be employed. However, urban dwellers should also be concerned about soil because the health of the soil determines the quality and quantity of food they will eat. If the soil is so abused that it can no longer grow crops, or if it is allowed to erode, causing air and water quality to be degraded, both urban and rural residents suffer. To understand how soil can be protected, we must first understand its properties and how it is formed.

Soil Formation

A combination of physical and biological events is involved in the formation of soil. Soil building begins with the physical fragmentation of the **parent material,** which consists of ancient layers of rock or more recent geologic deposits from lava flows or glacial activity. The kind of parent material and the climate determine the kind of soil formed. Factors that can bring about fragmentation or chemical change of the parent material are known as **weathering.** Temperature changes and abrasion are two primary agents of **mechanical weathering.**

Heating a large rock can cause it to fracture because rock does not expand evenly. Pieces of the rock flake off and can be further reduced in size by other processes, such as the repeated freezing and thawing of water. Water that has seeped into rock cracks and crevices expands as it freezes, causing the cracks in the rock to widen. The subsequent thawing of the water allows it to fill the widened cracks, which are enlarged further by another period of freezing. Alternating freezing and thawing results in the fragmentation of large rock pieces into smaller fragments. (See figure 13.1.) The roots of plants growing in cracks can also exert enough force to cause rock to break.

Forces that cause rock particles to move and rub against each other also cause the physical breakdown of rock. A glacier causes rock particles to grind against one another, resulting in smaller fragments and smoother surfaces. These particles are carried by glaciers and are deposited when the ice melts. In many parts of the world, the parent material from which soil is formed consists of glacial deposits. Wind and moving water also cause small particles to collide, resulting in further weathering. The smoothness of rocks and pebbles in a stream or on the shore is evidence that moving water has caused them to rub together, removing their sharp edges. Similarly, particles carried by wind collide with objects, resulting in the fragmentation of both the object and the wind-driven particles.

In addition, wind and moving water remove small particles and deposit them at new locations, exposing new surfaces to the weathering process. For example, the landscape of the Painted Desert was created by a combination of wind and moving water that removed some easily transported particles, while rocks more resistant to weathering remained. (See figure 13.2.)

In addition to the forces of wind, moving water, glaciers, and changing temperature, certain chemical activities also alter the size and composition

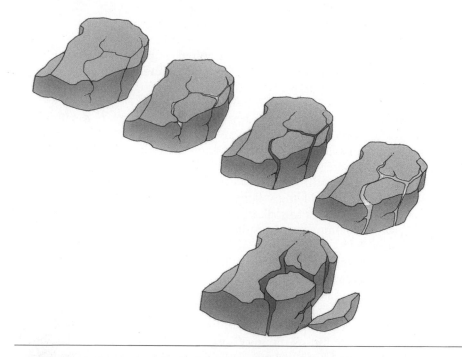

FIGURE 13.1

Physical Fragmentation by Freezing and Thawing. The cracks in the rock fills with water. As the water freezes and becomes ice, the pressure of the ice enlarges the crack. The ice melts, and water again fills the crack. The water freezes again and widens the crack. Alternating freezing and thawing splits the rock into smaller fragments.

FIGURE 13.2

The Painted Desert. This landscape was created by the action of wind and moving water. The particles removed by these forces were deposited elsewhere and may have become part of the soil in that new location.

of parent material and participate in the soil-building process. This process is called **chemical weathering.** Small rock fragments exposed to the atmosphere may be oxidized; that is, they combine with oxygen from the air and chemically change to different compounds. Other kinds of rock may combine with water molecules in a process known as hydrolysis. Often, the oxidized or hydrolyzed molecules are more readily soluble in water and, therefore, may be removed by rain or moving water. Since rain is normally slightly acid, the acid content assists in dissolving rocks.

The first organisms to gain a foothold in this modified parent material also contribute to the process of fragmentation. Lichens often form a pioneer community that traps small particles and chemically alters the underlying rock. As other kinds of organisms, like plants and small animals, become established, they contribute increasing amounts of organic matter, which are incorporated with the small rock fragments to form soil. Decaying organic matter called **humus** becomes mixed with the top layers of rock particles and constitutes a very important ingredient of soil. Humus supplies some of the nutrients needed by plants and also increases the acidity of the soil so that inorganic nutrients, which are more soluble under acidic conditions, become more available to plants. For example, wheat and corn grow best in soils with a pH between 5.5 and 7.0. A soil with a pH above 7.0 would be more productive if humus were added to increase the acidity. In an acidic soil, many nutrients are more soluble and can be carried away by water if there is heavy rainfall or if there are few plants to capture the nutrients. In addition to affecting nutrient availability, humus is also important in modifying soil texture. It helps to create a loose, crumbly texture that allows water to soak in and permits air to be incorporated into the soil, while a compact soil results in the water running off the soil and poor aeration.

Burrowing animals, soil bacteria, fungi, and the roots of plants are also part of the biological process of soil formation. One of the most important burrowing animals is the earthworm. One hectare of soil may support a population of five hundred thousand earthworms that can process as much as 9 metric tons of soil a year. These animals literally eat their way through the soil, resulting in further mixing of organic and inorganic material, which increases the amount of nutrients available for plant use. They often bring nutrients from the deeper layers of the soil up into the area where plant roots are more concentrated, thus improving the soil's fertility. Soil aeration and drainage are also improved by the burrowing of earthworms and other small soil animals. They also help to incorporate organic matter into the soil by collecting dead organic material from the surface and transporting it into burrows and tunnels.

Fungi and bacteria are important decomposers and serve as important links in many mineral cycles. (See chapter 4.) They, along with animals, reduce organic material to smaller particles and, therefore, improve the quality of the soil.

Over a period of time, this complex array of physical, chemical, and biological processes formed the soils we have today. Under ideal climatic conditions, soft parent material may develop into a centimeter of soil within fifteen years. Under poor climatic conditions, a hard parent material may require hundreds of years to develop into an equivalent amount of soil. In any case, soil formation is a slow process.

Soil Properties

Soil properties include soil texture, structure, atmosphere, moisture, biotic content, and chemical composition. **Soil texture** is determined by the size of the rock particles within the soil. The largest soil particles are gravel, which consists of fragments larger than 2.0 millimeters in diameter. Particles between 0.05 and 2.0 millimeters are classified as sand. Silt particles range from 0.002 to 0.05 millimeters in diameter, and the smallest particles are clay particles, which are less than 0.002 millimeters in diameter.

Large particles, such as sand and gravel, have many tiny spaces between them, which allow both air and water to flow through the soil. Water drains from this kind of soil very rapidly, often carrying valuable nutrients to lower soil layers, where they are beyond the reach of plant roots. Clay particles tend to be flat and are easily packed together to form waterproof layers. Soils with a lot of clay do not drain well and are poorly aerated. Because water does not flow through clay very well, clay soils tend to stay moist for longer periods of time and do not easily lose minerals to infiltrating water.

Human Influences on Ecosystems

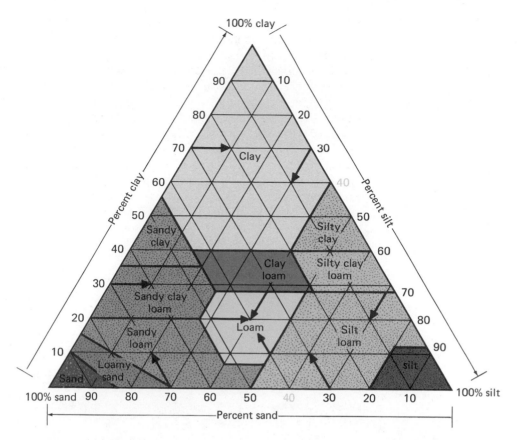

FIGURE 13.3

Soil Texture. Texture depends upon the percentage of clay, silt, and sand particles in the soil. Soils with the best texture for most crops are loams. As shown in the illustration, if a soil were 40 percent sand, 40 percent silt, and 20 percent clay, it would be a loam.

Source: Soil Conservation Service.

Rarely does a soil consist of a single size of particle. Various particles are mixed in many different combinations, resulting in many different soil classifications. (See figure 13.3.) An ideal soil is a **loam,** which combines the good aeration and drainage properties of large particles with the nutrient-retention ability of clay particles.

Soil structure is different from its texture. **Soil structure** refers to the way various soil particles clump together. Sand particles do not clump, and therefore, sandy soils lack structure, whereas clay soils tend to stick together in large clumps. A good soil forms small clumps that crumble easily. This ability to crumble is known as the soil's **friability.** The friability is determined by the soil structure and its moisture content. Sandy soils are very friable, while clay soils are not. If clay soil is worked when it is too wet, it can stick together in massive chunks that will not break up for years.

A good soil will crumble and has spaces to allow air and water to mix with the soil. In fact, the air and water content of the soil depends upon the presence of these spaces. (See figure 13.4.) In good soil, about two-thirds of the spaces contain air after the excess water has drained. The air in these spaces provides a source of oxygen for plant root cells. Water for the roots occupies the remaining soil space. The relationship between the amount of air and water is not fixed. After a heavy rain, all of the spaces may be filled with water. If some of the excess water does not drain from the soil, the plant roots may die from lack of oxygen. They are literally drowned. Also, if there is not enough soil moisture, the plants wilt from lack of water. Soil moisture and air are also important in determining the numbers and kinds of soil organisms.

Protozoa, nematodes, earthworms, insects, bacteria, and fungi are typical inhabitants of soil. (See figure 13.5.) The role of protozoa in the soil is not firmly established, but they seem to act as parasites on other forms

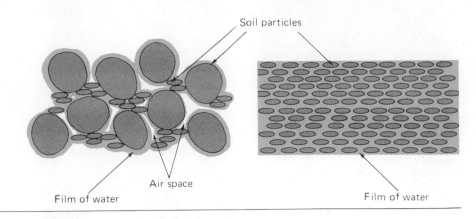

FIGURE 13.4

Pore Spaces and Particle Size. The soil on the left, which is composed of particles of various sizes, has spaces for both water and air. The particles have water bound to their surfaces (represented by the colored halo around each particle), but some of the spaces are so large that an air space is present. The soil on the right, which is composed of uniformly small particles, has less space for air. Since roots require both air and water, the soil on the left would be better able to support crops than that on the right.

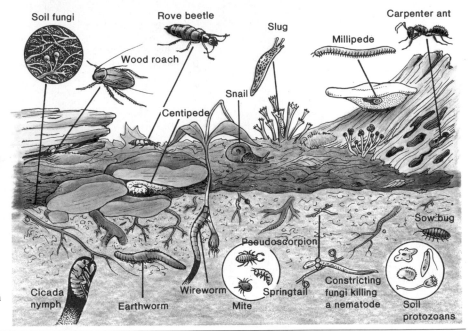

FIGURE 13.5

Soil Organisms. All of these organisms occupy the soil and contribute to it by rearranging soil particles, participating in chemical transformation, and recycling dead organic matter.

of soil organisms and, therefore, help to regulate the population size of other soil organisms. Nematodes, which are often called wireworms or roundworms, may aid in the recycling of dead organic matter. Some nematodes are parasitic on the roots of plants. Even though insects contribute to the soil by forming burrows and recycling organic materials, they are also major crop pests that feed on plant roots. Bacteria and fungi are particularly important in the decay and recycling of materials. Their chemical activities change complex organic materials into nutrients that can be used by plants. For example, these microorganisms can convert the nitrogen contained in the protein component of organic matter into ammonia or nitrate, nitrogen compounds that can be utilized by plants. The amount of nitrogen produced will vary with the type of organic matter, type of microorganisms, drainage, and temperature. All of these organisms are active within distinct layers of the soil, known as the soil profile.

Soil Profile

The **soil profile** is a series of horizontal layers of different chemical composition, particle size, and amount of organic matter. Each recognizable

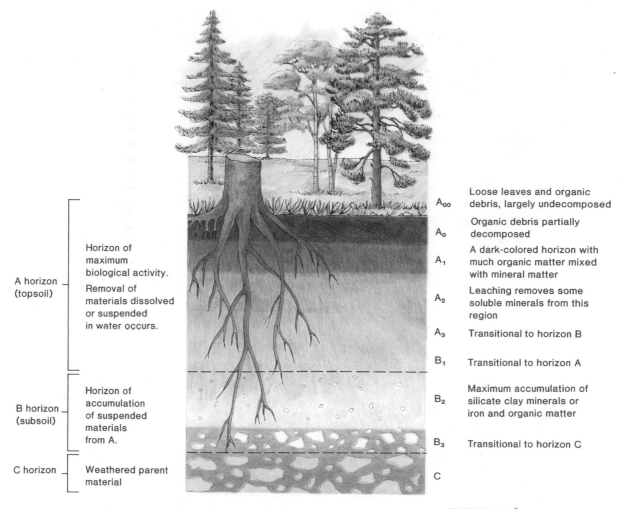

A horizon (topsoil)
- Horizon of maximum biological activity.
- Removal of materials dissolved or suspended in water occurs.

B horizon (subsoil)
- Horizon of accumulation of suspended materials from A.

C horizon
- Weathered parent material

A_{00} — Loose leaves and organic debris, largely undecomposed

A_0 — Organic debris partially decomposed

A_1 — A dark-colored horizon with much organic matter mixed with mineral matter

A_2 — Leaching removes some soluble minerals from this region

A_3 — Transitional to horizon B

B_1 — Transitional to horizon A

B_2 — Maximum accumulation of silicate clay minerals or iron and organic matter

B_3 — Transitional to horizon C

C

FIGURE 13.6

Soil Profile. A soil has layers that differ physically, chemically, and biologically. The top layer is known as the A horizon and contains most of the organic matter. The B horizon accumulates minerals and particles as water carries dissolved minerals downward from the A to the B horizon. The B horizon is often called the subsoil. Below the B horizon is a C horizon of weathered parent material.

layer is known as a **horizon.** (See figure 13.6.) The uppermost layer of the soil contains more nutrients and organic matter than the deeper layers. This layer is known as the A horizon or topsoil. The thickness of the A horizon may vary from less than a centimeter on steep mountain slopes to over a meter in the rich grasslands of the central United States. The majority of the living organisms and nutrients are found near the top of the A horizon. The lower portions of the A horizon often contain few nutrients because water flowing through the soil from the top to deeper layers dissolves and transports nutrients to the B horizon. This process is known as **leaching.** The B horizon, which is often called the subsoil, contains less organic material, fewer organisms, and accumulations of nutrients that were leached from higher levels. Because of this, the B horizon in many soils is a valuable source of nutrients for plants, and such subsoils support a well-developed root system. Because the amount of leaching is dependent on the available rainfall, areas of the world where rainfall is low may have a poorly developed B horizon.

The area below the subsoil is known as the C horizon, and it consists of weathered parent material. This parent material contains no organic materials, but it does contribute to some of the soil's properties. The chemical composition of the C horizon helps to determine the pH of the soil. The C horizon may also influence the soil's rate of water absorption and retention.

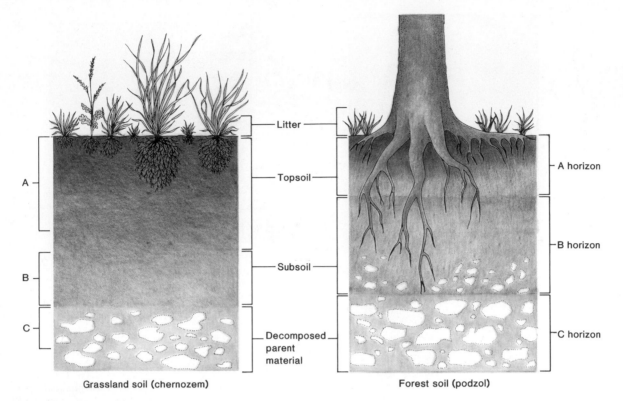

Litter

Topsoil

Subsoil

Decomposed parent material

A

B

C

Grassland soil (chernozem)

A horizon

B horizon

C horizon

Forest soil (podzol)

FIGURE 13.7

Major Soil Types. There are thousands of different soil types, but many of them can be classified into two broad categories. Soils formed in grasslands are known as chernozem soils and have a deep *A* horizon. The shallow *B* horizon does not have sufficient nutrients to support root growth. In forest soils, known as podzol soils, the *A* horizon is thinner, and leaching results in many nutrients in the *B* horizon. Thus roots are found in both the *A* and *B* horizons.

Soil profiles and the factors that contribute to soil development are extremely varied. Over fourteen thousand separate soil types have been classified in the United States. However, most of the cultivated land in the world can be classified as either grassland soil or forest soil. (See figure 13.7.)

Grassland soils usually have a deep *A* horizon. The low amount of rainfall in grassland areas limits the amount of leaching from the topsoil. As a result, the *A* horizon is deep and supports most of the root growth. This lack of leaching also results in a thin layer of subsoil, which is low in mineral and organic content and supports little root growth.

Forest soils develop in areas of more abundant rainfall. The rainfall results in a layer of topsoil that is usually not as thick as grassland soil, but the material leached from the topsoil forms a subsoil that supports substantial root growth. One of the materials that accumulates in the *B* horizon is clay. In some soils, particularly forest soils, clay or other minerals may accumulate and form an impermeable "hardpan" layer that limits the growth of roots and may prevent water from reaching the soil's deeper layers.

Desert soils have very poorly developed horizons since there is little rainfall to leach materials from upper layers to lower layers. In addition, the low amount of plant growth results in little contribution of organic matter to the soil. In cold, wet climates, typical of the northern parts of Europe and Canada, there may be considerable accumulations of organic matter since the rate of decomposition is reduced. Furthermore, the extreme acidity of these soils also reduces the rate of decomposition. Hot, humid climates also tend to have poorly developed soil horizons since the organic matter decays very rapidly, and soluble materials are carried away by the abundant rainfall.

In addition to the differences caused by the kind of vegetation and rainfall, topography influences the soil profile. (See figure 13.8.) On a relatively flat area, the topsoil formed by soil-building processes will collect in place

Human Influences on Ecosystems

| A horizon | } 15 cm |
| B horizon | } 25 cm |

| A horizon | }150 cm |
| B horizon | } 15 cm |

FIGURE 13.8
The Effect of Slope on a Soil Profile.
The topsoil formed on a large area of the hillside is continuously transported down the slope by the flow of water. It accumulates at the bottom of the slope and results in a thicker *A* horizon. The resulting "bottomland" is highly productive because it has a deep, fertile layer of topsoil, while the soil on the slope is less productive.

and gradually increase in depth. The topsoil formed on rolling hills or steep slopes is often transported down the slope as fast as it is produced. On such slopes, the accumulation of topsoil may not be sufficient to support a cultivated crop. The topsoil removed from these slopes is eventually deposited in the flat floodplains. These regions serve as collection points for topsoil that was produced over extensive areas. As a result, these river-bottom and delta regions have a very deep topsoil layer and are highly productive agricultural land.

Soil Erosion

Erosion is the wearing away and transportation of soil by water or wind. The Grand Canyon of the Colorado River, the floodplains of the Nile in Egypt, the little gullies on hillsides, and the deltas that develop at the mouths of rivers all attest to the ability of water to move soil. Anyone who has seen muddy water after a rainstorm has observed soil being moved by water. (See figure 13.9.) The force of moving water allows it to carry large amounts of soil. Each year, the Mississippi River transports over 325 million metric tons of soil from the central region of the United States to the Gulf of Mexico. This is equal to the removal of a layer of topsoil approximately 1 millimeter thick from the entire region. This movement of soil by water is repeated by every stream and river in the world. Dry Creek, a small stream in California, has only 500 kilometers of mainstream and tributaries; however, each year, it removes 180,000 metric tons of soil from a 340-square-kilometer area.

Badly eroded soil has lost all of the topsoil and some of the subsoil and is no longer productive farmland. Most current agricultural practices result in the loss of soil faster than it is being replaced. Farming practices that reduce erosion, such as contour farming and terracing, are discussed later in the chapter.

In addition to water, wind is an important mover of soil. Dust blown by the wind is soil movement. Under certain conditions, wind can move large amounts of soil. (See figure 13.10.) Wind erosion may not be as evident as water erosion, since it does not leave gullies on the surface of the soil. Nevertheless, it has the potential to be a serious problem. Wind erosion is most common in dry, treeless areas where the soil is left exposed.

FIGURE 13.9
Water Erosion. The force of moving water is able to pick up soil particles and remove them. The volume and speed of the moving water determine the size of the particles moved. Muddy streams show erosion in action.

FIGURE 13.10

Wind Erosion. The dry, unprotected topsoil from this field is being blown away. The force of the wind is capable of removing all of the topsoil and transporting it several thousand kilometers.

FIGURE 13.11

Wind Erosion in the Sahel. The semiarid region just south of the Sahara Desert is in an especially vulnerable position. The rainfall is unpredictable, which often leads to crop failures. In addition, population pressure forces people in this region to try to raise crops in marginal areas. This often results in increased wind erosion.

In the Sahel region of Africa, much of the land has been denuded of vegetation because of drought, overgrazing, and improper farming practices. This has resulted in extensive wind erosion of the soil. (See figure 13.11.) In the Great Plains region of the United States, there have been four serious periods of wind erosion since the area was settled in the 1800s. If this area receives less than 30 centimeters of rain per year, there is not enough moisture to support crops. When this occurs for several years in a row, it is called a drought. Farmers plant crops, hoping for rain. When the rain does not come, they plow their fields again to prepare it for another crop. Thus, the loose, dry soil is left exposed, and wind erosion results. Because of the large amounts of dust in the air during those times, the region is known as the Dust Bowl. During the 1930s, wind destroyed 3.5 million hectares of farmland and seriously damaged an additional 30 million hectares in the Dust Bowl. (See figure 13.12.)

Human Influences on Ecosystems

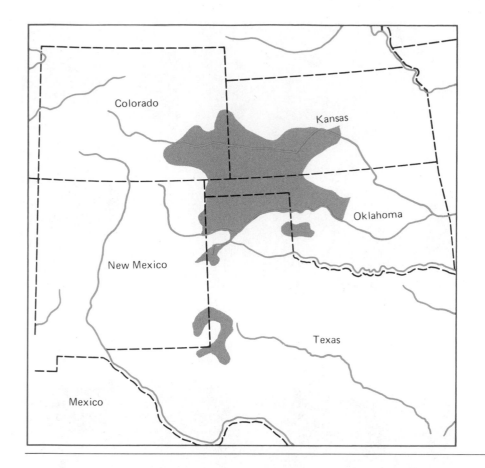

FIGURE 13.12
Areas Damaged by Wind Erosion during the Dust Bowls. Although wind erosion is a factor in most of the world, certain dry, treeless areas are more susceptible. In the Great Plains of the United States over 30 million hectares of land were damaged by wind erosion. These areas are indicated on the map.

Soil Conservation Practices

Whenever soil is lost by water or wind erosion, the topsoil, the most productive layer, is the first layer to be removed. When the topsoil is lost, the soil's fertility decreases, and expensive fertilizers must be used to restore the fertility that was lost. This has the effect of raising the cost of the food we buy. In addition, the movement of the excessive amounts of soil lost from farmland into streams has several undesirable effects. First, a dirty stream is less aesthetically pleasing than a clear stream. Second, a stream laden with sediment affects the fish population. Fishing may be poor because of unwise farming practices hundreds of kilometers upstream. Third, the soil carried by a river is eventually deposited somewhere. In many cases, this soil must be removed by dredging to clear shipping channels. We pay for dredging with our tax money, and it is a very expensive operation.

For all of these reasons, proper soil conservation measures should be employed to minimize the loss of topsoil. Figure 13.13 contrasts poor soil conservation practices with proper soil protection. When soil is not protected from the effects of running water, the topsoil is removed and gullies result. This can be prevented by slowing the flow of water over sloping land.

The kinds of agricultural activity that land can be used for are determined by soil structure, texture, friability, drainage, fertility, rockiness, slope of the land, amount and nature of rainfall, and other climatic conditions. The United States has a relatively large proportion of its land suitable for

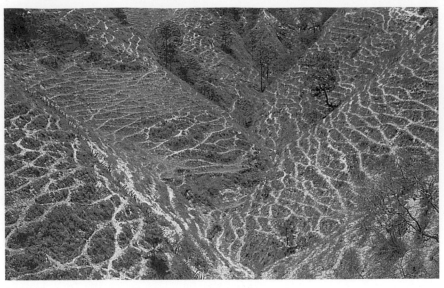

a.

b.

FIGURE 13.13

Poor and Proper Soil Conservation Practices. (*a*) This land is no longer productive farmland since erosion has removed the topsoil. (*b*) This rolling farmland shows strip contour farming to minimize soil erosion by running water. It should continue to be productive farmland indefinitely.

crop cultivation. (See figure 13.14.) Over 50 percent of the land can be used for raising crops. However, only 2 percent of that land does not require some form of soil conservation practice. This means that nearly all of the soil in the United States must be managed in some way to reduce the effects of soil erosion by wind or water.

Not all parts of the world are as well supplied with land that has agricultural potential. (See table 13.1.) For example, worldwide, approximately 11 percent of the land surface is suitable for crops, and an additional 24 percent is in permanent pasture. In the United States, about 21 percent is cropland, and 26 percent is in permanent pasture. This can be contrasted with the continent of Africa, in which only 6 percent is suitable for crops,

Human Influences on Ecosystems

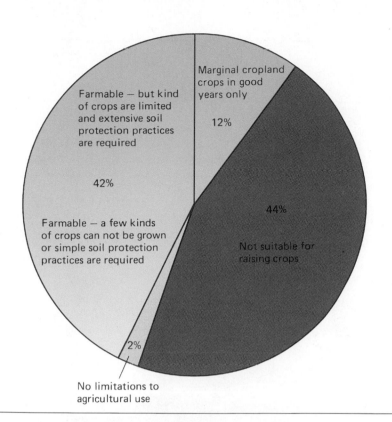

Farmable — but kind
of crops are limited
and extensive soil
protection practices
are required

42%

Marginal cropland
crops in good
years only

12%

44%

Not suitable for
raising crops

Farmable — a few kinds
of crops can not be grown
or simple soil protection
practices are required

2%

No limitations to
agricultural use

FIGURE 13.14
U.S. Land Suitable for Raising Crops.
Only 2 percent of the land in the United
States can be cultivated without some
soil conservation practices. This 2
percent is primarily flatland, which is not
subject to wind erosion. On 42 percent
of the remaining land, some special
considerations for protecting the soil are
required, or the kinds of crops are
limited. Twelve percent of the land is
marginal cropland that can only provide
crops in years when rainfall and other
conditions are ideal. Forty-four percent is
not suitable for cultivating crops but may
be used for other purposes, such as
grazing cattle or as forests.

TABLE 13.1
Percentage of land suitable for agriculture.

Country	Percent cropland	Percent pasture
World	11	24
Africa	6	26
Egypt	2.5	0
Ethiopia	12.7	41
Kenya	4	6.6
South Africa	11	65
North America	13	16.8
Canada	5	2.6
United States	21	26
South America	8	26
Argentina	13	52
Venezuela	4.3	19.7
Asia	17	24
China	10.8	30.6
Japan	13	1.6
Europe	30	18
USSR	10.4	16.8

Source: Data modified from *World Resources* 1987, published by the International Institute for
Environment and Development and the World Resources Institute.

and an additional 26 percent can be used for pasture. Europe has the highest
percentage of cropland with 30 percent, but it has only 18 percent in permanent pasture.

Few reserves of arable land remain in the temperate regions of Europe
and North America. Most undeveloped land that is suitable for cultivation
and that receives enough rainfall for agriculture is found in the rain forests
of South America and Africa.

BOX 13.1
Land Capability Classes

Not all land is suitable for raising crops, grazing, or urban building. Such factors as the degree of slope, soil characteristics, rockiness, erodibility, and other characteristics determine the best use for a parcel of land. In an attempt to encourage people to use land wisely, the U.S. Soil Conservation Service has established a system to classify land-use possibilities. The table shows the eight classes of land and lists the characteristics and capabilities of each.

Unfortunately, many of our homes and industries are located on type I and II land, which has the least restrictions on agricultural use. This does not make the best use of the land. Zoning laws and land-use management plans should consider the land-use capabilities and institute measures to assure that land will be used to its best potential.

Contour Farming

Contour farming, which involves tilling at right angles to the slope of the land, is one of the simplest methods to prevent soil erosion. This practice is useful on gentle slopes and produces a series of small ridges at right angles to the slope. (See figure 13.15.) Each ridge acts as a dam to hold water from running down the slope. This allows more of the water to soak into the soil. Contour farming reduces soil erosion by as much as 50 percent and, in drier regions, increases crop yields by conserving water.

Strip Farming

When a slope is too steep or too long, contour farming alone may not be an effective method of preventing soil erosion. However, a combination of contour and strip farming may still allow this land to be used for raising

	Land class	Characteristics	Capability	Conservation measures
Land suitable for cultivation	I	Excellent, flat, well-drained land	Cropland	None
	II	Good land; has minor limitations, such as slight slope, sandy soil, or poor drainage	Cropland Pasture	Strip cropping Contour farming
	III	Moderately good land with important limitations of soil, slope, or drainage	Cropland Pasture Watershed	Contour farming Strip cropping Terraces Waterways
	IV	Fair land with severe limitations of soil, slope, or drainage	Pasture Orchards Urban Industry Limited cropland	Crops on a limited basis Contour farming Strip cropping Terraces Waterways
Land not suitable for cultivation	V	Use for grazing and forestry; slightly limited by rockiness, shallow soil, or wetness	Grazing Forestry Watershed Urban Industry	No special precautions if properly grazed or logged; must not be plowed
	VI	Moderate limitations for grazing and forestry because of moderately steep slopes	Grazing Forestry Watershed Urban Industry	Grazing or logging may be limited at times
	VII	Severe limitations for grazing and forestry because of very steep slopes vulnerable to erosion	Grazing Forestry Watershed Recreation Wildlife Urban Industry	Careful management is required when used for grazing or logging
	VIII	Unsuitable for grazing and forestry because of steep slope, shallow soil, lack of water, or too much water	Watershed Recreation Wildlife Urban Industry	Not to be used for grazing or logging; steep slope and lack of soil present problems

FIGURE 13.15

Contour Farming. Tilling at right angles to the slope creates a series of ridges that slows the flow of the water and prevents soil erosion. This soil conservation practice is useful on gentle slopes.

FIGURE 13.16
Strip Farming. On rolling land, a combination of contour and strip farming prevents excessive soil erosion. The strips are planted at right angles to the slope, with bands of closely sown crops, such as wheat or hay, alternating with bands of row crops, such as corn or soybeans.

a.

b.

FIGURE 13.17
Terraces. Since the construction of terraces requires the movement of soil and the protection of the steep slope between levels, terraces are expensive to build. (*a*) The terraces seen here are extremely important for people who live in countries that have little flatland available. They require much energy and hand labor to maintain but make effective agricultural use of the land without serious erosion. (*b*) This modification of the terracing concept allows the use of the large farm machines typical of farming practices in Canada, Europe, and the United States.

crops. **Strip farming** consists of alternating strips of closely sown crops like hay, wheat, or other small grains with strips of row crops like corn, soybeans, cotton, or sugar beets. (See figure 13.16.) The closely sown crops retard the flow of water, which reduces soil erosion and allows more water to be absorbed into the ground. The type of soil, steepness, and length of slope dictate the width of the strips and determine whether strip or contour farming is practical.

Terracing

On very steep land, the only practical method of preventing soil erosion is to construct terraces. **Terraces** are level areas constructed at right angles to the slope to retain water and greatly reduce the amount of erosion. (See figure 13.17.) Terracing has been used for centuries in nations with a shortage of level farmland. The type of terracing seen in figure 13.17*a* requires the use of small machines and considerable hand labor and is not

a.

b.

FIGURE 13.18
**Protection of Waterways Prevents
Erosion.** (*a*) An unprotected waterway
has been converted into a gully. (*b*) A
well-maintained waterway. The waterway
is not cultivated; a strip of grass retards
the flow of water and protects the
underlying soil from erosion.

suitable for the mechanized farming typical in much of the world. Terracing is an expensive method of controlling erosion since it requires the movement of soil to construct the level areas, protection of the steep areas between terraces, and constant repair and maintenance. Many factors, such as length and steepness of slope, type of soil, and amount of precipitation, determine whether terracing is feasible.

Waterways

Even with such soil conservation practices as contour farming, strip farming, and terracing, protected channels for the movement of water often must be provided. **Waterways** are depressions on sloping land where water collects and flows off the land. When not properly maintained, these areas are highly susceptible to erosion. (See figure 13.18.) If a waterway is maintained with a permanent sod covering, the speed of the water is reduced, and soil erosion is decreased.

Soil and Its Use

301

a.

b.

FIGURE 13.19

Windbreaks. (*a*) In sections of the Great Plains, trees provide protection from wind erosion. The trees along the road are planted in a north-south direction to protect the land from the prevailing westerly winds. (*b*) In this field, temporary strips of vegetation serve as windbreaks.

Windbreaks

Contour farming, strip farming, terracing, and maintaining waterways are all important ways of reducing water erosion, but wind is also a problem with certain soils, particularly in relatively dry areas of the world. Wind erosion can be reduced if the soil is protected from the wind. The best protection is a layer of vegetation over the surface of the soil. However, the process of preparing soil for planting and the method of planting often leave the soil exposed to wind. **Windbreaks** are plantings of trees or other plants that protect bare soil from the full force of the wind. Windbreaks reduce the velocity of the wind, thereby decreasing the amount of soil that it can carry away. (See figure 13.19.) In some cases, rows of trees are planted at right angles to the prevailing winds to reduce the force of the wind, while in other cases, a kind of strip farming is practiced in which strips of hay or grains are alternated with row crops that leave large amounts of the soil exposed. In some areas of the world, the only way to protect the soil is not to cultivate it at all, but to leave it in a permanent cover of grasses.

Conservation Tillage

Conventional tillage methods in much of the world require extensive use of farm machinery to prepare the soil for planting and to control weeds. Typically, a field is plowed and then disked or harrowed one to three times before the crop is planted. The plowing, which turns the soil over, has several desirable effects: Any weeds or weed seeds are buried, thus reducing the weed problem in the field. Crop residue from previous crops is incorporated into the soil where it will decay faster and contribute to soil structure. Nutrients that had been leached to deeper layers of the soil are brought near the surface. And the dark soil is exposed to the sun so that it warms up faster. This last effect is most critical in areas with short growing seasons. In many areas, fields are plowed in the fall of the year, after the crop has been harvested, and the soil is left exposed all winter.

After plowing, the soil is worked by disks or harrows to break up any clods of earth, kill remaining weeds, and prepare the soil to receive the seeds. After the seeds are planted, there may still be weed problems. Farmers often must cultivate row crops to kill the weeds that begin to grow between the rows of the crop. Each trip over the field costs the farmer money, while at the same time increasing the amount of time the soil is exposed to wind or water erosion.

In recent years, several innovations in chemical herbicides and farm equipment have led to the development of tilling practices that protect the soil by reducing the amount of time the soil is exposed to erosion forces. Reduced tillage uses less cultivation to control weeds and to prepare the soil to receive seeds. Selective herbicides are used to kill unwanted vegetation prior to planting of the new crop and to control weeds after the crop has been planted. Several different variations of conservation tillage are used:

1. Plowing followed by reduced secondary tillage. This usually involves plowing followed immediately by planting.
2. Strip tillage, a method that involves tilling only in the narrow strip that is to receive the seeds. The rest of the soil and the crop residue from the previous crop is left undisturbed.
3. No-till farming involves special planters that place the seeds in slits cut in the soil that still has on its surface the crop residue from the previous crop.

All of these types of reduced tillage use less fuel and less time but may require the use of more herbicides. (See table 13.2.) For many kinds of crops, yields are comparable to that produced by conventional tillage methods.

Other positive effects of reduced tillage, in addition to reducing erosion, are:

1. The amount of winter food and cover available for wildlife increases, which can lead to increased wildlife populations.
2. Since there is less runoff, the amount of siltation in streams and rivers is reduced. This results in clearer water for recreation and a reduction in the need to dredge waterways to keep them open for shipping.
3. Reduced tillage also allows planting of row crops on hilly land that cannot be converted to such crops under conventional tilling methods. This allows a farmer to convert low-value pasture land into cropland that gives a greater economic yield.

TABLE 13.2
Comparison of various tilling methods.

Tillage method	Hours required per 100 hectares	Liters of fuel per 100 hectares	Cost of herbicide per 100 hectares (1973 $U.S.)
Conventional	200	2,915	$2,717
Reduced secondary tillage	125	1,390	2,717
Strip tillage	95	1,020	2,717
No tillage	62	375	4,000

Source: Data from J. E. Beuerlein and S. W. Bone, "Selecting a Tillage System" Extension Publication, Ohio State University; and D. H. Doster, "Economics of Alternative Tillage Systems," in *Bulletin of the Entomological Society of America* 22 (1976):297.

4. Since fewer trips are made over the field, petroleum is saved, even if the feedstocks necessary to produce the herbicides are taken into account.

5. Reduced tillage may also allow for the growing of two crops on a field in areas that had previously been restricted to growing one crop per field per year. In some areas, immediately after harvesting wheat, farmers have used reduced tillage methods to plant soybeans in the wheat stubble.

6. Because reduced tillage reduces the number of trips made over the field by farm machinery, the soil does not become compacted as quickly.

However, there are also some drawbacks to the use of reduced-tillage methods:

1. The residue from previous vegetation may delay the warming of the soil, which may, in turn, delay planting some crops for several days.

2. The crop residue reduces evaporation from the soil and the upward movement of water and soil nutrients from deeper layers of the soil, which may retard the growth of plants.

3. The accumulation of plant residue can harbor plant pests and diseases that will require increased expenditures for insecticides and fungicides.

Reduced tillage is not the complete answer to soil erosion problems but may be useful in reducing soil erosion on well-drained soils. It also requires that farmers pay close attention to the condition of the soil and the pests to be dealt with.

In 1972, 12 million hectares in the United States were under some form of conservation tillage. About 1.3 million hectares were being farmed using no-till methods. By 1982, this had risen to 40 million hectares of conservation-tilled land, of which 4.2 million hectares were being farmed using no-till methods. It is estimated that, by the year 2010, 95 percent of U.S. cropland will be under some form of reduced-tillage practice.

Uses of Nonfarm Land

Not all land is suitable for crops or continuous cultivation. Some must never be plowed for use as cropland, but it can still be used by people to provide for some of their needs. By using appropriate soil conservation practices,

Human Influences on Ecosystems

FIGURE 13.20

Noncrop Use of Land to Raise Food.
As long as it is properly protected and managed, this land can produce food through grazing, but it should never be plowed to plant crops because the topsoil is too shallow and the rainfall is too low.

FIGURE 13.21

Forest and Recreational Use. Although this land is not capable of producing crops or supporting cattle, it furnishes lumber, a habitat for wildlife, and recreational opportunities.

much of the land not usable for crops can be used for grazing, wood production, wildlife production, or scenic and recreational purposes. Figure 13.20 shows land that is not suitable for cultivation but that is capable of producing grass for cattle or sheep. Such land, if properly used, will produce food for many generations.

The land shown in figure 13.21 is not suitable for either crops or grazing. However, it is still a valuable and productive piece of land since it can be used to furnish lumber, wildlife habitats, and recreational opportunities.

All land is not equal. Each section has its own soil characteristics, climate, and degree of slope. When these are all taken into consideration, a proper use can be determined for each portion of the planet. Wise planning and careful husbandry of the soil is necessary if the land and its soil are to be able to provide food and other necessities of life.

Summary

Soil is an organized mixture of minerals, organic material, living organisms, air, and water. Soil formation begins with the physical breakdown of the parent material by such physical processes as changes in temperature, freezing and thawing, and movement of particles by glaciers, flowing water, or wind. Oxidation and hydrolysis can chemically alter the parent material. Organisms also affect soil building by burrowing and mixing the soil, by releasing nutrients, and by their decomposition.

Topsoil contains a mixture of humus and inorganic material, both of which supply soil nutrients. Soil fertility is determined by the inorganic matter, organic matter, water, and air spaces in the soil. The mineral portion of the soil consists of various mixtures of sand, silt, and clay particles.

A soil profile consists of the A horizon, which is rich in organic matter; the B horizon, which accumulates materials leached from the A horizon; and the C horizon, which consists of slightly altered parent material. Forest soils typically have a shallow A horizon and a deep, nutrient-rich B horizon with much root development. Grassland soils usually have a thick A horizon and very few nutrients in the thin B horizon. Therefore, most of the roots of the grasses are in the A horizon.

Soil erosion is the wearing away and transportation of soil by water or wind. Proper use of such conservation practices as contour farming, strip farming, terracing, waterways, windbreaks, and conservation tillage can reduce soil erosion. Misuse of soil reduces the soil's fertility and causes air- and water-quality problems. Land unsuitable for crops may still be used for grazing, lumber, wildlife habitats, or recreation.

1. How are soil and land different?
2. Name the five major components of soil.
3. Describe the process of soil formation.
4. Name five physical and chemical processes that break parent material into smaller pieces.
5. In addition to fertility, what other characteristics determine the usefulness of soil?
6. How does soil particle size affect texture and drainage?
7. Describe a soil profile.
8. Define erosion.
9. Describe three soil conservation practices that help to reduce soil erosion.
10. Besides cropland, what are other possible uses of soil?
11. What kinds of information are important in determining land use?

CHAPTER FOURTEEN
Agricultural Methods and Pest Management

Objectives

After reading this chapter, you should be able to:

Explain how the invention of new farm machinery encouraged the development of a monoculture type of farming.

List the advantages and disadvantages of monoculture farming.

Explain why chemical fertilizers are used.

Understand that, while fertilizer does replace soil nutrients, it does not furnish organic materials for the soil, and the soil characteristics are altered.

Explain why modern agriculture makes extensive use of pesticides.

Differentiate between hard pesticides and soft pesticides.

Explain how chemicals can be used to delay or accelerate the harvesting of a crop.

List four problems associated with pesticide use.

Define biological amplification.

Define organic farming.

Explain why integrated pest management depends upon a complete knowledge of the pest's life history.

Chapter Outline

Key Terms

auxin
bioaccumulation
biocide
biological amplification
carbamate
chlorinated hydrocarbon
fungicide
hard pesticide
herbicide
insecticide
integrated pest
   management
macronutrient

micronutrient
monoculture
nontarget organism
organophosphate
persistent pesticide
pest
pesticide
pheromone
rodenticide
soft pesticide
target organism
weed

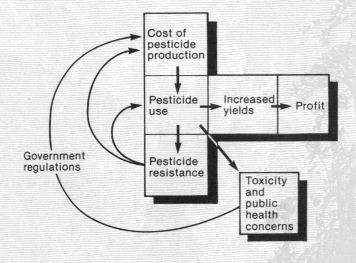

FIGURE 14.1

Slash-and-Burn Agriculture. In many areas of the world where the soils are poor, crops can be raised if only small areas of the ecosystem are disturbed. The burning of vegetation releases nutrients that can be used by crops for one or two years before the soil is exhausted. The return of the natural vegetation prevents erosion and repairs the damage done by temporary agricultural use.

Differing Agricultural Methods

Although many countries have highly mechanized methods of farming, there are still many places in the world where this is not true. In many parts of the world with poor soil and low populations, a form of agriculture known as "slash and burn" is practiced. This involves the felling of the forest in a small area, followed by the burning of the trees and other vegetation on the site. This releases nutrients that are tied up in the biomass and allows a crop or two to be raised before the soil is exhausted. Once the soil is no longer suitable for the raising of crops, the site is abandoned. The surrounding forest quickly recolonizes the area, and the area returns over a period of time to the original forest through the process of succession. This method is particularly useful on thin tropical soils and on sites where the slope may be steep. The small size of the opening in the forest and its temporary existence prevent widespread damage to the soil, and erosion is minimized. The gardens that are planted are typically a mixture of plants, rather than a regular set of rows containing single species of plants. (See figure 14.1.)

The traditional practices of the people who engage in slash-and-burn agriculture have been developed over hundreds of years and often are found to be more effective than other methods of gardening. For example, the mixing of plants may have a beneficial effect, since shade-requiring species may be helped by taller plants, or nitrogen-fixing legumes may provide needed nutrients for species that have a nitrogen requirement. In addition, the mixing of species may reduce insect pest problems because molecules produced by some plants are natural insect repellants. The small, isolated, temporary nature of the gardens also reduces the likelihood of insect plagues. Obviously, this form of agriculture is limited to a very small number of people in specialized habitats and is not the most efficient way to farm all kinds of habitats. Slash-and-burn farming also requires a long recovery time before the forest can be cleared for another cycle of agriculture.

In many other parts of the world, extensive areas are farmed with a great deal of manual labor. Two different situations favor this kind of farming: (1) when the growing site does not allow for mechanization, and (2) when the kind of crop does not allow it. Soils or terrain that require that fields

FIGURE 14.2
Labor-Intensive-Agriculture. In many of the less-developed countries of the world the extensive use of hand labor allows for impressive rates of production with a minimal input of fossil fuels and fertilizers. This kind of agriculture is also necessary in areas that have only small patches of land suitable for farming.

be small discourage the development of mechanization, since large tractors and other machines cannot be used efficiently on small, oddly shaped fields. Many mountainous areas of the world fit into this category. In addition, some crops require such careful handling in planting, weeding, or harvesting that large amounts of hand labor are required. The planting of paddy rice and the harvesting of many fruits and vegetables are examples.

However, the primary reason for large amounts of labor-intensive farming in the world is economic. Many densely populated countries have a large number of small farms that can be effectively managed with human labor, supplemented by that of draft animals and a few small gasoline-powered engines. (See figure 14.2.) In addition, in many of the less developed regions of the world, the cost of labor is low, which encourages the use of hand labor to do planting, weeding, and other activities rather than mechanization. Mechanization requires large tracts of land that could only be accumulated by the expenditure of large amounts of money or the development of larger cooperative farms from many small units. Even if social and political obstacles to the establishment of such large land holdings could be overcome, there is still the problem of obtaining the necessary capital to purchase the machines necessary for mechanization. Large parts of the developing world fit into this category, including much of Africa, many areas in Central and South America, and many areas in Asia.

Mechanized agriculture is typical of North America, much of Europe, the USSR, South America, and many other parts of the world where money and land are available to support this form of agriculture. In large measure, machines and fossil-fuel energy replace the energy formerly supplied by human and animal muscles. Mechanization requires large expanses of fairly level land for the machines to operate efficiently. In addition, large expanses of land must be planted in the same crop for efficiency of planting, cultivating, and harvesting. Such large tracts of land planted in the same crop is known as **monoculture.** (See figure 14.3.) Small sections of land

FIGURE 14.3

Monoculture. This wheat field is an example of monoculture, a kind of agriculture that is highly mechanized and requires large fields for the efficient use of machinery. In this kind of agriculture, machines and fossil-fuel energy have replaced the energy of humans and draft animals.

with many kinds of crops grown on them require many changes of farm machinery, which takes time. Also, many crops interspersed with one another reduces the efficiency of farming operations because farmers must skip certain parts of the field, which increases travel time and uses expensive fuel.

Even though monoculture is an efficient method of producing food, it is not without serious drawbacks. With mechanized monoculture, large farm machines are used to plow, plant, and harvest vast expanses of land at the same time. When large tracts of land are not covered by vegetation, soil erosion increases. (See chapter 13.)

Mechanized farming methods also result in the removal of much of the organic matter each year. When the crop is harvested, a major portion of the plant typically is removed for use as food or some other purpose. This means that little organic matter is returned to the soil unless the farmer specifically plants a crop that is later plowed under to increase the soil's organic content.

To ensure that a crop can be planted, tended, and harvested efficiently, farmers rely on hybrid seed that provides uniform plants with the characteristics suitable for mechanized farming. These hybrid plants have little genetic variety. When all the farmers in an area plant the same varieties, pest control becomes a serious problem. If diseases or pests begin to spread, the magnitude of the problem becomes devastating because all the plants have the same characteristics and, thus, are susceptible to the same diseases. If genetically diverse crops are planted, or crops are rotated from year to year, this problem is not as great.

Because farm equipment is expensive, farmers tend to specialize in a few crops. This means that the same crop may be planted in the same field for several years in a row. This lack of crop rotation may deplete certain essential soil nutrients, thereby requiring special attention be paid to soil chemistry. In addition, planting the same crop repeatedly encourages the growth of insect and fungus pest populations because they have a huge food supply at their disposal. This requires the frequent use of insecticides and fungicides or some other method of pest control.

Even though problems are associated with mechanized, monoculture agriculture, it has greatly increased the amount of food available to the world

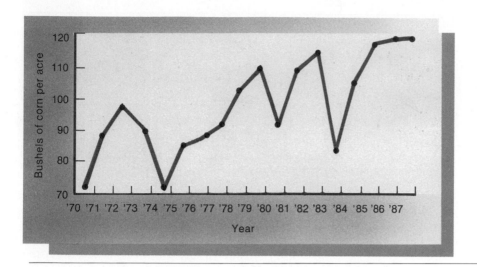

FIGURE 14.4

Increased Yields Resulting from Modern Technology. The increased yields in many parts of the world are the result of a combination of factors, including better genetic qualities in the seed, improved agricultural methods, the application of fertilizers and pesticides, and more efficient machinery.

Source: Data from Department of Agriculture, *Agriculture Statistics.*

over the last one hundred years. Yields per hectare of land being farmed have increased over much of the world and particularly in the developed world, which includes the United States. (See figure 14.4.) This increase has been the result of a combination of factors, including improved varieties of crops, better farming methods, the use of agricultural chemicals, more efficient machines, and the use of energy-intensive technology as opposed to labor-intensive processes.

Energy Versus Labor

The development of mechanized agriculture has resulted in the substitution of the energy stored in petroleum products for the labor of humans. For example, in the United States in 1913, it required 135 hours of labor to produce 2,500 kilograms of corn. In 1980, it required about 15 hours of labor to produce 2,500 kilograms of corn. The energy supplied by petroleum products replaced the equivalent of 120 hours of labor. Energy is needed for tilling, planting, harvesting, and pumping irrigation water. In addition, much of the energy is in the form of manufactured fertilizers, herbicides, fungicides, and insecticides that encourage crop growth while inhibiting weeds and reduce the impact of pests that feed on the crop or cause disease. For example, about 5 metric tons of fossil fuel are required to produce about 1 metric ton of fertilizer. Since the developed world is dependent on oil to provide energy and organic molecules needed to manufacture pesticides to support its agriculture, any change in the availability or cost of oil will have a major impact on the world's ability to feed itself.

The Impact of Fertilizer

Table 14.1 shows that, in some respects, fertilizer can be used to replace human labor. The United States uses twice as much fertilizer as either Brazil or India and has a cereal grain yield twice that of Brazil or India with only 3 percent of its work force in agriculture. The yields are similar for India and Brazil. However, Brazil uses about one-third less labor to accomplish the same yield. In this case, fertilizer appears to directly substitute for labor, but the use of fertilizers also has brought about a significant increase in yields even when reduced labor is taken into account.

TABLE 14.1
Fertilizer and food production.

Country	Percentage of the work force in agriculture	Fertilizer used (kilograms/hectare)	Yield in 100 kilograms/hectare (cereal grains)
United States	3	93	44
Brazil	39	49	18
India	64	52	16

Source: Data from Population Reference Bureau, Inc.

Approximately 25 percent of the world's crop yield is estimated to be directly attributed to the use of chemical fertilizers. Thus, if the world stopped using chemical fertilizers, food production would decline by 25 percent. At present, the world demand for fertilizer is doubling every ten years. Since fertilizer production relies on oil, the price and availability of chemical fertilizers is strongly influenced by the price of oil on the world market. If the price of oil increases, the price of fertilizer goes up, as does the cost of food. This will be felt most acutely in parts of the world where money is in short supply, since the farmers will be unable to buy fertilizer, and crop yields will fall accordingly.

Chemical fertilizers are valuable because they replace the soil nutrients removed by plants. Some plant nutrients, such as carbon, hydrogen, and oxygen, are easily replaced by carbon dioxide from the air and water from the soil, but others are less easily found. The three primary soil nutrients often in short supply are potassium, phosphorus, and nitrogen compounds. They are often referred to as **macronutrients** and are the common ingredients of chemical fertilizers. Their replacement is important because, when the crop is harvested, the chemical elements that are a part of the crop are removed from the field. Since many of those elements originated from the soil, they need to be replaced if another crop is to be grown. Certain other elements are necessary in extremely small amounts and are known as **micronutrients.** Examples are boron, zinc, manganese, and many others. As an example of the difference between macronutrients and micronutrients, harvesting a metric ton of potatoes removes 10 kilograms of nitrogen (a macronutrient) but only 13 grams of boron (a micronutrient). When the same crop is grown repeatedly in the same field, certain micronutrients may be depleted, resulting in reduced yields. These necessary elements can be returned to the soil in sufficient amounts by having them incorporated into the fertilizer the farmer applies to the field.

Although chemical fertilizers replace inorganic nutrients, they do not replace soil organic matter. Organic material is important because its decay produces humus, which modifies the structure of the soil, preventing compaction and maintaining pore space, which is necessary for the movement of water and air to the roots. Organic matter is also important to maintain proper soil chemistry because it tends to lower the pH of the soil, which helps to release certain soil nutrients for use by plants. Soil bacteria and other organisms use organic matter as a source of energy. Since these organisms serve as important links in the carbon and nitrogen cycles, the presence of organic matter is important to their function. Thus, total dependency upon chemical fertilizers can change the physical, chemical, and biotic properties of the soil.

As water moves through the soil, it gathers soil nutrients and carries them into streams and lakes, where they may encourage the growth of un-

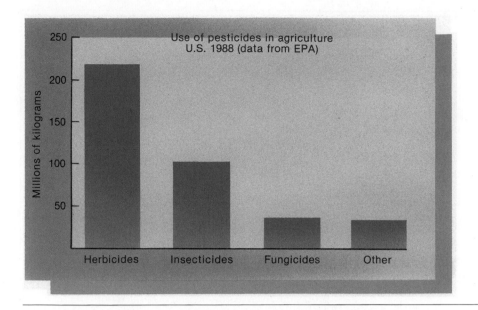

FIGURE 14.5
Pesticide Use in the United States. The most widely used pesticides are herbicides to control weeds and facilitate harvest of crops. Insecticides are also widely used.

Data from Environmental Protection Agency.

wanted plants and algae. This is particularly true when fertilizers are applied at the wrong time of the year, just before a heavy rain, or in such large amounts that the plants cannot efficiently remove them from the soil before they are lost. These ideas are covered in greater detail in chapter 15 on water pollution.

Pesticides

In addition to chemical fertilizers, mechanized monoculture requires the use of large amounts of other agricultural chemicals, such as pesticides, growth regulators, and preservatives. These chemicals have specific scientific names but are usually categorized into broad groups based on their effects. A **pesticide** is any chemical used to kill or control the populations of unwanted fungi, animals, or plants, often called **pests.** The term *pest* is not a scientific term but refers to any organism that is unwanted. Insects that feed on crops are pests, while others, like bees, are beneficial for pollinating plants. Unwanted plants are generally referred to as **weeds.**

Pesticides can be subdivided into several categories based on the kinds of organisms they are used to control. **Insecticides** are used to control insect populations by killing them. The growth of unwanted fungal pests that can weaken plants or destroy fruits is controlled by **fungicides.** Mice and rats are killed by **rodenticides,** and plant pests are controlled by **herbicides.** A perfect pesticide would be one that kills or inhibits the growth of the specific pest organism causing a problem. The specific pest is often referred to as the **target organism.** However, most pesticides are not very specific and kill many **nontarget organisms** as well. For example, most insecticides kill both beneficial and pest species, rodenticides kill other kinds of animals as well as rodents, and most herbicides kill a variety of different kinds of plants, both pests and nonpests. Therefore, the various pesticides do not just kill pests, but can kill a large variety of living things, including humans. Because of this, these chemicals might be more appropriately called **biocides,** since they kill many kinds of living things. (See figure 14.5.)

BOX 14.1
Regulation of Pesticides

The Environmental Protection Agency (EPA) is charged by Congress to protect the nation's land, air, and water systems. Under a mandate of national environmental laws focused on air and water quality, solid waste management, the control of toxic substances, pesticides, noise, and radiation, the agency strives to formulate and implement actions that lead to a compatible balance between human activities and the ability of natural systems to support and nurture life.

To fulfill this mandate, the EPA is charged with the enforcement of various federal laws and acts. One such act governs the registration of pesticides. Section 3 (c) (1) of the Federal Insecticide, Fungicide, and Rodenticide Act (FIFRA) states:

Procedure for Registration

1. *Statement required. Each applicant for registration of a pesticide shall file with the Administrator a statement which includes—*
   A. *the name and address of the applicant and of any other person whose name will appear on the labeling;*
   B. *the name of the pesticide;*
   C. *a complete copy of the labeling of the pesticide, a statement of all claims to be made for it, and any directions for its use;*
   D. *except as otherwise provided in subsection (c) (2) (D) of this section, if requested by the Administrator, a full description of the tests made and the results thereof upon which the claims are based, or alternately, a citation to data that appears in the public literature or that previously had been submitted to the Administrator and that the Administrator may consider in accordance with the following provisions;*
   E. *the complete formula of the pesticide; and*
   F. *a request that the pesticide be classified for general use, or restricted use, or for both.*

If a corporation wants to manufacture and market a pesticide within the United States, the pesticide must not adversely affect the environment. There is no specific format to test a product's effects on the environment, but some of the tests conducted include the following:

1. **Degradation**—A determination of the physical and chemical methods of decay and an identification of the decay products.

2. **Metabolism**—A determination of the organisms that act on the pesticide and the metabolic products released.
3. **Mobility**—A determination of where the molecules are likely to go and what route they will follow in the ecosystem.
4. **Accumulation**—Do the pesticide molecules accumulate in living tissue? If so, in what form and in what quantities?
5. **Hazard**—Evaluation of the possible hazard to plants, microorganisms, wildlife, and humans, whether target or nontarget organisms.

After receiving the information submitted in the procedure for registration, the Administrator of EPA may register the pesticide for use under the provisions of Section 3 (c) (5) of FIFRA:

Approval of Registration
The Administrator registers a pesticide if he or she determines that, when considered with any restrictions imposed under subsection (d),

1. *its composition is such as to warrant the proposed claims for it;*
2. *its labeling and other material required to be submitted comply with the requirements of this Act;*
3. *it will perform its intended function without unreasonable adverse effects on the environment; and*
4. *when used in accordance with widespread and commonly recognized practice, it will not generally cause unreasonable adverse effects on the environment.*

The Administrator also publishes all applications and supporting data so that various governmental agencies and the public have an opportunity to comment.

The FIFRA is designed to protect the environment from being damaged by the use of pesticides. The law places the burden upon the manufacturer to prove that a pesticide is safe. One source estimates that eight to twelve years and $50 million are needed to bring a major new pesticide from discovery to first registration. These figures *do not* include the capital cost of constructing a facility to manufacture the pesticide.

Insecticides

For centuries, insects consumed a large proportion of the crops produced by farmers. In small garden plots, insects can be controlled by manually removing them and killing them. However, in large fields, this is not practical, and people have sought other ways to control pest insects. In addition, many insects harm humans because they spread diseases, such as sleeping sickness, bubonic plague, and malaria. Mosquitoes are known to carry over thirty diseases harmful to humans. The discovery of chemicals that could kill insects was a major advance in the control of disease and the protection of crops from insect damage.

Nearly three thousand years ago, the Greek poet Homer mentioned the use of sulfur to control insects. For centuries, it was known that natural plant products were able to repel or kill insect pests. Plants known to have insect-repelling abilities were interplanted with crops to help control insect pests. Nicotine from tobacco, rotenone from tropical legumes, and pyrethrum from chrysanthemums are extracted and still used today to control insects. However, because they are difficult to apply and their effects are short-lived, other compounds were sought. In 1867, the first synthetic inorganic insecticide, Paris green, was formulated. It was a mixture of acetate and arsenide of copper and was used to control Colorado potato beetles.

The first synthetic organic insecticide to be used was DDT. It was originally thought to be the perfect insecticide. It was long-lasting, relatively harmless to humans, and very deadly to insects. During the first ten years of its use (1942–1952), DDT is estimated to have saved five million lives, primarily because of its use to control disease-carrying mosquitoes. However, after a period of use, many mosquitoes and other insects became tolerant of DDT. In 1965, mosquitoes were estimated to be responsible for causing the spread of diseases like malaria and sleeping sickness to one hundred million people. The early success of DDT and its loss of effectiveness caused a search for additional synthetic organic compounds that could be used to replace DDT in the control of insect pests. Over sixty thousand different compounds have been synthesized that have potential as insecticides. However, most of these have never been put into production because of cost, human health effects, or other drawbacks that make them unusable. Several different categories of these compounds have been developed. Three that are currently used are chlorinated hydrocarbons, organophosphates, and carbamates.

Chlorinated Hydrocarbons

Chlorinated hydrocarbons are a group of pesticides of complex, stable structure that contain carbon, hydrogen, and chlorine. DDT (**d**ichloro**d**iphenyl**t**richloroethane) was the first such pesticide manufactured, but several other compounds have also been developed. The chemical structure of DDT is shown in figure 14.6. Other forms of chlorinated hydrocarbons are chlordane, aldrin, heptachlor, dieldrin, and endrin. It is not fully understood how these compounds work, but they are believed to affect the nervous systems of insects, resulting in their death.

One of the major characteristics of these pesticides is that they are very stable chemical compounds. This is both an advantage and a disadvantage. Since they are long-lasting **persistent pesticides,** they can be applied once and be effective for long periods of time. They are sometimes known as

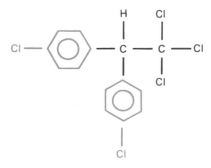

FIGURE 14.6
The Chemical Structure of DDT. This diagram shows the arrangement of the atoms in a molecule of DDT. The two chlorophenyl portions of the molecule are shown in blue. The other chlorines are shown in red, and the remainder of the ethane molecule appears black.

hard pesticides as well. However, since they do not break down easily, they tend to accumulate in the environment and affect many nontarget organisms, not just the insects that were the original target. In temperate regions of the world, DDT has a half-life (the amount of time required for half of the chemical to decompose) of from ten to fifteen years. This means that, if 1 metric ton of DDT were sprayed over an area, 500 kilograms would still be present in the area ten to fifteen years later, and thirty years from the date of application, 250 kilograms would still be present. The half-life varies, depending on soil type, temperature, the kinds of soil organisms present, and other factors. In tropical parts of the world, the half-life may be as short as six months. Persistent pesticides tend to accumulate in the soil and in the bodies of animals in the food chain. An additional complication is that hard pesticides may break down into products that are still harmful.

Because of the negative effects, most of the chlorinated hydrocarbons have been removed from use. DDT, aldrin, dieldrin, toxaphene, chlordane, and heptachlor have been banned in the United States and some other countries. However, many developing countries still rely on chlorinated hydrocarbons for insect control for public health reasons and to protect crops.

Organophosphates and Carbamates

Because of the problems associated with hard pesticides, a new generation of **soft pesticides** that decompose to harmless products in a few hours or days have been developed for use. However, like other pesticides, these are not species specific and kill beneficial insects as well as harmful ones. They are also much more toxic to humans and other vertebrates than the older hard pesticides. Although the short half-life is important in preventing the accumulation of toxic material in the environment, it is a disadvantage to farmers, since it requires more frequent applications to control the pest insects. This requires more labor and fuel and, therefore, is more expensive.

Both **organophosphates** and **carbamates** are soft pesticides that work by interfering with the ability of the nervous system to conduct impulses normally. Under normal conditions, a nerve impulse is conducted from one nerve cell to another by means of a chemical known as acetylcholine. When this chemical is produced at the end of one nerve cell, it causes an impulse to be passed to the next cell, thereby transferring the nerve message. As soon as this transfer is completed, an enzyme known as cholinesterase destroys acetylcholine, so the second nerve cell in the chain is only stimulated for a short period of time. Organophosphates and carbamates attach to the enzyme cholinesterase, preventing cholinesterase from destroying acetylcholine. This results in nerve cells being continuously stimulated, causing uncontrolled spasms of nervous activity and uncoordination that result in death.

Although these soft pesticides have the advantage of being less persistent in the environment than chlorinated hydrocarbons, they are generally much more toxic to humans and other vertebrates. Persons who use such pesticides must use special equipment and should receive special training because improper use of soft pesticides can result in death. Since organophosphates bind more strongly to cholinesterase than carbamates, they are considered more dangerous and have been replaced for many applications by carbamates.

Common organophosphates are malathion, parathion, and diazinon. Malathion is widely used for such projects as mosquito control, but parathion is a restricted organophosphate because of its high toxicity to humans. Diazinon is widely used in gardens. Sevin, aldicarb, and propoxur are examples of carbamates. Table 14.2 lists a variety of pesticides, what they are used for, and how they act to control the pest. This table also includes other categories of insecticides and their uses.

Herbicides

Herbicides constitute another major class of chemical control agents. In fact, about 60 percent of the pesticides used in U.S. agriculture are herbicides. (See figure 14.5.) They are widely used to control unwanted vegetation along power-line rights-of-way, railroad rights-of-way, and highways, as well as on lawns and cropland, where they are commonly referred to as weed killers. Weeds are plants growing in the wrong place. Bluegrass in a lawn is desirable, but bluegrass growing in a cornfield is a weed, so many herbicides are designed to kill beneficial as well as harmful plants.

Weed control is extremely important for agriculture since the weeds take nutrients and water from the soil, thus making them unavailable to the crop species. In addition, weeds may shade the crop species and prevent it from getting the sunlight it needs for rapid growth. At harvest time, weeds reduce the efficiency of the harvesting machines. Also, the weeds generally must be sorted from the crop before it can be sold, which adds to the time and expense of the harvesting process.

Traditionally, much energy has been expended in the control of weeds. Initially, weeds were eliminated with manual labor and the hoe. Tilling of the soil also helps to control weeds. Once the crop is planted, row crops like corn or soybeans may be cultivated to remove weeds from between the rows. All of these activities are expensive in time and fuel. Therefore, selective use of herbicides can have a tremendous economic impact on a farmer's profits.

TABLE 14.2
Various kinds of insecticides and their uses.

Type of insecticide	Examples	Target organisms	Action on target organism
Chlorinated hydrocarbons	Aldrin Chlordane DDT Dieldrin Endosulfan Endrin Heptachlor Kepone® Lindane Mirex® Strobane Toxaphene	Various insect species	Interferes with nervous system by changing sodium and potassium ion balance

Essentially all chlorinated hydrocarbons have been banned or discontinued in the United States and many other countries because of their persistence and ability to accumulate in food chains. Lindane and Endosulfan have limited use in the United States, but others are used in many less developed countries to control disease-spreading insects.

Organophosphates			
General use	Azinphosmethyl (Guthion®) Chlorpyrifos (Dursban®, Lorsban®) Diazinon (Spectracide®) Malathion Dimethoate (Cygon®) Parathion Mevinphos (Phosdrin®) Trichlorfon (Dylox®) Dicrotophos (Bidrin®) Acephate (Orthene®) Methamidophos (Monitor®) Tetrachlorvinphos (Gardona®, Rabon®) Crotoxyphos (Ciodrin®) Naled (Dibrom®) Propetamphos (Safrotin®) Sebutos (Apache®, Rugby®) Profenofos (Curacron®) Sulprofos (Bolstar®) Isofenphos(Amaze®, Pryfon®) Penitrothion (Accothion®) Isazophos (Brace®) Fenthion (Baytex®) Fensulfothion (Dasanit®) Famphur (Bash®, Warbex®) Methidathion (Supracide®) Phosmet (Imidan®) Pirimiphos-methyl (Actellic®) Phosalone (Zolone®)	Various insect and mite species	Interferes with cholinesterase enzyme of nervous system, causing uncontrollable muscle twitching
Flea collars and pest strips	Dichlorvos (Vapona®)	Fleas and ticks, household and barn insects	Vapors kill
Plant systemics	Demetron (Systox®) Dicrotophos (Bidrin®) Disulfoton (Di-Syston®) Mevinphos (Phosdrin®) Oxydemetonmethyl (Meta Systox®) Phosphamidon (Demecron®, Swat®) Phorate (Thimet®)	Insects that suck plant juices	Incorporates into plants; taken up when juices sucked from plant
Animal systemics	Crufomate (Ruelene®) Cythioate (Cyflee®)	Cattle grubs, fleas, ticks on dogs and cats	Taken up by pest from body tissues of host

TABLE 14.2
(Continued)

Type of insecticide	Examples	Target organisms	Action on target organism
Carbamates			
General use	Carbaryl (Sevin®)	Many insects	Cholinesterase poisons
	Propoxur (Baygon®)		
	Methomyl (Lannate®, Nudrin®)		
	Thiocarb (Larvin®)		
	Aminocarb (Matacil®)		
	Bendiocarb (Ficam®, Turcam®)		
	Trimethacarb (Broot®)		
	Mexacarbate (Zectran®)		
	Fenoxycarb (Logic®)		
Plant systemics	Methomyl (Lannate®, Nudrin®)	Plant-sucking insects	Cholinesterase poisons
	Aldicarb (Temik®)		
	Carbonfuran (Furadan®)		
	Oxamyl (Vydate®)		
	Aldoxycarb (Standak®)		
Formamidines	Formetanate (Carzol®)	Insect eggs, mites, ticks, and young caterpillars	Nervous system poisons
	Amitraz (Taktic®, Mitac®)		
Organosulfurs	Aramite®	Mites	Nervous system poisons
	Genite®		
	Ovex (Ovotran®)		
	Tetradifon (Tedion®)		
Plant products	Nicotine	Various insects	Nerve poison inhibits ATP paralysis
	Rotenone		
	Pyrethins		
Synthetic pyrethrins	Fenvalarate (Pydrin®, Tribute®)	Many insect pests	Interferes with nervous system by changing sodium and potassium balance
	Permethrin (Ambush®, Dragnet®, Pramex®, Pounch®, Torpedo®)		
	Bifenthrin (Capture®, Talstar®)		
	Cyhalothrin (Karate®, Grenade®)		
	Cypermmethrin (Ammo®, Cymbush®, Cynoff®)		
	Cyfluthrin (Baythroid®)		
	Deltamethrin (Decis®)		
	Esfenvalerate (Asana XL®)		
	Fenpropathrin (Danitol®)		
	Flucythrinate (Payoff®)		
	Tefluthrin (Force®)		
	Fluvalinate (Mavrik®, Spur®)		
	Tralomethrin (Scout®)		

Source: Data from George W. Ware, *The Pesticide Book*, 3d edition, 1989, Thomson Publications.

Many of the recently developed herbicides can be very selective if used appropriately. Some are used to kill weed seeds in the soil before the crop is planted, while others are used after the weeds and the crop begin to grow. In some cases, a mixture of herbicides can be used to control weed species. Figure 14.7 shows the result when two different herbicides were used together. One stops photosynthesis, and the other prevents germination of the seeds. Several major types of herbicides are in current use. One type consists of synthetic plant-growth regulators that mimic natural-growth regulators known as **auxins.** Two common synthetic plant-growth regulators of this type are 2, 4-dichlorophenoxyacetic acid (2, 4-D) and 2, 4, 5-trichlorophenoxyacetic acid (2, 4, 5-T). When applied to broadleaf plants, these chemicals cause the metabolism of the plants to exceed their food-producing potential. The plants grow so rapidly that they run out of

FIGURE 14.7
The Effect of Herbicides. The grasses in this photograph have been treated with herbicides. The soybeans are unaffected and grow better without competition from grasses.

FIGURE 14.8
Herbicide Use to Maintain Rights-of-Way. Power-line rights-of-way are commonly maintained by herbicides that kill the woody vegetation, which might grow so tall that it interferes with power lines.

stored food and die. These kinds of herbicides can be used to control broadleaf weeds in grain crops, weed species in coniferous forests, and brush along rights-of-way. (See figure 14.8.)

A second kind of herbicide is responsible for disrupting photosynthetic activity of plants, causing their death. Others inhibit enzymes, precipitate proteins, stop cell division, or destroy cells directly. Depending on the concentration of the herbicide used, some are toxic to all plants, while others are very selective as to which species of plants they affect. One such herbicide is diuron. In proper concentrations and when applied at the appropriate time, it can be used to control annual grasses and broadleaf weeds in over twenty different crops. However, at higher concentrations, it kills

During its involvement in the conflict in Southeast Asia, the United States used more than 105 million metric tons of herbicides to defoliate large areas so that the North Vietnamese and Viet Cong soldiers could not use these areas as hiding places. Herbicides were also used to kill crops that supplied food for these soldiers.

The most commonly used herbicide, Agent Orange, was an equal mixture of 2, 4-D (2, 4-dichlorophenoxy-acetic acid) and 2, 4, 5-T (2, 4, 5-trichlorophenoxya-cetic acid). The herbicide got its name from the bright orange stripes on the steel drums in which it was contained. The United States first sprayed Agent Orange in 1965 and eventually sprayed over 2,400,000 hectares of land. Despite the vast amounts of money and effort expended, the defoliation program was a failure in most respects. It did have a detrimental effect on the native people, however, because their crops were destroyed.

Some of the 2, 4, 5-T was contaminated with an extremely toxic compound known as dioxin (2, 3, 7, 8-tetrachlorodibenzo-p-dioxin), also called TCDD. Exposure to Agent Orange and its contaminants has been linked with skin disorders, psychological problems, liver damage, cancer, and birth defects. By the time the United States stopped using Agent Orange in 1970, between fifty and sixty thousand members of the U. S. Armed Forces were exposed to the chemical mixture, as were thousands of Vietnamese citizens.

Despite dioxin's adverse effect on laboratory animals, it has not been conclusively proven that dioxin-contaminated Agent Orange has the same effect on humans. In humans, it has not been found to be directly responsible for ailments more serious than chloracne, a disfiguring skin problem. But after a 1976 explosion at an Italian chemical plant, which spread dioxin over a village and resulted in chloracne among humans and widespread death of animals, many

Vietnam veterans became convinced that Agent Orange was responsible for a variety of ailments they had developed.

After many years of litigation, the issue reached the federal court in 1984. Before the case was to be heard, the seven chemical companies who produced Agent Orange agreed to place $180 million into a fund that would be used to compensate victims and their families. The fund is expected to last twenty-five years. The chemical companies denied any liability for the veterans' illnesses; their position was that Agent Orange did not cause the health problems and that they had merely manufactured the defoliant according to the instructions of the military. The companies also maintained that, if herbicides are correctly formulated, no contamination by dioxin will occur. They said further that if herbicides are used according to manufacturers' specifications, they are safe and valuable agricultural chemicals.

all vegetation in an area. Fenuron is a herbicide that kills woody plants. In low concentrations, it is used to control woody weed plants in cropland. In high concentrations, it is used on noncroplands, such as power-line rights-of-way. Table 14.3 lists several kinds of herbicides, their target species, and how they work.

Other Agricultural Chemicals

In addition to herbicides, certain other agrochemicals have become available for use in special applications. A synthetic auxin sprayed on cotton plants prior to harvest causes the leaves to drop off, which facilitates the harvesting process by reducing clogging of the mechanical cotton picker.

TABLE 14.3
Various kinds of herbicides and their uses.

Type of herbicide	Examples	Target organisms	Action on target organisms
Acetanilides	Metolachlor (Dual®) Alachlor (Lasso®) Butachlor (Machete®) Diethatyl (Antor®) Propachlor (Ramrod®) Acetochlor (Harness®)	Preemergent for many kinds of weeds	Meristematic growth inhibitor
Petroleum oils	Motor oil, gasoline, etc.	All plants	Disrupts cell membranes
Arsenic compounds	Arsonic acid, MSMA, arsinic acid, DSMA, cacodylic acid	All plants	Inhibits enzymes
Phenoxy herbicides	2,4-D; 2,4,5-T; 2,4-DB; MCPA, silvex	Broadleaf weeds	Stimulates uncontrolled growth
Substituted amides	Diphenamid (Dymid®, Enide®)	Preemergent for all plants	Inhibits seedlings
	Propanil (Stampede®)	Postemergent for many weeds	Inhibits photosynthesis
	Napropamide (Devrinol®) Naptalam (Alanap®) Pronamide (Kerb®) Bensulide (Betasan®, Prefar®)	Many weeds	Inhibits growth
Nitroanilines	Trifluralin (Treflan®) Benefin (Balfin®, Quilan®) Isopropalin (Paarlan®) Oryzalin (Surflan®) Pendimethalin (Prowl®)	Preemergent for all plants	Inhibits root and shoot growth
Substituted ureas	Diuron (Karmex®, Krovar®) Linuron (Lorox®) Fluometuron (Cotoran®) Fenuron-TCA (Dozer®) Siduron (Tupersan®) Tebuthiuron (Spike®)	Preemergent	Inhibits photosynthesis
Carbamates	Propham (Chem Hoe®) Chlorpropham (Furloe®) Barban (Carbyne®) Asulam (Asulox®) Phenmedipham (Betanal®, Spin-Aid®)	All plants	Stops cell division
Pyridinoxy and picolinic acids	Picloram (Tordon®) Triclopyr (Garlon®) Fluroxypyr (Starane®) Clopyralid (Lonrel®)	Broadleaf weeds	Stimulates growth
Aliphatic acid	TCA, Dalapon	Quack grass, Bermuda grass	Precipitates proteins
Arylaliphatic acids	Dicamba (Banvel®) Chloramben (Amiben®) Fenac (Fenatrol®)	Preemergent against germinating seeds and seedlings	Inhibits growth
Phenol derivatives	DNOC, dinoseb	All plants	Prevents ATP formation
	Petachlorophenol	Defoliant	Destroys cells
Bipyriodyliums	Diquat, paraquat	All plants	Destroys cell membranes
Benzonitriles	Dichlobenil (Casoron®) Bromoxynil (Brominal®, Buctril®)	Contact against broadleaf weeds	Poisons respiration

TABLE 14.3
(Continued)

Type of herbicide	Examples	Target organisms	Action on target organisms
Benzothiadiazoles	Bentazon (Basagran®)	Postemergent against broadleaf weeds	Inhibits photosynthesis
Diphenyl ethers	Nitrofen Acifluorfen (Blazer®) Lactofen (Cobra®) Oxyfluorfen (Goal®)	Contact against many weeds	Inhibits respiration
Imidazoles	Imazapyr (Arsenal®) Imazaquin (Scepter®, Image®) Imazethapyr (Pursuit®) Imazamethabenz-methyl (Assert®)	Broadleaf weeds	Inhibits growth
Oxyphenoxy acid esters	Fluazifop-butyl (Fusilade®) Fenoxaprop-ethyl (Whip®) Diclofop (Hoelon®) Quizalofop-ethyl (Assure®) Cycloxydim (Focus®, Laser®) Haloxyfop-methyl (Verdict®, Gallant®)	Grasses	Inhibits growth
Phosphono amino acids	Glyphosate (Roundup®) Glufosinate (Ignite®) Fosamine ammonium (Krenite®) Sulfosate (Touchdown®)	Grasses	Inhibits growth
Phthalic acids	Chlorothal (Dacthal®) Endothall (Aquathol®, Endothal®)	Many weeds	Inhibits growth
Pyridazinones and pyridinones	Amitrole Pyrazon (Pyramin®) Fluridone (Sonar®) Norflurazon (Evital®, Solicam®) Dimethazone (Command®) Oxadiazon (Ronstar®)	All plants	Prevents chlorophyll formation
Sulfonylureas	Chlorsulfuron (Glean®) Chlorimuron-ethyl (Classic®) Metsulfuron-methyl (Ally®, Escort®) Thiameturon-methyl (Pinnacle®, Harmony®) Sulfometuron-methyl (Oust®)	Broadleaf weeds	Inhibits growth
Thiocarbamates	EPTC (Eptam®) Pebulate (Tillam®) Molinate (Ordram®) Cycloate (Ro-Neet®) Thiobencarb (Bolero®, Saturn®) Vernolate (Vernam®) Butylate (Surtan®)	Preemergent against grasses and some broadleaf weeds	Inhibits seedling growth
Triazines	Atrazine (Aatrex®) Simazine (Princep®) Cyanazine (Bladex®) Prometon (Pramitol®) Propazine (Milogard®) Metribuzin (Lexone®, Sencor®) Hexazinone (Velpar®)	All plants	Inhibits photosynthesis
Uracils	Lenacil (Venzar®) Bromacil (Hyvar®) Terbacil (Sinbar®)	All plants	Inhibits photosynthesis

Source: Data from George W. Ware, *The Pesticide Book,* 3d edition, 1989, Thomson Publications.

FIGURE 14.9

Chemical Loosening of Cherries to Allow Mechanical Harvesting. By using chemicals and machinery, this farmer can rapidly harvest the cherry crop. This practice reduces the amount of labor required to pick the cherries but requires the application of chemicals to loosen the fruit.

NAA (naphthaleneacetic acid) is used by fruit growers to prevent immature apples from dropping from the trees and being damaged. This chemical can keep the apples on the trees for up to ten extra days, which allows for a longer harvest period and fewer apples lost.

Under other conditions, it may be valuable to get fruit to fall more easily. Cherry growers use ethephon to promote loosening of the fruit so that the cherries will fall more easily from the tree when shaken by the mechanical harvester. This method lowers the cost of harvesting the fruit. (See figure 14.9.)

Fungicides and Rodenticides

Fungus pests can be divided into two different categories. Some are natural decomposers of organic material. However, when the organic material being destroyed happens to be a crop or other product useful to humans, the fungus is a pest. Other fungi are parasites on crop plants and weaken or kill the plants, thereby reducing the yield. Fungicides are used as fumigants to protect agricultural products from spoilage, as sprays and dusts to prevent the spread of diseases among plants, and as seed treatments to protect seeds from rotting in the soil before they have a chance to germinate. Methylmercury is often used on seeds to protect them from spoilage prior to germination. However, since methylmercury is extremely toxic to humans, these seeds should never be used for food. To reduce the chance of a mix-up, treated seeds are usually dyed a bright color.

Like fungi, rodents are harmful because they destroy food supplies. In addition, they can carry disease and damage crops in the field. In many parts of the world, such as India, the government pays a bounty to people who kill rats because this is an inexpensive way to protect the food supply. Several kinds of rodenticides have been developed to control rodents. One of the most widely used is warfarin, a chemical that causes internal bleeding in animals that consume the rodenticide. It is usually incorporated into a food substance so that rodents eat warfarin along with the bait. Because it is effective in all mammals, including humans, it must be used with care to prevent nontarget animals from having access to the chemical. As with many kinds of pesticides, some populations of rodents have become tolerant of warfarin, while others avoid baited areas. In many cases, rodent problems can be minimized by building storage buildings that are rodent proof, rather than relying on rodenticides to control them.

Problems with Pesticide Use

A perfect pesticide would have the following characteristics:

1. It would be inexpensive.
2. It would affect only the target organism.
3. It would have a short half-life.
4. It would break down into harmless materials.

However, the perfect pesticide has not been invented. Many of the more recently developed pesticides have fewer drawbacks than the early hard pesticides, but none are without problems.

Although there has been a trend away from using hard pesticides in the United States, some are still allowed for special purposes, and they are still in common use in other parts of the world. Because of their stability, these chemicals have become a long-term problem. They may be transported from the place they were originally applied to other parts of the world by wind or ocean currents. Hard (persistent) insecticides become attached to small soil particles, which are easily moved by wind or water to any part of the world. Hard pesticides and other pollutants have been discovered in the ice of the poles and are present in detectable amounts in the body tissues of animals, including humans, throughout the world. Thus, chemicals originally sprayed on a cornfield in Iowa or a sugarcane field in Brazil may become a problem in Arctic or Antarctic ecosystems.

Another problem associated with persistence is that pesticides may accumulate in the bodies of animals without killing the animals. If an animal receives small quantities of hard pesticides in its food and is unable to eliminate them, the concentration within the animal increases. Many hard pesticides and other persistent pollutants fall into this category. This process of accumulating higher and higher amounts of material within the body of an animal is called **bioaccumulation.** Many of the hard pesticides and their breakdown products build up in the fat of animals since these molecules are fat soluble. Because such an affected animal can be eaten by a carnivore, these toxins may build up in food chains, causing disease or death of the carnivores at higher trophic levels, even though lower-trophic-level organisms are not injured. This phenomenon of increasing levels of a substance in the bodies of higher-trophic-level organisms is known as **biological amplification.**

The well-documented case of DDT is an example of how biological amplification occurs. Since DDT is not very soluble in water, it is usually dissolved in an oil or fatty compound, which is then mixed with water before application. This mixture is then sprayed over an area and falls on plants that the insect population uses for food, or it may fall directly on the insect. The insect takes the DDT into its body, where the DDT interferes with normal metabolic activities. If the insect obtains a high enough concentration, it dies. If small amounts are taken in, the insect digests and breaks down the DDT as it would any other organic chemical compound. Since DDT is soluble in fat or oil, the insect stores DDT or its breakdown products in the fatty tissues of its body.

If an area has been lightly sprayed with DDT, some insects die, but others will be able to tolerate the low DDT levels. Some insects may contain as much as one part per billion of DDT in their tissues. This is not very much DDT, but it can have a tremendous effect on the animals that feed on the insects.

If an aquatic habitat is sprayed with a small concentration of DDT, or receives DDT from runoff, small aquatic organisms may accumulate a concentration of DDT that is up to 250 times greater than the concentration of DDT in the surrounding water. These organisms are eaten by shrimp, clams, and small fish, which are, in turn, eaten by larger fish. DDT concentrations of large fish can be as much as two thousand times the original concentration sprayed on the area. (See figure 14.10.)

What was a very small initial concentration has now become so high that it could be fatal to animals at higher trophic levels. This has been of particular concern for birds, since DDT interferes with the production of eggshells, making them much more fragile. This problem is more common in carnivorous birds because they are at the top of the food chain. Although all birds of prey have probably been affected to some degree, those that rely on fish for food seem to have been affected most severely. Eagles, osprey, cormorants, and pelicans are particularly susceptible species.

Although DDT is a well-known case of biological amplification in ecosystems, other persistent molecules are known to behave in similar fashion. Mercury, aldrin, chlordane, and other chlorinated hydrocarbons, such as polychlorinated biphenols (PCB's) used as insulators in electric transformers, are all known to accumulate in ecosystems.

Because of their persistence, their effects on organisms at higher trophic levels, and concerns about long-term human health problems, most chlorinated hydrocarbon pesticides have been banned from use in the United States and some other countries. In the early 1970s, the use of DDT was prohibited in the United States. Aldrin, dieldrin, heptachlor, and chlordane have also been prohibited from use on crops, although heptachlor, and chlordane were still used for termite control until recently. In 1987, Velsicol Chemical Corporation agreed to stop selling chlordane in the United States.

Another problem associated with insecticides is the ability of insect populations to become resistant to pesticides. All organisms within a given species are not identical. Each individual has a slightly different genetic composition from other members of the same species. If an insecticide is used for the first time on a particular kind of insect, it kills all the insects that are genetically susceptible. However, individuals with a genetic component that allows them to tolerate the insecticide may live to reproduce.

If, in a population of insects, only 5 percent of the individuals possess genes that make them resistant to an insecticide, the first application of the insecticide would, therefore, kill 95 percent of the population and would be of great benefit in controlling the size of the insect population. However, the surviving individuals that are tolerant of the insecticide would then constitute the majority of the breeding population. Since most of these individuals possess genetic characteristics for tolerating the insecticide, so will their offspring. Therefore, in the next generation, the number of individuals able to tolerate the insecticide will increase, and the second use of the insecticide will not be as effective as the first. Since some species of insect pests can produce a new generation each month, this process of selecting those individuals capable of tolerating the insecticide can result in resistant populations in which 99 percent of the individuals are able to tolerate the insecticide within five years. As a result, that particular insecticide

Human Influences on Ecosystems

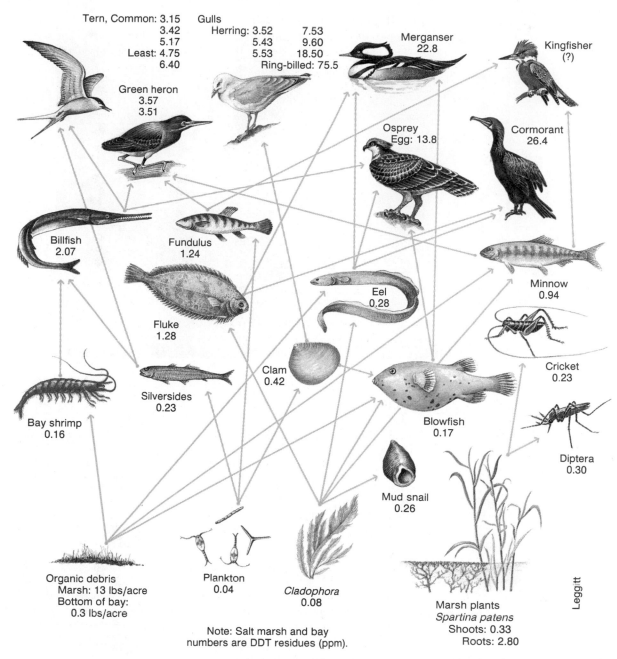

Tern, Common: 3.15
3.42
5.17
Least: 4.75
6.40

Gulls
Herring: 3.52   7.53
5.43   9.60
5.53   18.50
Ring-billed: 75.5

Merganser
22.8

Kingfisher
(?)

Green heron
3.57
3.51

Osprey
Egg: 13.8

Cormorant
26.4

Billfish
2.07

Fundulus
1.24

Minnow
0.94

Eel
0.28

Fluke
1.28

Clam
0.42

Cricket
0.23

Silversides
0.23

Blowfish
0.17

Bay shrimp
0.16

Diptera
0.30

Mud snail
0.26

Organic debris
Marsh: 13 lbs/acre
Bottom of bay:
0.3 lbs/acre

Plankton
0.04

Cladophora
0.08

Marsh plants
Spartina patens
Shoots: 0.33
Roots: 2.80

Leggitt

Note: Salt marsh and bay
numbers are DDT residues (ppm).

FIGURE 14.10

Biological Amplification. Note how
the concentration of DDT increases as it
passes through the food chain. At each
successive trophic level, the amount of
DDT increases because the animals are
accumulating the DDT from the bodies
of the animals they are eating as food.

is no longer as effective in controlling insect pests, and increased dosages
or more frequent spraying may be necessary. Figure 14.11 indicates that
over 450 species of insects have populations resistant to insecticides.

Cotton growers have become extremely dependent on the use of insecticides. In the United States, about 40 percent of the insecticides used
are to control pests in cotton. Because of this intensive use of insecticides,
many populations of pest insects have become resistant to the commonly
used insecticides. Table 14.4 shows the effect of continual use of insecticides on two pests, the bollworm and the tobacco budworm. The size of
the dose necessary to kill the pest increased greatly in a five-year period.
In some cases, the dose increased by ten times and in others a hundred
times or more. By 1965, these insecticides were no longer able to control
bollworm or tobacco budworm.

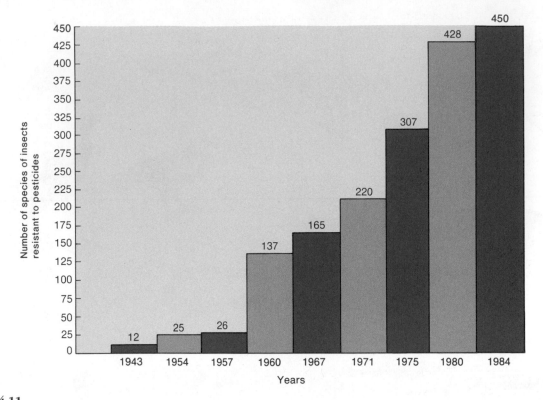

FIGURE 14.11

Resistance to Insecticides. The continued use of insecticides has increased the number of resistant species in the world. This reduces the effectiveness of insecticides and requires higher dosages and more frequent applications of insecticide or the costly development of new compounds.

TABLE 14.4
Average dose necessary to kill two cotton pests.

Compound	Average dose necessary to kill (milligrams per gram of larva)			
	Bollworm		*Tobacco budworm*	
	1960	*1965*	*1961*	*1965*
DDT	0.03	1000+	0.13	16.51
Endrin	0.01	0.13	0.06	12.94
Carbaryl	0.12	0.54	0.30	54.57
Strobane and DDT	0.05	1.04	0.73	11.12
Toxaphene and DDT	0.04	0.46	0.47	3.52

From P. L. Adkisson, "Controlling Cotton's Insect Pests: A New System" in *Science,* 216: 19–22, April 1982. Copyright © 1982 American Association for the Advancement of Science Washington, DC. Reprinted by permission of the publisher and author.

A third problem associated with pesticide use is that most pesticides are not specific to certain pest species but kill many beneficial species as well. Predator and parasitic insects that normally control herbivorous insects may be killed, which allows pest species to increase rapidly because there are no natural checks to population growth. This means that insecticides are necessary to prevent the pest population from rebounding to levels even higher than the initial one. Therefore, once the decision is made to use pesticides, it often becomes an irreversible tactic, because stopping their use would result in extensive crop damage.

A related concern is that the use of insecticides may cause an insect that is not a problem to become one. For example, when synthetic organic insecticides came into common use with cotton in the 1940s, the insect parasites and predators of the bollworm and tobacco budworm were eliminated, and the bollworm and tobacco budworm became major pests. The use of insecticides had caused a different pest problem to develop.

A fourth problem associated with pesticide use is their short-term and long-term health effects. If properly applied, most pesticides are safe. However, in many cases, people applying the pesticide are unaware of how the pesticide works and what precautions should be used in its application. In many parts of the world, small subsistence farmers may lack the ability to read the caution labels on the packages. Consequently, deaths from the misuse of pesticides are common.

However, for most people, the most critical health problems related to pesticide use involve inadvertent exposure to very small quantities of pesticides. Many pesticides have been proven to cause mutations, produce cancers, or cause abnormal births in experimental animals. There are questions about the effects of chronic, minute exposures to pesticide residues in food or through contamination of the environment. Although the risks are very small, many people find pesticides unacceptable and seek to prohibit their sale and use.

For example, in 1986, the U.S. Environmental Protection Agency banned the use of dinoseb, a herbicide, because tests by the German chemical company Hoechst AG indicated that dinoseb causes birth defects in rabbits. Other studies indicate that dinoseb causes sterility in rats. Similarly, in 1987, Velsicol Chemical Corporation, in an agreement with the U.S. EPA, agreed to stop producing and distributing chlordane in the United States. Chlordane had been banned previously for all applications except for termite control. After it was shown that harmful levels of chlordane could exist in treated homes, its use was finally curtailed by this agreement. Also in 1987, the Dow Chemical Company announced that it would stop producing the controversial herbicide 2,4,5-T. In both of these latter cases, other pesticides have been developed to replace the older types, so that the discontinuance of the older pesticides did not result in an economic hardship to farmers or other consumers.

Why Are Pesticides So Widely Used?

If pesticides have so many drawbacks, why are they used so extensively? There are three primary reasons. First, the use of pesticides has increased, at least in the short term, the amount of food that can be grown in many parts of the world. In the United States, pests are estimated to consume 33 percent of the crops grown. On a worldwide basis, approximately 35 percent of crops are consumed by pests. This represents an annual loss of $18.2 billion in the United States alone. Farmers, grain-storage operators, and the food industry continually seek to reduce this loss. Therefore, a retreat from dependence on pesticides would certainly result in a reduction in the amount of food that could be produced. Agricultural planners in most countries are not likely to suggest changes in pesticide use that would result in malnutrition and starvation for many of their inhabitants.

A second reason pesticides are used so extensively is economic. The cost of pesticides is more than offset by increased yields for the farmer and an increase in profits. In addition, the production and distribution of pesticides is big business. Companies that have spent millions of dollars developing a new pesticide are going to argue very strongly for its continued use. Since farmers and agrochemical interests have a very strong voice in government, they have been successful in preventing certain pesticides from being prohibited.

TABLE 14.5		
Yield difference between organic and conventional farming.		
Crop	Metric tons per hectare	
	Organic farming	*Conventional farming*
Corn	6.45	7.00
Soybeans	2.44	2.57
Wheat	1.88	3.28

Source: Data from William Lockeretz, Georgia Shearer, and Daniel H. Kohl, "Organic Farming in the Corn Belt," *Science* 211 (6 February 1981): 540–47.

A third reason for extensive pesticide use is that many special health situations are currently impossible to control without insecticides. This is particularly true in areas of the world where insect-borne diseases would cause widespread public health consequences if insecticide use were discontinued.

Organic Farming

Before the invention of synthetic fertilizers, insecticides, herbicides, fungicides, and a wide variety of other agrochemicals, all farming was organic. Animal manure and crop rotation provided soil nutrients; a mixture of crops prevented regular pest problems; and manual labor killed insects and weeds. With the development of mechanization, larger areas could be farmed without draft animals, and many farmers changed from mixed agriculture, in which animals were an important ingredient, to monoculture. Chemical fertilizers replaced manure as a source of soil nutrients, and crop rotation was no longer important since hay and grain were no longer grown for draft animals and cattle. The larger fields of crops like corn, wheat, and cotton presented opportunities for pest problems to develop. Chemical pesticides "solved" this problem.

As a result of the problems and costs associated with the use of chemical fertilizers and pesticides, some farmers are seeking to return to an earlier form of agriculture in which they use traditional methods to produce crops without the use of pesticides and chemical fertilizers. They are even willing to accept lower yields because they do not have to pay for fertilizers and pesticides. Thus, they can still make a profit. In addition, organic farmers often receive premium prices for products that are "organically grown."

A study by Lockeretz, Shearer, and Kohl lists some comparisons between conventional and organic farming. (See table 14.5.) Conventional farms had a higher yield than organic farms and also netted $384 per hectare, as compared to $333 per hectare for organic farms. Organic farms used only 40 percent of the energy used on conventional farms, thus reducing their costs and raising their profit margin.

This transition to organic farming does present some problems, however. The use of legumes, such as beans, soybeans, clover, or alfalfa, in crop rotation reduces the amount of land available for cash crops and requires that cattle be a part of the farmer's operation. Crop rotation also requires a greater investment in farm machinery, since certain crops require specialized equipment. Also, the raising of cattle requires additional expenditures for feed supplements and veterinary care. Critics say that organic farming cannot produce the amount of food required for today's population and that it can only be successful in specific cases. Proponents disagree and

Human Influences on Ecosystems

Food additives are chemicals added to food before its sale. They have several different purposes:

1. To prolong the storage life of the food
2. To make the food more attractive by adding color or flavor
3. To modify nutritive value

Many different molecules are added to foods to prolong shelf life. Calcium propionate is added to baked goods to increase shelf life because baked goods can become contaminated with airborne spores of molds and bacteria, which grow on the food and spoil it. BHT, TBHQ, and other commonly seen alphabetic mixtures have a similar function.

Sometimes, additives are used just to make the food appear more attractive. For years, Red Dye II was used to color a variety of foods. Its use was discontinued when it was found to be carcinogenic. Other food colorings are still widely used to increase the appeal to the consumer. Many kinds of artificial flavors are added to products as well. Commonly used flavor enhancers are monosodium glutamate, table salt, and citric acid.

Other food additives are used to modify the nutritive value of the food product. Iodine in table salt is a good example. Iodine is a trace element required for proper thyroid functioning. Individuals suffering from a lack of iodine often develop an enlargement of the thyroid gland known as goiter. The addition of iodine to table salt has eliminated goiter in the United States. Most cereals and baked goods and many other products have various vitamins and minerals added to improve the food's nutritive value. Some additives, like Nutrasweet™, are added to reduce the number of calories present in the food while giving the consumer the sensation of tasting something sweet.

Some additives are an unavoidable residue of some step of the food production industry and could be more properly called contaminants. Pesticide residues are an example. The pesticides are used in the growth of foods, but they are also used to eliminate pests in the storing, processing, and transportation steps of the food industry.

Diethylstilbestrol (DES) was at one time used in the poultry industry to produce fatter birds. Because there were indications that DES is carcinogenic, the U.S. Food and Drug Administration banned the use of DES in chickens and declared that it was potentially hazardous to humans. Further studies were conducted to determine if DES was safe to use in connection with raising beef. It was used with cattle because an animal gains weight more rapidly with DES in its feed. Because DES was eventually linked to breast cancer in women, it has been banned from all animal feed use in the United States and Europe.

stress that organic farming reduces the release of chemicals into the environment and reduces soil erosion. Regardless, as fuel and chemical costs increase, organic farming or some modification of it may allow for a reduction in the cost of producing food.

Integrated Pest Management

Another way of reducing dependence on pesticides is to use integrated pest management. **Integrated pest management** uses a variety of methods to control pests, rather than relying on pesticides alone. It depends on a complete understanding of all ecological aspects of the crop and the particular pests to which it is susceptible. Information about the metabolism of the crop plant, the biological interactions among pests and their predators or parasites, the climatic conditions that favor certain pests, and techniques for encouraging beneficial insects are all involved, as well as the selective use of pesticides. Much of the information necessary to make integrated pest management work goes beyond the knowledge of the typical farmer. The metabolic and ecological studies necessary to pinpoint weak points in the life cycle of pests can usually only be carried out at universities or government research institutions. These kinds of studies are expensive and must be completed for each kind of pest, since each pest has a unique biology. Once a viable technique has been developed, an educational program is necessary to interest farmers in using integrated pest management rather than the "spray and save" techniques that they understand and that pesticide salespersons continually encourage.

Several different methods are employed in integrated pest management. These include sex attractants, male sterilization, the release of natural predators or parasites, the development of resistant crops, the use of natural pesticides, the modification of farming practices, and the selective use of pesticides. In some species of insects, a chemical called a **pheromone** is released by females to attract males. Males of some species of moths can detect the presence of a female from a distance of up to 3 kilometers. Since many moths are pests, synthetic odors can be used to control them. Spraying an area with the pheromone confuses the males and prevents them from finding a female, which results in a reduced moth population the following year. In a similar way, a synthetic sex attractant molecule known as Gyplure is used to lure male gypsy moths into traps, where they become stuck. Since the females cannot fly and the males are inside the traps while the females are outside the traps, the reproductive rate drops, and the insect population may be controlled.

Another technique that reduces reproduction is male sterilization. In the southern United States and Central America, the screwworm fly weakens or kills large grazing animals, such as cattle, goats, and deer. The female screwworm fly lays eggs in open wounds on these animals, where the larvae feed. However, it was discovered that the female screwworm fly only mates once in her lifetime. Therefore, the fly population can be controlled by raising and releasing large numbers of sterilized male screwworm flies. Any female that mates with a sterile male fails to produce fertilized eggs and cannot reproduce. In Curacao, an island 65 kilometers west of Venezuela, a program of introducing sterile male screwworm flies eliminated this disease from the twenty-five thousand goats on the island. In parts of the southwestern United States, the sterile male technique has also been very effective. The screwworm fly has been eliminated from the United States and northern Mexico, and much of Central America may become free of them as well. In 1990, sterile males were released in Libya to begin eliminating screwworm flies that had been introduced with a South American cattle shipment.

During an epidemic of Mediterranean fruitflies in the early 1980s, a similar technique was used. Unfortunately, the X-ray sterilization technique used was ineffective, and most of the flies released were not sterile, which made the problem worse rather than better. Pesticides were eventually used to control the fruitflies. Recent concern about the Mediterranean fruitfly in California has resulted in the controversial aerial spraying of malathion.

The manipulation of predator-prey relationships can also be used to control pest populations. For instance, the ladybird beetle, commonly called a ladybug, is a natural predator of aphids and scale insects. (See figure 14.12.) Artificially increasing the population of ladybird beetles leads to reduced aphid and scale populations. In California during the late 1800s, scale insects on orange trees greatly affected the health of the trees and reduced crop yields. The introduction of a species of ladybird beetle from Australia quickly brought the pests under control. Years later, when chemical pesticides were first used in the area, so many ladybird beetles were accidentally killed that scale insects once again became a serious problem. When

FIGURE 14.12
Beneficial Insect. The ladybird beetle is a predator of many kinds of pest insects, including aphids.

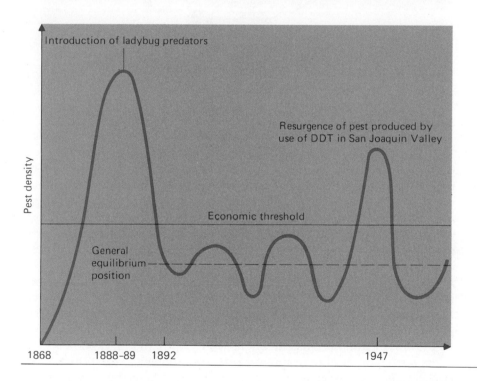

FIGURE 14.13

Insect Control with Natural Predators. In 1889, the introduction of ladybird beetles (ladybugs) brought the cottony cushion scale under control in the orange groves of the San Joaquin Valley. In the 1940s, DDT reduced the ladybird beetle population and the cottony cushion scale population increased. Stopping the use of DDT allowed the ladybird population to increase, reducing the pest population and allowing the orange growers to make a profit.

pesticide use was discontinued, ladybird beetle populations rebounded, and the scale insects were once again brought under control. (See figure 14.13.)

In 1961, grape growers in California's San Joaquin Valley were troubled by a grape leafhopper. To combat the pest, growers applied DDT. However, the leafhopper quickly developed resistance to DDT, and other insecticides had to be employed. Grape growers spent over $8 million to control

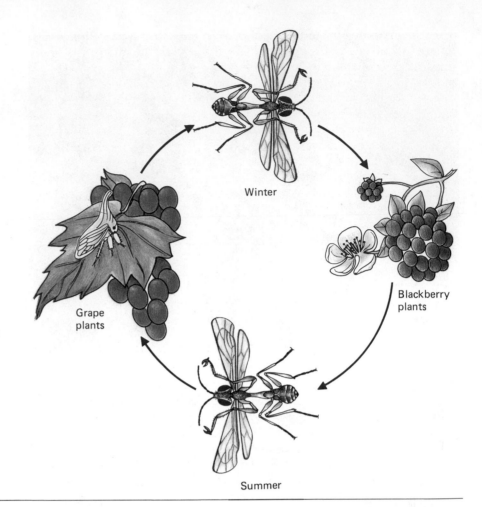

Winter

Grape plants

Blackberry plants

Summer

FIGURE 14.14

Life Cycle of a Parasitic Wasp. During the summer, the population of leafhoppers that feed on the grape leaves is controlled by a parasitic wasp. These wasps migrate to blackberry bushes to overwinter but return to the grapes during the summer. If the distance is too great, the wasps are unable to migrate between grapes and the blackberry bushes. As a result, the grape leafhoppers are unchecked, and they damage the grape plants. Therefore, the removal of blackberry patches to clear land for the growing of grapes increases the amount of damage that the leafhoppers inflict on the grapes.

leafhoppers but created a resistant pest instead. Research into the biology of leafhoppers revealed that a particular species of parasitic wasp is a natural control of the leafhopper. The female wasp deposits an egg on the egg of a leafhopper. When the wasp egg hatches, the larva uses the leafhopper egg as a source of food. In one growing season, the wasp produces nine or ten generations, compared to three for the leafhopper. This is sufficient to keep the leafhopper under control.

However, it was also discovered that the leafhopper spent its entire life in the vineyards, while wasps only lived in the vineyard during the summer. In the winter, the wasp required an alternate host, a noneconomically important leafhopper that normally lived on blackberry bushes. (See figure 14.14.) In a natural ecosystem where wild grapes and blackberries were interspersed, the leafhoppers never reached problem levels. With the establishment of large vineyards, most of the blackberry bushes were destroyed, and the remaining bushes were so far from the grapes that the wasps were unable to migrate from their winter habitat to their summer habitat. Thus, the wasps were no longer able to keep the leafhopper population under control.

The use of specific strains of the bacterium *Bacillus thuringiensis* to control mosquitoes and moths is another example of the use of one organism to control another. A crystalline material produced by the bacterium causes the destruction of the lining of the gut of the feeding insect,

Human Influences on Ecosystems

resulting in its death. One strain of *B. thuringiensis* is used to control mosquitoes while another is primarily effective against the caterpillars of leaf-eating moths, including the gypsy moth.

An area receiving increasing attention is that of producing resistant crop species through selective breeding or genetic engineering. These techniques seek to develop characteristics in major crop plants that allow them to resist fungi, insects, and other pests. For a long time, wheat rust, a fungus parasite that weakens wheat and reduces yields, was a major problem. Modern strains of wheat and other crops have now been developed that are able to resist such common plant pests. One development involves strains of crop plants with shorter growing seasons that often go through their entire life cycle before pest populations become large enough to cause problems. Another recent development could allow herbicide-resistant genes to be introduced into crop plants by genetic engineering techniques. This would allow herbicides to be used on the crop without harming it, while the competing weeds would be destroyed.

Naturally occurring pesticides found in plants also can be used to control pests. For example, marigolds are planted to reduce the number of soil nematodes, and garlic plants are used to check the spread of Japanese beetles.

Often, modification of farming practices can reduce the impact of pests. In some cases, all crop residues are destroyed to prevent insect pests from finding overwintering sites. For example, shredding and plowing under the stalks of cotton in the fall reduces overwintering sites for boll weevils and reduces their numbers significantly, thereby reducing the need for expensive insecticide applications. Many farmers are also returning to crop rotation, which tends to prevent the buildup of specific pests that typically occurs when the same crop is raised in the same field year after year.

Integrated pest management also makes selected use of pesticides. Identifying the precise time when the pesticide will have the greatest effect at the lowest possible dose has several advantages: It reduces the cost to the farmer, reduces the amount of pesticide used, and may still allow the parasites and predators of pests to survive. This often requires the assistance of a trained pest professional who can correctly identify the pests, measure the size of the population, and time pesticide applications for maximum effect. In several instances, pheromone-baited traps capture insect pests from fields, and an assessment of the number of insects caught can be a guide to when insecticides should be applied.

Several of these techniques were used successfully in an integrated pest management system in the Canete Valley in Peru. The Canete Valley consists of 22,000 hectares of cultivated land surrounded by an arid region. Since 1920, about 60 percent of the land has been planted in cotton. As a result of the monoculture approach, cotton pests increased rapidly. Until 1949, heavy metals and natural organic insecticides were used to control the insects. Beginning in 1949, synthetic organic pesticides were used to control the insects; the results were disastrous. The widespread use of synthetic pesticides reduced the number of beneficial insects and created resistant strains of pest species. At first, the chemicals were applied every fifteen days; this was then shortened to every eight days, and then to every three days. This dependency on synthetic organic pesticides had become very expensive. In addition, insects that had not previously been present in large enough numbers to be pests began to threaten the crop. In spite of the extensive use of pesticides, in 1956 the cotton yield dropped to its lowest level in a decade.

Since the late 1960s, a number of carnivorous birds at the end of long food chains were discovered to be producing eggs with thin shells. Some shells were so thin that they collapsed when they were incubated. Analysis of fat in such eggs revealed the presence of high levels of DDE (dichloro-diphenyl-dichloroethane), a chemical closely related to DDT. It is now known that DDE is a metabolite of DDT. DDE interferes with the action of a liver enzyme that causes calcium to mobilize from the birds' bones and become incorporated into eggshells. The greater the quantity of DDE, the less effective the enzyme and the thinner the eggshells.

In May 1968, two researchers from the Smithsonian Institute visited Anacapa Island, located just offshore of Ventura, California. Although May is the most active nesting time for brown pelicans, they found no young in an area that should have contained fifteen hundred nests with up to three young per nest. Within a short time, other scientists became aware of nesting failure in other colonies of pelicans. In 1969, many people visited Anacapa Island to survey pelican nesting and found only six hundred nests, out of which there were only two surviving young.

Analysis of the fat in the pelican eggs on Anacapa Island revealed levels of DDE as high as 2,500 parts per million (ppm). Information was also collected

Location of island	ppm DDE	Eggshell thickness
Anacapa Island	1,223	0.32 mm
West Coronado Island	1,158	0.34 mm
San Martin Island	429	0.45 mm
San Benitos Island	128	0.51 mm
Gulf of California	13	0.57 mm

from other islands at various distances from the Los Angeles area. The map and table relate eggshell thickness (0.57 millimeters is normal), DDE concentration in ppm, and distance from Los Angeles. Pelicans that bred farther from Los Angeles produced eggs with less DDE and thicker shells.

As more data accumulated, it seemed that the waters around Los Angeles may have contained abnormally high levels of DDE. A search for the source determined that, in this highly urbanized area, agricultural runoff was not a significant factor. When municipal sewage outfalls were checked between San Francisco and San Diego, only one showed high levels of DDE. Los Angeles' outfall, however, was discharging about 50 kilograms per day. By monitoring the sewer lines that flowed to the ocean outfall, the source of the DDE was traced to a chemical company a few kilometers inland

To combat this problem, the farmers of the Canete Valley organized to implement an integrated pest management system, which included the following elements:

1. Strict planting and harvesting dates were enforced to deprive the pests of a source of food during part of the year.
2. Marginal farmland was taken out of production so that only the most healthy plants, capable of resisting pests, would be grown.
3. Irrigation practices were altered to hinder pest population growth.
4. Synthetic pesticides were banned or severely reduced.
5. Natural predators and parasites of pests were encouraged.

After six years of this integrated pest management scheme, cotton production increased almost 30 percent.

Integrated pest management will become increasingly popular as the cost of pesticides increases and knowledge about the biology of specific pests becomes available. However, as long as humans raise crops, there will be pests that will outwit the defenses we develop. Integrated pest management is just another approach to a problem that began with the dawn of agriculture.

Pelican nesting success on Anacapa Island, California.		
Year	Number of active nests	Number of young
1968	No active nesting	No young
1969	600 nests	2 young
1970	552 nests	1 young
(April 1970, dumping of DDE into sewer system banned)		
1971	600 nests	7 young
1972	Nests not counted	56 young
1973	597 nests	134 young
1974	1,286 nests	1,185 young
1975		250 young
1980	1,258 nests	
1981	2,150 nests	1,505 young
1984	848 nests	584 young
1985		6,497 young
1986		4,601 young
1987		4,398 young
1988	3,000 ± nests	2,400 young

Source: Data from Ray E. Williams, Rio Hondo College, Whittier, California.

that was manufacturing DDT. Following a court order issued in April 1970, the chemical company stopped releasing its wastes into the sewer system and began trucking them to a nearby landfill. The other table shows nesting success beginning with 1968 and the subsequent recovery of pelican nesting on Anacapa Island following the termination of DDE release in 1970. While there is considerable variation, it is clear that nesting success has greatly improved since 1970.

Has the pelican problem been solved?
Does trucking wastes to a landfill solve the problem?

Summary

Although small slash-and-burn garden plots are common in some parts of the world, most of the food in the world is raised on more permanent farms. In countries where population is high and money is in short supply, much of the farming is labor-intensive, making use of human labor for many of the operations necessary to raise crops. However, the majority of the food of the world is grown on large, mechanized farms that use energy rather than human muscle to produce crops. Energy is needed for tilling, planting, and harvesting crops and for the production and application of fertilizers and pesticides.

Monoculture involves planting large areas of the same crop year after year. This causes problems with plant diseases, pests, and soil depletion. Although chemical fertilizers can replace soil nutrients that are removed when the crop is harvested, they do not replace the organic matter necessary to maintain soil texture, pH, and biotic richness.

Mechanized monoculture is heavily dependent on the control of pests by chemical means. Hard pesticides are stable and persist in the environment, where they may be amplified in ecosystems. Consequently, many of the older hard pesticides have been replaced by soft pesticides that decompose much more quickly and present less of an environmental hazard. However, most soft pesticides are more toxic to humans and must be handled with greater care than the older hard pesticides.

Pesticides can be divided into several categories based on the organism they are used to control. Insecticides are used to control insects, herbicides are used for plants, fungicides for fungi, and rodenticides for rodents. Because of the problems of biological amplification, resistance of pests to pesticides, and human health concerns, many people are seeking pesticide-free alternatives to raising food.

Integrated pest management makes use of a complete understanding of an organism's ecology to develop pest-control strategies that do not use pesticides or severely reduce the need for pesticides.

Review Questions

1. What is monoculture?
2. List three reasons why fossil fuels are essential for mechanized agriculture.
3. Describe why pesticides are commonly used in mechanized agriculture.
4. Why are fertilizers used?
5. How do hard and soft pesticides differ?
6. What is biological amplification?
7. How do organic farms differ from conventional farms?
8. Name three nonchemical methods of controlling pest populations.
9. What are the advantages and disadvantages of integrated pest management?
10. List three uses of food additives.

Human Influences on Ecosystems

CHAPTER FIFTEEN
Water Management

Objectives

After reading this chapter, you should be able to:

Explain how water is cycled through the hydrologic cycle.

Explain the significance of groundwater, aquifers, and runoff.

Explain how land use affects infiltration and surface runoff.

List the various kinds of water use and the problems associated with each.

List the problems associated with water impoundment.

List the major sources of water pollution.

Define biochemical oxygen demand (BOD).

Explain how nutrients cause water pollution.

Differentiate between point and nonpoint sources of pollution.

Explain how heat can be a form of pollution.

Differentiate between primary, secondary, and tertiary sewage treatments.

Describe some of the problems associated with storm-water runoff.

List sources of groundwater pollution.

Explain how various federal laws control water use and prevent misuse.

List the problems associated with water-use planning.

Explain the rationale behind the federal laws that attempt to preserve specific water areas and habitats.

List the problems associated with groundwater mining.

Explain the problem of salinization associated with large-scale irrigation in arid areas.

List the kinds of water-related services provided by local governments.

Chapter Outline

Key Terms

activated sludge sewage
  treatment
agricultural products
agricultural runoff
aquiclude
aquifer
aquitard
artesian aquifer
biochemical oxygen
  demand (BOD)
confined aquifer
domestic water
groundwater
groundwater mining
hydrologic cycle
industrial uses
in-stream uses
irrigation
landfills
limiting factor

nonpoint source
point source
potable waters
primary sewage
  treatment
runoff
secondary sewage
  treatment
septic tanks
storm-water runoff
surface impoundments
tertiary sewage treatment
thermal pollution
transpiration
unconfined aquifer
underground storage
  tanks
water diversion
water table

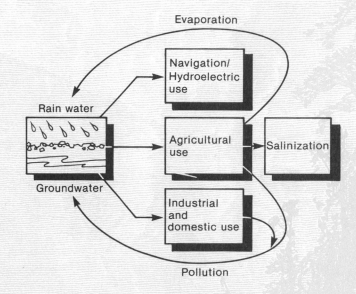

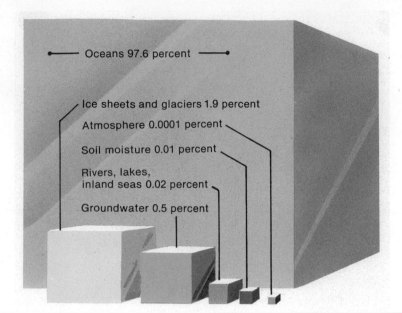

Oceans 97.6 percent

Ice sheets and glaciers 1.9 percent

Atmosphere 0.0001 percent

Soil moisture 0.01 percent

Rivers, lakes,
inland seas 0.02 percent

Groundwater 0.5 percent

FIGURE 15.1

Free-Water Storage on the Earth. The majority of the world's supply of water is in the oceans. The readily available fresh water is found as groundwater in porous rock beds. Although ice sheets and glaciers hold a great deal of fresh water, their turnover is too slow to be usable.

The Water Issue

All living organisms are composed of cells that contain at least 60 percent water. Organisms can only exist where there is access to adequate supplies of water. Water, however, is also a unique and necessary resource because it has remarkable physical properties. Water's ability to act as a solvent and its capacity to store heat are perhaps its most useful properties. As a solvent, water can dissolve and carry substances ranging from nutrients to industrial and domestic wastes. A glance at any urban sewer will quickly point out the value of water in dissolving and transporting wastes.

Water's ability to store or contain heat is another unusual and important physical property. Because water heats and cools more slowly than most other substances, it is used in large quantities for cooling purposes in electric power generation plants and in other industrial processes. Water's ability to retain heat modifies local climatic conditions in areas near large bodies of water. These areas do not have the wide temperature changes characteristic of other areas.

For most human uses, as well as some commercial ones, the quality of the water is as important as its quantity. Water must be substantially free of salinity, plant and animal waste, and bacterial contamination to be suitable for human consumption. Unpolluted freshwater supplies are known as **potable waters.** Early human migration routes and settlement sites were influenced by the availability of drinking water. Today, despite advances in drilling, irrigation, and purification, the location, quality, quantity, ownership, and control of potable waters remains an important human concern.

Over 70 percent of the earth's surface is covered by water. The vast majority of this water, however, is ocean salt water, which has limited use. (See figure 15.1.) Salt water cannot be consumed by humans or used for many kinds of industrial processes. Clean freshwater supplies have always been considered inexhaustible. Only now is it understood that we will probably exhaust our usable water supplies in some areas because of both human and natural factors. Human factors that have affected usable water supplies include a steadily increasing demand for fresh water for industrial, agricultural, and personal needs.

Human Influences on Ecosystems

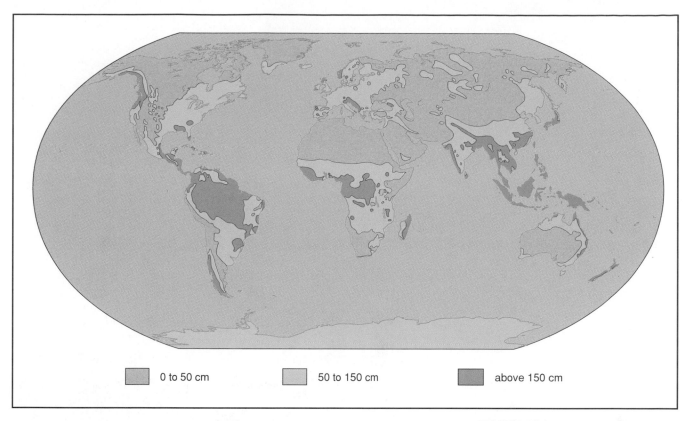

| | 0 to 50 cm | | 50 to 150 cm | | above 150 cm |

FIGURE 15.2
Global Average Annual Precipitation.
The world's precipitation is not
uniformly distributed. Although some
areas of the world, such as Southeast
Asia and Central America, have
extremely high levels of precipitation,
other areas, such as northern Africa and
the Middle East, have very little annual
precipitation.

Shortages of potable water throughout the world can also be directly attributed to human abuse in the form of pollution. Water pollution has negatively affected water supplies in almost all of the world's densely populated industrialized nations, including Japan, Europe, the Soviet Union, and North America.

Unfortunately, the outlook for the world's freshwater supply is not very promising. According to studies by the United Nations and the International Joint Commission, many sections of the world will experience shortages of potable water by the year 2000. Although the world's supply of water is continually being replenished by rainfall, this rainfall varies significantly. (See figure 15.2.) Parts of the world, particularly Mexico, India, Europe, and sections of Africa, continue to suffer massive droughts because of long periods of extremely limited rainfall. In other parts of the world, floods result from too much rainfall.

The Hydrologic Cycle

All water is locked into a constant recycling process called the **hydrologic cycle.** (See figure 15.3.) Solar energy evaporates water from the ocean surface. Additional quantities of water are evaporated from the soil and from the surfaces of plants. The water loss from plants is called **transpiration.** The air containing water vapor moves across the surface of the earth. As this warm, moist air cools, water droplets form and fall to the land as precipitation. Although some precipitation may simply stay on the surface until it evaporates, most will either sink into the soil or flow downhill until it eventually returns to the ocean. The water that infiltrates the soil may be

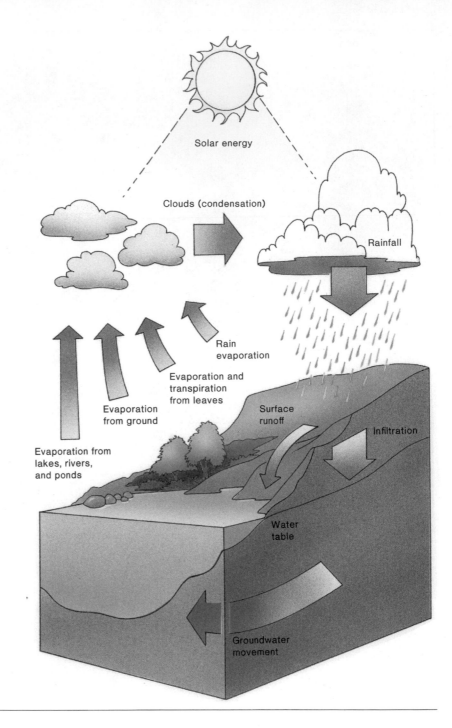

FIGURE 15.3

Hydrologic Cycle. This cycle is powered by solar energy. Water evaporates from the oceans, collects in the atmosphere, and then is moved by the wind. When the air cools, the water molecules condense into droplets that fall as precipitation. Some of it falls on the land and runs off the surface. This water eventually returns to the sea.

stored for long periods in underground reservoirs. This water is called **groundwater,** as opposed to the surface water that enters a river system as **runoff.**

The way in which land is used has significant impact on evaporation, runoff, and the rate of infiltration. When water is used for cooling purposes in power plants or for the irrigation of crops, the rate of evaporation is increased. Water impounded in reservoirs also evaporates rapidly. This rapid evaporation can affect local atmospheric conditions. Likewise, water infiltration is not constant but is greatly influenced by human activity. Urban complexes with paved surfaces and storm sewers increase runoff and reduce

Human Influences on Ecosystems

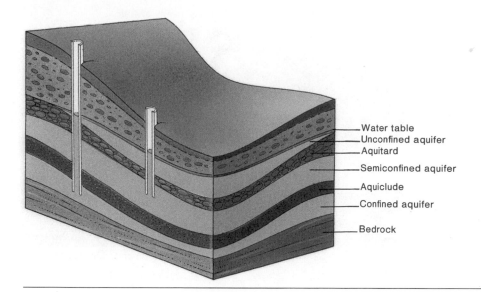

Water table
Unconfined aquifer
Aquitard
Semiconfined aquifer
Aquiclude
Confined aquifer
Bedrock

FIGURE 15.4

Types of Aquifers. The figure depicts the vadose zone (unsaturated zone or zone of aeration), aquitard and aquiclude layers, and water table (unconfined) and confined aquifers.

infiltration. Demand for underground water in urban areas is usually high, but urban developments actually increase the gap between supply and demand.

Water that enters the soil and is not picked up by plant roots moves slowly downward until it reaches an impervious layer of rock. This water accumulates in porous strata called an **aquifer.** There are two basic kinds of aquifers: **unconfined** and **confined.** An unconfined aquifer usually occurs near the land's surface and may be called a water-table aquifer because the upper surface of its water is the **water table.** The lower boundary is an impermeable layer of clay or rock. Unconfined aquifers are replenished (recharged) primarily by rain that falls on the ground directly above the aquifer and percolates down to the water table. The water in such aquifers is at atmospheric pressure and flows in the direction of the water table's slope.

A confined aquifer, also known as an **artesian aquifer,** is bounded on the top and bottom by confining layers and is saturated with water under greater-than-atmospheric pressure. The artesian aquifer is primarily replenished by rain and surface water from a recharge zone that may be many kilometers from where the aquifer is tapped for use. If water cannot pass through the confining layer (impermeable) of an artesian aquifer, the layer is called an **aquiclude.** If water can pass in and out of the confining layer (permeable), the layer is called an **aquitard.** The vadose zone (also known as the unsaturated zone or zone of aeration) is an unsaturated area below the ground surface. (See figure 15.4.)

Aquifers are extremely important in supplying water for industrial, agricultural, and municipal use. Most of the larger urban areas in the western part of the United States depend upon underground water for their water supply. This groundwater supply exists as long as it is not used faster than it can be replaced. Determining how much water can be used and what the uses should be is often a problem.

Water uses can be measured by either the amount withdrawn or the amount consumed. Water withdrawn for use is diverted from its natural course. It may be withdrawn and then later returned to its source so it can be used again in the future. For example, when a factory removes water from a river for cooling purposes, it returns the water to the river so it can

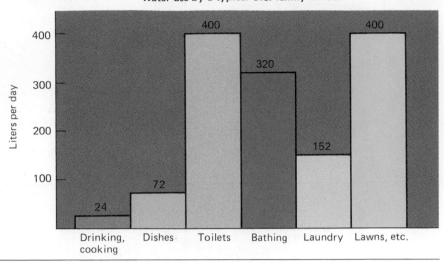

FIGURE 15.5
Urban Domestic Water Uses. Over 100 billion liters of water are used each day for urban domestic purposes in the United States.

be used later. Water that is incorporated into a product or lost to the atmosphere through evaporation and transpiration cannot be reused in the same geographic area and so is said to be consumed. Much of the water used for irrigation evaporates. Some of this water is removed with the crop when it is harvested. Therefore, irrigation must be considered to both withdraw and consume water.

Kinds of Water Use

The use of water can be classified into four categories: (1) domestic use, (2) agricultural use, (3) in-stream use, and (4) industrial use.

Domestic Use of Water

Many rural residents in North America still obtain safe water from untreated private wells, but urban residents are usually supplied with water from complex and costly water purification facilities. Extensions and merges of urban communities have created problems in the development, transportation, and maintenance of quality water supplies.

Domestic activities in highly developed nations require a great deal of water. This domestic use includes drinking, air conditioning, bathing, washing clothes, washing dishes, flushing toilets, and watering lawns and gardens. On average, each person in a North American home uses 300–400 liters of water each day. Most of this **domestic water** is used as a solvent to carry away wastes, with only a small amount used for drinking. (See figure 15.5.) Yet all water that enters the house has been purified and treated to make it safe for consumption. Until recently, the cost of water in almost every community has been so low that there was very little incentive to conserve, but increasing purification costs have raised the price of domestic water.

Natural processes cannot cope with the highly concentrated wastes typical of a large urban area. The unsightly and smelly results present a potential health problem for the municipality. Cities and towns must provide for both the domestic water supplies and the treatment of the wastewater following its use, and both processes are expensive and require trained personnel.

Human Influences on Ecosystems

FIGURE 15.6
Urban Expansion. Maintaining a suitable supply of water for growing metropolitan areas can pose major problems, especially in arid areas such as the southwestern United States.

The major problem associated with domestic use of water is maintaining an adequate, suitable supply for growing metropolitan areas. (See figure 15.6.) Demand for water in urban areas sometimes exceeds the immediate supply, particularly when domestic supply consists of local surface water. During the summer, water demand is high, and precipitation is often low.

More domestic water is wasted than consumed. This loss, nearly 20 percent of the water withdrawn from public supplies (mainly through leaking water pipes and water mains), is amazingly large. Another major cause of water loss has been that of public attitudes. As long as water is considered a limitless, inexpensive resource, there will be little effort to conserve. As the cost of water rises and attitudes toward water change, so will usage and efforts to conserve.

Agricultural Use of Water

The major consumptive use of water in North America is for agricultural purposes and principally for **irrigation.** In the 1980s, for example, irrigation accounted for nearly 80 percent of all the water consumed in the United States. The amount of water used for irrigation and livestock continues to increase throughout the world. Future agricultural demand for

FIGURE 15.7

Irrigation. Large areas of the United States require irrigation to be farmed economically. The fields in the top photograph have a long pipe, which releases water while slowly rotating about a central pivot. In this way, the field is automatically irrigated. In contrast to the rotating pipe system, trickle irrigation delivers the water directly to the roots of the plants, thus reducing water loss.

water will depend on the cost of water for irrigation; the demand for agricultural products, food, and fiber; governmental policies; and the development of new technology.

In some areas, irrigation is a problem because there is not a supply of water nearby. This is particularly true in the western United States, where about 14 million hectares of land are irrigated. (See figure 15.7.) In some places, water must be piped hundreds of kilometers for irrigation.

Because most of the world's consumptive use of water results from irrigation, it is becoming increasingly important to modify irrigation practices to use less water. Water loss from irrigation can be reduced in many ways. Increasing the cost of water will stimulate conservation of water by farmers just as it does home owners. Another method is to reduce the amount of water-demanding crops grown in dry areas, or change from high water-demanding to lower water-demanding crops. Planting wheat or soybeans instead of potatoes or sugar beets reduces the amount of water required. Switching to trickle irrigation also reduces water loss. With trickle

Human Influences on Ecosystems

FIGURE 15.8
Dams Interrupt the Flow of Water.
The flow of water in most large rivers is controlled by dams. Most of these dams provide electricity. In addition, they prevent flooding and provide recreational areas. However, dams destroy the natural river system.

irrigation, a series of pipes are placed on the ground with openings strategically placed so that when water flows through the pipe, it delivers a particular quantity of water to the individual plants. This method delivers the water directly to the roots of the plants, rather than flooding entire fields. Although used extensively in greenhouses, trickle irrigation is generally too costly for large agricultural operations. Methods that do not use as much water as flooding irrigation and that are not as expensive in terms of labor and equipment as the trickle method include furrow irrigation, corrugation irrigation, overhead irrigation, and subirrigation. Each of these methods has its drawbacks and advantages as well as conditions under which it works well.

Irrigation requires a great deal of energy. Estimates indicate that 40 percent of the energy devoted to agriculture in Nebraska is used for irrigation. Increasing energy costs may force some farmers to reduce or discontinue irrigation. In addition, much of western Nebraska relies on groundwater for irrigation, and the water table is dropping rapidly. If a water shortage develops, land values will decline. Land use and water use are interrelated and cannot be viewed independently.

In-stream Use of Water

When the flow of water in streams is interrupted or altered, the value of the stream is changed. Major **in-stream uses** of water are for hydroelectric power, recreation, and navigation. Electricity from hydroelectric power plants is an important energy resource. Presently, hydroelectric power plants produce about 13 percent of the total electricity generated in the United States. Hydroelectric power plants do not consume water and do not add waste products to it. However, the dams needed for hydroelectric power plants have definite disadvantages, including the high cost of construction and the resultant destruction of the natural habitat in streams and surrounding lands. While dams reduce the amount of flooding, they do not eliminate it. (See figure 15.8.) In fact, the building of a dam often encourages people to develop the floodplain. As a result, when flooding occurs, the potential loss of property and lives is greater.

FIGURE 15.9
Recreational Use of Water. Marinas provide recreation; however, wetlands destruction and large dredging operations may be necessary to build them.

The sudden discharge from a dam of the impounded water also can seriously alter the downstream environment. If the discharge is from the top of the reservoir, the stream temperature rapidly increases. Discharging the colder water at the bottom of the reservoir causes a sudden decrease in the stream's water temperature. Either of these changes is harmful to aquatic life in the stream.

The impoundment of water also reduces the natural scouring action of a flowing stream. If water is allowed to flow freely, the silt accumulated in the river is carried downstream during times of high water. This maintains the river channel and carries nutrient materials to the river's mouth. But if a dam is constructed, the silt deposits behind the dam, eventually filling the reservoir with silt.

In addition, impounded water has a greater surface area, which increases the amount of evaporation. In areas where water is scarce, the amount of water lost through such evaporation can be serious. This is particularly evident in hot climates. Furthermore, flow is often intermittent below the dam, which alters the water's oxygen content and interrupts fish migration. The populations of algae and other small organisms are also altered. Therefore, dam construction requires careful prior planning.

Water tends to be a focal point for recreational activities. (See figure 15.9.) Sailing, waterskiing, swimming, fishing, and camping all require water of reasonably good quality. Water is used for recreation in its natural setting and often is not physically affected. Even so, it is necessary to plan for recreational use, because overuse or inconsiderate use can degrade water quality. For example, waves generated by powerboats can accelerate shoreline erosion and cause siltation.

Human Influences on Ecosystems

Dam construction creates new recreational opportunities because reservoirs provide new sites for boating, camping, and related recreation. However, this is at the expense of a previously free-flowing river. Some recreational opportunities, such as river fishing, have been lost.

Most major rivers are used for navigation. North America currently has more than 40,000 kilometers of commercially navigable waterways. Waterways used for navigation must have sufficient water to ensure passage of transport vessels. Canals, locks, and dams are employed to guarantee this. Often, dredging is necessary to maintain the proper channel depth. Dredging can resuspend in the water contaminated sediments that had previously been covered over. In addition, the flow within the hydrologic system is changed, which, in turn, affects the water's value for other uses.

Most large urban areas rely on water to transport needed resources. During recent years, the inland waterway system has carried about 10 percent of the goods, such as grain, coal, ore, and oil. In the United States, expenditures for the improvement of the inland waterway system have totalled billions of dollars.

In the past, almost any navigation project was quickly approved and funded, regardless of the impact on other uses. Today, however, such decisions are not made until the impacts on various other uses are carefully analyzed.

Industrial Use of Water

Water for **industrial use** accounts for more than half of total water withdrawals. Ninety percent of the water used by industry is for cooling. Most industrial processes involve heat exchanges. Water is a very effective liquid for carrying heat away from these processes. For example, electric power generating plants use water to cool steam so that it changes back into water. If the water heated in an industrial process is dumped directly into a watercourse, it significantly changes the stream's water temperature. This affects the aquatic ecosystem by increasing the metabolism of the organisms and reducing the water's ability to hold dissolved oxygen.

Industry also uses water to dissipate and transport waste materials. In fact, many streams are now overused for this purpose, especially watercourses in urban centers. The use of watercourses for waste dispersal degrades the quality of the water and may reduce its usefulness for other purposes. This is especially true if the industrial wastes are toxic.

During the past thirty years, many nations have passed laws that severely restrict industrial discharges of wastes into watercourses. In the United States, the federal role in maintaining water quality began in 1948 with the passage of the Federal Water Pollution Control Act. This act provided federal funds and technical assistance to strengthen local, state, and interstate water-quality programs. Through amendments to the act in 1956, 1965, 1972, and 1987, the federal role in water-pollution control was increased to include establishing water-quality standards, financing area-wide waste-treatment management plans, and establishing the framework for a national program of water-quality regulation.

Kinds and Sources of Water Pollution

Water pollution occurs when the use by one segment of society interferes with the health and well-being of other members. In an industrialized society, maintaining unpolluted water in all drains, streams, rivers, and lakes is probably impossible. Water quality is related to the use intended of the water. Adding material to water may cause the water to become unfit for some uses but may not affect other uses. If silt is added to a lake, the water may still be drinkable, but the lake may no longer be an acceptable place to swim. If salts are added to a lake, the water may then be less acceptable as drinking water, but the salts may not interfere with the lake's recreational value. It may not be necessary to maintain absolutely pure water.

There are also economic considerations. The cost of removing the last few percentages of some materials from the water may not be justified. This is certainly true of organic matter, which is biodegradeable. However, radioactive wastes and toxins that may accumulate in living tissue are a different matter. Attempting to remove these materials is often justified because of their potential harm to humans and other living organisms.

Municipal Water Pollution

Municipalities are faced with the double-edged problem of providing suitable drinking water for the population and disposing of wastes. These wastes consist of storm-water runoff, wastes from industry, and wastes from homes and commercial establishments.

Wastes from homes consist primarily of organic matter from garbage, food preparation, cleaning of clothes and dishes, and human wastes. Human wastes are mostly undigested food material and a concentrated population of bacteria, such as *Escherichia coli* and *Streptococcus faecalis*. These particular bacteria normally grow in the human large intestine, where they are responsible for some food digestion and for the production of vitamin K. Low numbers of these bacteria in water are not harmful to healthy people. Because they can be easily identified, their presence in the water is used to indicate the amount of pollution from human waste. The number of this type of bacteria present in water is directly related to the amount of human waste entering the water. When human wastes are disposed of in water systems, other potentially harmful bacteria from the human large intestine may be present in amounts too small to detect by sampling. The greater the amount of wastes deposited in the water, the more likely that there will be small populations of disease-causing bacteria. Therefore, the harmless bacteria are indicators that other organisms may also be present.

The nonliving organic matter in sewage presents a different kind of pollution problem because it decays in the water. Microorganisms use oxygen dissolved in the water when they degrade the organic material. As the microorganisms metabolize the organic matter, they use up the available oxygen. The amount of oxygen required to decay a certain amount of organic matter is called the **biochemical oxygen demand (BOD).** (See figure 15.10.) Measuring the BOD of a body of water is one way to determine how polluted it is. If too much organic matter is added to the water, all of the available oxygen will be used up. Then anaerobic (nonoxygen-requiring) bacteria begin to break down wastes. Anaerobic respiration produces chemicals that have a foul odor and an unpleasant taste and that generally interfere with the well-being of humans. Therefore, they are pollutants.

Although the wastewater from cleaning dishes and clothing may contain some organic material, the more important group of contaminants found in this water are soaps and detergents. Soaps and detergents are useful because one end of the molecule dissolves in dirt or grease and the other end dissolves in water. When the soap or detergent molecules are rinsed away by the water, the dirt or grease goes with them.

Until recently, many detergents contained phosphates as a part of their chemical structure. A **limiting factor** is a necessary material that is in short

Human Influences on Ecosystems

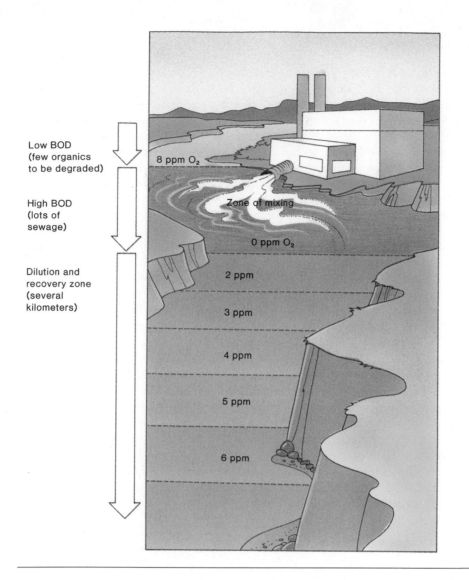

Low BOD
(few organics
to be degraded)

8 ppm O$_2$

High BOD
(lots of
sewage)

Zone of mixing

0 ppm O$_2$

Dilution and
recovery zone
(several
kilometers)

2 ppm

3 ppm

4 ppm

5 ppm

6 ppm

FIGURE 15.10

Effect of Organic Wastes on Dissolved Oxygen. Sewage contains a high concentration of organic materials. When these are degraded by organisms, oxygen is removed from the water. Therefore, there is an inverse relationship between sewage and oxygen in the water. The greater the BOD, the less desirable the water is for human use.

supply, and because of the lack of it, an organism cannot reach its full po tential. Phosphates are plant nutrients and therefore are a limiting factor of plant growth. Thus, when phosphates from detergents are added to the surface water, they act as a fertilizer and promote the growth of undesirable algae populations. Algae and larger plants may interfere with the use of the water by fouling boat propellors, clogging water intake pipes, changing the taste and odor of water, and causing the buildup of organic matter on the bottom. As this organic matter decays, oxygen levels decrease, and fish and other aquatic species die.

Because these problems are all associated with the addition of plant nutrients such as phosphates to the water, many states have banned the sale of detergents with high-phosphate content. Detergent manufacturers point out that nutrients from **agricultural runoff** are probably more significant than detergents in adding phosphate to lakes and streams. While such runoff is important in adding to the total phosphate load of surface water, this defense does not alter the fact that detergents also add to this surface water load. Most domestic wastewater goes through a sewage treatment plant, so it is easier to measure and control the phosphate content. In those areas where agriculture is uncommon, detergents can be the major source of phosphate pollution.

Industrial Water Pollution

Frequently, a factory or industrial complex disposes of some or all of its wastes into a municipal sewage system. Depending on the type of industry involved, these wastes are likely to be a combination of organic materials, petroleum products, metals, acids, and so forth. The organics and oil add to the BOD of the water. The metals, acids, and other ions need special treatment, depending on their nature and concentration. As a result, municipal sewage treatment plants must be designed with their industrial customers in mind. In most cases, it is preferred that industries take care of their own wastes. This allows the industry to segregate and control toxic wastes and design a wastewater facility that meets its specific needs.

Most companies, when they remodel their facilities, include wastewater treatment as a necessary part of an industrial complex. However, many older facilities continue to pollute. These companies discharge acids, particulates, heated water, and noxious gases into the water. Generally, these plants are easily detected because the pollution is from a single effluent pipe or series of pipes. This pollution is said to come from a **point source.** Diffuse pollutants, such as agricultural runoff, road salt, and acid rain, are said to come from **nonpoint sources.** Economic pressure and adverse publicity can affect those companies that continue to pollute from point sources. However, pollutants that come from nonpoint sources are very difficult to control. In addition, pollution legislation relating to them is very difficult to enforce.

Amendments to the U.S. Federal Water Pollution Control Act of 1972 (PL 92-500) have mandated changes in how industry treats water. Industries are no longer allowed to use water and return it to its source in poor condition. One of the standards regulates the temperature of the water that is returned to its source. Because many industries use water for cooling, thermal pollution has become a problem.

Thermal Pollution

Thermal pollution occurs when an industry removes water from a source, uses the water for cooling purposes, and then returns the heated water to its source. In the United States alone, industry uses more than 225 billion liters of water annually for cooling purposes. Electric generating plants use 75 percent. Most of the remainder is used for generating steam, cooling steel, and manufacturing plastics.

Power plants heat water to convert it into steam, which drives the turbines that generate electricity. For steam turbines to function efficiently, the steam must be condensed into water after it leaves the turbine. This condensation is usually accomplished by taking water from a lake or stream to absorb the heat. This heated water is then discharged.

Cooling water used by industry does not have to be released into aquatic ecosystems. There are three other methods of discharging the heat. One method is to construct a large, shallow pond. Hot water is pumped into one end of the pond, and cooler water is removed from the other end. The heat is dissipated from the pond into the atmosphere and substrate.

A second method is to use a cooling tower. In a cooling tower, the heated water is sprayed into the air and cooled by evaporation. The disadvantage of cooling towers and shallow ponds is that large amounts of water are lost by evaporation. The release of this water into the air can also produce localized fogs.

The third method of cooling, the dry tower, does not release water into the atmosphere. In this method, the heated water is pumped through tubes, and the heat is released into the air. This is the same principle used in an automobile radiator. The dry tower is the most expensive to construct and operate.

While the least expensive and easiest method of discharging heated water is to return the water to the aquatic environment, this can create problems for the inhabitants of the area. Although an increase in temperature of only a few degrees may not seem significant, some aquatic ecosystems are very sensitive to minor temperature changes. The spawning behavior of many fish is triggered by temperature changes. Lake trout will not spawn in water above 10° C; therefore, if a lake has a temperature of 8° C, the lake trout population will reproduce. But an increase of 3° C would prevent spawning of this species and result in their eventual elimination from that lake.

Ocean estuaries are very fragile. The discharge of heated water into an estuary may alter the type of plant food available. As a result, animals with specific food habits may be eliminated because the warm water supports different kinds of food organisms. The entire food web in the estuary may be altered by only slight temperature increases.

Wastewater Treatment

Because water must be cleaned before it is released, most companies and municipalities maintain wastewater treatment facilities. Treatment of sewage is usually classified as primary, secondary, or tertiary. **Primary sewage treatment** removes larger particles by filtering through large screens and settling in ponds or lagoons. Water is removed from the top of the settling lagoon and released. Water thus treated does not have any sand or grit, but it still carries a heavy load of organic matter, dissolved salts, bacteria, and other microorganisms. The organisms use the organic material for food, and as long as there is sufficient oxygen, they will continue to grow and reproduce. If the receiving body of water is large enough and the organisms have enough time, the organic matter will be degraded. In crowded areas, where several municipalities take water and return it to a lake or stream within a few kilometers of each other, primary water treatment is not adequate.

Secondary sewage treatment usually follows primary treatment and involves holding the wastewater until the organic material has been degraded by the bacteria and other microorganisms. To encourage this action, the wastewater is mixed with large quantities of highly oxygenated water, or the water is aerated directly, as in a trickling filter system. In this system, the wastewater is sprayed over the surface of rock to increase the amount of dissolved oxygen. The rock also provides a place for the bacteria and other microbes to attach so they are exposed simultaneously to the organic material and oxygen. Most secondary treatment facilities are concerned with promoting the growth of microorganisms. These microorganisms feed on the dissolved organic matter and small suspended particles, which then become incorporated into their bodies as part of their cell structure. The bodies of the microorganisms are larger than the dissolved and suspended organic matter, so this process concentrates the organic wastes into particles that are large enough to settle out. The sludge that settles consists of living and dead microorganisms and their waste products. In **activated sludge sewage treatment** plants, some of the sludge is returned to aeration tanks, where it is mixed with incoming wastewater. This kind of process uses less land than a trickling filter. (See figure 15.11.) Both processes produce a sludge that settles out of the water.

a.

b.

FIGURE 15.11

Primary and Secondary Wastewater Treatment. Primary treatment is physical; it includes filtrating and settling of wastes. Photograph (*a*) is a settling tank in which particles settle to the bottom. Secondary treatment is mostly biological and includes the concentration of dissolved organics by microorganisms. Two major types of secondary treatment are the trickling filter and activated sludge methods. Photograph (*b*) shows an activated sludge system.

The sludge that remains is concentrated and often dried before disposal. Sludge disposal is a major problem in large population centers. In the San Francisco Bay area, 2,500 metric tons of sludge are produced each day. Most of this is carried to landfills and lagoons, and some amount is composted and returned to the land as fertilizer.

Primary and secondary facilities are the most common types of sewage treatment in North American cities. The water discharged from these sewage treatment plants must be disinfected. The least costly method of disinfection is chlorination of the wastewater after it has been filtered and the organic materials have been allowed to settle. Using ultrasonic energy to mechanically break down waste may be less harmful and more effective, but it is also more expensive than chlorination.

A growing number of larger sewage treatment plants use additional processes called tertiary sewage treatment. **Tertiary sewage treatment** involves a variety of different techniques to remove dissolved pollutants left after primary and secondary treatments. The tertiary treatment of wastewater removes phosphorus and nitrogen that could increase aquatic plant growth. Tertiary treatment is very costly because it requires specific chemical treatment of the water to eliminate specific problem materials. (See table 15.1.) Certain industries are beginning to maintain their own secondary or tertiary wastewater facilities because of the specific nature of their waste products.

Table 15.1
Tertiary treatment.

Kind of tertiary treatment	Problem chemicals	Methods
Biological	Phosphorus and nitrogen compounds	1. Large ponds are used to allow aquatic plants to assimilate the nitrogen and phosphorus compounds from the water before the water is released. 2. Columns containing denitrifying bacteria are used to convert nitrogen compounds into atmospheric nitrogen.
Chemical	Phosphates and industrial pollutants	1. Water can be filtered through calcium carbonate. The phosphate substitutes for the carbonate ion, and the calcium phosphate can be removed. 2. Specific industrial pollutants, which are nonbiodegradable, may be removed by a variety of specific chemical processes.
Physical	Primarily industrial pollutants	1. Distillation. 2. Water can be passed between electrically charged plates to remove ions. 3. High-pressure filtration through small-pored filters. 4. Ion-exchange columns.

Runoff

Storm-water runoff from streets and buildings is often added directly to the sewer system and sent to the municipal wastewater treatment facility. During heavy precipitation or spring thaws, the sewage treatment plant may be unable to handle the volume of wastewater, so some of it might be discharged directly to the surface water without treatment. Modern wastewater treatment facilities have holding tanks to contain storm runoff and domestic wastes during heavy rains until they can be treated. Many plants lack adequate storage and must release untreated wastewater directly into receiving waters.

Agricultural runoff and mine drainage are nonpoint sources of pollution and thus are difficult to detect and control. One of the largest water-pollution problems is associated with agricultural runoff from large expanses of open fields. Precipitation dissolves materials in these areas and carries the materials away. Either the water runs over the surface and carries away exposed topsoil and nutrients (which are deposited in drains, streams, and rivers), or water seeps into the soil and carries dissolved nitrogen and phosphorus compounds into the groundwater.

Farmers can reduce runoff in several ways. One is to leave a zone of undisturbed land near drains or stream banks. This retards surface runoff because soil covered with vegetation tends to slow the movement of water and allows the silt to be deposited on the surface of the land rather than in the streams. This is costly because farmers may need to remove valuable cropland from cultivation. To retard leaching, farmers can keep the soil

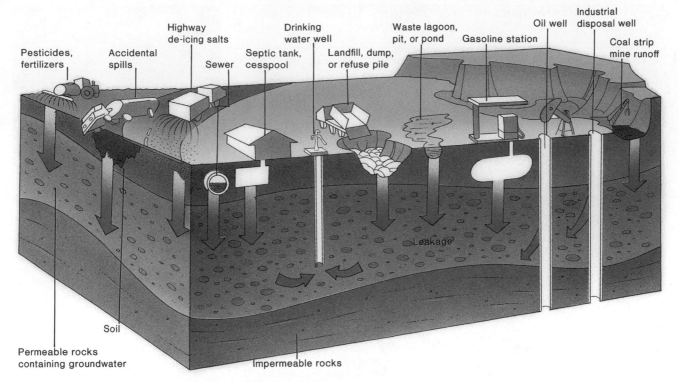

Pesticides,
fertilizers

Accidental
spills

Highway
de-icing salts

Sewer

Septic tank,
cesspool

Drinking
water well

Landfill, dump,
or refuse pile

Waste lagoon,
pit, or pond

Gasoline station

Oil well

Industrial
disposal well

Coal strip
mine runoff

Leakage

Soil

Permeable rocks
containing groundwater

Impermeable rocks

FIGURE 15.12

**Sources of Groundwater
Contamination.** A wide variety of
activities have been identified as sources
of groundwater contamination.

Source: Environmental Protection Agency.

covered with a crop as long as possible and carefully control the amount
and the timing of fertilizer application. This makes good economic sense
because any fertilizer that runs off or leaches out of the soil has to be replaced.

Another nonpoint source of water pollution is associated with mining.
When coal is mined, water that drains from these mines is often very acidic.
In addition, fine coal-dust particles are suspended in the water, which make
the water chemically and physically less valuable as a habitat. Dissolved
ions of iron, sulfur, zinc, and copper also are present in mine drainage.
Control involves the containment of mine drainage so that it does not mix
with surface water.

Groundwater Pollution

A wide variety of activities, some once thought harmless, have been identified as potential sources of groundwater contamination. In fact, possible
sources of human-induced groundwater contamination span every facet of
social, agricultural, and industrial activities. (See figure 15.12.) Major sources
include:

Agricultural products Pesticides contribute to unsafe levels of organic
contaminants in groundwater. The EPA has detected seventy-three
different pesticides in the groundwater of thirty-four states.
Accidental spills or leaks of pesticides pollute groundwater sources
with ten to twenty additional pesticides. Other agricultural practices
contributing to groundwater pollution include animal feeding
operations, fertilizer applications, and irrigation practices.

Underground storage tanks Up to 350,000 of the 1.4 million underground storage tanks containing gasoline and other hazardous substances that are covered by U.S. inventory rules may already be leaking. Four liters of gasoline can contaminate the water supply of a community of fifty thousand people.

Landfills According to the EPA, approximately 90 percent of the active U.S. landfills have no liners to stop leaks to underlying groundwater, and 96 percent have no system to collect the leachate that seeps from the landfill. Sixty percent of landfills place no restrictions on the waste accepted, and many landfills are not inspected even once a year.

Septic tanks Poorly designed and inadequately maintained septic systems have contaminated groundwater with nitrates, bacteria, and toxic cleaning agents. Over twenty million septic tanks are in use, and up to a third have been found to be operating improperly.

Surface impoundments Over 190,000 pits, ponds, and lagoons are used in the United States to store or treat wastes. Seventy-one percent are unlined, and only 1 percent use a plastic or other synthetic, nonsoil liner. Ninety-nine percent of these impoundments have no leak-detection systems. Seventy-three percent have no restrictions on the waste placed in the impoundment. Sixty percent of the impoundments are not even inspected annually. Many of these ponds are located near groundwater supplies.

Other sources of groundwater contamination identified by the U.S. Office of Technology Assessment include: mining wastes, salting for snow control, land application of treated wastewater, open dumps, cemeteries, radioactive disposal sites, urban runoff, construction excavation, and animal feedlots.

Water-Use Planning Issues

In the past, wastes were discharged into waterways with little regard to the costs imposed on other users by the resulting decrease in water quality. With today's increasing demands for high-quality water, unrestrained waste disposal could lead to serious conflicts about water uses and cause social, economic, and environmental losses.

Water-use planning will need to deal with a number of different issues, such as the following:

Increased demand for water will force increased reuse of existing water supplies.

In many areas where water is used for irrigation, both the water and the soil become salty because of evaporation. When this water returns to a stream, the quality of the water is lowered.

In some areas, wells provide water for all categories of use. If the groundwater is pumped out faster than it is replaced, the water table is lowered.

In coastal areas, seawater may intrude into the aquifers and ruin the water supply.

The demand for water-based recreation is increasing dramatically and requires high-quality water, especially for water recreation involving total body contact, such as swimming.

BOX 15.1
Marine Oil Pollution

Marine oil pollution has many sources. One source is accidents, such as oil-drilling blowouts or oil-tanker accidents. Examples include the Torrey Canyon supertanker that broke up off the coast of England in 1967 and the Amoco Cadiz supertanker that broke up off the coast of France in 1978, resulting in a spill of more than 254 million liters of oil.

In 1989, the Exxon Valdez ran aground in Prince William Sound, Alaska, causing an oil spill of over 42 million liters of oil. The Exxon Valdez spill affected nearly 1,500 kilometers of Alaskan coastline and caused extensive damage to native wildlife. Long-term effects on such species as the salmon are still uncertain. Salmon are of particular concern in Alaska because of the over $700 million-a-year salmon-fishing industry.

By the end of 1990, Exxon Corporation had spent close to $2.5 billion in an effort to clean up the worst-ever oil spill in the United States. In addition, in March 1991, Exxon agreed to an additional settlement of $1 billion, $900 million to complete the cleanup and $100 million to pay criminal fines. How successful the cleanup was is open to debate. Some argue that a greater cleanup effort is needed, while others contend that the problems resulting from the spill have, for the most part, been resolved. Whatever the outcome of the debate, one fact is certain: Cleaning up an oil spill the size of the Alaskan spill will never be a simple task and court challenges continue.

The greatest sources of all marine oil pollution are not accidental. Nearly two-thirds of all human-caused marine oil pollution comes from three sources: (1) runoff from streets, (2) improper disposal of lubricating oil from machines or automobile crankcases, and (3) intentional oil discharges that occur during the loading and unloading of tankers.

The oil spill resulting from the *Exxon Valdez* in 1989 had a wide-ranging environmental impact in Alaska.

With regard to the latter, pollution occurs when the tanks are cleaned or oil-contaminated ballast water is released. Oil tankers use seawater as ballast to stabilize the craft after they have discharged their oil. This oil-contaminated water is then discharged back into the ocean when the tanker is refilled.

As the number of offshore oil wells and the number and size of oil tankers grew, the potential for increased oil pollution also grew. Many methods for controlling marine oil pollution have been investigated and tried. Some of the more promising methods include recycling and reprocessing of used oil and grease from automobile service stations and industries, and enforcing stricter regulations on the offshore drilling, refining, and shipping of oil.

Preserving Scenic Water Areas and Wildlife Habitats

As mentioned earlier, the use of land influences water quality. This is particularly true along shorelines and riverbanks, where the land and water meet. Some bodies of water have unique scenic value. To protect these resources, the way in which the land adjacent to the water is used must be consistent with preserving these scenic areas.

The U.S. Federal Wild and Scenic Rivers Act of 1968 established a system to protect wild and scenic rivers from development. All federal agencies must consider the wild, scenic, or recreational values of rivers in planning for the use and development of rivers and adjacent land resources. The process of designating a river or part of a river as wild or scenic is complicated. It often encounters local opposition from businesses dependent on growth. Following reviews by state and federal agencies, rivers may be des-

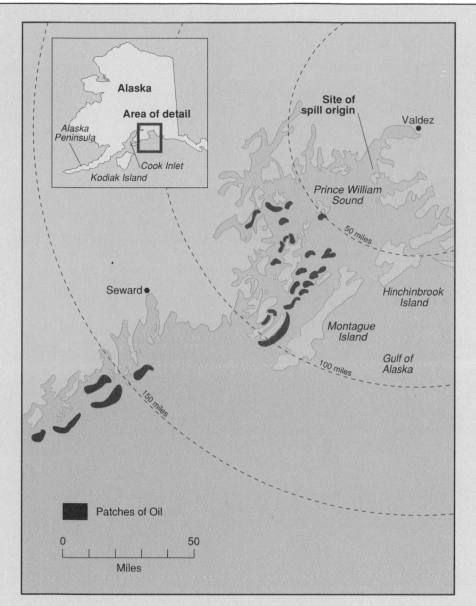

Ways to Clean Up Oil Spills

Mechanical

- Floating booms contain the spill near the source or block it from sensitive areas
- Skimmer boats herd the oil together to be vacuumed up onto collection barges
- Strings of absorbent pads soak up oil on beaches and in water too shallow for skimmer boats

Fire

- Oil must be fairly concentrated to burn, and crude is hard to ignite
- Chemical burning agents and lasers improve effectiveness

Chemical

- Coagulating agents cause floating crude to gather together for easier pickup, or sink to the bottom where it does less harm
- Dispersing agents break slicks
- Disadvantage: Chemical agents are pollutants just as the oil is

Natural

- Wind and wave action emulsify some oil into the water (like shaking salad dressing)
- Bacteria consume some crude over time

(Map labels: Alaska, Area of detail, Alaska Peninsula, Cook Inlet, Kodiak Island, Site of spill origin, Valdez, Prince William Sound, 50 miles, Seward, Hinchinbrook Island, Montague Island, 100 miles, Gulf of Alaska, 150 miles, Patches of Oil, 0 50 Miles)

ignated as wild and scenic by action of either Congress or the secretary of the interior. Sections of over sixty-five streams in the United States have been designated as wild or scenic.

Many unique and scenic shorelands have also been protected from future development. Until recently, U.S. estuaries and shorelands have been subjected to significant physical modifications, such as dredging and filling, which may improve conditions for navigation and construction but destroy fish and wildlife habitats. Recent actions throughout North America have attempted to restrict the development of shorelands. Development has been restricted in some particularly scenic areas, such as Cape Cod National Seashore in Massachusetts.

Historically, poorly drained areas were considered worthless. Subsequently, many of these wetlands were filled or drained. The natural and

BOX 15.2

Is It Too Late for the Everglades?

Close to a million people visit the Everglades Park each year to view a landscape that is part African veldt and part tropical swamp. While the Everglades may look healthy, parts of its finely tuned ecosystem are breaking down. According to some scientists, the park is dying.

Years of mismanagement of the water supply, with agriculture and urban development increasingly diverting the flow that sustains the park, have desiccated much of the park's wetlands. Water policies favoring farmers and urban dwellers have forced wading birds to relocate their colonies and cut their population in the southern Everglades from three hundred thousand to fifteen thousand. Alligators, whose population in the drying park is also drastically shrinking, are cannibalizing their young. Exotic plants, thriving on nutrients from agricultural runoff, are forcing out natural vegetation. Bass and catfish in the park should not be eaten because they are laced with natural mercury leached into ponds from soil dried to dust. Panthers, snail kites, and wood storks, three of the park's thirteen endangered wildlife species, are on the verge of extinction.

After Everglades Park was established in 1947, about 800,000 hectares of wetlands to the north were drained for farms and urban development. South Florida boomed. Some 4.5 million people now live in the horseshoe crescent around the Everglades region, and six hundred new residents arrive each day. The once-clear waters of Lake Okeechobee, which is north of the park, are now being polluted with dairy-farm runoff and choked with algae. Cattails that proliferate in the phosphorus-laden runoff from the sugar plantations are choking the waterways. Some 175,000 hectares of drained Everglades are in sugar production. Since the water is free, South Florida's farmers take as much as they need and have little incentive to conserve.

There are plans to help the Everglades. The Kissimmee River, which flows into Lake Okeechobee, was channelized into an arrow-straight river in the late 1960s. This project destroyed the marshes and allowed for nutrients to pollute the lake and thus the park. Beginning in 1990, the Kissimmee was being returned to its natural state, with twisting oxbow curves and extensive wetlands. In 1989, Congress approved the purchase of 43,000 hectares to be added to the east section of the park. The state of Florida has obtained an additional 60,000 hectares as a buffer zone for the park. In addition, there is modification of canals, spillways, and pumping stations to assure a steady supply of cleaner water for the park. This ambitious attempt to put a little wild back in the wilderness of the Everglades project has a good chance to succeed—if it is not too late.

economic importance of wetlands has been recognized only recently. In addition to providing the necessary spawning and breeding habitat for many species of wildlife, wetlands act as natural filtration systems by trapping nutrients and pollutants and preventing them from entering adjoining lakes or streams. Wetlands also slow down the force of floodwaters and permit nutrient-rich particles to settle out. In addition, wetlands can act as reservoirs and release water slowly into lakes, streams, or aquifers, thereby preventing floods. Coastal estuarine zones and adjoining sand dunes also provide significant natural flood control. Sand dunes act as barriers and absorb damaging waves caused by severe storms.

Groundwater Mining

Groundwater mining means that water is removed from an aquifer faster than it is replaced. When this practice continues over a long period of time, the water table eventually declines. Groundwater mining is common in areas

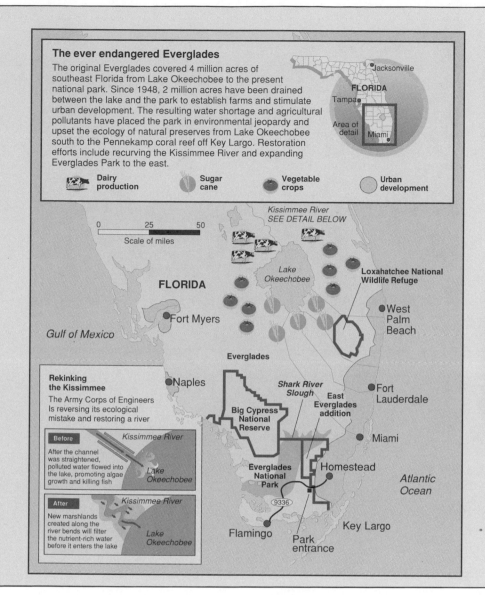

The ever endangered Everglades

The original Everglades covered 4 million acres of southeast Florida from Lake Okeechobee to the present national park. Since 1948, 2 million acres have been drained between the lake and the park to establish farms and stimulate urban development. The resulting water shortage and agricultural pollutants have placed the park in environmental jeopardy and upset the ecology of natural preserves from Lake Okeechobee south to the Pennekamp coral reef off Key Largo. Restoration efforts include recurving the Kissimmee River and expanding Everglades Park to the east.

Dairy production   Sugar cane   Vegetable crops   Urban development

FLORIDA
Jacksonville
Tampa
Area of detail
Miami

Kissimmee River
SEE DETAIL BELOW

0   25   50
Scale of miles

FLORIDA

Gulf of Mexico

Fort Myers

Lake Okeechobee

Loxahatchee National Wildlife Refuge

West Palm Beach

Everglades

Naples

Rekinking the Kissimmee

The Army Corps of Engineers Is reversing its ecological mistake and restoring a river

Before   Kissimmee River

After the channel was straightened, polluted water flowed into the lake, promoting algae growth and killing fish

Lake Okeechobee

After   Kissimmee River

New marshlands created along the river bends will filter the nutrient-rich water before it enters the lake

Lake Okeechobee

Shark River Slough

Fort Lauderdale

East Everglades addition

Big Cypress National Reserve

Miami

Everglades National Park

Homestead

Atlantic Ocean

9336

Flamingo   Park entrance

Key Largo

Construction threatens to destroy the fragile ecosystem of the Everglades.

Everglades wading birds once numbered more than a million, but drought and pollution have decimated their number.

of the western United States due to growth of cities and the use of increased amounts of water for irrigation. In aquifers with little or no recharge, virtually any withdrawal constitutes mining, and sustained withdrawals will eventually exhaust the supply. This problem is particularly serious in communities that depend heavily upon groundwater for their domestic needs.

Groundwater mining can also lead to problems of settling or subsidence of the ground surface when withdrawal of groundwater exceeds its replenishment rate by rainfall. Removal of the water allows the ground to become compacted, and large depressions may result. For example, in the San Joaquin Valley of California, groundwater has been withdrawn for irrigation and cultivation since the 1850s. In the last forty years, groundwater levels have fallen over 100 meters. More than 1,000 hectares of ground have subsided, some as much as 6 meters. Currently, the ground surface in that area is sinking 30 centimeters per year. In 1981, in Winter Park, Florida, a large area of subsidence—a "sinkhole"—occurred because of excessive water withdrawal. (See figure 15.13.)

FIGURE 15.13

Development of a "Sinkhole." If the water table is lowered as a result of groundwater mining or a severe drought, the land surface can subside. The space formerly occupied by water can collapse and create sinkholes. This photo shows only one of many sinkholes that occurred in Florida in 1981 as the result of the combined effects of excessive groundwater use and a long drought.

BOX 15.3
Groundwater Mining in Garden City, Kansas

For twenty million years, much of the precipitation that fell on the Great Plains infiltrated the sand and gravel aquifer surrounding Garden City, Kansas. For nearly one hundred years, wheat (a low water-demanding crop) was the predominant crop. But in 1960, the farmers began to tap the groundwater in the Ogallala Aquifer, and because of this availability of water, corn (which requires more water than wheat) quickly replaced wheat as the main crop. For $4.50, a farmer could distribute 1 meter of water over 1 hectare of land.

An economic boom resulted when the farmers began to irrigate and grow corn. Today, nearly twenty-five thousand wells are irrigating 1.4 million hectares of land, and the corn is being fed to feedlot cattle. Five large packing plants in the area process enough cattle in a single day to feed a million people.

The amount of irrigation in the area is the subject of much controversy. Although irrigation has been in

operation for over thirty years, the cost of pumping enough water to cover 1 hectare of land with 1 meter of water could increase to $250. The increase is a result of higher fuel costs, inflation, and the decreased amount of water in the aquifer. In over thirty years of pumping, the water level in the aquifer has dropped 4 meters. At the present rate of pumping, the aquifer is estimated to be dry by the year 2000.

Soil conservationists predict that, when the water is no longer available from the aquifer, the farmers will not be able to produce a crop, and that the conditions will be similar to those of the dust bowls of the 1930s. Farmers, cattle growers, and slaughterhouse owners are resisting any attempts to limit the amount of water allowed for irrigation. They are certain that new sources of water will be found in unexploited aquifers or by diverting water from water-rich regions like the Great Lakes.

Groundwater mining is a serious problem in western Texas. This area depends on irrigation for agriculture. The population of this area has also increased dramatically over the last twenty-five years. The ever-increasing demand for groundwater has led to its rapid depletion. Precipitous declines in agricultural production are forecast within the next ten years. As the groundwater is depleted, the land values will decline. Land use then directly affects water use, and water use and availability directly influence land use. London, Mexico City, Venice, Houston, and Las Vegas are some other cities that have experienced subsidence as a result of groundwater withdrawal.

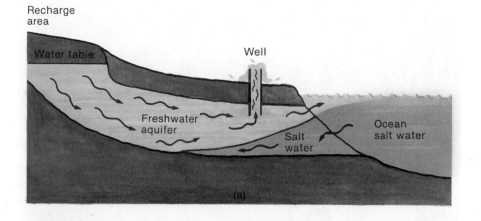

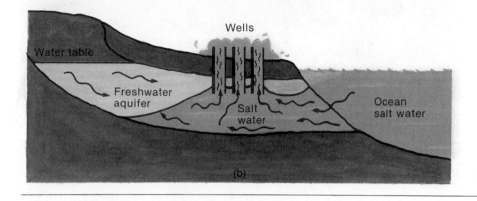

FIGURE 15.14
Saltwater Intrusion. When saltwater intrudes on fresh groundwater, the groundwater becomes unusable for human consumption and for many industrial processes.

Groundwater mining poses a special problem in coastal areas. As the fresh groundwater is pumped from wells along the coast, the saline groundwater moves inland, replacing fresh groundwater with unusable salt water. (See figure 15.14.) Saltwater intrusion is a serious problem in heavily populated coastal areas, such as New York, New Jersey, southern California, and Florida.

Salinization

Another water-use problem results from the salinity caused by increasing salt concentrations in soil. When plants extract the water they need, the salts present in all natural waters become concentrated. Irrigation of arid farmland makes this problem more acute because so much water is lost due to evaporation. Irrigation is most common in hot, dry areas, which normally have high rates of evaporation. This results in a concentration of salts in the soil and in the water that runs off the land. (See figure 15.15.) Every river increases in salinity as it flows to the ocean. The salinity of the Colorado River water increases twenty times as it passes through irrigated cropland between Grand Lake in north central Colorado and the Imperial Dam in southwest Arizona. The problem of salinity will continue to increase as irrigation increases.

Water Diversion

Water diversion is the physical process of transferring water from one area to another area. Early examples of water diversion were the aqueducts of ancient Rome. Thousands of diversion projects have been constructed

FIGURE 15.15

Salinization. As water evaporates from the surface of the soil, the salts it was carrying are left behind. In some areas of the world, this has permanently damaged cropland. Other areas must flush the soil to rid it of salt to maintain its fertility.

BOX 15.4
Death of a Sea

The Soviet Union's Aral Sea, which was once larger than any of the Great Lakes except Superior, is disappearing. Since the 1920s, Soviet agricultural planners have used up the Aral Sea, diverting its waters to irrigate cotton. The two rivers feeding the Aral were drawn off to irrigate millions of hectares of cotton. The irrigation canal, the world's longest, stretches over 1,300 kilometers, paralleling the boundaries of Afghanistan and Iran. The cotton production plan worked, and by 1937, the Soviet Union became a net exporter of cotton. The success of the cotton program, however, was to spell the end for the Aral Sea.

For a long time, the ecological impact on the sea and surrounding area was largely hidden from public view. Since the 1960s, however, the Aral has lost about 40 percent of its surface area, or almost 20,000 square kilometers of what are now largely dry, salt-encrusted wastelands. The once-thriving fishing industry that depended on the water is all but gone. Another apparent consequence is a host of human illnesses. A high rate of throat cancer is attributed to dust from the drying sea. In the northwest part of the republic of Uzbekistan, the infant-mortality rate is the highest in the Soviet Union.

The former fishing center of the sea was a town named Muynak. The town is now landlocked more than 30 kilometers from the water. Less than twenty-five years ago, Muynak was a seaport. In 1990, the mayor of Muynak and its last harbormaster commented:

Desperate fishermen make a last-ditch attempt to hold onto the sea. A dried-up canal to the retreating shoreline of the Aral Sea is strewn with abandoned boats.

The water continued to go away while the salinity increased. The weather changed for the worse, with the summers getting hotter and the winters colder. The people feel salt on their lips and in their eyes all the time. It's getting hard to open your eyes here.

With the advent of glasnost in the Soviet Union, controversy between the old-line economic managers and the country's fledgling environmental movement is growing.

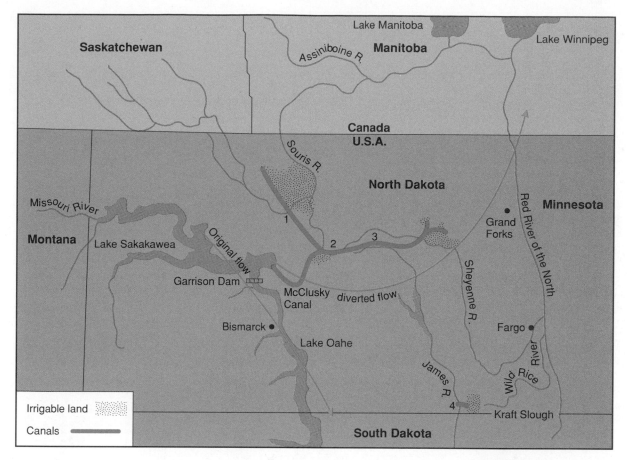

FIGURE 15.16

The Garrison Diversion Unit. This unit would divert water that flows into the Missouri River through the McClusky Canal to the Sheyenne River and eventually into Hudson Bay.

since then. New York City, for example, diverts water from Pennsylvania (250 kilometers away), and Los Angeles obtains part of its water supply from the Colorado River.

While diversion is necessary in many parts of the world, it can be misused or detrimentally affect an area. An example of this is the proposed Garrison Diversion Unit, which is designed to divert water from the Missouri River system for use in irrigation. (See figure 15.16.) This action would have enormous environmental consequences. Wildlife refuges, native grasslands, forests, and waterfowl breeding marshes would be damaged or destroyed. In addition, hundreds of kilometers of streams could be seriously damaged.

There is considerable opposition to this plan, both by private groups and governmental agencies. In fact, the Canadian government has fought against the project. In contrast, the people who need the water for irrigation see the Garrison Diversion Unit plan as beneficial. If this plan was implemented, they could raise corn (a more profitable crop), rather than the dryland crop of wheat. The corn could then be used for cattle feed.

The initial plan to irrigate this portion of the Great Plains began during the Dust Bowl era of the 1930s. The Federal Flood Control Act of 1944 authorized the construction of the Garrison Diversion Unit, but controversy concerning it continues fifty years after its proposal.

BOX 15.5
Is It Safe to Drink the Water?

Roughly one thousand contaminants have been detected in the public water supply in the United States, and virtually every major water source is vulnerable to pollution. About half the U.S. population relies on surface water—from rivers, lakes, and reservoirs that may contain industrial wastes and pesticides washed off fields by rain. The other half uses groundwater that may be tainted by chemicals slowly seeping in from toxic-waste dumps. In some areas where groundwater supplies are being gradually depleted, the chemical pollutants are becoming more concentrated.

Most pollutants are probably not present in large enough concentrations to pose significant health hazards; however, there are exceptions. The most widespread danger in water is lead, which can cause high blood pressure and an array of other health problems. Lead is especially hazardous to children, since it impairs the development of brain cells. The

U.S. EPA estimates that at least forty-two million Americans are exposed to unacceptably high levels of lead, and the U.S. Public Health Service estimates that perhaps nine million children are at least slightly affected by it.

The contamination comes from old lead piping and solder that have been used in plumbing for years. These materials are gradually being replaced in homes and water systems. Individuals may want to have their water tested for lead by an official lab. If the level is too high, they can investigate ways to deal with the problem or switch to bottled water for drinking and cooking. Even then, caution is called for: Some bottled waters contain many of the same contaminants that tap water does.

The other four types of contamination in the U.S. water supply, along with their source and risk, are shown in the table. Regardless of the problems, however, the water supply in the United States is among the cleanest in the world.

Toxins from the tap

Five types of contamination in the U.S. water supply:

Substance	Source	Risk
Chlorinated solvents	Industrial pollution; used for chemical degreasing, machine maintenance, and as intermediaries in the manufacture of other chemicals	Cancer
Trihalomethanes	Produced by chemical reactions in water that has been disinfected with chlorine	Liver and kidney damage, possibly cancer
Lead	Old piping and solder in public water distribution systems, homes, and other buildings	Nerve problems, learning disabilities in children, birth defects, possibly cancer
PCBs	Wastes from many outmoded manufacturing operations	Liver damage, possibly cancer
Pathogenic bacteria and viruses	Leaking septic tanks, overflowing sewer lines	Acute gastrointestinal illness, more serious diseases like meningitis

Managing Urban Water Use

Providing water services for metropolitan areas is another serious water-planning issue. Metropolitan areas must provide three basic water services:

1. Water supply for human and industrial needs
2. Wastewater collection and treatment
3. Storm-water collection and management

Water for human and industrial use must be properly treated and purified. It is then pumped through a series of pipes to consumers. After the water is used, it flows through a network of sewers to a wastewater treatment plant before it is released. Metropolitan areas must also deal with great volumes of excess water during storms. Because urban areas are paved and little rainwater can be absorbed into the ground, management of storm water is a significant problem. Cities often have severe local flooding because the water is channeled along streets to storm sewers. If these sewers are overloaded or blocked with debris, the water cannot escape and flooding occurs.

Many cities have a single system to handle both sewage and storm-water runoff. During heavy runoff, the flow is often so large that the wastewater treatment plant cannot handle the volume. The wastewater is then diverted directly into the receiving body of water without first being treated. Some cities have areas in which to store this excess water until it can be treated. This is expensive and, therefore, is not usually done unless government grants are available.

All water services provided by metropolitan areas are expensive. These services must be provided with an understanding that water supplies are limited. Also, water's ability to dilute and degrade pollutants is limited. Proper land-use planning is essential if these objectives are to be met.

In pursuing these objectives, city planners encounter many obstacles. Large metropolitan areas often have hundreds of local jurisdictions (governmental and bureaucratic areas) that divide responsibility for management of basic water services. The Chicago metropolitan area is a good example. This area is composed of six counties and approximately two thousand local units of government. It has 349 separate water-supply systems and 135 separate wastewater disposal systems. Efforts to implement a water-management plan when so many layers of government are involved are complicated and frustrating.

Summary

Water is a renewable resource that circulates continually between the atmosphere and the earth's surface. The energy for the hydrologic cycle is provided by the sun. Water loss from plants is called transpiration. Water that infiltrates the soil and is stored in underground reservoirs is called groundwater, as opposed to surface water that enters a river system as runoff. The two basic kinds of aquifers are unconfined and confined. The way in which land is used has a significant impact on rates of evaporation, runoff, and infiltration.

The four human uses of water are domestic, agricultural, in-stream, and industrial. Water use is measured by either the amount withdrawn or the amount consumed. Domestic water is in short supply in many metropolitan areas. Most domestic water is used for waste disposal and washing, with only a small amount used for drinking. The largest consumptive use of water is for agricultural irrigation. Major in-stream uses of water are for hydroelectric power, recreation, and navigation. Most industrial uses of water are for cooling and for dissipating and transporting waste materials.

CONSIDER THIS CASE STUDY
The California Water Plan

The management of fresh water is often a controversial subject, involving social, ecological, and economic aspects. A good example is the California Water Plan.

In the early 1900s, it became clear that the growth of Los Angeles, which was then a small coastal town, would be encouraged by irrigating the surrounding land. Los Angeles looked to the Owens Valley, 400 kilometers north, for a source of water. The Los Angeles Aqueduct connecting these two areas was completed in 1913.

Since the Owens Valley project, California has developed a statewide water program known as the California Water Plan. This water plan was necessary because most of the state's population and irrigated land are found in the central and southern regions, but most of the water is in the north. The plan details the construction of aqueducts, canals, dams, reservoirs, and power stations to transport water from the north to the south. In addition to supplying water for southern regions, the aqueducts provide irrigation for the San Joaquin Valley. Eventually, the new land made available for agriculture will amount to about 400,000 hectares.

The California Water Plan has been one of the most controversial programs ever undertaken in California. Adoption of the plan instigated a sectional feud between the moist "north" and the dry "south." Southern California was accused of trying to steal northern water, but because the population concentration in the southern part of the state carried the vote, the plan was adopted. Environmentalists still claim that the project has irreparably scarred the countryside and upset natural balances of streams, estuaries, vegetation, and wildlife. They argue that providing water to southern California promotes population growth, which leads to further urbanization and land development.

Many questions and controversies center on whether the water is really needed. Ninety percent of the water used in southern California is for irrigation; there is evidently abundant water for domestic and industrial use. A few of the crops raised in the San Joaquin Valley account for a large percentage of the irrigation water. Most notable is rice. Rice is not a native crop and demands intensive irrigation. Although the cost is borne by all the rate payers, most of whom are urban, California is now one of the nation's most productive agricultural areas.

Allocation of water resources is a matter of economics as well as a matter of technology. The California water project has been criticized for using public funds to increase the value of privately held farmland. Furthermore, technological advances in desalination plants may give a new dimension (unforeseen when the water plan was devised) to the problem of water resources. Although more aqueducts, canals, and pumping plants are planned, whether they will be completed is uncertain.

Southern California has been in a drought emergency for several years. By 1991, Los Angeles was in its eighth straight year without adequate rainfall. A number of unusual proposals for emergency supplies of water were reviewed, including:

Bringing in Colorado water by train
Importing Canadian water by barge
Building a desalination plant for Pacific Ocean water
Using an oil pipeline to carry water from northern California

In addition to new sources of water, conservation practices must be enforced. In Santa Barbara, water use decreased 40 percent in three months once the city raised water rates and banned landscape irrigation. In Los Angeles, 1 million homeowners had to cut water use by 10 percent, whereas residents of Marin County, north of San Francisco, were limited to 200 liters of water per day. Santa Monica mandated a 25 percent cutback for residents and businesses.

Water for irrigation was also targeted for big cutbacks, with some farmers receiving little, if any, water for crops. In Marin County, irrigation accounts were reduced 85 percent. By the summer of 1991, nearly a dozen California counties had proclaimed drought disasters and were seeking aid from the state and federal governments. Most California communities had water rationing, conservation programs, and depleted reservoirs.

What are the major advantages of having a water plan?
What problems develop when water is transported to arid regions?
Should water from northern California be sent to southern California?
Should crops like rice be raised in California?
Aside from lack of rainfall, why do you think southern California is in a water crisis?

Shasta Lake

Lake Oroville

Sacramento River

Feather River

Lake Tahoe

Sacramento

North Bay Aqueduct

San Francisco

South Bay Aqueduct

California Aqueduct

Los Angeles Aqueduct

San Luis Reservoir

San Joaquin Valley

Coastal Branch

Tehachapi Mts.

Silverwood Lake

Lake Perris

Colorado River Aqueduct

Castaic Lake

Los Angeles

Colorado River

Salton Sea

San Diego

CALIFORNIA AQUEDUCT
STATE WATER PROJECT
DEPARTMENT OF WATER RESOURCES

Water can accept large amounts of use without permanent damage. However, adding material to water may make the water unfit for some uses but not for others.

Major sources of water pollution are municipal sewage, industrial wastes, and agricultural runoff. Organic matter in water requires oxygen for its decomposition and therefore has a large biochemical oxygen demand (BOD). Oxygen depletion can result in fish death and changes in the normal algae community, which lead to visual and odor problems. Nutrients, such as nitrates and phosphates from detergents and agricultural runoff, enrich water and stimulate algae and aquatic plant growth.

Point sources of pollution are easy to identify and resolve. Nonpoint sources of pollution, such as agricultural runoff and mine drainage, are more difficult to detect and control than those from municipalities or industries.

Thermal pollution occurs when an industry returns heated water to its source. Temperature changes in water can alter the kinds and numbers of plants and animals that live in that area. The methods of controlling thermal pollution include cooling ponds, cooling towers, and dry cooling towers.

Wastewater treatment consists of primary treatment, a physical settling process; secondary treatment, biological degradation of the wastes; and tertiary treatment, chemical treatment to remove specific components. Two major types of secondary wastewater treatments are the trickling filter and the activated sludge sewage methods.

Groundwater pollution comes from a variety of sources, including agriculture, landfills, and septic tanks. Marine oil pollution results from oil drilling and oil-tanker accidents, runoff from streets, improper disposal of lubricating oil from machines and car crankcases, and intentional discharges from oil tankers during loading or unloading.

Reduced water quality can seriously threaten land use and in-place water use. In the United States and other nations, legislation helps to preserve certain scenic water areas and wildlife habitats. Shorelands and wetlands provide valuable services as buffers, filters, reservoirs, and wildlife areas. Water management concerns of growing importance are groundwater mining, increasing salinity, water diversion, and the management of urban water use. Urban areas face several problems, such as providing water suitable for human use, collecting and treating wastewater, and handling stormwater runoff in an environmentally sound manner. Water planning involves many governmental layers, which makes effective planning difficult.

Review Questions

1. Describe the hydrologic cycle.
2. List several uses of water that are nonconsuming and nonwithdrawing.
3. What is the major industrial use of water?
4. What are the similarities between domestic and industrial water use? How are they different from in-stream use?
5. How is land use related to water quality and quantity?
6. Is the definition of water pollution related to its intended use? Give some examples to support your answer.
7. What is biochemical oxygen demand? How is it related to water quality?
8. How is water pollution related to agricultural activities?
9. Differentiate between point and nonpoint sources of water pollution.
10. How are most industrial wastes disposed of?
11. What is thermal pollution? How can it be controlled?
12. Describe primary, secondary, and tertiary sewage treatment.
13. What are the types of wastes associated with agriculture?
14. Why is storm-water management more of a problem in an urban area than in a rural area?
15. What is the Federal Wild and Scenic Rivers Act? Why is it important?
16. Define groundwater mining.
17. How does irrigation increase salinity?
18. What are the three major water services provided by metropolitan areas?

PART FIVE
Pollution and Policy

The quality of the environment is something that most people have strong feelings about. However, these feelings vary greatly because people do not agree on what is good or bad for environmental quality. Economics, perception of risk, and politics have become interwoven into the environmental quality equation. There are great differences of opinion about what environmental quality should be or what the current quality of the environment is. The degree of pollution can often be quantified, and pollution may have different degrees of seriousness depending on a variety of local conditions. Although pollution, by definition, is harmful, different amounts of pollution can be tolerated, depending on the specific situation.

Chapter 16 provides background information on risk assessment and economic principles useful in evalu-

ating the costs of pollution and the value gained from eliminating specific kinds of pollution.

Chapters 17, 18, and 19 discuss the specific problems of air pollution, solid waste disposal, and hazardous and toxic wastes, respectively. These chapters also deal with the approaches used in dealing with the specific problems. Throughout this section, economic and political realities are discussed as a part of the pollution equation.

Chapter 20 discusses how environmental decisions get made. It is a complex process, involving the integration of public input, economic demands, political posturing, and legal requirements. As complex as the process is, the quality of our environment has been significantly improved over the past several decades.

CHAPTER SIXTEEN
Risk and Cost: Elements of Decision Making

Objectives

After reading this chapter, you should be able to:

Describe why the analysis of risk has become an important tool in environmental decision making.

Understand the difference between risk assessment and risk management.

Describe the issues involved in risk management.

Understand the difference between true and perceived risks.

Define what an economic good or service is.

Understand the relationship between the available supply of a commodity or service and its price.

Understand how and why cost-benefit analysis is used.

Understand the concept of sustainable development.

Understand environmental external costs and the economics of pollution prevention.

Understand the market approach to curbing pollution.

Chapter Outline

Key Terms

carcinogens
cost-benefit analysis
demand
external costs
negligible risk
pollution costs
pollution prevention
probability

resources
risk assessment
risk management
subsidy
supply
supply/demand curve
sustainable development

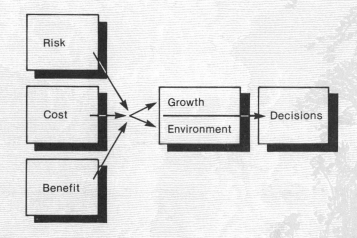

Measuring Risk

Two primary factors are important in many decisions in life: risk and cost. Such questions as "How likely is it that someone will be hurt?" and "What is the cost of this course of action?" are common. This also applies to the area of environmental decision making. If a new air-pollution regulation is contemplated, industry will be sure to point out that it will cost industry a considerable amount of money to put these controls in place and will reduce profitability. Citizens will point out that there will be an additional governmental bureaucracy to support with their money. On the other side will be advocates who will point out the reduced risk of specific illnesses and the reduced cost of health care for people who live in areas of high air pollution.

Risk analysis has become an important decision-making tool at all levels of society. In the area of environmental concerns, the assessment of risk and the management of risk are important in determining what environmental policies are appropriate. The analysis of risk generally involves a probability statement. **Probability** is a mathematical statement about how likely it is that something will happen. Probability is often stated in terms like "The probability of developing a specific illness is 1 in 10,000," or "The likelihood of winning the state lottery is 1 in 5,000,000." It is important to make a distinction between *probability* and *possibility.* When we say something is *possible,* we are just saying that it could occur. It is a very inexact term. *Probability* defines how likely *possible* events are.

Other important considerations are the consequences of an event happening. If the presence of a specific disease is likely to make 50 percent of the population ill (the probability of becoming ill is 50 percent) but no one dies, that is very different from analyzing the safety of a dam, which if it failed would cause the deaths of thousands of people downstream from it. We would certainly not accept a 50 percent probability that the dam would fail. Even a 1 percent probability in that case would be unacceptable. The assessment and management of risk involve an understanding of probability and the consequences of decisions. (See figure 16.1.)

Risk assessment involves the analysis of risk to determine the probability of an adverse effect. Risk management is a much broader concept that involves risk assessment and the consequences of risks in the decision-making process.

Risk Assessment

Environmental **risk assessment** is the use of facts and assumptions to estimate the probability of harm to human health or the environment that may result from exposures to specific pollutants, toxic agents, or management decisions. What risk assessment provides for environmental decision makers is an orderly, clearly stated, and consistent way to deal with scientific issues when evaluating whether a hazard exists and what the magnitude of the hazard may be.

Calculating the hazardous risk of a particular activity, chemical, or technology to humans is difficult. Probabilities based on past experience are used to estimate risks associated with well-known technologies. For example, the risk of developing black lung disease from coal dust in mines is well established. However, much less accurate statistical probabilities, based on models rather than real-life experiences, must be used to predict the risks associated with new technology.

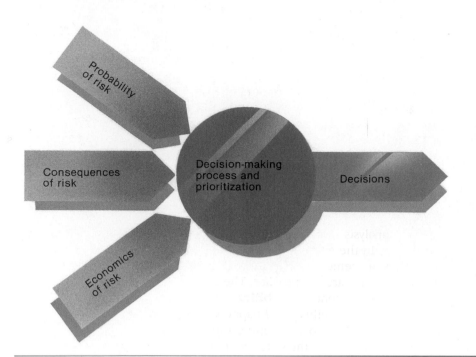

FIGURE 16.1
Decision-Making Process. The assessment, cost, and consequences of risks are all important to the decision-making process.

Risks associated with new chemicals are difficult to accurately quantify. While animal tests are widely accepted in predicting whether or not a chemical will cause cancer in humans, their use in predicting how many cancers will be caused in a group of exposed people is still very controversial. Most risk assessments are *estimates* of the probability that a person who is exposed to certain chemicals will develop cancer or other negative effects. (See figure 16.2.)

Such estimates typically are based on broad assumptions to ensure that a lack of complete knowledge does not result in an underestimation of the risk. For example, people may be more or less sensitive to the effects of certain chemicals than the laboratory animals studied. Also, people vary in their sensitivity to cancer-causing compounds. Thus, what may present no risk to one person may be a high risk to others. Persons with breathing difficulties are more likely to be adversely affected by high levels of air pollutants than are young, healthy individuals. In addition, the estimate of human risk is based on extrapolation from animal tests using high, chronic doses. Human exposure is likely to be lower or infrequent. Because of all these uncertainties, government regulators have decided to err on the side of safety to protect the public health. That approach has been criticized by those who say it carries protection to the extreme, usually at the expense of industry.

Over the past decade, risk assessment has had its largest impact in regulatory practices involving cancer-causing chemicals called **carcinogens.** In the United States, for example, the decisions to continue registration of pesticides, to list substances as hazardous air pollutants under the Clean Air Act, and to regulate water contaminants under the Safe Drinking Water Act depend to a large degree on the risk assessments for the substances in question.

Risk assessment analysis is also being used to help set regulatory priorities and support regulatory action. Those chemicals or technologies that have the highest potential to cause damage to health or the environment

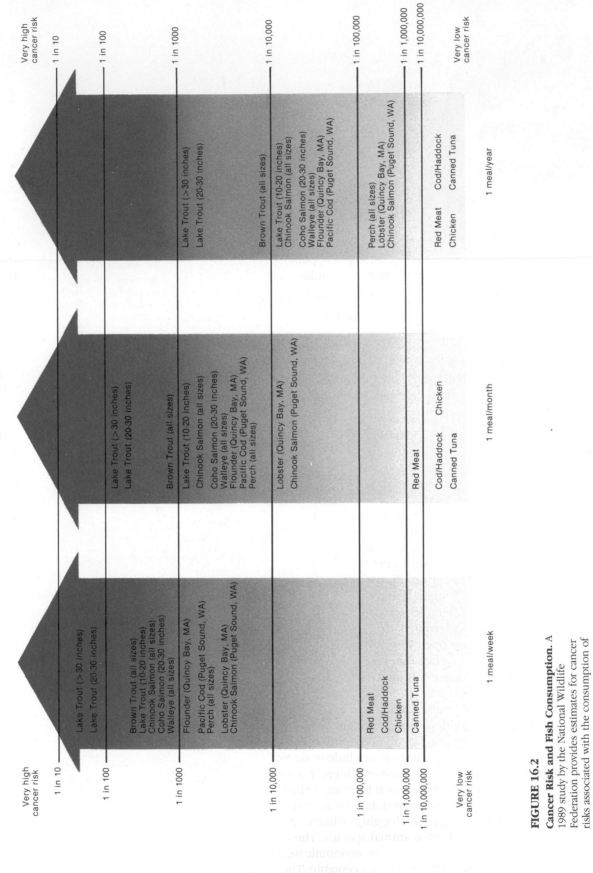

Estimates of cancer risks for Lake Michigan fish and other foods

FIGURE 16.2

Cancer Risk and Fish Consumption. A 1989 study by the National Wildlife Federation provides estimates for cancer risks associated with the consumption of sport fish.

receive attention first, while those perceived as having minor impacts receive less immediate attention. Medical waste is perceived as having high risk, and laws have been enacted to minimize the risk, while the risk associated with the use of fertilizer on lawns is considered minimal and is not regulated.

The science supporting environmental regulatory decisions is complex and rapidly evolving. Many of the most important threats to human health and the environment are highly uncertain. Risk assessment quantifies risk and states the uncertainty that surrounds many environmental issues. This can help institutions to make research and planning decisions in a way that is consistent with both scientific and public concern for environmental protection.

Risk Management

Risk management refers to a decision-making process involving risk assessment, technological feasibility, economic impacts, public concerns, and legal requirements. Risk management concerns such issues as:

1. Deciding which risks should be given the highest priority
2. Deciding how much money will be needed to reduce each risk to an acceptable level
3. Deciding where the greatest benefit would be realized by the spending of limited funds
4. Deciding how much risk is acceptable
5. Deciding how the plan will be enforced and monitored

Risk management raises several issues. With environmental concerns such as acid rain, ozone depletion, and hazardous waste, the scientific basis for regulatory decisions is often controversial. As was previously mentioned, hazardous substances can be tested, but only on animals. Are animal tests appropriate for making determinations about impacts on humans? There is not an easy answer to this question. The questions of global warming, ozone depletion, and acid rain require projecting into the future and estimating the magnitude of future effects. Will sea level rise? How many lakes will become acidified? How many additional skin cancers will be caused by depletion of the ozone layer? Estimates from equally reputable sources vary widely. Which ones do we believe?

The politics of risk management frequently focus on the adequacy of the scientific evidence. The scientific basis can be thought of as a kind of problem definition. Science determines that some threat or hazard exists, but because scientific facts are open to divergent interpretations, there is controversy. For example, it is a fact that dioxin is a highly toxic material known to cause cancer in laboratory animals. It is also very difficult to prove that human exposure to dioxin has led to the development of cancer, although high exposures have resulted in acne in exposed workers.

This is why problem definition is so important. Defining the problem helps to determine the rest of the policy process (making rules, passing laws, or issuing statements). If a substance poses little or no risk, then policy action is unnecessary. Some observers see many threats from chemicals that need to be addressed. Others see little threat from most chemicals; instead, they see scare tactics and government regulations as unnecessary attacks on businesses. Decisions relating to the logging of forests pose risks of soil erosion and the possible loss of some animal species. The timber industry sees these risks as minimal and as a threat to its economic well-being, while many environmentalists consider the risks unacceptable. These and similar

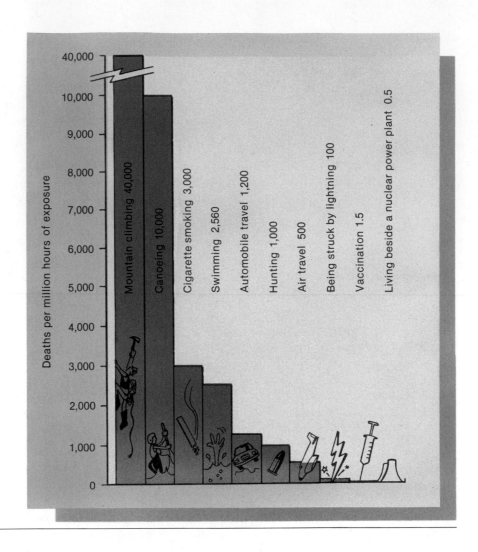

FIGURE 16.3

Risks of Death. In many cases, historical data can be used to predict the risk of exposure to specific activities, chemicals, or technologies.

problems are often a serious public-relations problem to both government and business because most of the public has a poor understanding of the risks they accept daily.

True and Perceived Risks

People often overestimate the frequency and seriousness of dramatic, sensational, well-publicized causes of death and underestimate the risks from more familiar, accepted causes that claim lives one by one. Risk estimates by "experts" and by the "public" on many environmental problems differ significantly. This problem and the reasons for it are extremely important because the public generally does not trust experts to make important risk decisions alone.

While public health and environmental risks can be minimized, eliminating all risks is impossible. Almost every daily activity—driving, walking, working—all involve some element of risk. (See figure 16.3.)

From a risk management standpoint, whether one is dealing with a site-specific situation or a national standard, the deciding question ultimately is: What degree of risk is acceptable? In general, we are not talking about a "zero risk" standard, but rather a concept of **negligible risk:** At what

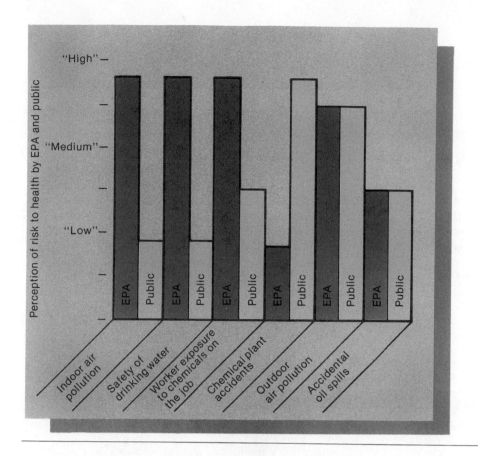

FIGURE 16.4
Perception of Risk. Professional regulators and the public do not always agree on what risks are.

point is there really no significant health or environmental risk? At what point is there an adequate safety margin to protect public health and the environment?

Risk management involves comparing the estimated true risk of harm from a particular technology or product with the risk of harm perceived by the general public. The public generally perceives involuntary risks, such as nuclear power plants or nuclear weapons, as greater risks than voluntary risks, such as drinking alcohol or smoking. In addition, the public perceives newer technologies, such as genetic engineering or toxic-waste incinerators, as greater risks than more familiar technologies, such as automobiles and dams. Many people are afraid of flying for fear of crashing; however, automobile accidents account for a far greater number of deaths—about forty-five thousand in the United States each year, compared to less than a thousand annually from plane crashes. (See figure 16.4.)

A fundamental problem facing governments today is how to satisfy people who are concerned about a problem that experts state presents less hazard than another less visible problem, especially when economic resources to deal with the problems are limited. Debate continues over the health concerns raised by asbestos, dioxin, radon, and Alar. (See box 16.1.)

Some researchers argue that the public is frequently misled by the politics of public health and environmental safety. This is understandable since many prominent people become involved in such issues and use their public image to encourage people to look at issues from a particular point of view.

In 1989, U.S. consumers experienced a public-health scare concerning the pesticide Alar (daminozide), a plant growth regulator that was used on apples to postpone fruit drop, enhance the fruit's color and shape, and extend storage life.

The public was alarmed to the point of panic. Worried mothers dumped apple juice down the drain, despite the EPA's repeated assurances that such measures were unnecessary. Apples were banished from school cafeterias. With each new event reported by the media, consumers were further unsettled and confused as points of disagreement between the EPA and outside groups on the risks of Alar were raised.

The initial panic concerning Alar has subsided somewhat, but an uneasy confusion remains, and consumers have been left with a lingering doubt about the safety of their food. In general, the public has limited patience with extended deliberations by scientists and regulators over data on a chemical and its potential harmful effects. While scientists and regulatory officials are concerned with questions of scientific uncertainty and statistical risk assessment, consumers, who generally do not speak the language of risk assessment, tend to ask very direct questions: Is it safe to eat apples? Is it safe for my child to consume apple products? Does Alar cause cancer?

Two questions are implicit in the question: Does Alar cause cancer? First, is Alar a known human carcinogen? The answer is no; scientists do not have direct evidence in humans that traces actual cancer cases to Alar exposure. In fact, comparatively few chemicals in the world have been demonstrated beyond doubt, on the basis of epidemiological data, to cause cancer in humans.

Second, does Alar cause cancer in laboratory animals? The answer is yes; Alar and its breakdown product called unsymmetrical dimethylhydrazine (UDMH) have increased the incidence of malignant tumors in mice.

It is difficult to understand cancer risks, or any kind of risk, without a meaningful frame of reference. For perspective, one of the key phrases consumers should keep in mind in the Alar case, and generally in cases of chemicals said to pose cancer risks, is "long term," In evaluating the risks of pesticides to consumers, the EPA uses the working assumption that dietary exposure to the pesticide occurs over a lifetime (seventy years). This is just one of a number of assumptions that the EPA factors into its chemical risk analysis.

The truth, however, is that hard evidence on the effects of pesticide chemicals is generally limited to the cases where short-term, highly concentrated pesticide exposure has caused acute toxic poisoning in humans or killed important nontarget organisms in significant numbers. Such acute toxic effects are immediately apparent.

Most risk scenarios are not so easy to assess, and this is especially true of chronic or delayed health effects, such as cancer, reproductive dysfunctions, or effects on the unborn. Such chronic or delayed effects do not become apparent for a long time, and when they do occur, it is almost always impossible to trace them with certainty to exposures to specific chemicals. Instead, the evidence at hand consists of the raw materials of risk assessment: animal data tabulations, cancer potency estimates based on animal study results, food consumption statistics, and exposure estimates. As the Alar case demonstrated, such data may be used selectively and inappropriately to make calculations that misrepresent pesticide risks.

Consumer education and risk communication will become increasingly necessary if pesticide decision making is to take place in an atmosphere that is relatively free of fear, confusion, and unnecessary economic disruption.

Whatever the specific issue, it is hard to ignore the will of the people, particularly when the sentiments are firmly held and not easily changed. A fundamental issue surfaces concerning the proper role of a democratic government and other organizations in a democracy when it comes to matters of risk. Should the government focus available resources and technology where they can have the greatest tangible impact on human and ecological well-being, or should it focus them on those problems about which the public is most upset? What is the proper balance? For example, adequate prenatal health care for all pregnant women would have a greater effect on the health of children than would requirements that asbestos be removed from all school buildings.

Obviously, there are no clear answers to these questions. However, experts and the public are both beginning to realize that they each have something to offer concerning how we view risk. Many risk experts who have been accustomed to looking at numbers and probabilities are now conceding that there is a clear rationale for looking at risk in broader terms. At the same time, the public is being supplied with more data to enable them to make more informed judgments.

Throughout this discussion of risk assessment and management have been numerous references to costs and economics. It is not economically possible to eliminate all risk. As risk is eliminated, the cost of the product or service increases. Many environmental issues are difficult to evaluate from a purely economic point of view, but economics is one of the tools useful to analyzing any environmental problem.

Economics and the Environment

It has been argued that environmental problems are primarily economic problems. While this may be somewhat of an overstatement, the fact is that it often is difficult to separate economics from environmental issues or concerns. Basically, economics deals with resource allocation. It is a description of how we value goods and services, and how we make decisions about the relative desirability of competing wants. We are willing to pay for things or services we value highly and are unwilling to pay for things we consider to have little value. For example, we will readily pay for a warm, safe place to live but would be offended if someone suggested that we pay for the air we breathe.

Economic Concepts

An economic good or service can be defined as anything that is scarce. Scarcity exists whenever the demand for anything exceeds its supply. We live in a world of general scarcity. With few exceptions, **resources,** which is the general term applied to anything that contributes to making desired goods and services available for consumption, are limited, relative to the desires of humans to consume. The **supply** is the amount of a good or service available to be purchased. **Demand** is the amount of a product that consumers are willing and able to buy at various possible prices. In economic terms, supply depends on:

1. The raw materials available to produce a good or service using present technology
2. The amounts of those materials available
3. The costs of extracting, shipping, and processing the raw materials
4. The degree of competition for those materials among users
5. The feasibility and cost of recycling already used material
6. The social and institutional arrangements that might have an impact (See figure 16.5.)

The relationship between available supply of a commodity or service and its price is known as a **supply/demand curve.** (See figure 16.6.) The price of a product or service reflects the strength of the demand for and the availability of the commodity. When demand exceeds supply, the price rises. Cost increases cause people to seek alternatives or to decide not to use a product or service, which results in lower demand.

For example, food production depends heavily on petroleum for the energy to plant, harvest, and transport food crops. In addition, petrochemicals are used for fertilizer and the production of chemical pest-control

FIGURE 16.5

Factors That Determine Supply. The supply of a good or service is dependent on the several factors shown here.

agents. As petroleum prices rise, farmers seek ways to reduce their petroleum use. Perhaps they farm less land or do not use as much fertilizer or pesticide. Regardless, the price of food must rise as the price of petroleum rises. As the price of certain foods rises, consumer habits change as consumers seek less costly forms of food.

When the supply of a commodity exceeds the demand, producers must lower their prices to get rid of the product, and eventually, some of the

Pollution and Policy

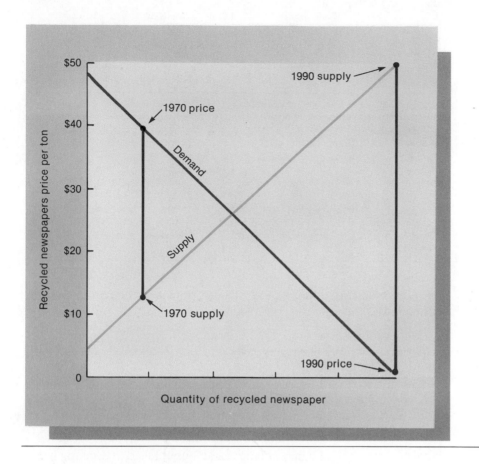

FIGURE 16.6
Supply and Demand. As a result of increased interest in recycling, the supply of recycled newspapers has increased substantially between 1970 and 1990. The demand changed very little; therefore, the price in 1990 was extremely low.

producers go out of business. Ironically, this happens to farmers when they have a series of good years. Production is high, prices fall, and some farmers go out of business.

Another way in which economic policy influences our future environment is through governmental subsidies. A **subsidy** is a gift from government to a private enterprise that is considered important to the public interest. This is particularly true when the private enterprise is having economic difficulty. Agriculture, transportation, space technology, and communication are frequently subsidized by governments. These gifts, whether loans, favorable tax situations, or direct grants, are all paid for by taxes on the public.

These subsidies are costly in two ways: First, the bureaucracy necessary to administer a subsidy costs money, and the subsidy is an indirect way of keeping the market price of a product low. The actual cost is higher because these subsidy costs must be added to the market price to arrive at the product's true cost. Second, in many cases, subsidies encourage activities that in the long term may be detrimental to the environment. For example, the transportation subsidies for highway construction encourage use of inefficient individual automobiles. Higher taxes on automobile use to cover the cost of building and repairing highways would encourage the use of more energy-efficient public transport.

Cost-Benefit Analysis

Cost-benefit analysis is concerned with determining whether a policy generates more social costs than social benefits, and if benefits outweigh

TABLE 16.1
Costs and benefits of improving air quality.

Costs	Benefits
Installation and maintenance of new technology 1. Scrubbers on smokestacks 2. Automobile emissions control	Reduced deaths and disease
Redesign of industries and machines	Reduced plant and animal damage
Additional energy costs to industry and public	Lower cleaning costs for industry and public
Unemployment as some industries go out of business	More clear, sunny days; better visibility
Retraining of employees to use new technology	Less eye irritation and fewer respiratory problems
Costs associated with monitoring and enforcement	Fewer odor problems

costs, how much spending would obtain optimal results. Steps in cost-benefit analysis include:

1. Identification of the project to be evaluated
2. Determination of all impacts, favorable and unfavorable, present and future, on all of society
3. Determination of the value of those impacts, either directly through market values or indirectly through price estimates
4. Calculation of the net benefit, which is the total value of positive impacts less the total value of negative impacts

For example, the cost of reducing the amount of lead in drinking water in the United States to acceptable limits is estimated to be about $125 million a year. The benefits to the nation's health from such a program are estimated at nearly $1 billion per year. Thus, the program would be economically sound under a cost-benefit analysis. Recycling of solid waste, which was once cost-prohibitive, is now cost-effective in many communities because the costs of landfills have risen dramatically. Table 16.1 gives examples of the kinds of costs and benefits involved in improving air quality. Although not a complete list, the table indicates the kinds of considerations that go into a cost-benefit analysis. Some of these are easy to measure in monetary terms; others are not.

Concerns about the Use of Cost-Benefit Analysis

Does everything have an economic value? Critics of cost-benefit analysis raise this point, along with others. Some argue that, if economic thinking pervades the whole of society, many simple noneconomic values like beauty or cleanliness can survive only if they prove to be "economic." (See figure 16.7.)

One particularly compelling critique of cost-benefit analysis is that for analysis to be applied to a specific policy, the analyst must decide which preferences count—that is, which preferences have "standing" in cost-benefit analysis. In theory, cost-benefit analysis should count all benefits and costs associated with the policy under review, regardless of who benefits or bears the costs. In practice, however, this is not always done. For example, if a cost is spread thinly over a great many people, it may not be recognized as a cost at all. The cost of air pollution in many parts of the

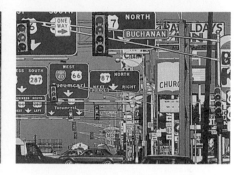

FIGURE 16.7
Does Everything Have an Economic Value? The use of water and land is often based on the economic benefits obtained. Are there other benefits that cannot be measured economically?

world could fall into such a category. Debates over how to count benefits and costs for future generations, inanimate objects such as rivers, and non-humans, such as endangered species, are also common.

A noted critic of cost-benefit analysis is E. F. Schumacher, who states in his book *Small Is Beautiful* (Harper and Row, 1973):

> Cost-benefit analysis is a procedure by which the higher is reduced to the level of the lower and the priceless is given a price. All it can do is lead to self-deception or the deception of others; for to undertake to measure the immeasurable is absurd and constitutes but an elaborate method of moving from preconceived notions to foregone conclusions; all one has to do to obtain the desired results is to impute suitable values to the immeasurable costs and benefits. The logical absurdity, however, is not the greatest fault of the undertaking; what is worse, and destructive of civilization, is the pretense that everything has a price or, in other words, that money is the highest of all values.

While Schumacher's objections to cost-benefit analysis may be somewhat idealistic, he does raise some issues that should be discussed. Is it possible to separate economic issues from environmental issues?

Economics and Sustainable Development

The past president of the Japan Economic Research Center, Saburo Okita, once stated that a central problem related to the economics of an improved environment involves the necessity of a slowdown of economic growth to prevent further deterioration of the environment. Whether or not a slowdown is necessary provokes sharp differences of opinion.

One school of thought argues that economic growth is essential to finance the investments necessary to prevent pollution and to improve the environment by a better allocation of resources. A second school of thought, which is also progrowth, stresses the great potential of science and technology to solve problems and advocates relying on technological advances to solve environmental problems. Neither of these schools of thought sees any need for fundamental changes in the nature and foundation of economic policy. Environmental issues are viewed mainly as a matter of setting priorities in the allocation of resources.

A newer school of economic thought is referred to as **sustainable development.** The former chairman of the World Commission on Environment and Development defined sustainable development as those "actions that address the needs of the present without compromising the ability of future generations to meet their own needs." Proponents of sustainable development believe that, if future generations are to have a high quality of life, and if poverty in the developing world is to be addressed, economic growth must be continued, but in a way that enhances, rather than exploits, the natural resource base.

Historically, rapid exploitation of resources has provided only short-term economic growth, and the environmental consequences in some cases have been incurable. For example, forty years ago, forests covered 30 percent of Ethiopia. Today, forest covers only 1 percent, and deserts are expanding. One-half of India once was covered by trees; today, only 14 percent of the land is in forests. As the Indian trees and topsoil disappear, the citizens of Bangladesh drown in India's runoff. (See figure 16.8.)

Sustainable development requires choices based on values. Both depend upon information and education, especially regarding the economics of decisions that affect the environment. A. W. Clausen, in his final address as President of the World Bank, noted the

> increasing awareness that environmental precautions are essential for continued economic development over the long run. Conservation, in its broadest sense, is not a luxury for people rich enough to vacation in scenic parks. It is not just a motherhood issue. Rather, the goal of economic growth itself dictates a serious and abiding concern for resource management.

High-income developed nations, such as the United States, Japan, and much of Europe, are in a position to promote sustainable development. They have the resources to invest in research and the technologies to implement research findings. Some believe that the world should not impose environmental protection standards upon poorer nations without also helping them move into the economic mainstream.

External Costs

Many of the important environmental problems facing the world today arise because modern production techniques and consumption patterns transfer waste disposal, pollution, and health costs to society. Such expenses, whether they are measured in monetary terms or in diminished environmental quality, and that are borne by someone other than the individuals who use a resource, are referred to as **external costs.** For example, consider an individual who attempts to spend a day of leisure fishing in a nearby lake. Transportation, food, fishing equipment, and bait for the day's activities can be purchased in readily accessible markets. But suppose that upon arrival at the lake, the individual finds a lifeless, polluted body of water. Let us further suppose that a chemical plant situated on the lakeshore is responsible for degrading the water quality on the lake to the point that all or most of the fish are destroyed. In this example, fishers would collectively

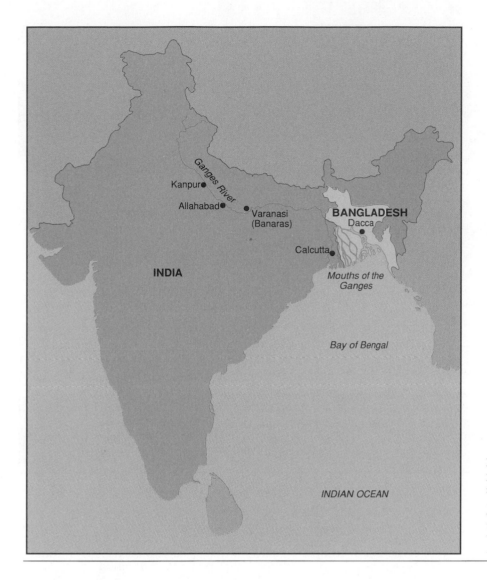

FIGURE 16.8
Indian Deforestation Causes Floods in Bangladesh. Because the Ganges River drains much of India and the country of Bangladesh is at the mouth of the river, deforestation and poor land use in India can result in devastating floods in Bangladesh.

bear the external costs of chemical production in the form of lost recreational opportunities.

Pollution-control costs include pollution-prevention costs and pollution costs. **Pollution-prevention** costs are those incurred either in the private sector or by government to prevent, either entirely or partially, the pollution that would otherwise result from some production or consumption activity. The cost incurred by local government to treat its sewage before dumping it into a river is a pollution-prevention cost; so is the cost incurred by a utility to prevent air pollution by installing new equipment.

Pollution costs can be broken down into two categories:

1. The private or public expenditures to avoid pollution damage once pollution has already occurred
2. The increased health costs and loss of the use of public resources because of pollution

The large cost of cleaning up oil spills, such as that from the Exxon Valdez and from the war in the Persian Gulf, is an example of a pollution cost, as is the increased health risk to humans from eating seafood contaminated from the oil.

The problems inherent in common ownership of resources were outlined by biologist Garrett Hardin in a now classic essay entitled "The Tragedy of the Commons." The original "commons" involved areas of pastureland in England that were provided free by the king to anyone who wished to graze cattle.

There are no problems on the commons as long as the number of animals is small in relation to the size of the pasture. From the point of view of each herder, however, the optimal strategy is to enlarge his herd as much as possible. If his animals do not eat the grass, someone else's will. Thus, the size of each herd grows, and the density of stock increases until the commons becomes overgrazed. The result is that everyone eventually loses as the animals die of starvation. The tragedy is that, even though the eventual result should be perfectly clear, no one acts to avert disaster. In a democratic society, there are few remedies to keep the size of herds in line.

The ecosphere is one big commons stocked with air, water, and irreplaceable mineral resources—a "people's pasture," but a pasture with very real limits. Each nation attempts to extract as much from the commons as possible while enough remains to sustain the herd. Thus, the United States and other industrial nations consume far more than their share of the total world resource harvest each year, much of it imported from less developed nations. The nations of the world compete frantically for all the fish that can be taken from the sea before the fisheries are destroyed. Each nation freely uses the commons to dispose of its wastes, ignoring the dangers inherent in overtaxing the waste-absorbing capacity of rivers, oceans, and the atmosphere.

The tragedy of the commons also operates on an individual level. Most people are aware of air pollution, but they continue to drive their automobiles without question. Many families claim to need a second or third car. It is not that these people are antisocial; most would be willing to drive smaller or fewer cars if everyone else did, and they could get along with only one small car if public transport were adequate. But people frequently get "locked into" harmful situations, waiting for others to take the first step, and many unwittingly contribute to tragedies of the commons. After all, what harm can be done by the birth of one more child, the careless disposal of one more beer can, or the installation of one more air conditioner?

Common Property Resource Problems

Economists have stated that, when everybody shares ownership of a resource, there is a strong tendency to overexploit and misuse that resource. Thus, common public ownership could be better described as effectively having no owner. (See box 16.2.)

For example, common ownership of the air makes it virtually costless for any industry or individual to dispose of wastes by burning them. The air-pollution cost is not reflected in the economics of the polluter but becomes an external cost to society. Common ownership of the ocean makes it inexpensive for cities to use the ocean as a dump for their wastes. (See figure 16.9.)

Similarly, nobody owns the right to harvest whales. If any one country delays in getting its share of the available supply of whales, other countries may beat that country to it. Thus, there is a strong incentive for overharvesting of the whale population and a consequent threat to the survival of the species. What is true on an international scale for whales is true also for other species within individual countries. Note that endangered species are wild and undomesticated; the survival of privately owned livestock is not a concern.

Finally, common ownership of land resources, such as parks and streets, is the source of other environmental problems. People who litter in public parks do not generally dump trash on their own property. The lack of enforceable property rights to commonly owned resources explains much of what economist John Kenneth Galbraith has termed "public squalor amid private affluence."

The Market Approach to Curbing Pollution

A growing number of political leaders are beginning to believe that many of our environmental problems will not be solved unless the forces of the economic marketplace are used. In the United States, one proposal would allocate "pollution rights" that would permit a business, such as a coal-burning utility, to pollute above the allowable federal limit. Only a limited number of these permits would be issued, but businesses would be able to trade them, thus establishing a market in the permits. The profit motive may then help reduce waste-disposal problems. The businesses responsible for pollution would have an incentive to internalize the external cost they were previously imposing on society: If they cleaned up their pollution sources, they could realize a profit by selling their permit to pollute. Once a business recognized the possibility of selling its permit, it would see that pollution is not costless. The creation of new markets (the permits) to reduce pollution reduces the external aspects to waste disposal and makes them internal costs, just like costs of labor and capital.

Critics of this concept argue that pollution permits are nothing more than taking public resources and turning them into something that can be treated as if they were property. Opponents also point out the philosophical argument that the permit system seems to approve of a given amount of pollution. However, even critics of the concept agree on one point: that anything that promises to make business a more willing partner in the fight against pollution is worth trying.

Economics, Environment, and Developing Nations

The greatest obstacle to economic and environmental improvements in developing countries is their large foreign debt. Collectively, the Third World owes $1.2 trillion to the banks and governments of industrialized countries. A World Bank report estimates that developing countries make net payments of $43 billion to the industrial nations, up from $38 billion in 1987. It is difficult for poor countries to launch environmental programs while struggling to pay off such loans.

One new approach to solving this problem is the so-called debt-for-nature swaps. Conceived in 1984 by Thomas Lovejoy of the Smithsonian Institution, these deals often involve the cooperation of governments, bankers, and conservation groups. In a 1989 debt-for-nature swap, the World Wildlife Fund, a nonprofit organization based in Washington, D.C., bought $1 million worth of Ecuadorian debt held by Bankers Trust at the discounted price of $354,000. The bank was happy to get the troublesome loan off its books, while the World Wildlife Fund gained the power to improve Ecuador's environment. The fund accomplished this by transferring the loan payments to Fundacion Natura, a conservation group in Ecuador. Fundacion Natura, in turn, uses the money to protect and maintain national parks and wildlife preserves.

Attitudes of banks in the industrialized nations also seem to be changing. For example, the World Bank, which lends money for Third World development projects, has long been criticized by environmental groups for backing large, ecologically unsound programs, such as a cattle-raising project in Botswana that led to overgrazing. During the past few years, however, the World Bank has been seeking to factor environmental concerns into its programs. One product of this new approach is an environmental action plan for Madagascar. The twenty-year plan, which will be drawn up jointly with the World Wildlife fund, is aimed at heightening public awareness of environmental issues, setting up and managing protected areas, and encouraging sustainable development.

As the previous discussion has shown, the economics of environmental problems is complex and difficult. The single most difficult problem to overcome is the assignment of an appropriate economic value to resources that have not previously been examined from an economic perspective. When air, water, scenery, and wildlife are assigned an economic value, they are looked at from an entirely different point of view.

CONSIDER THIS CASE STUDY
Shrimp and Turtles

Each year, shrimp boats trawling the South Atlantic and Gulf of Mexico accidentally catch an estimated forty-five thousand sea turtles in their nets. More than twelve thousand of them drown, including hundreds of the endangered Kemp's ridley sea turtle, as well as threatened loggerhead and endangered green sea turtles.

The National Marine Fisheries Service, looking for some technological innovation to stop the accidental netting and killing of the turtles, developed a device to keep turtles out of shrimp nets. The turtle excluder device, or TED, attaches to standard shrimp nets. Data compiled by the National Marine Fisheries Service show that TEDs can reduce turtle captures in shrimp nets by 97 percent.

Beginning in 1982, conservationists and shrimp industry representatives joined forces in a program to encourage shrimpers to use TEDs voluntarily. TEDs, however, were not accepted widely by the shrimpers. Subsequent to the failure of the voluntary plan, the U.S. Department of Commerce established the mandatory use of TEDs. Congressional delegations from states bordering the Gulf of Mexico are constantly pressuring for a review or change of the regulations. Many shrimpers have willingly broken the law by refusing to use TEDs.

Shrimpers who oppose the TEDs claim that the devices are expensive and dangerous, and cut down on their catches. They also claim that, since Mexican shrimpers are not required in Mexican waters to use TEDs, the U.S. shrimpers cannot compete in the marketplace. Redesigning the TEDs has led to a model that is less bulky and does not reduce catch as much; however, even this version severely worries the shrimpers.

Biologists argue that any delay in bringing TEDs into widespread use will mean the continued destruction of tens of thousands of turtles and bring the Kemp's ridley even closer to extinction. Shrimpers argue that the U.S. shrimp industry could be in danger of financial ruin. Mexico, the shrimpers claim, will soon dominate the shrimp industry because Mexican shrimpers are not required to use TEDs.

Trawling for shrimp also kills several endangered and threatened species of turtles in the South Atlantic and Gulf of Mexico.

Are the benefits of saving the turtles greater than the costs to the shrimpers?

Should the U.S. government put pressure on Mexico to require its shrimpers to use TEDs?

Can you arrive at a solution to this issue? Are there opportunities for compromise?

Summary

Risk assessment is the use of facts and assumptions to estimate the probability of harm to human health or the environment that may result from exposures to specific pollutants, toxic agents, or management decisions. While it is difficult to calculate risks, risk assessment is used in risk management, which analyzes various factors in decision making. The politics of risk management focus on the adequacy of scientific evidence, which is often open to divergent interpretations. In assessing risk, people frequently overestimate new and unfamiliar risks, while underestimating familiar risks.

To a large degree, environmental problems can be viewed as economic problems. Economic policies and concepts, such as supply/demand and subsidies, play important roles in environmental decision making. Another important economic tool is cost-benefit analysis. Cost-benefit analysis is concerned with whether a policy generates more social benefits than social costs. Criticism of cost-benefit analysis is based on the question of whether everything has an economic value. It has been argued that, if economic thinking dominates the whole of society, then even noneconomic values, like beauty, can survive only if a monetary value is assigned to them.

A newer school of economic thought is referred to as sustainable development. Sustainable development has been defined as actions that address the needs of the present without compromising the ability of future generations to meet their own needs. Sustainable development requires choices based on values.

Pollution is extremely costly. When the costs are imposed on society, they are referred to as external costs. Costs of pollution control include pollution-prevention costs and pollution costs. Prevention costs are less costly, especially from a societal perspective.

Economists have stated that, when everyone shares ownership of a resource, there is a strong tendency to overexploit and misuse the resource. This concept was developed by Garrett Hardin in "The Tragedy of the Commons."

Recently, a market approach to curbing pollution has been proposed that would assign a value to not polluting, thereby introducing a profit motive to pollution reduction. Critics of this approach argue that pollution "permits" are nothing more than an acceptance of pollution of public resources.

Economic concepts are also being applied to the debt-laden developing countries. One such approach is the debt-for-nature swap. This program, which involves transferring loan payments for land that is later turned into parks and wildlife preserves, is gaining popularity.

Review Questions

1. How is risk assessment used in environmental decision making?
2. What is incorporated in a cost-benefit analysis?
3. What are some of the concerns about the use of cost-benefit analysis in environmental decision making?
4. What concerns are associated with sustainable development?
5. What are some examples of environmental external costs?
6. Define what is meant by pollution-prevention costs.
7. Define the problem in common property resource ownership.

CHAPTER SEVENTEEN
Air Pollution

Objectives

After reading this chapter, you should be able to:

Explain why air can accept and disperse significant amounts of pollutants.

List the major sources and effects of the five primary pollutants.

Describe how photochemical smog is formed and how it affects humans.

Explain how PCV valves, APC valves, catalytic converters, scrubbers, precipitators, filters, and changes in fuel types reduce air pollution.

Explain how acid rain is formed.

Understand that humans can alter the atmosphere in such a way that the climate may change.

Understand the link between chlorofluorocarbon use and ozone depletion.

Recognize that enclosed areas can trap air pollutants that are normally diluted in the atmosphere.

Chapter Outline

The Atmosphere
Primary Air Pollutants
    Carbon Monoxide (CO)
    Box 17.1 Air Pollution in Mexico City
    Hydrocarbons (HC)
    Particulates
    Sulfur Dioxide (SO_2)
    Oxides of Nitrogen (NO and NO_2)
Photochemical Smog
    Box 17.2 The 1990 Clean Air Act
Control of Air Pollution
Acid Deposition
    Box 17.3 Acid Rain: Canada Versus the
      United States
Global Warming
Ozone Depletion
    Box 17.4 Aesthetic Pollution
Indoor Air Pollution
    Box 17.5 Noise Pollution
    Box 17.6 Radon
Consider This Case Study: International Air Pollution

Key Terms

acid deposition
acid rain
carbon dioxide (CO_2)
carbon monoxide (CO)
carcinogens
greenhouse effect
hydrocarbons (HC)
oxides of nitrogen (NO and NO_2)

ozone
particulates
photochemical smog
primary pollutants
radon
secondary pollutants
sulfur dioxide (SO_2)
thermal inversion

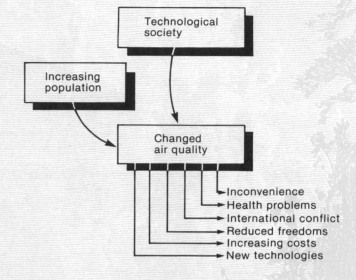

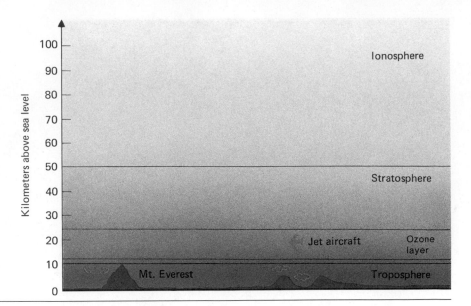

The Atmosphere

The atmosphere is normally composed of 79 percent nitrogen, 20 percent oxygen, and a 1 percent mixture of carbon dioxide, water vapor, and small quantities of several other gases. Most of the atmosphere is held close to the earth by the pull of gravity. The atmosphere gets thinner with increasing distance from the earth. (See figure 17.1.)

Even though gravity keeps the air near the earth, the air is not static. As it absorbs heat from the earth, it expands and rises. When its heat content is radiated into space, the air cools, becomes more dense, and flows toward the earth. As the air circulates due to heating and cooling, it also moves horizontally over the surface of the earth because the earth rotates on its axis. The combination of all air movements creates the specific wind patterns characteristic of different regions of the world. (See figure 17.2.)

As discussed in chapter 11, pollution is material produced by humans that interferes with our well-being. Because we cause pollution, we may be able to do something to prevent it. In this text, volcanic ash and gases are not considered air pollution because humans do not cause the problem and cannot control it. Automobile emissions and odors and factory smoke are considered air pollution, however, and we will focus on ways that we influence our air quality.

If it were possible to live as single-family units on farms where each family would grow its own food, air pollution would not be a problem. The amount of waste put into the atmosphere would be diluted, so no one's air quality would be seriously altered. However, in an urbanized, industrialized civilization with a growing population and a history of increasing use of fossil fuels and technological aids, we manufacture the products needed by this concentrated population, and we also release by-products into our environment.

Gases or small particles released into the atmosphere are likely to be mixed, diluted, and circulated, but they are likely to stay near the earth due to gravity. When we put a material into the air, we do not get rid of it; we just dilute it and move it out of the immediate area. When people lived in small groups, the smoke from fires was diluted; the smoke was in such low

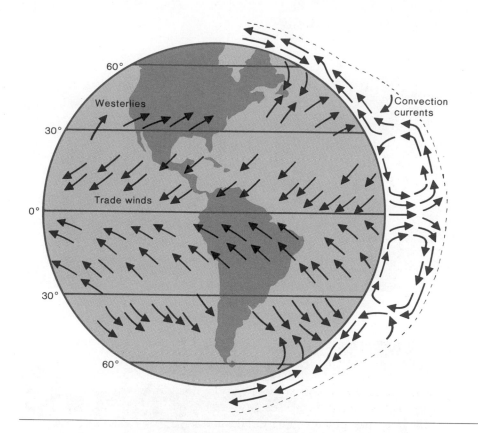

FIGURE 17.2

Global Wind Patterns. Wind is the movement of air caused by temperature differences and the rotation of the earth. Both of these contribute to the patterns of world air movement. In North America, most of the winds are westerlies (from the west to the east).

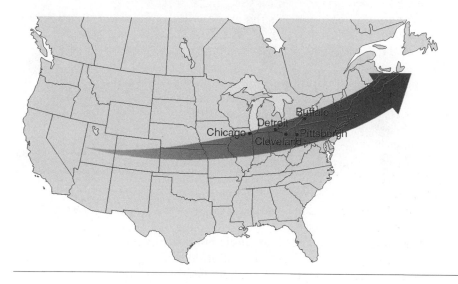

FIGURE 17.3

Accumulation of Pollutants. As an air mass moves across the continent from west to east, each population center adds its pollutants to the total load in the atmosphere.

concentrations that it did not interfere with neighboring groups down-wind. In industrialized urban areas, the pollutants cannot always be diluted before the air reaches another city. The polluted air from Chicago is further polluted by Gary, Indiana, supplemented by the wastes of Detroit and Cleveland, and finally moves over southeastern Canada and New England to the ocean. (See figure 17.3.) While not every population center adds the same kind or amount of waste, each adds to the total load carried.

In addition to aesthetic problems resulting from dirty air, there are also health problems associated with air pollution. Hundreds of deaths have

a.

FIGURE 17.4

Air Pollution. (*a*) Donora, Pennsylvania, was the scene of a serious air-pollution incident. The pollutants from industry were trapped by an inversion, and almost half of the population became ill. (*b*) The major source of carbon monoxide, hydrocarbons, and nitrogen oxides is the internal combustion engine, which is used to provide most of our transportation. The more concentrated the number of automobiles, the more concentrated the pollutants. Carbon monoxide concentrations of a hundred parts per million are not unusual in rush-hour traffic in large metropolitan areas. These concentrations are high enough to cause fatigue, dizziness, and headaches.

b.

been directly related to poor-quality air in cities—for example, London in 1952 and 1956, and New York City in 1965. A well-documented case of pollution that was harmful to human health occurred in Donora, Pennsylvania. (See figure 17.4*a*.) The city of Donora is located in a valley. In October 1948, the pollutants from a zinc plant and steel mills became trapped in the valley, and a dense smog formed. Within five days, seventeen people died, and 5,910 persons became ill. The polluted atmosphere affected nearly 50 percent of the city's 12,300 inhabitants. The technology that provided jobs for the people was also killing them. Deaths from air pollution occur primarily among the elderly, infirm, and the very young. Bronchial inflammations, allergic reactions, and irritation of the mucous membranes of the eyes and nose all indicate that air pollution must be reduced.

TABLE 17.1
Sources of primary air pollutants.

Pollutant	Sources
Carbon monoxide	Incomplete burning of fossil fuels Tobacco smoke
Hydrocarbons	Incomplete burning of fossil fuels Tobacco burning Chemicals
Particulates	Burning fossil fuels Farming operations Construction operations Industrial wastes Building demolition
Sulfur dioxide	Burning fossil fuels Smelting ore
Nitrogen compounds	Burning fossil fuels

Primary Air Pollutants

Five major types of materials are released directly into the atmosphere in their unmodified forms and in sufficient quantities to pose a health risk and to be considered **primary air pollutants:** carbon monoxide, hydrocarbons, particulates, sulfur dioxide, and nitrogen compounds. (See table 17.1.) Furthermore, these may interact with one another to form new **secondary air pollutants** in the presence of an appropriate energy source.

Carbon Monoxide (CO)

Carbon monoxide (CO) is produced when organic materials, such as gasoline, coal, wood, and trash, are incompletely burned. The automobile is responsible for most of the carbon monoxide produced in cities. (See figure 17.4b.) The number of automobiles is a great part of this problem. Although increased fuel efficiency and the use of catalytic converters have reduced carbon monoxide emissions, carbon monoxide remains a problem because the number of automobiles has increased.

The next largest source of carbon monoxide is smoking tobacco. Currently, there is a great deal of pressure to restrict the areas where smoking is permitted to minimize exposure to secondhand cigarette smoke. Restaurants designate nonsmoking sections (some even advertise themselves as smoke-free), public buildings have designated nonsmoking areas, and corporations find that their employees are requesting smoke-free lounge or break areas. In 1976, Finland adopted comprehensive legislation to control smoking. Smoking is forbidden in all public places in that country except where it is expressly allowed. No advertisement of smoking tobacco is allowed, and health warnings are published along with the quantity of tar and nicotine contained in the product. Smoking is decreasing in the industralized world today, but in the developing nations, smoking retains its image of glamour and sophistication.

Exposure to air containing 0.001 percent of carbon monoxide for several hours can cause death. Because carbon monoxide remains attached to hemoglobin for a long time, even small amounts tend to accumulate and reduce the blood's oxygen-carrying capacity. The amount of carbon monoxide produced in heavy traffic can cause headaches, drowsiness, and

Mexico City has been labeled the city with the worst air pollution ever recorded. The air over Mexico City exceeded ozone limits set by the World Health Organization on over three hundred days in one year. With twenty million people, Mexico City is the largest city in the world and grows larger each day. Open sewers and garbage dumps contribute dust and bacteria to the atmosphere. The city contains about thirty-five thousand factories and 2.5 million automobiles. Most of these are older models that are poorly tuned and, therefore, pollute the air with a mixture of hydrocarbons, carbon monoxide, and nitrogen oxides. Estimates are that 80 percent of the air pollution is the result of automobiles. The high altitude (over 2,000 meters) results in even greater air pollution from automobiles because automobile engines do not burn fuel efficiently at such high altitudes. Mexico City's location in a valley also allows for conditions suitable for thermal inversions during the winter.

Many foreign companies and governments give special "hazard pay" for working in Mexico City because of the air pollution. Pediatricians estimate that 85 percent of childhood illnesses are related to air pollution and say that the only way to improve the health of many of the children is to get them out of the city. In an attempt to respond to these conditions, officials have cancelled school during the middle of the winter, hoping that the children will be able to leave the area when air pollution is at its worst. An effort to actually reduce the air pollution is the "day without car" program. This program encourages people to use alternative transportation or to ride in a car pool to reduce the traffic in the city. Widespread compliance and some minor improvements in air quality seem to be associated with this program. In 1991, Mexico's president closed the largest oil refinery in the city in an effort to improve air quality. However, the problem is far from being solved. Most of the people are poor and cannot afford new, less-polluting automobiles, and they rely on the polluting industries for jobs. This appears to trap Mexico City in a pollution cycle for some time to come.

blurred vision. A heavy smoker in congested traffic is doubly exposed and may experience a severely impaired reaction time as compared to non-smoking drivers.

Fortunately, carbon monoxide is not a persistent pollutant. Natural processes convert carbon monoxide to other compounds that are not harmful. Therefore, the air can be cleared of its carbon monoxide if no new carbon monoxide is introduced into the atmosphere.

Hydrocarbons (HC)

In addition to carbon monoxide, automobiles emit a variety of **hydrocarbons (HC)**. Hydrocarbons are a group of organic compounds consisting of carbon and hydrogen atoms. They are either evaporated from fuel supplies or are remnants of the fuel that did not burn completely. Undoubtedly, the internal combustion engine is the major culprit, although refineries and other industries do add hydrocarbons to the total atmospheric burden. Hydrocarbons in the atmosphere should be no great problem. Most of them are washed out of the air when it rains and run off into surface water. They cause an oily film on surfaces, but hydrocarbons do not generally cause more than nuisance problems, except when they react to form secondary pollutants.

Many modifications to automobile engines have reduced the loss of hydrocarbons to the atmosphere. Recycling some gases through the engine, using higher oxygen concentrations in the fuel-air mixture and using valves to prevent the escape of gases are three of these modifications. In addition, catalytic converters result in the more complete burning of exhaust gases so that fewer hydrocarbons leave the tail pipe.

Particulates

Particulates, small pieces of solid materials dispersed into the atmosphere, constitute the third largest category of air pollutants. Smoke particles from fires, bits of asbestos from brake linings and insulation, dust particles, and ash from industrial plants contribute to the particulate load. Particulates cause problems ranging from the annoyance of soot settling on a backyard picnic table to the **carcinogenic** (cancer-causing) effects of asbestos. Particulates frequently get attention because they are so readily detected by the general public. Heavy black smoke from a factory can be seen without expensive monitoring equipment and generally causes an outcry, whereas the production of colorless gases like carbon monoxide and sulfur dioxide goes unnoticed.

Particulates cause most of their health effects by acting as centers for the deposition of moisture and gases from the atmosphere. As we breathe air containing particulates, we come in contact with concentrations of other potentially more harmful materials that have accumulated on the particulates. Sulfuric, nitric, and carbonic acids, which irritate the lining of our respiratory system, frequently form on particulates.

Sulfur Dioxide (SO_2)

Sulfur dioxide (SO_2) is a compound containing sulfur and oxygen that is produced when sulfur-containing fossil fuels are burned. Coal and oil were produced from organisms that had sulfur as one of the component parts of their living structure. When the coal or oil was formed, some of the sulfur was incorporated into the fossil fuel, and it is released as sulfur dioxide when the fuel is burned. Sulfur dioxide has a sharp odor and irritates respiratory tissue. It also reacts with water, oxygen, and other materials in the air to form sulfur-containing acids. The acids can become attached to particles, which, when inhaled, are very corrosive to lung tissue. In 1306, Edward I of England banned the burning of "sea coles," coal found on the sea shore, in the city of London. This coal was high in sulfur content and was, in part, responsible for the city's noxious odors. This might very well be some of the earliest environmental legislation concerning air quality.

London, England, was also the site of one of the earliest killer fogs. In 1952, London was covered with a dense fog for several days. During this time, the air over the city failed to mix with the layers of air in the upper atmosphere due to the temperature conditions. The factories continued to release smoke and dust into this stagnant layer of air, and the air became so full of the fog, smoke, and dust that people got lost in familiar surroundings. This combination of smoke and fog has become known as smog. Many individuals who lived in London developed such symptoms as respiratory discomfort, headache, and nausea. Four thousand people died in the next few weeks. Their deaths have been associated with the high levels of sulfur compounds in the smog. Thousands of others suffered from severe bronchial irritation, sore throats, and chest pains. The 1948 Donora, Pennsylvania, incident already mentioned also involved symptoms related to the particles and sulfur dioxide in the air.

Oxides of Nitrogen (NO and NO_2)

Oxides of nitrogen (NO and NO_2) are also major primary air pollutants. A variety of different compounds contain nitrogen and oxygen; however, nitrogen oxide (NO) and nitrogen dioxide (NO_2) are the most common.

FIGURE 17.5
Photochemical Smog. The interaction
among hydrocarbons, oxides of nitrogen,
and sunlight produces new compounds
that are irritants to humans. The visual
impact of this smog is shown in these
photographs.

When combustion takes place in air, the nitrogen and oxygen molecules
from the air may react with each other and oxides of nitrogen result:

$N_2 + O_2 \rightarrow 2NO$ (nitrogen oxide)
$2NO + O_2 \rightarrow 2NO_2$ (nitrogen dioxide)

A mixture of nitrogen oxide and nitrogen dioxide is called NO_x. The nitrogen dioxide in the mixture reacts with other compounds to produce
photochemical smog, discussed in the next section.

The primary source of nitrogen oxides is the automobile engine. Catalytic converters reduce the amount of nitrogen oxides released from the
internal combustion engine, but increased automobile traffic has resulted
in significant levels of NO_x in many metropolitan areas. Nitrogen oxides
are probably most noteworthy because they are involved in the production
of secondary pollutants.

Photochemical Smog

Secondary air pollutants are compounds that result from the interaction of
various primary air pollutants with one another. **Photochemical smog** is
a mixture of pollutants resulting from the interaction of nitrogen oxide and
nitrogen dioxide with ultraviolet light. (See figure 17.5.) The two most de-

The 1990 Clean Air Act

In November 1990, U.S. President George Bush signed the 1990 Clean Air Act, the first major revision to the Clean Air Act of 1977. This bill is surprisingly strong, considering that several powerful interest groups lobbied Congress intensively to weaken provisions of the bill that would apply to them. Although electrical utilities, the automobile industry, and certain urban areas with severe air-pollution problems won special concessions, the bill signed by President Bush mandates major changes in the amounts of air pollutants that can be released. The major consequences are economic. Ultimately, consumers will be asked to pay more for products and services as the producers pass on their air-pollution control costs in the prices of their products. The table lists some of the major provisions and expected consequences of the 1990 Clean Air Act.

Major provisions	Consequences
1. Reduce urban smog by 15 percent by 1996 and 3 percent per year until federal air-quality standards are met. Some cities with severe smog problems, like Los Angeles, have up to twenty years to comply.	Many changes needed, which will require improved technology and additional costs.
2. Utilities must reduce release of sulfur dioxide by one-half by the year 2000. 3. Utilities must reduce release of nitrogen oxides by one-third by the year 2000, starting in 1992.	Utilities will need to spend $3 billion a year for low-sulfur coal and pollution-control devices.
4. Utilities can buy and sell "pollution credits." Dirty plants can purchase permits to pollute from utilities that are not using their full allotment of pollution credits.	Economic incentives will exist for utilities to clean up their emissions. In special cases, they may be able to prolong the use of a dirty plant by purchasing permits to pollute.
5. Passenger cars must emit 60 percent less nitrogen oxide and 40 percent less hydrocarbons by the year 2003. 6. Auto-emission controls must last for 100,000 miles. 7. On-board canisters required to capture vapors during refueling.	Increased cost of fuel. Increased cost for new cars.
8. Cleaner-burning fuels will be required in the most polluted cities (Baltimore, Chicago, Hartford, Houston, Los Angeles, Milwaukee, New York, Philadelphia, and San Diego). 9. One million special low-emission vehicles (primarily in fleets) must be present by the year 2000. Three hundred thousand private low-emission vehicles must be in California by 1999.	Oil companies will need to develop new fuels. New automotive technology will need to be developed. Costs will be passed on to the consumer.
10. Toxic emissions must be reduced by 90 percent by the year 2000. Factories must install maximum-achievable control technology. This includes small businesses.	Many small businesses may need help to meet this standard. Additional costs will be passed on to the consumer as higher prices.
11. Production of chlorofluorocarbons (CFCs), halons, and carbon tetrachloride will be banned by the year 2000, methyl chloroform by the year 2002. Recycling of chemicals will be required.	Alternative chemicals will need to be developed. Additional costs will be passed on to the consumer for recycling equipment and higher-priced substitutes.

structive kinds of materials formed are ozone (O_3) and peroxyacetylnitrates. Both of these materials are excellent oxidizing agents, which means that they react readily with many other compounds, including those found in living things, causing destructive changes. Ozone is particularly harmful because it destroys chlorophyll and injures lung tissue. Peroxyacetylnitrates, in addition to being oxidizing agents, are eye irritants. Both ozone and peroxyacetylnitrates are secondary air pollutants that result from chemical reactions assisted by light.

A typical smog incident involves an interesting series of events. Morning rush-hour traffic results in the production of large amounts of nitrogen oxide (NO):

$$N_2 + O_2 \rightarrow 2NO$$

The nitrogen oxide reacts with molecular oxygen (O_2) from the atmosphere to form nitrogen dioxide (NO_2). The nitrogen dioxide in the atmosphere gives photochemical smog its reddish-brown haze.

$$2NO + O_2 \rightarrow 2NO_2$$

Later in the morning, nitrogen dioxide reacts with ultraviolet light to form atomic oxygen (O):

$$NO_2 \xrightarrow[\text{light}]{\text{ultraviolet}} NO + O$$

Abundant molecular oxygen in the atmosphere reacts with atomic oxygen to form ozone:

$$O_2 + O \rightarrow O_3$$

Hydrocarbons in the atmosphere react with ozone to produce peroxyacetylnitrates. Various pollution-control devices now reduce the amount of hydrocarbons escaping to the atmosphere, thereby decreasing the amount of peroxyacetylnitrates formed. As the ozone and peroxyacetylnitrates react with living things, they cause damage and are converted to less reactive molecules, and the smog eventually clears.

Due to their climate and the geographic features of their locations, such large metropolitan areas as Los Angeles, Salt Lake City, Phoenix, and Denver have more trouble with photochemical smog than East Coast metropolitan areas. Each of these cities is located within a ring of mountains. The prevailing winds are from the west. As cool air flows into a valley, it pushes the warm air upward. This warm air becomes sandwiched between two layers of cold air and acts like a lid on the valley, a condition known as a **thermal inversion.** The air is trapped in the valley. (See figure 17.6.) The warm air cannot rise further because it is covered by a layer of cooler air pushing down on it. It cannot move out of the area because of the ring of mountains. Without normal air circulation, smog accumulates. Harmful chemicals continue to increase in concentration until a major weather change causes the warm air to move up and over the mountains. Then the underlying cool air can begin to circulate, and the polluted air is diluted.

Smog problems could be substantially decreased by reducing the use of internal combustion engines (perhaps eliminating them completely) or by moving population centers away from the valleys that produce thermal inversions. Both of these solutions would require expenditures of billions of dollars; therefore, people will probably continue to live with the problem.

Control of Air Pollution

Eliminating photochemical smog would require large-scale changes in lifestyle and culture, but other pollutants are much more readily controlled. Government regulations have pressured the automobile industry to engage in air-pollution control research. The positive crankcase ventilation valve (PCV) and gas caps with air-pollution control valves (APC) reduce hydrocarbon loss. Catalytic converters reduce carbon monoxide, oxides of nitrogen, and hydrocarbons in emissions and necessitate the use of lead-free

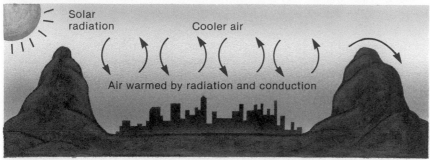

Normal situation

Solar radiation

Cooler air

Air warmed by radiation and conduction

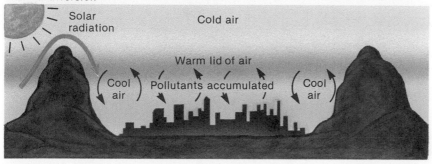

Thermal inversion

Solar radiation

Cold air

Warm lid of air

Cool air   Pollutants accumulated   Cool air

FIGURE 17.6

Thermal Inversion. Under normal conditions, the air at the earth's surface is heated by the sun and rises to mix with the cooler air above it. When a thermal inversion occurs, a layer of warm air is formed above the cooler air at the surface. The cooler air is then unable to mix with the warm air above and cannot escape because of surrounding mountains. The cool air is trapped, sometimes for several days, and accumulates pollutants. If the thermal inversion continues, the levels of pollution can become dangerously high.

FIGURE 17.7

Control of Particulates. The installation of proper pollution-control devices can significantly reduce the amount of particulate material released into the air. These photographs illustrate the difference before and after the installation of control devices.

fuel. This lead-free fuel requirement, in turn, significantly reduces the amount of lead (and other metal additives) in the atmosphere.

Particulates are produced primarily by burning. In industry, their release can be controlled with scrubbers, precipitators, and filters. (See figure 17.7.) These devices are effective, but expensive. They can be retrofitted to the smokestack, or they can be designed into the combustion system. They filter the products of combustion and remove some, while allowing others to be released into the atmosphere. Electric charge differentials may be used to remove the unwanted materials from the emission.

Industry and automobiles are not the only producers of particulate pollutants. Many people in the world use wood as their primary source of fuel for cooking and heat. In developed countries like the United States and Canada, fireplaces and wood-burning stoves are used by some as a primary source of heat, but most use them for supplemental heat or for aesthetic purposes. However, the use of large numbers of wood-burning stoves and fireplaces can generate a significant air-pollution problem, called a brown cloud problem. Many municipalities, such as Boise, Idaho; Salt Lake City, Utah; and Denver, Colorado (and some entire states), use fines to enforce a ban on wood burning during severe air-pollution episodes. Many other communities issue pollution alerts and request that people not use wood-burners. Many of these communities have regulations about the number and efficiency of wood-burning stoves and fireplaces. Some communities, such as Castle Rock, Colorado, prohibit the construction of houses with fireplaces or wood-burning stoves. Many people readily switch to gas fireplaces or high-efficiency wood-burners when they understand that their enjoyment of a rustic stove may be leading to a degraded environment, but most need to be forced to comply by official regulations and the threat of fines. Obviously, accumulations of many small sources of air pollution may cause as big a problem as one large emission source and are frequently more difficult to control.

To control sulfur dioxide, which is produced primarily by electric power generating plants, several possibilities are available. One alternative is to change from high-sulfur fuel to low-sulfur fuel. Switching from a high-sulfur coal to a low-sulfur coal reduces the amount of sulfur released into the atmosphere by 66 percent. Switching to oil, natural gas, or nuclear fuels would reduce sulfur dioxide emissions even more. However, this is not a long-term solution because low-sulfur fuels are in short supply, and nuclear power plants present a new set of pollution problems.

A second alternative is to remove the sulfur from the fuel before the fuel is used. Chemical or physical treatment of coal before it is burned can remove nearly 40 percent of the sulfur. This is technically possible, but it increases the cost of electricity to the rate payer.

Scrubbing the gases emitted from a smokestack is a third alternative. The technology is available, but, of course, these control devices are costly to install, maintain, and operate. As with auto emissions, governments have required the installation of these devices, but when industries install them, the cost of construction and operation is passed on to the consumer. The cost of installing scrubbers on a typical power plant is about $200 million. Large amounts of lime are used in a scrubber, and a great deal of waste material is produced. In the United States, an estimated 20+ million metric tons of sulfur oxides are released into the atmosphere each year. Installing scrubbers on just fifty of the largest coal-burning plants would reduce this amount by over one-third. The sulfur can then be sold, and some of the cost of its removal recovered.

There is resistance to taking these steps, however. Large utility companies (prime sources of atmospheric sulfur dioxide) question the effect of this pollutant and also deny that they are the main source of sulfur emissions.

In the past, a common solution was to build taller smokestacks. Tall stacks release their gases above the inversion layers and, therefore, add the sulfur dioxide to the upper atmosphere, where it is diluted before it comes in contact with the population downwind. This works as long as the stack is tall enough and as long as not so much pollution is added to the air that it cannot be diluted to an acceptable level. However, because the sulfur dioxide content of the upper atmosphere has increased, sulfur dioxide reacts more frequently with oxygen and dissolves in the water in the atmosphere to form sulfuric acid. This acid is washed from the air when it rains or snows and damages plants and animals and increases corrosion of building materials and metal surfaces.

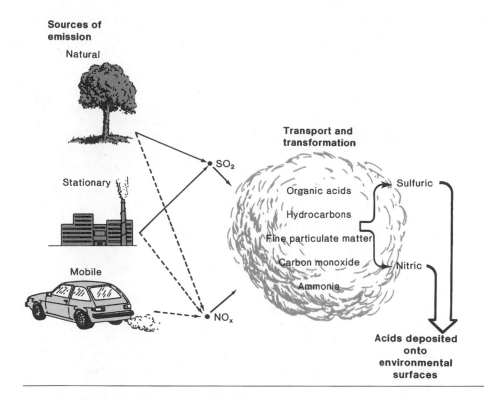

Sources of emission

Natural

Stationary

Mobile

SO_2

NO_x

Transport and transformation

Organic acids

Hydrocarbons

Fine particulate matter

Carbon monoxide

Ammonia

Sulfuric

Nitric

Acids deposited onto environmental surfaces

FIGURE 17.8
Acid Deposition. Molecules from natural sources, power plants, and internal combustion engines react to produce the chemicals that are the source of acid deposition.

Acid Deposition

Acid deposition is the accumulation of potential acid-forming particles on a surface. Acids result from natural causes, such as vegetation, volcanoes, and lightning; and from human activities, such as coal burning and use of the internal combustion engine. (See figure 17.8.) These combustion processes produce sulfur dioxide (SO_2) and oxides of nitrogen (NO_x). Oxidizing agents, such as ozone, hydroxyl ions, or hydrogen peroxide, along with water, are necessary to convert the sulfur dioxide or nitrogen oxides to sulfuric or nitric acid. Various reactive hydrocarbons (HC) encourage the production of oxidizing agents.

The acid-forming reactants are classified as wet or dry. Wet reactions occur in the atmosphere and come to earth as some form of precipitation: acid rain, acid snow, or acid dew. Dry deposition occurs with the settling of the precursors of the acid on a surface. An acid does not actually form until these materials mix with water. Even though the acids are formed and deposited in several different ways, all of these processes usually are referred to as **acid rain.**

Acid rain is a worldwide problem. Reports of high acid-rain damage have come from Canada, England, Germany, France, Scandinavia, and the United States. Rain is normally slightly acidic, with a pH between 5.6 and 5.7 due to atmospheric carbon dioxide that dissolves to produce carbonic acid. But acid rains sometimes have a concentration of acid a thousand times higher than normal. In 1969, New Hampshire had a rain with a pH of 2.1. In 1974, Scotland had a rain with a pH of 2.4. The average rain in much of the northeastern part of the United States and adjoining parts of Ontario has a pH between 4.0 and 4.5.

Acid rain can cause damage in several different ways. Buildings and monuments are often made from materials that contain limestone (calcium carbonate, $CaCO_3$) because limestone is relatively soft and easy to work.

BOX 17.3
Acid Rain: Canada Versus the United States

"U.S. foot-dragging and interference in the development of scientific information has reached frustrating proportions." These words, spoken by the Honorable John Roberts, Canadian Minister of the Environment, sum up the confrontation between Canada and the United States regarding the question of acid deposition. Although the Canadians do release large amounts of sulfur dioxide and oxides of nitrogen into the atmosphere, they have long contended that much of the acid deposition in their country originates in the United States. They are very concerned because 2.5 million square kilometers of Canada are highly susceptible to acid deposition.

With the signing of the Memorandum of Intent on Transboundary Air Pollution in 1980, the United States and Canada took the first step to cooperatively reduce the amount of acid deposition. The memorandum created scientific groups to study the problems of air pollution. After two years of study, there was still no accord. The Canadians accused the Reagan administration of delaying the studies, and the United States accused the Canadians of acting too rapidly.

In fact, the dispute reached such proportions that in 1983 the U.S. Department of Justice ruled that a Canadian-produced film on acid rain had to be labeled as political propaganda before it could be shown in the United States. The U.S. Department of Justice also required that the names of U.S. groups viewing the film be reported to the Justice Department. This attitude prompted the Canadian Minister of the Environment to observe, "It sounds like something you would expect from the Soviet Union, not the United States."

In 1982, Canada suggested a mutual 50 percent reduction of sulfur dioxide by 1990. Citing a lack of research, the United States did not agree with the plan. This prompted Minister Roberts to state, "Always the constant refrain rings out from the administration that nothing is proven, and that an indefinite amount of further study is needed, not prompt action. Well, we can't wait. Our lakes and forests are literally dying."

In his 1984 State of the Union Address, President Reagan affirmed that the United States would take no direct action regarding the question of acid deposition other than to continue to research the problem. Later that year, the Canadians announced a goal of reducing acid deposition by 50 percent and trusted that the United States would join them. However, the possibility seemed remote, for in May 1984, the House subcommittee voted against a bill to reduce the emissions of sulfur dioxide by 10 million metric tons by 1993. This killed any U.S. action regarding reduced acid deposition for 1984.

During the Bush administration, there has been a softening of attitudes toward transboundary air pollution. The 1990 Clean Air Act will significantly lower the amount of acid precipitation and move the United States toward a more progressive position. As evidence of this change in attitude, President Bush and Prime Minister Mulroney signed an agreement in 1991 to cooperate in reducing transboundary air pollution.

Sulfuric acid (H_2SO_4), a major component of acid rain, converts limestone to gypsum ($CaSO_4$), which is more soluble and is eroded over many years of contact with acid rain. (See figure 17.9.) Metal surfaces can also be attacked by acid rain.

The effects of acid rain on ecosystems are often more difficult to quantify. Intense sulfur dioxide pollution around smelters is known to cause the death of many kinds of trees and other vegetation. But this is an extreme case and may not be directly comparable to less intense acid rain. However, in many parts of the world, acid rain is suspected of causing the death of many forests and reducing the vigor and rate of growth of others. (See figure 17.10.) In Central Europe, many forests have declined significantly, resulting in the death of about 6 million hectares of trees. Northeastern North America has also seen significant tree death and reduction in vigor, particularly at higher elevations. Some areas have had 50 percent mortality of red spruce trees.

A clear link between the decline of the forests and acid rain is difficult to establish, but several hypotheses have been formulated. Molecules like sulfur dioxide and ozone are known to be associated with air pollution and acid rain and to cause direct damage to plants. Also, as soil becomes acidic,

Pollution and Policy

FIGURE 17.9

Damage Due to Acid Deposition.
Sulfuric acid (H_2SO_4), which is a major component of acid deposition, reacts with limestone ($CaCO_3$) to form gypsum ($CaSO_4$). Since gypsum is water-soluble, it washes away with rain. The damage to this monument is the result of such acid reacting with the stone.

FIGURE 17.10

Forest Decline. Many forests at high elevations in northeastern North America have shown significant decline, and dead trees are common.

aluminum is released from binding sites and may interfere with the plant roots' ability to absorb nutrients. Reduction in the pH of the soil may also cause changes in the kind of bacteria in the soil and reduce the availability of nutrients for the plant. While none of these alone would necessarily result in plant death, each could add to the stresses on the plant and may allow other factors, such as insect infestations, extreme weather conditions (particularly at high elevations), or drought, to further weaken trees and ultimately cause their death.

The effects of acid rain on aquatic ecosystems are much more clear-cut. In several experiments, lakes were purposely converted to acid lakes and the changes in the ecosystems recorded. The experiments showed that, as lakes become more acidic, there is a progressive loss of many kinds of organisms. The food web becomes less complicated, many organisms fail to

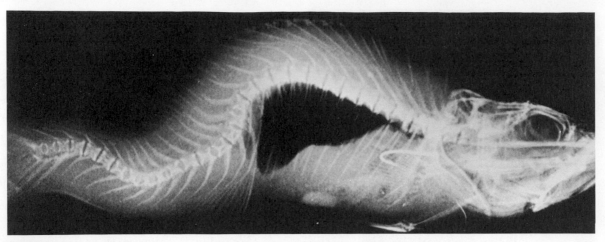

FIGURE 17.11
Effects of Acid Deposition on Organisms. The low pH of the water in which this fish lived caused the abnormal bone development that ultimately resulted in the death of the fish.

reproduce, and many others die. Most healthy lakes have a pH above 6. At a pH of 5.5, many desirable species of fish have been eliminated; at a pH of 5, only a few starving fish may be found, and none are reproducing. Lakes with a ph of 4.5 are nearly sterile.

There are several reasons for these changes. Many of the early reproductive stages of insects and fish are more sensitive to acid conditions than the adults. In addition, the young are often in shallow water, which is most affected by a flood of acid into lakes and rivers during the spring snowmelt. The snow and its acids have accumulated over the winter, and the snowmelt releases large amounts of acid all at once. Crayfish need calcium to form their external skeleton. As the pH of the water decreases, the crayfish eventually are unable to form new skeletons and die. Reduced calcium availability also results in some fish with malformed skeletons. (See figure 17.11.) As mentioned earlier, increased acidity also results in the release of aluminum, which impairs the function of a fish's gills.

The extent to which lakes have been acidified is very large. About fourteen thousand lakes in Canada and eleven thousand in the United States have been seriously altered by becoming acidic. Many lakes in Scandinavia are similarly affected. The extent to which acid deposition affects an ecosystem depends on the nature of the bedrock in the area and the ecosystem's proximity to acid-forming pollution sources. (See figure 17.12.) Parent material derived from igneous rock is not capable of buffering the effects of acid deposition, while soils derived from sedimentary rocks such as limestone release bases that neutralize the effects of acids. Because of this, eastern Canada and the U.S. Northeast are particularly susceptible to acid rain. These areas have high amounts of granite rock and are downwind from the major air-pollution sources of North America. Scandinavian countries have a similar geology and receive pollution from industrial areas in the United Kingdom and Europe. Thousands of kilometers of streams and up to two hundred thousand lakes in eastern Canada and the northeastern United States are estimated to be in danger of becoming acidified because of their location and geology.

Global Warming

During the 1980s, scientists, governments, and the public became concerned about the possibility that the world may be getting warmer. In fact, though, over the decade of the 1980s, there were some warm years and some cool years, but no convincing proof of global warming. Several gases,

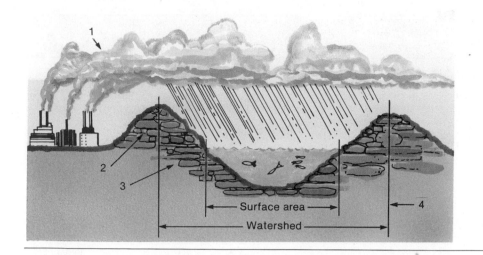

FIGURE 17.12

Factors That Contribute to Acid Rain Damage. In an aquatic ecosystem, the following factors increase the risk of damage from acid deposition: (1) a lake is located downwind from a major source of air pollution; (2) the area around the lake is hard, insoluble bedrock covered with a layer of thin infertile soil; (3) the soil has a low buffering capacity; and (4) there is also a low watershed to lake surface area ratio.

Source: From EPA's "Acid Rain."

however, could cumulatively produce that effect over longer periods of time. This is enough to convince many that action must be taken.

If global warming occurs, there could be major changes in world climate. Sea level could rise because the warmer water in the seas would expand, and water stored in polar glaciers would melt, adding to the amount of water in the oceans. Temperate regions of the world could shift northward, and many areas that are currently temperate could be converted to hot, dry regions. Many of the predictions made by scientists are based on computer models of climate and weather patterns and have, therefore, been criticized as being inaccurate and unable to predict outcomes from sketchy data. For example, many different computer models predict an increase in the average annual temperature if the carbon dioxide in the atmosphere doubles, but the models differ by as much as 3.5° C as to what the increase would be. However, the consequences of a global warming would be so great that many are suggesting that we should alter our life-styles regardless to prevent major world climate changes.

An appreciation of the problem requires careful examination of several facts. Several gases in the atmosphere are transparent to light but absorb infrared radiation. These allow sunlight to penetrate the atmosphere and be absorbed by the earth's surface. This energy is reradiated as infrared radiation (heat), which is absorbed by the gases. Because the effect is similar to what happens in a greenhouse (the glass allows light to enter but retards the loss of heat), these gases are called greenhouse gases and the warming thought to occur from their increase is called the **greenhouse effect.** The most important greenhouse gases are carbon dioxide (CO_2), chlorofluorocarbons (primarily CCl_3F and CCl_2F_2), methane (CH_4), and nitrous oxide (N_2O). Each of these gases is currently increasing in amount as a result of human activity. Table 17.2 lists the relative contribution of each of these gases to the potential for global warming.

Carbon dioxide (CO_2) is the most abundant of the greenhouse gases and is present as a natural consequence of respiration. However, much larger quantities are put into the atmosphere as a waste product of energy production. Measurement of carbon dioxide levels at the Mauna Loa Observatory in Hawaii show that the carbon dioxide level has increased from about 315 ppm (parts per million) in 1958 to about 352 ppm in 1990. (See figure 17.13.) Coal, oil, natural gas, and biomass are all burned to provide heat and electricity for industrial processes, home heating, and cooking. These sources are causing an increase in the amount of carbon dioxide in the atmosphere.

TABLE 17.2
Major greenhouse gases.

Gas	Contribution to global warming (percentage)
Carbon dioxide	57
Chlorofluorocarbons	25
Methane	12
Nitrous oxide	6

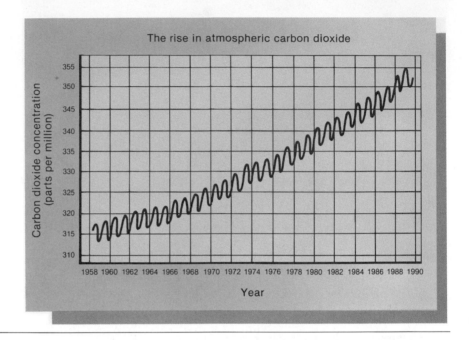

FIGURE 17.13

Change in Atmospheric Carbon Dioxide. Since the establishment of a carbon dioxide monitoring station at Mauna Loa Observatory in Hawaii, a steady increase in carbon dioxide levels has been observed. The fluctuations are caused by the increased uptake of carbon dioxide during summer in the Northern Hemisphere.

A major step toward slowing global warming would be to increase the efficiency of energy utilization. This would also be of value in conserving the continually shrinking supplies of energy resources. It makes sense to increase energy efficiency, thus reducing carbon dioxide production, even if global warming is not a concern. One way to stimulate a move toward greater efficiency would be the imposition of a carbon tax. A carbon tax would increase the cost of any fuels used by taxing the amount of carbon put into the atmosphere by their use. This would increase the demand for fuel efficiency because the cost of fuel would rise, stimulate the development of alternative fuels with a lower carbon content, and generate funds for research in many aspects of fuel efficiency and alternative fuel technologies.

Another approach to the problem is to increase the amount of carbon dioxide removed from the atmosphere. If enough biomass is present, the excess carbon dioxide can be utilized by vegetation during photosynthesis, thereby reducing the impact of carbon dioxide released by fossil-fuel burning. Australia, the United States, and several other countries have announced plans to plant billions of trees to help remove carbon dioxide from the atmosphere. Many critics argue that this approach will only provide a short-term benefit since, eventually, the trees will mature and die, and their decay will release carbon dioxide into the atmosphere at some later time.

TABLE 17.3
Worldwide uses of chlorofluorocarbons, 1985.

Use	Share of total (percentage)
Aerosols	25
Rigid-foam insulation	19
Solvents	19
Air conditioning	12
Refrigerants	8
Flexible foam	7
Other	10

Source: Data from Daniel F. Kohler, et al., *Projection of Consumption of Products Using Chlorofluorocarbons in Developing Countries,* Rand Corporation, 1987, Santa Monica, CA.

An associated concern is the destruction of vast areas of rain forest in tropical regions of the world. These ecosystems are extremely efficient at removing carbon dioxide and storing the carbon atoms in the structure of the plant. The burning of tropical rain forest to provide farm or grazing land not only adds carbon dioxide to the atmosphere but also reduces the ability to remove carbon dioxide from the atmosphere, since the grasslands or farms created do not remove carbon dioxide as efficiently as the original rain forest. Furthermore, the grazing lands and farms in such regions of the world are often abandoned after a few years and do not return to their original forest condition.

Chlorofluorocarbons are entirely the result of human activity. They are used as refrigerant gases in refrigerators and air conditioners, as cleaning solvents, as propellants in aerosol containers, and as expanders in foam products. Although they are present in the atmosphere in minute quantities, they are extremely efficient as greenhouse gases (about fifteen thousand times more efficient at retarding heat loss than carbon dioxide). Since all chlorofluorocarbons are made by people for specific purposes, their levels can be easily controlled. Table 17.3 lists several chlorofluorocarbons and their uses.

The use of chlorofluorocarbons as propellants in aerosol spray cans or as expanders in foam products is not necessary. Other more benign materials, such as hydrocarbons, could be used. Since the 1970s, when chlorofluorocarbons were linked to the depletion of the ozone layer in the upper atmosphere, their use in aerosol cans has been banned in the United States, Canada, Norway, and Sweden, and the European Economic Community agreed to reduce use of chlorofluorocarbons in aerosol cans. However, worldwide, chlorofluorocarbons are still widely used as aerosol propellants. In foam and solvents, care can be taken to recover the chlorofluorocarbons for reuse rather than to allow them to escape into the atmosphere. Alternative products are available for use as refrigerants, and care also could be taken to recycle refrigerant gases rather than just venting them into the air. In January 1991, DuPont announced the development of a group of new refrigerants that would not harm the ozone layer. These will be installed in new refrigerators and air conditioners in the future. All of these options will need to be exploited if chlorofluorocarbon production is to be reduced.

In 1987, several industrialized countries, including Canada, the United States, the United Kingdom, Sweden, Norway, The Netherlands, the Soviet Union, and West Germany, agreed to freeze production of chlorofluorocarbons at present levels and reduce production by 50 percent by the year

2000. This document, known as the Montreal Protocol, was ratified by the U.S. Senate in 1988. In May 1989, eighty-two nations signed the Helsinki Declaration, pledging to phase out the use of most chlorofluorocarbons by the year 2000. Several companies in the United States and Japan (General Motors, AT&T, Nissan) have announced plans to either phase out chlorofluorocarbon use or reduce use significantly. In 1990 in London, international agreements were reached to further reduce the use of chlorofluorocarbons. A major barrier to these negotiations involved a commitment from the developed countries of the world to establish a fund to assist less developed countries in implementing technologies that would allow them to obtain refrigeration and air conditioning without the use of chlorofluorocarbons. Regardless of all of these agreements, however, large quantities of chlorofluorocarbons are currently in use and will eventually find their way into the atmosphere. So even if all chlorofluorocarbon production were stopped today, atmospheric levels of these chemicals would still increase for several years.

Methane enters the atmosphere primarily from biological sources. Several kinds of bacteria that are particularly abundant in wetlands and rice fields release methane into the atmosphere. Methane-releasing bacteria are also found in large numbers in the guts of termites and various kinds of ruminant animals like cattle. Some methane enters the atmosphere from fossil-fuel sources. Control of methane sources is unlikely since the primary sources involve agricultural practices that would be very difficult to change. For example, it would involve the conversion of rice paddies to other forms of agriculture and a drastic reduction in the number of animals used for meat production. Neither is likely to occur since food production in most parts of the world needs to be increased, not decreased.

Nitrous oxide, a minor component of the greenhouse gas picture, primarily enters the atmosphere from fossil fuels and fertilizers. It could be reduced by more careful use of nitrogen-containing fertilizers.

Ozone Depletion

In the 1970s, various sectors of the scientific community became concerned about the possible reduction in the ozone layer surrounding the earth in the upper atmosphere. In 1985, it was discovered that a significant thinning of the ozone layer over the Antarctic occurred during the Southern Hemisphere spring. Some regions of the ozone layer showed 95 percent depletion. Ozone depletion also has been found to be occurring farther north than previously. These findings have caused several countries to become involved in efforts to protect the ozone layer.

The presence of ozone in the outer layers of the atmosphere, approximately 12–25 kilometers from the earth's surface, shields the earth from the harmful effects of ultraviolet light radiation. **Ozone** is a molecule that consists of three atoms of oxygen (O_3). This molecule absorbs ultraviolet light and is split into an oxygen molecule and an oxygen atom:

$$O_3 \xrightarrow{\text{Ultraviolet light}} O_2 + O$$

Oxygen molecules are also split by ultraviolet light to form oxygen atoms:

$$O_2 \xrightarrow{\text{Ultraviolet light}} 2O$$

Recombination of oxygen atoms and oxygen molecules allows ozone to be formed again and to be available to absorb more ultraviolet light.

Pollution and Policy

Visual pollution is a sight that offends us. This is highly subjective and is, therefore, very difficult to define or control. To most people, an open garbage dump is a form of visual pollution. A dilapidated home or building may also be offensive, especially if located in an area of higher-priced homes. A badly littered highway or street is aesthetically offensive, and litter along a wilderness trail is even more unacceptable. To many, roadside billboards are offensive, but they are helpful to the advertiser and to travelers looking for information. This conflict is typical of aesthetic forms of pollution. People do not agree on what is offensive. Therefore, control or regulation of many forms of aesthetic pollution is difficult.

Some of the chemicals discharged into our waterways have a definite taste. If the chemical does no biological injury but has an unpleasant taste, it is a taste pollutant. If minute quantities of certain chemicals enter drinking water or food, we may taste them. In fish, a concentration of 100 ppm of phenol can be tasted. Although this small amount of chemical is not harmful biologically, it does make the fish very unappetizing.

Odor pollution may originate from several sources. Various airborne chemicals have a specific odor that may be offensive. People who live near stockyards, paper mills, chemical plants, steel mills, and many other types of industry may be offended by the odors

originating from these sources. Many of these industries discharge waste materials into the water that decompose or evaporate and cause odor pollution. People who are constantly exposed to an odor are not as offended by it as are people who are newly exposed to it. When an odor is constantly received, the brain ceases to respond to the stimulus. In other words, the person is not aware of the odor.

Aesthetic pollutants, such as sights, tastes, and odors, are extremely difficult to define. Each of us has our own idea of what is offensive, which makes it very difficult to establish aesthetic pollution standards. This type of pollution can be controlled by educating the public so that generally unacceptable levels of aesthetic pollution are eliminated.

$$O_2 + O \longrightarrow O_3$$

This series of reactions results in the absorption of 99 percent of the ultraviolet light energy coming from the sun and prevents it from reaching the earth's surface. Reduction in the amount of ozone in the upper atmosphere would result in more ultraviolet light reaching the earth's surface, causing increased skin cancers and cataracts in humans and increased mutations in all living things.

Chlorofluorocarbons are strongly implicated in the ozone reduction in the upper atmosphere. Chlorine reacts with ozone in the following way to reduce the quantity of ozone present:

$$Cl + O_3 \rightarrow ClO + O_2$$

$$ClO + O \rightarrow Cl + O_2$$

These reactions both destroy ozone and reduce the likelihood that it will be formed because atomic oxygen (O) is removed as well.

The use of chlorofluorocarbons in air conditioners, refrigerators, and as propellants has resulted in the release of large amounts into the atmosphere. Because chlorofluorocarbons are implicated in both global warming and ozone depletion, considerable international attention is focused on controlling their manufacture and release.

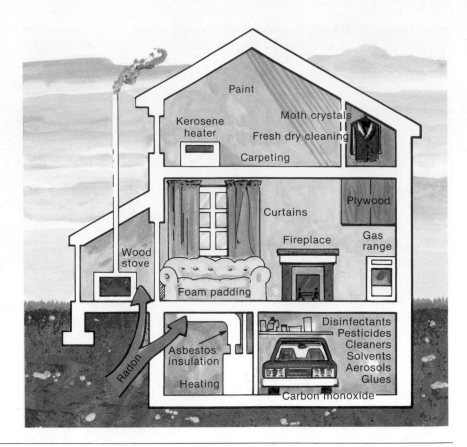

FIGURE 17.14
Sources of Indoor Air Pollution. In tightly sealed modern buildings, many small sources of air pollution can become a problem.

Indoor Air Pollution

A growing body of scientific evidence indicates that the air within homes and other buildings can be more seriously polluted than outdoor air in even the largest and most industrialized cities. Many indoor air pollutants and pollutant sources are thought to have an adverse effect on human health. These pollutants include: asbestos; formaldehyde, which is associated with many consumer products, including certain wood products and aerosols; airborne pesticide residues; chloroform; perchloroethylene (associated particularly with dry cleaning); paradichlorbenzene (from mothballs and air fresheners); and many disease-causing microorganisms. (See figure 17.14.) Smoking is the most important air pollutant source in the United States in terms of human health. In the United States, the Surgeon General estimates that 350,000 people die each year from emphysema, heart attacks, strokes, lung cancer, or other diseases caused by tobacco smoking. Banning smoking probably would save more lives than any other pollution-control measure.

A recent contributing factor to the concern about indoor air pollution is the weatherizing of buildings to reduce heat loss and save on fuel costs. In most older homes, there is a complete exchange of air energy every hour. This means that fresh air leaks in around doors, windows, cracks, and holes in the building. In a weatherized home, a complete air exchange may occur only once every five hours. Such a home is more energy efficient, but it also tends to trap air pollutants.

BOX 17.5
Noise Pollution

Noise is referred to as unwanted sound. However, noise can be more than just unpleasant sound. Research has shown that exposure to noise can cause physical, as well as mental, harm to the body. The loudness of the noise is measured by decibels (db). Decibel scales are logarithmic, rather than linear. Thus, the change from 40 db (a library) to 80 db (a dishwasher or garbage disposal) represents a ten-thousandfold increase in sound loudness.

The frequency or pitch of a sound is also a factor in determining its degree of harm. High-pitched sounds are the most annoying. The most common sound pressure scale for high-pitched sounds is the A scale, whose units are written "dbA." Hearing loss begins with prolonged exposure (eight hours or more) to 80 to 90 dbA levels of sound pressure. Sound pressure becomes painful at around 140 dbA and can kill at 180 dbA. (See the table.)

In addition to hearing loss, noise pollution is linked to a variety of other ailments, ranging from nervous tension headaches to neuroses. Research has also shown that noise may cause blood vessels to constrict (which reduces the blood flow to key body parts), disturbs unborn children, and sometimes causes seizures in epileptics. The EPA has estimated that noise causes about forty million U.S. citizens to suffer hearing damage or other mental or physical effects. Up to sixty-four million people are estimated to live in homes affected by aircraft, traffic, or construction noise.

The Noise Control Act of 1972 was the first major attempt made in the United States to protect the public health and welfare from detrimental noise. This act also attempted to coordinate federal research and activities in noise control, to set federal noise emission standards for commercial products, and to provide information to the public. Subsequent to the passage of the Noise Control Act, many local communities in the United States enacted their own noise ordinances. While such efforts are a step in the right direction, the United States is still controlling noise less than are many European and Scandinavian countries. Several European countries have developed quiet construction equipment in conjunction with strongly enforced noise ordinances. For example, in the Soviet Union, factory noise levels above 85 dbA are prohibited, as are noise levels above 30 dbA in residential areas. The Germans and Swiss have also established maximum day and night noise levels for certain areas. Regarding noise-pollution abatement, North America has much to learn from European countries.

Intensity of noise.	
Source of sound	**Intensity in decibels**
Jet aircraft at takeoff	145
Pain occurs	140
Hydraulic press	130
Jet airplane (160 meters overhead)	120
Unmuffled motorcycle	110
Subway train	100
Farm tractor	98
Gasoline lawn mower	96
Food blender	93
Heavy truck (15 meters away)	90
Heavy city traffic	90
Vacuum cleaner	85
Hearing loss after long exposure	85
Garbage disposal unit	80
Dishwasher	65
Window air conditioner	60
Normal speech	60

Even though we spend almost 90 percent of our time indoors, the movements to reduce indoor air pollution lag behind regulations governing outdoor air pollution. In the United States, the Environmental Protection Agency is conducting research to identify and rank the human health risks that result from exposure to individual indoor pollutants or mixtures of multiple indoor pollutants.

BOX 17.6
Radon

In 1985, the clothing worn by an engineer at the Limerick Nuclear Generating Station in Pottstown, Pennsylvania, registered a high radiation level. Initially, the generating station was believed to be the source of the radiation. However, subsequent studies indicated that radon 222 from the engineer's home was the source. Following this incident, there has been an increased interest in the source of radon and its effects.

The source of radon is uranium 238, a naturally occurring element that makes up about three parts per million of the earth's crust. Uranium 238 goes through fourteen steps of decay before it becomes stable, nonradioactive lead 206. **Radon,** an inert radioactive gas having a half-life of 3.8 days, is one of the products formed during this process.

Since radon is an inert gas, it does not enter into any chemical reactions. The health hazards are not from radon but from the "daughters" of radon, those products formed as radon undergoes radioactive decay. These decay products of radon—polonium 218, which has a three-minute half-life; lead 214, which has a twenty-seven-minute half-life; bismuth 214, which has a twenty-minute half-life; and polnium 214, which has a millisecond half-life—are solid materials and chemically active. Therefore, as the radon gas decays, it produces materials that are solids. These solids can be inhaled and may become attached to lung tissue.

Increased incidence of lung cancer is the only known health effect associated with radon decay products. It is estimated that the decay products are responsible for five thousand to twenty thousand cases of lung cancer annually in the United States.

Some four million homes in the United States may have higher-than-acceptable levels of radon. Mining areas or areas with high amounts of igneous and metamorphic rocks are most likely to be affected. Uranium mining regions of Colorado and sections of Pennsylvania, New York, New Jersey, and Florida are the areas most affected in the United States.

As the radon gas is formed in the rocks, it usually diffuses up through the rocks and soil and escapes harmlessly into the atmosphere. It can also diffuse into groundwater. Radon usually enters a home through an open space in the foundation. A crack in the basement floor or the foundation, the gap around a water or sewer pipe, or a crawl space allows the radon to enter the home. It may also enter in the water supply from wells.

Usually, the natural ventilation within a house allows air to escape through loose-fitting doors or windows or other openings. Thus, radon that enters a house can escape as a result of this ventilation. Problems could arise, however, as people make their homes more airtight in an attempt to save energy.

Only 10 percent of the homes in the United States have a potential radon problem. However, the increased publicity about radon has many people worried. In addition, in 1988, the Environmental Protection Agency and the U.S. Surgeon General recommended that all Americans (other than those living in apartment buildings above the second floor) test their home for radon. People who are concerned about radon should contact their state's public health department or environmental protection agency. These agencies can recommend self-testing kits to detect radon levels. The cost of these kits ranges from fifteen to thirty-five dollars. Consumer frauds that prey upon people's fear of radon are increasing, so consumers should be wary. Two excellent books on radon may be obtained free: *Radon Reduction Methods, a Homeowner's Guide* and *Radon Reduction Methods for Detached Houses.* The address is Environmental Protection Agency, 230 South Dearborn Street, Chicago, Illinois 60604.

Radon risk evaluation chart

pCi/L	WL	Estimated lung-cancer deaths due to radon exposure (out of 1,000)	Comparable exposure levels	Comparable risk
200	1.0	440–770	One thousand times average outdoor level	More than sixty times nonsmoker risk
100	0.5	270–630	One hundred times average indoor level	Four-pack-a-day smoker
				Two thousand chest X rays per year
40	0.2	120–380		
				Two-pack-a-day smoker
20	0.1	60–120	One hundred times average outdoor level	
				One-pack-a-day smoker
10	0.05	30–120	Ten times average indoor level	Five times nonsmoker risk
4	0.02	13–50		
				Two hundred chest X rays per year
2	0.01	7–30	Ten times average outdoor level	Nonsmoker risk of dying from lung cancer
1	0.005	3–13	Average indoor level	
0.2	0.001	1–3	Average outdoor level	Twenty chest X rays per year

Note: Measurement results are reported in one of two ways: (1) pCi/L (plcocuries per liter) — Measurement of *radon gas*, or (2) WL (working levels) — Measurement of *radon decay products*.

Source: Office of Air and Radiation Programs, U.S. Environmental Protection Agency.

CONSIDER THIS CASE STUDY
International Air Pollution

Acid rain originates as a form of air pollution, but it may damage the environment in the form of water pollution. Also, air pollution that originates in one country may result in water pollution in another country. A particular nation may have stringent environmental controls within its own boundaries but have environmental problems because of the actions of a neighboring country.

Recently, the phenomenon of acid rain has underscored the need for international cooperation in dealing with various environmental concerns. For example, an estimated 56 percent of the acid rain falling in Sweden originates outside of that country. The main sources of the pollutants are Germany and the United Kingdom because the west coast of Sweden receives winds from these countries. When these industrialized countries release more sulfur dioxide into the atmosphere, the result is more acid rain in Sweden. Since the 1930s, the lakes in western Sweden have become more acidic by a value of two pH

units. Ten thousand lakes have a pH below 6.0, and five thousand lakes have a pH below 5.0. A pH below 5.5 is too acidic for many species of fish.

In the 1930s, the United Kingdom initiated a program to reduce air pollution. This program has been successful in that the air in the United Kingdom has become cleaner. But one of their solutions was to build taller smokestacks. As a result, more of the pollutants from the United Kingdom are transported to Sweden.

Should the United Kingdom be permitted to disperse air pollutants in a manner that damages the Swedish environment?

Should there be a series of international agreements to control and regulate the movement of airborne pollutants across international boundaries?

In addition to air pollutants, are there any other forms of pollutants that can naturally be transported across international boundaries?

Summary

The atmosphere has a tremendous ability to accept and disperse pollutants. Carbon monoxide, hydrocarbons, particulates, sulfur dioxide, and nitrogen compounds are the primary air pollutants. They can cause a variety of health problems.

Photochemical smog is a secondary pollutant formed when hydrocarbons and oxides of nitrogen are trapped by thermal inversions and react with each other in the presence of sunlight to form peroxyacetylnitrates and ozone. Elimination of photochemical smog requires changes in technology, such as scrubbers, precipitators, and filter removal of sulfur from fuels.

Acid rain is caused by emissions of sulfur dioxide and oxides of nitrogen in the upper atmosphere, which form acids that are washed from the air when it rains or snows. Direct effects of acid rain on terrestrial ecosystems are difficult to prove, but changes in many forested areas are suspected of being partly the result of additional stresses caused by acid rain. The effect of acid rain on aquatic ecosystems is easy to quantify. As waters become more acidic, the complexity of the ecosystem decreases, and many species fail to reproduce.

Currently, many are concerned about the damaging effects of greenhouse gases: carbon dioxide, methane, and chlorofluorocarbons. Chlorofluorocarbons are also thought to lead to the destruction of ozone in the upper atmosphere. Many commonly used materials release gases into closed spaces where they cause potential health risks.

1. List the five primary air pollutants commonly released into the atmosphere and their sources.
2. Define *secondary air pollutant,* and give an example.
3. List three health effects of air pollution.
4. Why is air pollution such a large problem in urban areas?
5. What is photochemical smog? What causes it?
6. Describe three actions that can be taken to control air pollution.
7. What causes acid rain? List three possible detrimental consequences of acid rain.
8. Why is carbon dioxide (a nontoxic normal component of the atmosphere) called a "greenhouse gas"?
9. What would the consequences be if the ozone layer surrounding the earth was destroyed?
10. How does energy conservation influence air quality?

CHAPTER EIGHTEEN
Solid Waste Management and Disposal

Objectives

After reading this chapter, you should be able to:

Explain why solid waste has become a problem throughout the world.

Understand that the management of municipal solid waste is directly impacted by economics, changes in technology, and citizen awareness and involvement.

Recognize that the management of municipal solid waste in the future will require an integrated approach.

Describe the various methods of waste disposal and the problems associated with each method.

Understand the difficulties in developing new sanitary landfills.

Define the problems associated with incineration as a method of waste disposal.

Describe some methods of source reduction.

Describe composting and how it fits into solid waste disposal.

List some benefits and drawbacks of recycling.

Chapter Outline

Key Terms

incineration

mass burn

municipal solid waste

recycling

sanitary landfill

source reduction

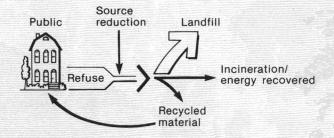

Introduction

In March 1987, a barge laden with 3,200 tons of garbage set out in search of a dump. The refuse had been turned away from a landfill in Islip, New York. The barge traveled 10,000 kilometers and stopped at several foreign ports, but found no one willing to accept its noxious load. The three-month odyssey took the barge to Mexico, Belize, and the Bahamas before it returned still fully loaded to New York. (See figure 18.1.) The futile voyage made headlines, giving many North Americans their first inkling of an impending crisis.

Lack of space for dumping solid waste has become a problem for many large metropolitan areas throughout the world. Communities are concerned about the increasing costs of waste disposal, possible hazards to groundwater, and maintaining air quality. In many areas, there is not suitable land available for new landfills. Problems with solid waste have increased dramatically over the past several decades because of population increases and an attitude that convenience is a very important part of the North American life-style.

The Disposable Decades

In 1955, *Life* magazine pictured a happy family in an article entitled "Throwaway Living." A disposable life-style was marketed as the wave of the future and as a way to cut down on household chores. "Use it once and throw it away" became a very popular advertising slogan in the 1950s.

The solid-waste problems facing us today have their roots during the economic boom following World War II. Marketing experts set to work trying new tactics to get consumers to buy and toss or, as economists would say, to "stimulate consumption." In the mid-1950s, marketing consultant Victor Lebow wrote an emotional plea for "forced consumption" in the *New York Journal of Retailing:*

> Our enormously productive economy demands that we make consumption a way of life, that we convert the buying and use of goods into rituals, that we seek our spiritual satisfactions in consumption. . . . We need things consumed, burned up, worn out, replaced, and discarded at an ever-growing rate.

Consumers were quick to adapt to the life-style that Lebow envisioned. In fact, it was not long until a disposable, throwaway life-style was seen as a consumer's right. The idea was to sell "convenience" to the prosperous postwar consumers. What was initially a convenience was soon to become a "necessity"; at least, that was what the advertisements were telling people.

A good example of this way of thinking is the TV dinner, which was first marketed in 1953. The food-packaging industry was evolving into a major service, closely allied with marketing and advertising. From the first TV dinner, we progressed rapidly, so that by the late 1980s, microwave meals were available. One such meal consisted of 340 grams of edible material and six separate layers of packaging, five of them plastic.

After nearly four decades of throwaway living, the disposable life-style may soon be changing. Cities and counties face a shortage of space in old landfills. Solid waste has become a major problem and, in a growing list of communities around the world, it has reached crisis proportions. (See figure 18.2.)

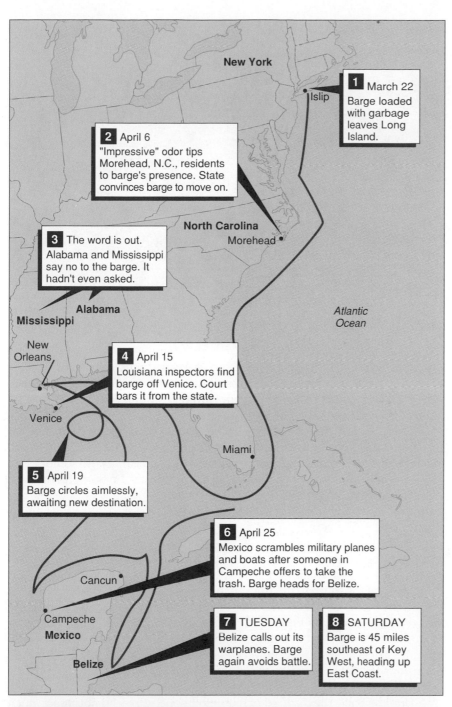

1 March 22
Barge loaded with garbage leaves Long Island.

2 April 6
"Impressive" odor tips Morehead, N.C., residents to barge's presence. State convinces barge to move on.

3 The word is out. Alabama and Mississippi say no to the barge. It hadn't even asked.

4 April 15
Louisiana inspectors find barge off Venice. Court bars it from the state.

5 April 19
Barge circles aimlessly, awaiting new destination.

6 April 25
Mexico scrambles military planes and boats after someone in Campeche offers to take the trash. Barge heads for Belize.

7 TUESDAY
Belize calls out its warplanes. Barge again avoids battle.

8 SATURDAY
Barge is 45 miles southeast of Key West, heading up East Coast.

New York

Islip

North Carolina
Morehead

Atlantic Ocean

Alabama
Mississippi
New Orleans
Venice

Miami

Cancun

Campeche
Mexico

Belize

FIGURE 18.1

Wandering Garbage. In 1987, a barge filled with garbage from New York traveled to four states and two countries looking for a place to unload its unwanted cargo. The barge was forced to return to New York to unload.

FIGURE 18.2
Disposable Life-Styles. The disposable life-style of the past four decades has created a shortage of space in landfills such as these. The development of new landfills is expensive, and in many areas, there is not available land appropriate for development. The photo of the rural dump was taken in a midwestern state in the summer of 1990.

The Nature of the Problem

As we approach the twenty-first century, the world must take steps to solve the ever-increasing burden of garbage—or as the professionals say, **municipal solid waste.** We are all part of the problem, but we can also be part of the solution.

The U.S. Environmental Protection Agency (EPA) estimates that the United States produces about 160 million tons of municipal solid waste each year. This equates to about 1.5 kilograms of trash per person per day. The volume of municipal solid waste has increased by 80 percent since 1960 and is expected to grow by an additional 20 percent by 2000. At this rate, the U.S. annual garbage heap could reach close to 200 million tons by the end of the century. (See figure 18.3.)

Nations with high standards of living and productivity tend to have more municipal solid waste per person than less developed countries. The United States and Canada, therefore, are world leaders in waste production. This high volume of solid waste and reliance on landfills for disposal have led to the problems we face today. For example, Toronto, Canada, is running out of places to put its municipal solid waste. With present methods, metropolitan Toronto's two landfill sites will be full by the mid-1990s. The outlying districts are not willing to be the site of a new landfill for Toronto. Solutions to such a problem will not be easy; however, solutions must be found.

Until recently, the disposal of municipal solid waste did not attract much public attention. From prehistory through the present day, the favored means of disposal was simply to dump solid wastes outside of the city or village limits. Frequently, these dumps were in a wetlands adjacent to a river or lake. To minimize the volume of the waste, the dump was often burned. This method of waste disposal was state of the art until only recently. Unfortunately, this method is still being used in remote or sparsely populated areas in North America and the world today. (See figure 18.4.)

As better waste-disposal technologies were developed and as values changed, more emphasis was placed on the environment and quality of life. Simply dumping and burning our wastes is no longer an acceptable practice from an environmental or health perspective. While the technology of waste disposal has evolved during the past several decades, our options are

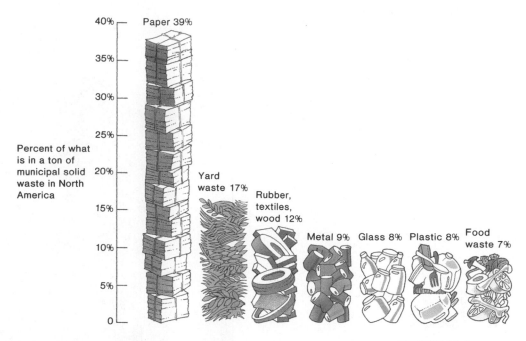

FIGURE 18.3

The Changing Nature of Trash. While food waste was the primary waste product in the past, changes in life-style and packaging have led to a change in the nature of trash. Most of what is currently disposed of could be recycled.

FIGURE 18.4

Burning Landfills. To minimize the volume of waste in landfills, the landfills were frequently burned. Waste is still being burned in sparsely populated areas in North America and other parts of the world today.

still limited. Realistically, there are no ways of dealing with waste that have not been known for many thousands of years. Essentially, four techniques are used: (1) dumping it, (2) burning it, (3) converting it into something that can be used again, and (4) minimizing the volume produced in the first place.

Methods of Waste Disposal

Our present waste-management systems are failing, and changes are needed. The U.S. Environmental Protection Agency suggests that each community adapt an integrated waste-management approach that combines the following four methods in a way best suited to local needs and capabilities: (1) landfills, (2) incineration, (3) source reduction, and (4) recycling.

What we do with our trash

Incinerated (3%)
Energy recovery (6%)

Recycled (11%)

Landfills (80%)

FIGURE 18.5

Landfill. The landfill is still the primary method of waste disposal. Historically, landfills have been the cheapest means of disposal. This may not be the case in the future.

Source: Data from Franklin Associates, Ltd., INFORM Inc.

Landfilling has historically been the primary method of waste disposal because it is the cheapest and most convenient method, and because the threat of groundwater contamination was not initially recognized. (See figure 18.5.) However, the landfill of today is far different from a simple hole in the ground into which garbage is dumped.

Landfilling

A modern **sanitary landfill** is typically a clay-lined depression in which each day's deposit of fresh garbage is covered with a layer of soil. Selection of modern landfill sites must be based on an understanding of groundwater geology, soil type, and sensitivity to local citizens' concerns. Once the site is selected, extensive construction activities are necessary to prepare the site for use. New landfills have complex bottom layers to trap contaminant-laden water leaking through the buried trash. In addition, monitoring systems are necessary to detect methane gas production and groundwater contamination. In some cases, methane produced by rotting garbage is collected and used to generate electricity. The water that leaches through the site must be collected and treated. As a result, new landfills are becoming increasingly more complex and expensive. They currently cost up to $1 million per hectare to prepare. (See figure 18.6.)

Today, almost 80 percent of municipal solid waste in North America goes into landfills, but this method is failing to handle the volume. For example, New York City's Fresh Kills Landfill on Staten Island is already standing 30 meters high and soon will be the sole recipient of the city's 27,000-ton daily trash load. Fresh Kills has been projected to reach a peak of 170 meters when it closes in a decade.

The problem New York City faces with the closing of its Fresh Kills Landfill is similar to what many communities will be facing in the near future. The EPA estimates that, by the year 2000, there will be only 3,250 landfills, compared to 18,500 landfills in 1979. (See figures 18.7 and 18.8.)

How a modern landfill works

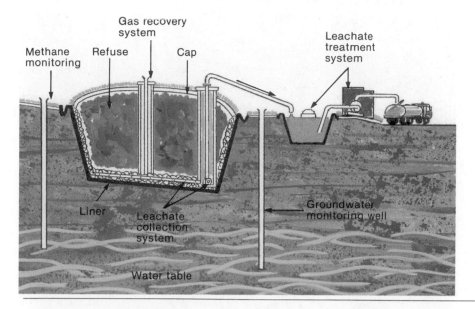

FIGURE 18.6

A Modern Well-Designed Landfill. A modern sanitary landfill is far different from a simple hole in the ground filled with trash. A modern landfill is more of a "cell," which is sealed when filled. Methane gas and groundwater are continuously monitored.

Source: National Solid Waste Management Association.

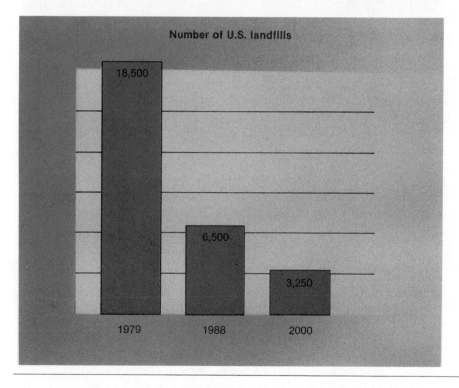

FIGURE 18.7

Reducing the Number of Landfills. The number of landfills in the United States is declining because they are filling up or because their design and operation do not meet environmental standards.

Source: Data from Environmental Protection Agency, 1990.

According to the EPA's 1984 report *The Solid Waste Dilemma: An Agenda for Action:*

> One-third of the nation's landfills will be full by 1993, which means that waste that is now disposed of in these facilities will have to be disposed of elsewhere. Many existing facilities are closing either because they are filled or because their design and operation do not meet federal or state standards for protection of human health and the environment.

A prolonged public debate over how to replace lost landfill capacity is developing where population density is high and available land is scarce.

FIGURE 18.8

Landfills Are Filling Up. The heavily populated northeastern United States will use up its existing landfill space faster than other parts of the country. But even in states like California that seemingly have plenty of room, individual cities are running out of places to bury their trash.

Source: Data from National Solid Wastes Management Association.

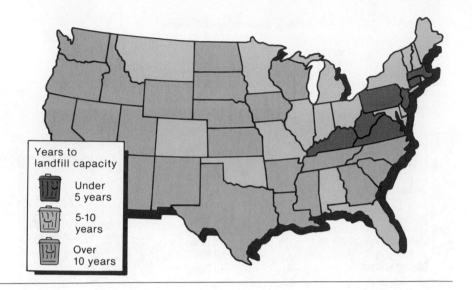

Years to landfill capacity

Under 5 years

5-10 years

Over 10 years

Siting new landfills in locations like Toronto, New York, and Los Angeles is extremely difficult because of local opposition, which is commonly referred to as the NIMBY, or "not-in-my-backyard," syndrome. Resistance by the public comes from concern over groundwater contamination, odors, and truck traffic. Public officials look for alternatives to landfills to avoid problems with the public over landfill sitings. Although siting a new landfill may be necessary, politicians are unwilling to take strong positions that might alienate their constituents. This is often called the NIMEY, or "not-in-my-election-year," syndrome.

Japan and many Western European countries have already moved away from landfilling as the primary method because of land scarcities and related environmental concerns. Sweden, Switzerland, and Japan dispose of 15 percent or less of their waste in landfills, compared to 80 percent in North America. Instead, recycling and incineration are the primary methods. In addition, the energy produced by incineration can be used for electric generation or heating.

Incineration

Incineration of refuse was quite common in the United States prior to 1940. However, many incinerators were eliminated because of aesthetic concerns, such as foul odors, noxious gases, and gritty smoke, rather than for reasons of public health. Today, about 13 percent of the municipal solid waste in the United States is incinerated. Most incinerators are not used just to burn trash. The heat derived from the burning is converted into steam and electricity. There are approximately 115 operating waste-to-energy incinerators and about 60 more contracted or in the advanced-planning stage. (See figure 18.9.)

Most incineration facilities burn unprocessed municipal solid waste, which is not as efficient as some other technologies. About one-fourth of the incinerators use refuse-derived fuel—collected refuse that has been processed into pellets prior to combustion.

The newest means of incineration, a European concept, is called **mass burn.** In the mass-burn technique, municipal solid waste is fed into a furnace, where it falls onto moving grates and is burned at temperatures up

FIGURE 18.9

Incineration. Incineration of municipal solid waste reduces its weight and volume significantly. However, there are concerns about air-quality problems and the toxicity and disposal of the ash.

to 1,130° C (2,400° F). The burning waste heats water, and the steam drives a turbine to generate electricity, which is sold to a utility.

Incinerators drastically reduce the amount of municipal solid waste—up to 90 percent by volume and 75 percent by weight. Primary risks of incineration, however, involve air-quality problems and the toxicity and disposal of the ash.

Though mass-burn technology works efficiently in Europe, the technology is not easily transferrable. North American municipal solid waste contains more plastic and toxic materials than European waste, thus creating air-pollution and ash-toxicity concerns. Modern incinerators have electrostatic precipitators, dual scrubbers, and fabric filters called baghouses; however, they still release small amounts of pollutants into the atmosphere, including certain metals, acid gases, and also classes of chemicals known as dioxins and furans, which have been implicated in birth defects and several kinds of cancer. The long-term risks from the emissions are still a subject of debate.

Ash from incineration is also a major obstacle to the construction of waste-to-energy facilities. Small concentrations of heavy metals are present both in the air emissions (fly ash) and residue (bottom ash) from these facilities. Because the ash contains lead, cadmium, mercury, and arsenic in varying concentrations from such items as batteries, lighting fixtures, and pigments, this ash may need to be treated as hazardous waste. The toxic substances are more concentrated in the ash than in the original garbage and can seep into groundwater from poorly sealed landfills. Many cities have had difficulty disposing of incinerator ash, and there is still considerable debate about what is the best method of disposal.

The cost and siting of new incinerators are also major concerns facing many communities. Incinerator construction is often a municipality's single largest bond issue. Incinerator construction costs in the United States in 1990 ranged from $30 million to $250 million, and the costs are not likely to decline.

Critics have argued that cities and towns have impeded waste reduction and recycling efforts by putting a priority on incinerators and committing resources to them. Proponents of incineration have been known to oppose source reduction. They argue that incinerators need large amounts

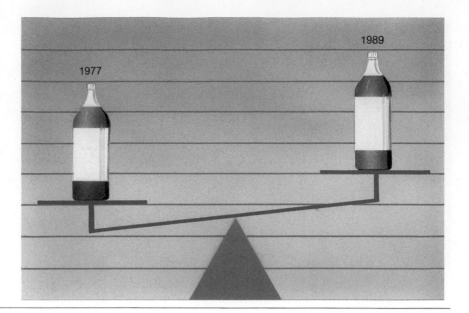

FIGURE 18.10
Source Reduction. Plastic soft-drink bottles are now 20 percent lighter than in 1977. The weight of aluminum cans has also been reduced by 35 percent since 1965.

of municipal solid waste to operate and that reducing the amount of waste generated makes incineration impractical. Many communities that have opposed incineration say that they support a vigorous waste-reduction and recycling effort.

Source Reduction

The most fundamental way to reduce waste is to prevent it from ever becoming waste in the first place. Examples of **source reduction** are using less material when making a product or converting from heavy packaging materials to lightweight ones. Some packagers have converted to lightweight aluminum and plastic by reducing the thickness of packages and thus the amount of packaging waste. Since the 2-liter soft-drink bottle was introduced in 1977, its weight has been reduced by more than 20 percent. Since 1965, aluminum cans have been reduced in weight by 35 percent. (See figure 18.10.)

Another way companies reduce waste is by making consumer products in concentrated form. These concentrated products can then be packaged in smaller containers. This approach, however, requires consumers to favor concentrates when making purchasing decisions.

Municipal composting is another source-reduction technique that could have substantial impact since 20 percent of material going into landfills in North America is yard waste. Building a compost pile is a popular practice with many home gardeners because it turns waste into a useful soil additive. (See figure 18.11.) Many communities require separate collection of such yard waste as grass clippings, leaves, and even discarded Christmas trees. Such waste is either composted or shredded into wood chips. The by-products can then be used for landscaping at city parks, schools, golf courses, and cemeteries. Preserving space in landfills by municipal composting is gaining in popularity because it is cost effective and substantially extends a landfill's useful life.

On an individual level, we can all attempt to reduce the amount of waste we generate. Every small personal commitment from each of us could have the cumulative result of a significant reduction in municipal solid waste.

FIGURE 18.11
Composting. Household composting of yard waste is a popular practice with many home gardeners. Yard waste accounts for up to 20 percent of waste going into landfills. Composting can have a significant impact on the life of a landfill.

Recycling

About 10 percent of the waste generated in the United States is handled through **recycling.** The U.S. goal is to have 25 percent of municipal solid waste recycled by the mid-1990s. The goal for Ontario, Canada, is to reduce the amount of garbage going to the landfill sites by 50 percent by the year 2000. Recycling, along with source reduction, is a major part of the Ontario plan.

Recycling initiatives have grown rapidly in North America during the past several years. In the United States, mandatory recycling laws for all materials have passed in ten states. Also, an estimated six hundred U.S. cities now have multimaterial recycling collection programs in place. In Canada, Toronto and Mississauga, Ontario, have comprehensive recycling programs.

Even with the growth of recycling programs, however, the United States recycles only a small percentage of the municipal solid waste generated. (See figure 18.12.) In contrast, Japan recycles about 45 percent of its municipal solid waste, including half the paper, about 55 percent of glass bottles, and 66 percent of food and beverage containers. In the best-run Japanese programs, residents separate waste into several categories, which are collected on different days.

Benefits of Recycling

Some benefits of recycling are readily recognizable, such as conservation of resources and pollution reduction. These perceived benefits provide the primary motivation for participation in recycling programs. The following examples illustrate the resource-conservation and pollution-reduction benefits of recycling:

One Sunday edition of the *New York Times* consumes sixty-two thousand trees. Currently, only about 20 percent of all paper used in the United States is recycled.

The United States imports 91 percent of its aluminum and throws away over $400 million worth of aluminum annually.

There are no sources of tin within the United States; however, 2.7 kilograms of ultrapure tin can be reclaimed from each 900 kilograms of metal food cans.

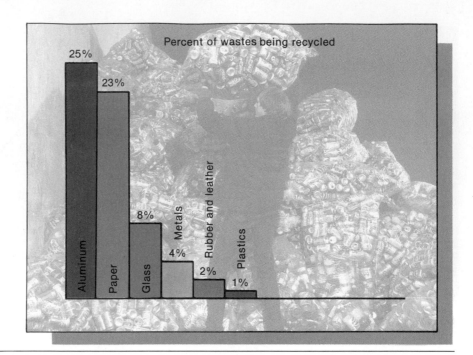

FIGURE 18.12

Recycling. Even with the growth of recycling programs during the past several years, the United States recycles only a small percentage of the municipal solid waste generated.

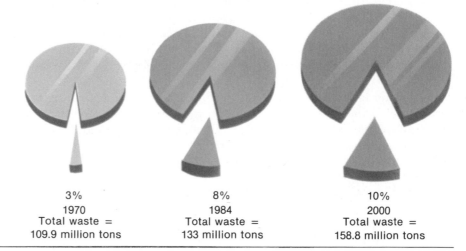

FIGURE 18.13

Increasing Amounts of Plastics in Trash. Plastics are a growing component of municipal solid waste in the United States. Increased recycling of plastics could reverse this trend.

3%	8%	10%
1970	1984	2000
Total waste =	Total waste =	Total waste =
109.9 million tons	133 million tons	158.8 million tons

Crushed glass (cullet) reduces the energy required to manufacture new glass by 50 percent. Cullet lowers the temperature requirements of the glass-making process, thus conserving energy and reducing air pollution.

Approximately 500 million barrels of oil were saved by recycling paper, metals, and rubber during 1990.

While recycling is a viable alternative to landfilling or the incineration of municipal solid waste, recycling does present several problems.

Recycling Concerns

Problems associated with recycling tend to be either technical or economic. Technical questions are of particular concern when recycling plastics. (See figure 18.13.) While the plastics used in packaging are recyclable, the technology to do so differs from plastic to plastic. There are many different types of plastic polymers. Since each type has its own chemical makeup, different plastics cannot be recycled together. In other words, a

BOX 18.1
Resins Used in Consumer Packaging

Thermoplastics, which account for 87 percent of plastics sold, are the most recyclable form of plastics because they can be remelted and reprocessed, usually with only minor changes in their properties. Thermoplastic resins are commonly used in consumer packaging applications:

1. Polyethylene terephthalate (PET) is used extensively in rigid containers, particularly beverage bottles for carbonated beverages.
2. Polyethylene is the most widely used resin. High-density polyethylene (HDPE) is used for rigid containers, such as milk and water jugs, household-product containers, and motor oil bottles.
3. Polyvinyl chloride (PVC) is a tough plastic often used in construction and plumbing. It is also used in some food, shampoo, oil, and household-product containers.
4. Low-density polyethylene (LDPE) is often used in films and bags.
5. Polypropylene (PP) is used in a variety of areas, from snack-food packaging to battery cases to disposable diaper linings. It is frequently interchanged for polyethylene or polystyrene.
6. Polystyrene (PS) is best known as a foam in the form of cups, trays, and food containers. In its rigid form, it is used in cutlery.
7. Other. These usually contain layers of different kinds of resins and are most commonly used for squeezable bottles (for example, ketchup).

Currently, HDPE and PET are the two most commonly recycled resins. Efforts to recycle PS are also being explored, especially within the fast-food industry.

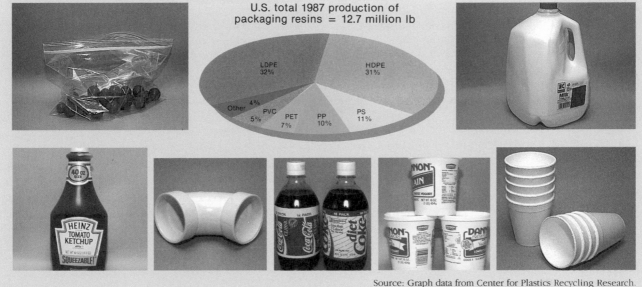

U.S. total 1987 production of packaging resins = 12.7 million lb

Source: Graph data from Center for Plastics Recycling Research.

milk container is likely to be high-density polyethylene (HDPE), while an egg container is polystyrene (PS) and a soft-drink bottle is polyethylene terephthalate (PET). (See box 18.1.)

Plastic recycling is still a relatively new field. Industry is researching new technologies that promise to increase the quantity of plastics recycled and that will allow mixing of different plastics. Until such technology is developed, separation of different plastic products before recycling will be necessary.

The economics of recycling are also a primary area of concern. The stepped-up commitment to recycling in many developed nations has produced a glut of certain materials on the market. Markets for collected ma-

BOX 18.2
Container Laws

In October 1972, Oregon became the first state to enact a "Bottle Bill." This statute required a deposit of two to five cents on all beverage containers that could be reused. It banned the sale of one-time-use beverage bottles and cans. The purpose of the legislation was to reduce the amount of litter. Beverage containers were estimated to make up about 62 percent of the state's litter. The bill succeeded in this respect: Within two years after it went into effect, beverage-container litter decreased by about 49 percent.

Those opposed to the bottle bill cited a loss of jobs that resulted from the enactment of the bill. In two small can-manufacturing plants, 142 persons did lose their jobs, but hundreds of new employees were needed to handle the returnable bottles. Other arguments against bottle bills focused on the "major" inconvenience to consumers who would need to return the bottles and cans to the store and on retailers' storage problems. Neither of these problems has proven to be serious. Nine other states—Vermont, Maine, Connecticut, New York, Iowa, Rhode Island, Michigan, Delaware, and California—have enacted legislation requiring deposits on returnable bottles, specifically beverage containers. Many states like New Jersey, Rhode Island, and California are turning more toward mandating recycling laws. In 1991, Maine expanded its bottle law to all nondairy beverage containers,

including all noncarbonated juice containers holding a gallon or less. Deposits are five cents, except on liquor and wine bottles, which carry a fifteen-cent deposit. Excluded are containers for milk and other dairy products, cough syrup, baby formula, soap, and vinegar.

At the present time, the amount of energy consumed in the United States is a major concern of many. Can the United States afford to use the equivalent of 125,000 barrels of oil a day to allow for the convenience of throwaway containers? Many argue that a national bottle bill is long overdue. A national bottle bill would reduce litter, save energy and money, and create jobs. It would also help to conserve natural resources. A national container bill could save approximately 7 million metric tons of glass, 2 million metric tons of steel, and 0.5 million metric tons of aluminum each year.

In 1989, more than fifty legislative bills were introduced in the U.S. Congress addressing ways to reduce, revise, and recycle the nation's growing volume of solid waste. One of the bills would have required a minimum five-cent refundable deposit on most beverage containers nationwide. While the bill received positive reviews, it did not survive the session of Congress. Proponents of the bill indicated that it would be reintroduced.

terials fill up just like landfills. Unless the demand for recycled products keeps pace with the growing supply, recycling programs will face an uncertain future. New uses and products need to be developed for recycled materials. Economic incentives to encourage the use of recycled materials would also be helpful. On an individual level, we can have an impact by purchasing products made from recycled materials. The demand for recycled products must grow if recycling is to succeed on a large scale.

Summary

Beginning with the post-World War II era, increased consumption of consumer goods became a way of life. Products were designed to be used once and then thrown away. By the 1980s, a disposable life-style began to cause problems. There simply were no places to dispose of waste. Barges filled with municipal solid waste from the metropolitan areas along the eastern United States were traveling the world trying to dispose of their unwanted cargo.

Municipal solid waste is managed by landfilling, incineration, waste reduction, and recycling. Landfilling is the primary means of disposal; however, a contemporary landfill is significantly more complex and expensive than the simple holes in the ground of the past. The availability of suitable landfill land is also a problem in large metropolitan areas.

Pollution and Policy

CONSIDER THIS CASE STUDY
Corporate Response to Environmental Concerns

In 1990, McDonald's Corporation announced that it would switch from polystyrene to paper for packaging its food products. In making this announcement, McDonald's stated that it was responding to consumer pressure to become more environmentally conscientious. Is using paper to wrap fast food better for the environment than using polystyrene? The answer is not a simple yes or no.

Advocates for the switch say that polystyrene takes up space in landfills and does not decompose. They also argue that the burning of polystyrene foam in incinerators might release harmful air pollutants.

Opponents of the switch argue that polystyrene can be recycled into useful products, such as insulation board or playground equipment. They further argue that using paper means cutting forests and that, since the paper used to wrap the food is coated with wax, it cannot be recycled.

Many grocery chains now offer several choices of carryout bags. Some encourage reuse of bags by deducting a small amount from the customer's bill. Others provide a choice of paper or plastic.

What are the pros and cons of paper and plastic? Which alternative do you prefer? Why? Is there an alternative to both?

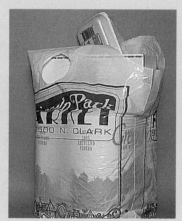

Which is better?

About 13 percent of the municipal solid waste in the United States is incinerated. The number of incinerators in the United States has declined over the past several decades. While incineration does reduce the volume of municipal solid waste, the problems of ash disposal and air quality continue to be major concerns.

The most fundamental way to reduce waste is to prevent it from ever becoming waste in the first place. Using less material in packaging, producing consumer products in concentrate form, and composting yard waste are all examples of source reduction. On an individual level, we can all attempt to reduce the amount of waste we generate.

About 10 percent of the waste generated in the United States is handled through recycling. In contrast, Japan recycles about 45 percent of its municipal solid waste. Recycling initiatives have grown rapidly in North America during the past several years. As a result, the markets for some recycled materials have become glutted. Recycling of municipal solid waste will only be successful if markets exist for the recycled materials. Another problem in recycling is the current inability to mix various plastics. The plastic industry is working on the development of a more universal plastic.

Future management of municipal solid waste will be an integrated approach involving landfilling, incineration, source reduction, and recycling. The degree to which any option will be used will depend on economics, changes in technology, and citizen awareness and involvement.

Review Questions

1. How is life-style related to our growing municipal solid waste problem?
2. What four methods are incorporated under integrated waste management?
3. Describe some of the problems associated with sanitary landfills.
4. What are four concerns associated with incineration?
5. Describe examples of source reduction.
6. Describe the importance of recycling household solid wastes.
7. Name several strategies that would help to encourage the growth of recycling.

Pollution and Policy

CHAPTER NINETEEN
Hazardous and Toxic Wastes

Objectives

After reading this chapter, you should be able to:

Explain what is meant by hazardous and toxic waste.

Explain the complexity in regulating hazardous wastes.

Describe the four characteristics by which hazardous wastes are identified.

Describe the environmental problems of hazardous and toxic wastes.

Understand the difference between a persistent and nonpersistent pollutant.

Describe the potential health risks associated with hazardous wastes.

Explain the problems associated with hazardous-waste dump sites and how such sites developed.

Describe how hazardous wastes are managed, and list five alternative technologies.

Describe the importance of source reduction with regard to hazardous wastes.

Chapter Outline

Key Terms

acute toxicity
air stripping
carbon absorption
chronic toxicity
Comprehensive
  Environmental
  Response,
  Compensation and
  Liability Act
  (CERCLA)
corrosiveness
fixation
hazardous waste
ignitability
LD_{50}
National Priority List
neutralization
nonpersistent pollutant

persistent pollutant
precipitation
reactivity
Resource Conservation
  and Recovery Act
solidification
source reduction
steam stripping
Superfund
synergism
thermal treatment
threshold levels
toxicity
toxic waste
waste destruction
waste immobilization
waste separation

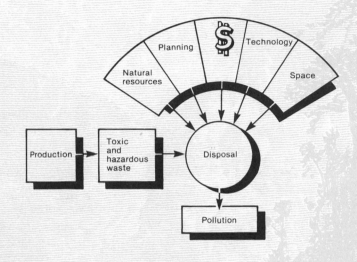

Hazardous and Toxic Wastes

At hazardous-waste sites around the world, toxic chemicals are contaminating the land, air, and water. The potential health hazards of these sites range from such minor discomforts as headaches and nausea to possibly life-threatening ailments as cancers that may not manifest themselves for years, in addition to potential birth defects. Today, names like Love Canal, New York, and Times Beach, Missouri, in the United States; Lekkerkerk in the Netherlands; Vac, Hungary; and Minamata Bay, Japan, are synonymous with the problems associated with the release of hazardous and toxic wastes into the environment. In each of these communities, residents were forced to leave their homes because the soil, surface water, and/or groundwater were contaminated with chemical wastes that had been improperly disposed of over the last few decades.

A question of benefits and risks marks the course of our chemical age. Events continue to reveal that "better living through chemistry" comes with serious costs, some of which have only recently come to light. Pesticides thought to degrade in soils turn up in rural drinking-water wells. Underground plumes of toxic chemicals emanating from abandoned waste sites contaminate city water supplies. A gas leak at a chemical production plant in Bhopal, India, killed more than two thousand people. Pesticides spilled into the Rhine River from a warehouse near Basel, Switzerland, destroyed a half million fish, disrupted water supplies, and caused considerable ecological damage. The collapse of an oil storage tank in Pennsylvania spilled over 2 million liters of oil into the Monongahela River and threatened the water supply for millions of residents. Toxic and hazardous products and by-products are becoming a major issue of our time.

Increasingly, governments and international agencies are attempting to control the growing problem of hazardous wastes. The problems stemming from such wastes, however, are complicated. Regulation of hazardous wastes is difficult in that they affect the air, water, and soil, and some can be hazardous in extremely low concentrations. In addition, definitions of hazardous and toxic wastes differ among countries, as do the concentrations of those wastes considered hazardous to human health.

While much progress has been made over the past decade to control the manufacture, distribution, and disposal of hazardous substances, we may have seen only the tip of the iceberg. Annual global amounts of hazardous wastes have been estimated, but there is no general acceptance of any of these estimates. The United States, however, leads the world in the generation of hazardous wastes, with an annual total of about 264 million metric tons. Many believe that the real problem lies in the countless amount of hazardous wastes that have been improperly or illegally disposed of in the past, with each burial dump a potential time bomb.

What Is Hazardous Waste?

The definition of hazardous waste varies from one country to another. One of the most widely used definitions, however, is contained in the U.S. **Resource Conservation and Recovery Act** of 1976 (RCRA). The RCRA considers wastes toxic and/or hazardous if they:

> cause or significantly contribute to an increase in mortality or an increase in serious irreversible, or incapacitating reversible, illness; or pose a substantial present or potential hazard to human health or the environment when improperly treated, stored, transported, disposed of, or otherwise managed.

This definition gives one an appreciation for the complexity of hazardous-waste regulation.

Working from the RCRA definition, the U.S. Environmental Protection Agency (EPA) compiled and proposed a list of hazardous wastes. Listing is the most common method for defining hazardous waste in European countries and in some state laws. The EPA has also proposed that a hazardous waste be identified by testing it to determine if it possesses any one of four characteristics. If it does, it is subject to regulation under RCRA. Three of the characteristics selected by the EPA produce acute effects likely to cause almost immediate damage; the fourth creates chronic effects most likely to appear over a longer time period. The four characteristics are:

1. **Ignitability,** which identifies wastes that pose a fire hazard during routine management. Fires not only present immediate dangers of heat and smoke but also can spread harmful particles over wide areas.
2. **Corrosiveness,** which identifies wastes requiring special containers because of their ability to corrode standard materials, or requiring segregation from other wastes because of their ability to dissolve toxic contaminants.
3. **Reactivity** (or explosiveness), which identifies wastes that, during routine management, tend to react spontaneously, to react vigorously with air or water, to be unstable to shock or heat, to generate toxic gases, or to explode.
4. **Toxicity,** which identifies wastes that, when improperly managed, may release toxicants in sufficient quantities to pose a substantial hazard to human health or the environment.

While the terms *toxic* and *hazardous* are often used interchangeably to refer to wastes, there is a recognized, though fine-line difference. **Toxic** commonly refers to a narrow group of poisonous substances that cause death or serious injury to humans and animals by interfering with normal body physiology. **Hazardous,** the broader term, refers to all dangerous wastes, including toxic ones, that present an immediate or long-term human health risk or environmental risk.

In the industrialized countries of Europe and North America, chemical and petrochemical industries are responsible for nearly 70 percent of all hazardous wastes; in developing countries, the figure is 50–66 percent. Most toxic and hazardous wastes come from chemical and related industries that produce plastics, soaps, synthetic rubber, fertilizers, medicines, paints, pesticides, herbicides, and cosmetics. (See figure 19.1.)

In attempting to regulate toxic and hazardous wastes, most countries simply draw up a list of specific substances for which there is sufficient scientific evidence linking them to adverse human health or environmental effects. However, since many potentially harmful chemical compounds have yet to be tested adequately, most lists only include the worst offenders. Countries contemplating regulation of chemical wastes must consider for each chemical not only how toxic it is but also how flammable, corrosive, and explosive it is, and whether it will produce mutations or cause cancer. Governmental agencies must look at not only the effects of one massive dose of a substance **(acute toxicity)**, but also at the effects of exposure to small doses over longer periods **(chronic toxicity)**. The latter may be lethal even though a single massive dose is not.

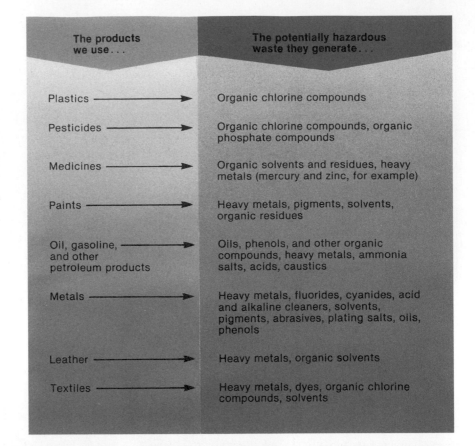

The products we use...	The potentially hazardous waste they generate...
Plastics	Organic chlorine compounds
Pesticides	Organic chlorine compounds, organic phosphate compounds
Medicines	Organic solvents and residues, heavy metals (mercury and zinc, for example)
Paints	Heavy metals, pigments, solvents, organic residues
Oil, gasoline, and other petroleum products	Oils, phenols, and other organic compounds, heavy metals, ammonia salts, acids, caustics
Metals	Heavy metals, fluorides, cyanides, acid and alkaline cleaners, solvents, pigments, abrasives, plating salts, oils, phenols
Leather	Heavy metals, organic solvents
Textiles	Heavy metals, dyes, organic chlorine compounds, solvents

FIGURE 19.1

Common Materials Can Produce Hazardous Wastes. Many commonly used materials can release toxic wastes if not properly disposed of.

Defining the quantities or concentrations at which wastes become hazardous is another problem facing governmental agencies. Nearly all substances are toxic in sufficiently high doses. The question is, "When does a chemical cross over from safe to toxic?" Even when concentrations are set, as in Belgium or the Netherlands, they may vary considerably. In the Netherlands, for example, 50 milligrams of cyanide per kilogram is considered hazardous; in neighboring Belgium, the toxicity standard is fixed at 250 milligrams per kilogram.

A further problem is the creation of new substances as the result of spontaneous reactions among chemicals placed together in dump sites. Most toxicological tests focus on a single compound, even though in waste dumps, the compounds are usually found in mixtures. Although the wastes may be relatively harmless as separate compounds, once mixed, they may

BOX 19.1
Exposure to Toxics

We are all exposed to materials that are potentially harmful. The question is, at what levels is such exposure harmful or toxic? One measure of toxicity is LD_{50}, the dosage of a substance that will kill (lethal dose) 50 percent of a test population. Toxicity is measured in units of poisonous substance per kilogram of body weight. For example, the deadly chemical that causes botulism, a form of food poisoning, has an LD_{50} in adult human males of 0.0014 milligrams per kilogram. This means that, if a dosage of only 0.14 milligrams—about the equivalent of a few grains of table salt—is consumed by one hundred adult human males, each weighing 100 kilograms, approximately fifty of them will die.

Lethal doses are not the only danger from toxic substances. During the past decade, concern has been growing over minimum harmful dosages, or threshold dosages, of poisons, as well as their sublethal effects.

The length of exposure further complicates the determination of toxicity values. Acute exposure refers to a single exposure lasting from a few seconds to a few days. Chronic exposure refers to continuous or repeated exposure for several days, months, or even years. Acute exposure usually is the result of a sudden accident, such as the tragedy at Bhopal, India,

mentioned earlier in the chapter. Acute exposures often make disaster headlines in the press, but chronic exposure to sublethal quantities of toxic materials presents a much greater hazard to public health. For example, millions of urban residents are continually exposed to low levels of a wide variety of pollutants. Many deaths attributed to heart failure or such diseases as emphysema may actually be brought on by a lifetime exposure to sublethal amounts of pollutants in the air.

become highly toxic. Often, the toxic end products escape adequate regulation because they do not fit into the definitional framework of hazardous wastes.

Environmental Problems of Hazardous Wastes

Hazardous wastes contaminate the environment in several different ways. Even before they are disposed of, hazardous wastes may pollute the air, water, or soil simply by how they have been stored or contained. Uncontrolled or improper incineration of hazardous wastes, whether on land or at sea, can contaminate the atmosphere and the surrounding environment. The discharge of hazardous substances into the sea or into lakes and rivers often kills fish and other aquatic life. Further, disposal on land in dumps that are later abandoned, or in improperly controlled landfills, can pollute both the soil and the groundwater as materials leach below the site.

Because most hazardous wastes are disposed of on or in land, the most serious environmental effect is contaminated groundwater. In the United States alone, an estimated 75,000 active industrial landfill sites may be possible sources of groundwater contamination, along with 200 special facilities for disposal of both liquid and solid hazardous wastes, and some 180,000 surface impoundments (ponds) for all types of waste. (See chapter 15.) Nearly 2 percent of North America's underground aquifers could be contaminated with such chemicals as chlorinated solvents, pesticides, trace metals, and PCBs. Once groundwater is polluted with hazardous wastes,

BOX 19.2
Toxic Chemical Releases

In 1987, industry had to report toxic chemicals released into the environment, as per a new EPA requirement. Any industrial plant that released 23,000 kilograms or more of toxic pollutants was required to file a report. Industrial plants that released under 23,000 kilograms were not required to file; thus, the data are incomplete. In 1988, 19,762 reports were filed, covering 332 toxic chemicals.

About 2 billion kilograms of toxic chemicals were reported released into the environment by industry in 1988, compared with nearly 2.3 billion in 1987. Another 800 million kilograms—not counted in the overall figure—were sent to municipal-waste treatment centers or private treatment and storage facilities.

According to the 1988 data, among the chemicals routinely emitted from industrial sources were seventy-seven carcinogens. The most widely released cancer-causing chemical—52 million kilograms—was dichloromethane, a chemical often used as an industrial solvent and paint stripper. Industry released a total of 128 million kilograms of carcinogens, including such chemicals as arsenic, benzene, and vinyl chloride. Chemical and allied industries accounted for about 60 percent of the total releases.

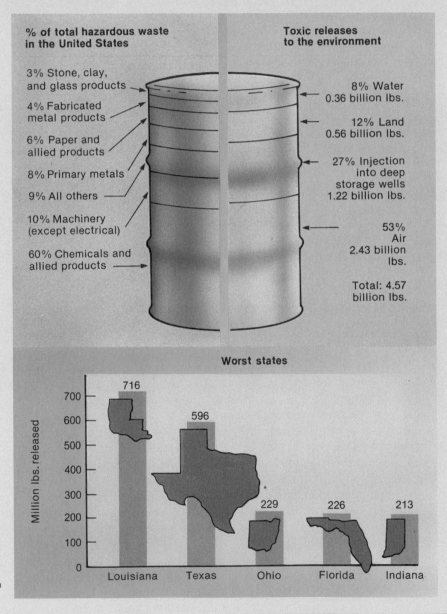

% of total hazardous waste in the United States

3% Stone, clay, and glass products
4% Fabricated metal products
6% Paper and allied products
8% Primary metals
9% All others
10% Machinery (except electrical)
60% Chemicals and allied products

Toxic releases to the environment

8% Water 0.36 billion lbs.
12% Land 0.56 billion lbs.
27% Injection into deep storage wells 1.22 billion lbs.
53% Air 2.43 billion lbs.

Total: 4.57 billion lbs.

Worst states

Louisiana 716
Texas 596
Ohio 229
Florida 226
Indiana 213

(Million lbs. released)

Source: Data from Environmental Protection Agency.

the cost of reversing the damage is prohibitive. In fact, if an aquifer is contaminated with organic chemicals, restoring the water to its original state is seldom physically or economically feasible.

Pollution Concerns

When dealing with the issue of pollution and hazardous wastes, a question that frequently arises is, how much is too much? Unfortunately, this is an extremely difficult question to answer, given the many unknown effects of pollution on living organisms. The effects of certain pollutants do not have to be immediate to be harmful. For example, lead has been used in paints for many years, but it was only recently discovered that it has harmful effects, particularly when paint chips are eaten by children. The **threshold levels** of pollutants must also be taken into consideration. Small quantities of a pollutant, such as fecal bacteria (one or two organisms per 100 milliliters), in drinking water may present no adverse health effects, but increased levels may result in illness in sensitive individuals. Threshold levels vary significantly among species, as well as among members of the same species. In addition, certain combinations of pollutants cause more serious problems than do individual pollutants. This is referred to as **synergism.** For example, uranium miners who smoke tobacco have unusually high incidences of lung cancer. Apparently, the radioactive gases found in uranium mines interact synergistically with the carcinogens found in tobacco smoke.

The methods used to deal with forms of pollution depend upon the pollutant's character and persistence. **Persistent pollutants** are those that remain in the environment for many years in an unchanged condition. Most of the persistent pollutants are human-made materials. An estimated thirty thousand synthetic chemicals are used in the United States. They are mixed in an endless variety of combinations to produce all types of products used in every aspect of daily life. They are part of our food, transportation, clothing, building materials, home appliances, medicine, recreational equipment, and many other items. Our way of life is heavily dependent upon synthetic materials.

An example of a persistent pollutant is DDT. It was used as an effective pesticide worldwide and is still used in some countries because it is so inexpensive. However, once released into the environment, it accumulates in the food chain and causes death when its concentration is high enough. (See chapter 14 for a discussion of DDT as a pesticide.)

Another widely used group of synthetic compounds of environmental concern are polychlorinated biphenyls (PCBs). PCBs are highly stable compounds that resist changes from heat, acids, bases, and oxidation. These characteristics make PCBs desirable for industrial use but also make them persistent pollutants when released into the environment. About half of the PCBs are used in transformers and electrical capacitors. Other uses include inks, plastics, tapes, paints, glues, waxes, and polishes. PCBs are harmful to fish and other aquatic forms of life because they interfere with reproduction. In humans, PCBs produce liver ailments and skin lesions. In high concentration, they can damage the nervous system, and they are suspected carcinogens. In 1970, PCB production was limited to those cases where satisfactory substitutes were not available.

In addition to synthetic compounds, heavy metals are used for many purposes. Mercury, beryllium, arsenic, lead, and cadmium are examples of heavy metals that are toxic. When released into the environment, they enter the food chain and become concentrated. In humans, these metals can produce kidney and liver disorders, weaken the bone structure, damage the

BOX 19.3
Lead and Mercury Poisoning

Lead and mercury are naturally present in the environment and probably have been a source of pollution for centuries. For example, the lead drinking and eating vessels used by the wealthy Romans may have caused many of their deaths. In 1865, when Alice in Wonderland was published, one of the characters was the Mad Hatter. At that period in history, mercury was widely used in the treatment of beaver skins for use in the making of hats. As a result of exposure to mercury, hat makers often suffered from a variety of mental problems; hence, the term "mad as a hatter."

In 1953, a number of physical and mental disorders in the Minamata Bay region of Japan were diagnosed as being caused by mercury; fifty-two people developed symptoms of mercury poisoning, seventeen died, and twenty-three became permanently disabled. In 1970, an outbreak of mercury poisoning in the United States was traced to mercury in the meat of swordfish and tuna. In both incidents, the toxic material was not metallic mercury but a mercurous compound, methylmercury. Metallic mercury is converted to methylmercury by bacteria in the water. Methylmercury enters the food chain and may become concentrated as the result of biological amplification. Sufficient amounts in humans can cause brain damage, kidney damage, or birth defects. Today, regulations reduce the release of mercury into the environment and set allowable levels in foods. The problem still persists, however, because it is impossible to eliminate the large amounts of mercury already present in the environment, and it is difficult to prevent the release of mercury in all cases. For example, burning coal releases 3,200 metric tons of mercury into the earth's atmosphere each year, and mercury is still "lost" when it is used for various industrial purposes.

Like mercury, lead is a heavy metal and has been a pollutant for centuries. Studies of the Greenland Ice Cap indicate a 1,500 percent increase in the lead content today as compared to 800 B.C. These studies reveal that the first large increase occurred during the Industrial Revolution and the second great increase occurred after the invention of the automobile. Oil companies added lead to gasoline to improve performance, and burning gasoline is a major source of lead pollution. Another source of lead pollution is older paints. Prior to 1940, indoor and outdoor paints often contained lead. In 1940, the use of lead in indoor

There are concerns about fish contamination and public health as a result of methylmercury.

paints was stopped, and in 1958, the use of lead in outdoor paints was reduced. The Lead Poisoning Prevention Act was passed in 1971.

There is still disagreement over what levels of lead and mercury can cause human health problems. Although ingested lead from paint can cause death or disability, such a strong correlation cannot be made for atmospheric lead.

Of the seventy thousand chemicals in use in North America, as many as thirty-five thousand are classified by the EPA as either definitely or potentially harmful to human health. A number of them, including some heavy metals (cadmium, arsenic) and certain organic compounds (carbon tetrachloride, toluene), are carcinogenic. Others, like mercury, are mutagenic, and many tend to induce brain and bone damage (mercury, copper, lead), kidney diseases (cadmium), neurological damage, and many other health problems. The risk resulting from exposure to more than one of these substances at the same time is not known. This uncertainty is one reason for the growing concern and public attention being directed at the problem of hazardous waste dumps.

As with most other forms of pollution, there is no agreement on how much is harmful. So it seems that we will continue to tolerate low levels of these materials in our environment and continue to disagree about the danger of these pollutants to humans.

central nervous system, cause blindness, and lead to death. Because these materials are persistent, they can accumulate in the environment even though only small amounts might be released each year. When industries use these materials in a concentrated form, it presents a hazard not found naturally. (See box 19.3.)

A **nonpersistent pollutant** does not remain in the environment for very long. Most nonpersistent pollutants are biodegradable. Others decompose as a result of inorganic chemical reactions. A biodegradable material is chemically changed by living organisms and often serves as a source of food and energy for decomposer organisms, such as bacteria and fungi. Waste from food-processing plants, garbage, human sewage, animal wastes, and other remains of organisms are examples of nonpersistent pollutants that are biodegradable.

In addition to the naturally occurring nonpersistent pollutants are synthetic nonpersistent pollutants. Many of these break down chemically by oxidation or hydrolysis. These include the "soft biocides." For example, organophosphates are a type of pesticide that usually decomposes within several weeks. As a result, organophosphates do not accumulate in food chains because they are pollutants for only a short period of time.

Health Risks Associated with Hazardous Wastes

Because most hazardous wastes are chemical wastes, controlling chemicals and their waste products is a major issue in most developed countries. Every year, roughly one thousand new chemicals join the nearly seventy thousand in daily use. Many of these hazardous chemicals are toxic, but they are little threat to human health unless they are improperly used or disposed of. Unfortunately, at the center of the hazardous-waste problem is the fact that the by-products of industry are often handled and disposed of improperly.

Establishing the medical consequences of exposure to toxic wastes is extremely complicated. The problem of linking a particular chemical or other hazardous waste to specific injuries or diseases is further compounded by the lack of toxicity data on most hazardous substances. (See table 19.1.)

Although assessing environmental contamination from toxic wastes and determining health effects is extremely difficult, what little is known is cause for concern. Most hazardous-waste dumps, for example, contain dangerous and toxic chemicals along with heavy metal residues and other hazardous substances. In 1984, the Washington, D.C.-based Conservation Foundation reported that a total of 444 toxic pollutants had been identified at over one thousand sites under investigation by the EPA. The most common substances found at these sites were lead, trichloroethylene, polychlorinated biphenyls, chloroform, arsenic, cadmium, and chromium. According to the Conservation Foundation:

> Some of these substances can cause immediate poisoning (e.g., arsenic and mercury). Virtually all of them can damage health as the result of low-level, long-term exposure, particularly through drinking water. Seven of the ten are thought to cause cancer; seven, birth defects; and five, genetic damage.

TABLE 19.1
Chronic effects of selected hazardous wastes.

Type of waste	Carcinogenic effects	Mutagenic effects	Teratogenic effects	Reproductive system damage
Pesticide wastes				
Halogenated organic pesticides	A	A	A	H
Methyl bromide				
Halogenated organic phenoxy herbicides	A	A	A	A
2,4-D				
Organophosphorous pesticides	A	A	A	
Organonitrogen herbicides (Paraquat® and Diquat®)	A	A	A	
Carbamate insecticides				
Dimethyldithiocarbonate fungicide compounds				
Aluminum phosphide				
Rotenone				
Polychlorinated biphenyls	A		A	
Cyanide wastes				
Toxic metals				
Zinc, copper, selenium, chromium, nickel	H			
Arsenic				
Organic lead compounds				
Mercury		H	H	
Cadmium	H			
Halogenated organics	H	H		
Nonhalogenated volatile organics	A	A		

Source: Adapted from Benjamin Goldman, et al., *Hazardous Waste Management: Reducing the Risk*, 1987. Council on Economic Priorities.

H = statistically verifiable effects on human beings. A = statistically verifiable effects on laboratory animals.

Hazardous-Waste Dumps—a Legacy of Abuse

Until only recently, the solution to hazardous-waste disposal was simply to bury or dump the wastes without any concern for potential environmental or health risks. Such uncontrolled sites included open dumps, landfills, bulk storage containers, and surface impoundments. The locations of these sites were usually selected for convenience and expedience. Many of these sites are located in environmentally sensitive areas, such as floodplains or wetlands. Rain and melting snow soak through the sites, carrying chemicals that contaminate underground waters, in which the chemicals are carried to nearby lakes and streams. When the sites become full or are abandoned, they are frequently left uncovered, thus increasing the likelihood of water

FIGURE 19.2
Toxic Chemical Storage. This is a site where toxic wastes were improperly stored. States often assume the problems and costs of correcting bankrupt companies' errors.

pollution from leaching or flooding, and increasing the chances of people having direct contact with the wastes. At some sites, specifically the uncovered ones, the air is also contaminated as toxic vapors rise from evaporating liquid wastes or from uncontrolled chemical reactions. (See figure 19.2.)

It is difficult, if not impossible, to determine how many of these sites have yet to be discovered. In the United States alone, the number of abandoned or uncontrolled sites is over twenty thousand, and the list grows yearly.

The costs involved in cleaning up the sites are high. Nearly all industrialized countries are faced with costly cleanup bills, some massive.

As is to be expected from the amount of toxic wastes generated each year, the United States has the highest number of hazardous-waste dumps needing immediate attention. Europeans are also paying a heavy price for their negligence. Every country in Europe (except Sweden and Norway) is plagued by an abundance of toxic-waste sites—both old and new—needing urgent attention. Holland is a good example. Authorities estimate that up to 8 million metric tons of hazardous chemical wastes may be buried in Holland. Estimates for cleaning up those wastes run as high as $6 billion. In the United States, the federal government has become the principal participant in the cleanup of hazardous-waste sites. The program that deals with the cleanup has popularly become known as **Superfund.**

Hazardous and Toxic Wastes

BOX 19.4
A Contaminated Community

"We don't know of any other community in the nation that has ever been contaminated like this one." This statement was made by a spokesperson from the U.S. Center for Disease Control in Atlanta, Georgia, about the six hundred residents of Triana, Alabama. Triana is located 25 kilometers from Huntsville.

From 1947 to 1970, the Olin Chemical Corporation manufactured DDT on a site leased from the nearby Redstone Arsenal. During the time of production, about 4,000 metric tons of DDT residue accumulated in the soil of an adjacent marsh. Today, the DDT from this soil is filtering into Indian Creek, which flows through Triana. The fish in this stream contain forty times the federal standard of allowable DDT. Because the people from Triana depend upon catfish from the stream as a source of food, they have accumulated large amounts of DDT in their bodies.

The average person in the United States contains 8 to 10 ppm of DDT in his or her body fat. People from Triana contain up to sixty times that amount. The older residents have higher levels of DDT than the younger members of the community, perhaps because the older people have bioaccumulated the DDT for a longer time.

A large portion of the community's income was earned from the sale of catfish. Because these fish are not suitable for sale, this income has been lost. The wildlife from nearby Wheeler National Wildlife Refuge also suffer from the effects of DDT. Mallard ducks have been found to contain 480 ppm DDT. A wintering population of thousands of double-crested cormorants once used the refuge, but they are now absent. A heron rookery that contained three hundred nests has disappeared from the area. These and other environmental changes will occur in Triana and the surrounding area as long as the DDT-contaminated soil remains in the marsh.

As yet, the only move toward a solution has been an allotment of $1.5 million by the federal government for a study to determine what to do about the problem. However, many residents feel that, regardless of the solution, the town will never be the same because of the problems caused by the DDT.

Superfund—Successes and Failures

Responding to public pressure to clean up hazardous-waste dumps and to protect the public against the dangers of such wastes, the U.S. Congress in 1980 enacted the **Comprehensive Environmental Response, Compensation and Liability Act (CERCLA).** CERCLA had several key objectives:

1. To develop a comprehensive program to set priorities for cleaning up the worst existing hazardous-waste sites.
2. To make responsible parties pay for those cleanups whenever possible.
3. To set up a $1.6 billion Hazardous Waste Trust Fund—popularly known as Superfund—to support the identification and cleanup of abandoned hazardous-waste sites.

TABLE 19.2
Common contaminants found at Superfund sites.

Chemical	Average occurrence (percent)	Average concentration (parts per billion)
Lead	51.4	309,000
Cadmium	44.7	2,185
Toluene	44.1	1,120,000
Mercury	29.6	1,379
Benzene	28.5	16,582
Trichloroethylene	27.9	103,000
Ethylbenzene	26.9	540,000
Benzo[a]anthracene	12.3	148,000
Bromodichloromethane	7.0	20
Polychlorinated biphenyls	3.9	128,000
Toxaphene	0.6	12,360

Source: Reprinted with permission from *Environmental Science and Technology*. Copyright © 1985 American Chemical Society.

4. To advance scientific and technological capabilities in all aspects of hazardous-waste management, treatment, and disposal.

A **National Priority List** of hazardous-waste dump sites requiring urgent attention was drawn up for Superfund action. The size of the list varies, however, depending on which governmental agency is contacted. The U.S. Office of Technology Assessment (OTA) estimates that ten thousand sites may eventually be placed on the National Priority List requiring Superfund cleanup. The OTA believes that cleaning up these hazardous dumps may take fifty years and cost up to $100 billion because of the complex mixture of contaminants in most sites. (See table 19.2.) The Government Accounting Office (GAO) believes that the National Priority List could reach more than four thousand sites with cleanup costs of around $40 billion. The EPA has the shortest list of sites (twenty-five hundred) that it says should be placed on the National Priority List at a cost of nearly $30 billion. Table 19.3 lists Superfund sites considered priority one.

After several years of congressional and administration argument, the Superfund program was reauthorized in 1986. The reauthorization increased the money available for hazardous-waste cleanup.

The 1986 Superfund legislation required the EPA to resume work on 375 priority sites, work that was delayed because of the delays in Congress to formulate cleanup standards. During the first five years after the passage of CERCLA, only thirteen sites were cleaned up and removed from the National Priority List, and there was controversy over the effectiveness of some of those cleanups. The 1986 legislation provided funds for research and development of new hazardous-waste disposal technologies. Such research is critical if we ever hope to effectively and safely deal with the growing volume of hazardous wastes being produced throughout the world. The bill was up for reauthorization in 1991. Major issues were how to pay for the cleanup of existing sites and what new sites should be added to the priority list. Appointed administrator of the EPA in 1988, William Reilly instituted a review of how the Superfund program was being carried out. The review resulted in several changes that revitalized the program: greater public participation, more efficient use of funding, and increased emphasis on site treatment technologies. The generators of the waste also had less control over the determination of the extent of the pollution and the cost to clean it up.

TABLE 19.3
Proposed national list of Superfund cleanup sites in the United States (priority one).

State/City/County	Site name	State/City/County	Site name
Alabama		New Hampshire	
Limestone and Morgan	Triana, Tennessee River	Epping	Kes-Epping
Arkansas		Nashua	Sylvester, Nashua
Jacksonville	Vertac, Inc.	Somersworth	Somersworth Landfill
California		New Jersey	
Glen Avon Heights	Stringfellow	Bridgeport	Bridgeport Rent. & Oil
Delaware		Fairfield	Caldwell Trucking
New Castle	Army Creek	Freehold	Lone Pine Landfill
New Castle County	Tybouts Corner	Gloucester Township	Gems Landfill
Florida		Mantua	Helen Kramer Landfill
Jacksonville	Pickettville Road Landfill	Marlboro Township	Burnt Fly Bog
Plant City	Schuylkill Metals	Old Bridge Township	CPS/Madison Industries
Indiana		Pittman	Lipari Landfill
Gary	Midco I	Pleasantville	Price Landfill
Iowa		New York	
Charles City	Labounty Site	Oswego	Pollution Abatement Services
Kansas		Oyster Bay	Old Bethpage Landfill
Cherokee County	Tar Creek, Cherokee County	Wellsville	Sinclair Refinery
Maine		Ohio	
Gray	McKin Company	Arcanum	Arcanum Iron & Metal
Massachusetts		Oklahoma	
Acton	W. R. Grace	Ottawa County	Tar Creek
Ashland	Nyanza Chemical	Pennsylvania	
East Woburn	Wells G&H	Bruin Boro	Bruin Lagoon
Holbrook	Baird & McGuire	Grove City	Osborne
Woburn	Industri-Plex	McAdoo	McAdoo
Michigan		Rhode Island	
Swartz Creek	Berlin & Farro	Coventry	Picillo Coventry
Utica	Liquid Disposal Inc.	South Dakota	
Minnesota		Whitewood	Whitewood Creek
Brainerd Baxter	Burlington Northern	Texas	
Fridley	FMC	Crosby	French, Ltd.
New Brighton/Arden	New Brighton	Crosby	Sikes Disposal Pits
St. Louis Park	Reilly Tar	Houston	Crystal Chemical
Montana		La Marque	Motco
Anaconda	Anaconda-Anaconda		
Silver Bow/Deer Lodge	Silver Bow Creek		

Source: Data from U.S. Environmental Protection Agency, "Proposed National Priorities List: As Provided for in Section 105(8) (b) of CERCLA,"
December 20, 1982, Washington, DC.

TABLE 19.4
Hazardous-waste management methods, United States.

Management method	Share of total waste managed (percent)
Land disposal[1]	67
Discharge to sewers, rivers, streams	22
Distillation for recovery of solvents	4
Burning in industrial boilers	4
Chemical treatment by oxidation	1
Land treatment of biodegradable waste	1
Incineration	1
Recovery of metals through ion exchange	less than 1
Total	100

Source: Data from U.S. Congressional Budget Office, *Hazardous Waste Management: Recent Changes and Policy Alternatives,* U.S. Government Printing Office, 1985, Washington, DC.
[1]Includes injection wells (25 percent of total), surface impoundments (19 percent), hazardous-waste landfills (13 percent), and sanitary landfills (10 percent).

Hazardous waste management methods, United States

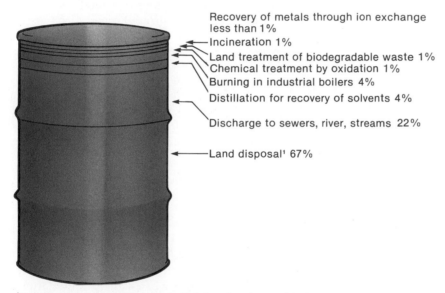

Recovery of metals through ion exchange less than 1%
Incineration 1%
Land treatment of biodegradable waste 1%
Chemical treatment by oxidation 1%
Burning in industrial boilers 4%
Distillation for recovery of solvents 4%
Discharge to sewers, river, streams 22%
Land disposal[1] 67%

[1]Includes injection wells (25 percent of total), surface impoundments (19 percent), hazardous waste landfills (13 percent), and sanitary landfills (10 percent).

Source: U.S. Congressional Budget Office, *Hazardous Waste Management: Recent Changes and Policy Alternatives* (Washington, DC: U.S. Government Printing Office, 1985).

Managing Hazardous Wastes— Various Technologies

Despite the attention given to hazardous wastes over the past decade, most industrial hazardous wastes still end up being disposed of on or in the land. (See table 19.4.) For over 80 percent of their hazardous wastes, the United States, Canada, and Europe still rely principally on six methods of disposal:

1. Deep-well injection into porous geological formations or salt caverns
2. Discharge of treated and untreated liquids into municipal sewers, rivers, and streams
3. Placement of liquid wastes or sludges in surface pits, ponds, or lagoons

Hazardous and Toxic Wastes

TABLE 19.5
Average amounts of hazardous wastes incinerated in selected countries in the 1980s.

Country	Amount incinerated (metric tons) On land	At sea
Denmark	circa 32,000	0
Belgium		10,000
France	400,000	10,000
Federal Republic of Germany	675,000	41,000
Netherlands	66,000	20,000
Switzerland	120,000	5,000
United Kingdom	80,000	3,500
United States	2,700,000*	0
Norway		8,000

Source: Data adapted from U.S. Congress, Office of Technology Assessment (OTA), *Ocean Incineration: Its Role in Managing Hazardous Waste* (OTA), August 1986, pp. 198–201, Washington, D.C.
*This large figure is only 1 percent of the total amount of hazardous wastes generated in the United States. In contrast, West Germany incinerates 15 percent of its total amount of hazardous wastes. The U.S. Congress, Office of Technology Assessment, estimates that 10–20 percent of all U.S. hazardous wastes could, in theory, be incinerated.

4. Storage of solid wastes in specially lined dumps covered by soil
5. Storage of liquid and solid wastes in underground caverns and abandoned salt mines
6. Dumping of wastes into sanitary landfills not designed for toxic or hazardous wastes

Methods of disposal of toxic and hazardous wastes have not noticeably changed over the past few decades. Several technologies, however, are under consideration as alternatives to land disposal. One way of categorizing these technologies is whether they destroy, immobilize, or separate the waste.

Waste Destruction Technology

"Destroying" hazardous waste means getting rid of most of it. Some harmful residues may still be left behind, however, and these must be properly disposed of. Thermal treatment and neutralization are the two most common types of **waste destruction.**

Thermal treatment involves heating waste in a flame-powered incinerator. Under controlled conditions, incineration can destroy 99.999 percent of organic wastes. Incineration accounts for the disposal of only about 2 percent of the hazardous wastes in the United States. In Europe, the amount of hazardous waste destroyed by incineration is higher, but still amounts to less than 50 percent. The relatively high costs of incineration (compared with landfills) and concerns for the safety of surrounding areas in case of accident have kept incineration from becoming a major form of disposal. (See table 19.5.)

Certain types of hazardous waste can be "neutralized." For example, an acid can be added to an excessively alkaline waste, or a base to an overly acidic waste. While **neutralization** is an effective means of destroying hazardous waste, it accounts for only a fraction of waste destroyed.

Pollution and Policy

Waste Immobilization Technology

Immobilizing a waste puts it into a solid form that is easier to handle and less likely to enter the surrounding environment. **Waste immobilization** is useful for dealing with wastes, such as certain metals, that cannot be destroyed. Once a waste has been immobilized, the material resulting from the process still must be properly disposed of.

Two popular methods of immobilizing waste are **fixation** and **solidification.** Engineers and scientists mix materials such as fly ash or cement with hazardous wastes. This either "fixes" hazardous particles, in the sense of immobilizing them or making them chemically inert, or "solidifies" them into a solid mass. Solidified waste is sometimes made into solid blocks that can be stored more easily than a liquid.

Waste Separation Technology

Waste separation entails either separating one hazardous waste from another or separating hazardous waste from a nonhazardous material that it has contaminated. Sometimes, this separation is achieved by changing the waste from one form (solid, liquid, or gas) to another. Examples of waste separation technology include air stripping, steam stripping, carbon absorption, and precipitation.

Air stripping is sometimes used to remove volatile chemicals from water. Volatile chemicals, which have a tendency to vaporize easily, can be forced out of liquid when air passes through it. **Steam stripping** works on the same general principle, except that it uses heated air to raise the temperature of the liquid and force out volatile chemicals that ordinary air would not.

Carbon absorption tanks contain specifically activated particles of carbon to treat hazardous chemicals in gaseous and liquid waste. The carbon chemically combines with the waste or catches hazardous particles just as a fine wire mesh catches grains of sand. Contaminated carbon must then be disposed of or cleaned and reused.

Precipitation involves adding special materials to a liquid waste. These bind to hazardous chemicals and cause them to precipitate out of the liquid and form large particles called "floc." Floc that settles can be separated as sludge; floc that remains suspended can be filtered.

Alternative Technologies

Alternative technologies for hazardous-waste management may be currently available, but there are often serious impediments to their use, including economic and marketplace uncertainties. One major uncertainty is the degree to which the public will accept a particular means of handling hazardous wastes; for example, this has been a special problem in the case of incineration.

Perhaps the wisest solution to the hazardous-waste problem is cutting off the flow of toxic wastes at the source—**source reduction.** While some industries have significantly reduced the amount of hazardous wastes generated, source reduction is still largely untried. The concept of reducing pollution at the source instead of at the end of a pipe, however, may soon become an economic necessity. (See table 19.6.)

TABLE 19.6
Industrial-waste management in Japan.

Waste disposition	Quantity (million tons)	Share of total (percent)
Total generated	220.5	100
Recycled and reused	112.7	51
Delivered off-site for reuse	(78.5)	(36)
Reused on-site	(34.2)	(15)
Reduced through treatment and incineration	68.9	31
Disposed of	38.9	18

Source: Data from Clean Japan Center, *Recycling '86: Turning Waste into RESOURCES* Tokyo, 1986.

CONSIDER THIS CASE STUDY
Love Canal

Love Canal, near Niagara Falls, New York, was constructed as a waterway in the nineteenth century. It was subsequently abandoned and remained unused for many years. In the 1930s, it became an industrial dump. The Hooker Chemical Company purchased the area in 1947 and used it as a burial site for 20,000 metric tons of chemicals.

Hooker then sold the property to the local government for one dollar. A housing development and an elementary school were constructed on the site. Soon after the houses were constructed, people began to complain about chemicals seeping into their basements. In 1978, eighty different chemicals were found in this seepage. Approximately one dozen probable carcinogens were identified among these chemicals.

As a result of these findings, $27 million in government funds were appropriated in 1978 to purchase homes and permanently relocate 237 families. The funds also provided for the construction of a series of ditches to contain the chemicals and a clay cap to prevent the fumes from entering the atmosphere. But the problems continued.

The remaining 710 families in the Love Canal area were not satisfied with the government's approach to the problem. They cited the fact that women in the area had a 50 percent higher rate of miscarriages. Of seventeen reported pregnancies in the area during 1979, two children were born normal, nine had defects, two were stillborn, and four ended in miscarriage. In addition to the abnormalities in birth, there are other biological problems in the Love Canal area.

Neurologists determined that the speed of the nerve impulses in thirty-seven residents who were examined were slower than normal. They stated that chemical exposure could have caused this damage. In 1980, the EPA released the findings of a study that found that eleven out of thirty-six residents tested in the Love Canal area had broken chromosomes, which are linked to cancer and birth defects. As a result, the federal government released $5 million to temporarily relocate Love Canal residents to motels or other quarters.

By 1990, a $150 million cleanup effort had sealed off the leaky dump, demolished 238 homes nearest the chemical graveyard, and scoured toxins from neighborhood storm sewers and streams (the two major sources of danger to area homes).

In early 1991, some families began to move back into the area. The Love Canal Area Revitalization Agency is planning to sell 236 homes in Love Canal (renamed Black Creek Village) by 1995. One of the incentives is that the houses can be purchased at a low cost. The families are confident that the houses are safe, but many environmental groups oppose them.

Who is responsible for providing treatment for the physical and mental problems experienced by the residents in this community?

Finally, there is a question that no one ever seems to ask. Why were permits ever awarded to construct a thousand-unit housing development on top of a site known to contain 20,000 metric tons of toxic wastes?

Laws and regulations dealing with solid- and hazardous-waste disposal are beginning to drive industrial behavior toward pollution prevention. In addition, because the costs of safe disposal are mounting, waste-handling firms—both private and public—are looking for better and cheaper ways to treat and dispose of hazardous wastes. Strong, enforceable laws have eliminated the economic incentives to pollute. Unfortunately, as strong laws have been enacted, some small industries that were unable or, more likely, unwilling to properly dispose of their wastes have turned to illegal night-time dumping.

The environmental costs of not managing hazardous wastes, as witnessed in virtually every industrialized country, are astronomical. And because major generators of hazardous wastes remain liable for past mistakes, economic and regulatory incentives for complying with hazardous-waste regulations should continue to encourage responsible management.

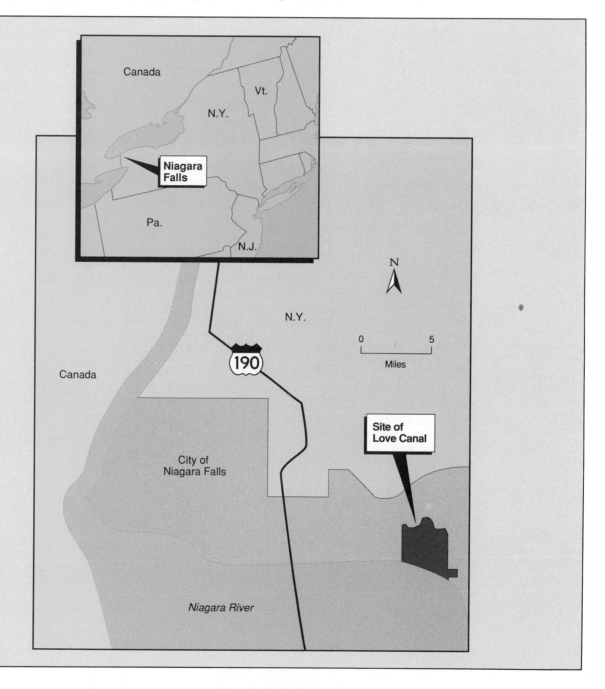

Summary

Public awareness of the problems of hazardous wastes is relatively recent. The industrialized countries of Europe and North America only began major regulation of hazardous waste during the past fifteen years, and most developing countries require little or no controls of such substances. As a result, many countries are living with serious problems from prior uncontrolled dumping practices, while current systems for management of hazardous and toxic waste remain incomplete and incapable of even identifying all hazardous waste.

A number of fundamental problems are involved in hazardous-waste management. Initially, there is no agreement as to what constitutes a hazardous waste. Moreover, little is known about the amounts of hazardous wastes generated throughout the world. The issue is further complicated by our limited understanding of the health effects of most hazardous wastes and the fact that large numbers of potentially hazardous chemicals are being developed faster than their health risks can be evaluated.

Hazardous-waste management must move beyond burying and dumping. Industries need to be encouraged to generate less hazardous waste in their manufacturing processes. Although toxic wastes cannot be entirely eliminated, technologies are available for destroying or recycling wastes. It is possible to enjoy the benefits of modern technology while avoiding the consequences of a poisoned environment. The final outcome rests with governmental and agency policymakers, as well as with an educated public.

Review Questions

1. Explain the problems associated with hazardous-waste dump sites and how such sites developed.
2. Distinguish between acute and chronic toxicity.
3. Give two reasons why there is difficulty in regulating hazardous wastes.
4. In what ways do hazardous wastes contaminate the environment?
5. Describe how hazardous wastes contaminate groundwater.
6. Why is there often a problem in linking a particular chemical or hazardous waste to a particular human health problem?
7. Describe what is meant by the National Priority List.
8. Describe five technologies for managing hazardous wastes.
9. What is meant by source reduction? Describe how source reduction could impact on the hazardous-waste problem.

CHAPTER TWENTY
Environmental Policy and Decision Making

Objectives

After reading this chapter, you should be able to:

Explain how the executive, judicial, and legislative branches of the U.S. government interact in forming policy.

Describe the forces that led to changes in environmental policy in the United States during the past three decades.

Understand the history of the major United States environmental legislation.

Understand what is meant by "Green" politics.

Describe why environmentalism is a growing factor in international relations.

Understand the factors which could result in "ecoconflicts."

Understand why it is not possible to separate politics and the environment.

Explain how citizen pressure can influence government environmental policies.

Chapter Outline

Politics in the United States
    Box 20.1 Farm Policy Legislation in the United States
U.S. Environmental Policy—Earth Day I
    Box 20.2 Is Environmental Quality a Right?
    Box 20.3 Earth Day 1970–1990
U.S. Environmental Policy—Earth Day II
    Box 20.4 The Greens
The Greening of Geopolitics
    Box 20.5 Eco-Terrorism
    Box 20.6 Environmental Disaster in Eastern Europe
International Environmental Policy
    Box 20.7 What *You* Can Do to Make the World a Better Place to Live In
Consider This Case Study: What Should We Expect of the EPA?

Key Terms

executive branch   legislative branch
"Green" politics   policy
judicial branch

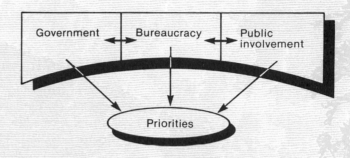

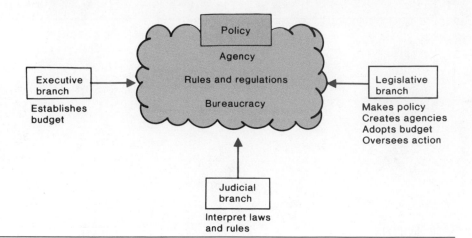

FIGURE 20.1
How Government Influences Policy.
The course of action with regard to a specific area of concern is planned by the legislative or executive branch of the government. These two branches influence the budgets of the agencies that will carry out the policy. The judicial branch is responsible for ensuring that the laws, rules, and regulations are upheld.

Politics in the United States

Government affects our lives in many ways. Food, air, and water quality are regulated by government. Individually, each of us supports the government by paying a variety of taxes. Even though not all governmental policies have been perceived as wise, many of them have helped to solve complex environmental problems. A knowledge of the political process and institutional structure of society is necessary before we can effectively influence public policies.

The government of the United States is structured into three separate branches: the legislative, judicial, and executive branches. Each of these branches has specific functions, and each is a check on the other two branches to ensure that the assigned responsibilities are fulfilled for the smooth operation of the whole system.

The **legislative branch** of the U.S. government is the Congress, composed of the House of Representatives and the Senate. A primary function of Congress is to develop and approve policy. **Policy** is a planned course of action on a question or topic. Congress generally supports policy by passing bills or acts that establish a government agency or by instructing an existing agency to take on new tasks or programs. (See figure 20.1.)

Another function of the legislative branch is to oversee the various agencies of the executive branch. Congress looks at how well the laws are being carried out and how effectively the agencies are spending their appropriations. An important input associated with the development, funding, and oversight of agencies is that concerned individuals and groups exert pressure to develop policy and to fund programs and policies that are favorable to them. These pressure groups are called lobbies, and they sometimes have tremendous influence on members of Congress. They can be a source of shortsighted legislative action, especially during an election year. (See figure 20.2.)

The **judicial branch** of the U.S. government is a complex and layered series of courts, ranging from local traffic courts to the Supreme Court. The courts interpret the several different categories of laws, such as constitutional law, statutory law, and administrative law, and specific courts deal

The legislative process

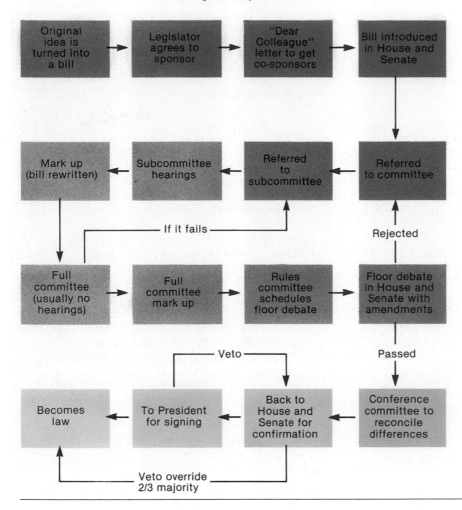

FIGURE 20.2

Passage of a Law. This figure illustrates the path of a bill from organization to becoming a law. As we can see, the process is not a quick one.

with specific areas of interpretation. Generally, statutory and administrative law are the areas involving the environment.

Federal agencies, such as the Environmental Protection Agency, have developed their own methods of establishing regulations that are as binding as laws passed by Congress. These regulations have the force of law because of the statutes establishing the agency, and the enforcement of their regulations falls under the category of statutory and administrative law.

The powers of the courts are generally less than the powers of the chief executive, the president. The powers of the **executive branch** come from the Constitution, but many additional powers have been assumed over the years by executives with particular personalities and priorities. Presidential powers are really permissions given to the office. When a president wants to gain support for programs or legislation, personality can be either an important asset or a serious liability. The powers of the chief executive when dealing with environmental issues include support for the development

BOX 20.1
Farm Policy Legislation in the United States

In 1985, the U.S. Congress faced the question of whether farm programs should be consistent with the nation's environmental goals. The outcome was a major piece of legislation: the 1985 Food Security Act, better known as the "Farm Bill." The Farm Bill addressed many environmental issues, such as groundwater and surface-water quality, wetlands protection, soil erosion, and pesticide residues.

Before 1985, U.S. federal farm programs too often encouraged negative environmental behavior—the draining of wetlands and the farming of erosion-prone soil. By passing two key conservation programs—the Conservation Compliance Program and the Conservation Reserve Program—in 1985, Congress made major strides toward reversing this trend. Unfortunately, each of these programs had loopholes, and the programs were not always enforced.

In 1990, a new Farm Bill was passed by Congress. The following environmental issues were addressed by the 1990 Farm Bill:

1. The amount of money in the budget for agricultural price supports was lowered by reducing the amount of land eligible for income support payments by 15 percent. Farmers now have 15 percent of their agricultural land exempted from farm support payments. This encourages farmers to take out of production marginal land that is most subject to degradation by being farmed.

2. An Office of Environmental Quality was established within the U.S. Department of Agriculture.

3. The bill encourages farmers to set aside high-quality wetlands by initiating a thirty-year Wetland Reserve program that pays farmers for not using wetlands for agricultural purposes.

4. The Conservation Reserve Program, which pays farmers to take highly erodible land out of production, now includes windbreaks, shelterbelts, and marginal pasturelands. This encourages farmers to preserve current shelterbelts and windbreaks, rather than to remove them to allow for more farmland. Some farmers may be encouraged to plant new windbreaks.

5. The Conservation Compliance Program, which requires farmers to use approved farming techniques on erodible land, has been made stronger by increasing the number of benefits a farmer could lose if U.S. Department of Agriculture recommendations are not followed.

The 1990 Farm Bill is the platform that is going to govern U.S. agricultural policy into the next century. It appears that environmental concerns will be a growing component of agricultural practices.

and funding of agencies and the budgeting of monies for existing environment-related agencies. The major environmental role of the executive branch is to lead members of Congress and the people they represent toward respect for and appreciation of quality in their surroundings. (See figure 20.3.)

Enactment of policies into law may be only the beginning. Relatively few statutes are "self-executing" in the sense that they can be put into effect immediately by existing agencies. Many require funding, if only in the form of new personnel to perform the investigatory services or the enforcement called for in the statute. In such cases, the appropriation process following submission of the budget serves as the second consideration of the issues addressed by the statute. Thus, the policy must be considered in light of its monetary ramifications. Once money has been both authorized and appropriated, the agency or program can spend the money. The agency is now a part of the executive branch and can implement policy.

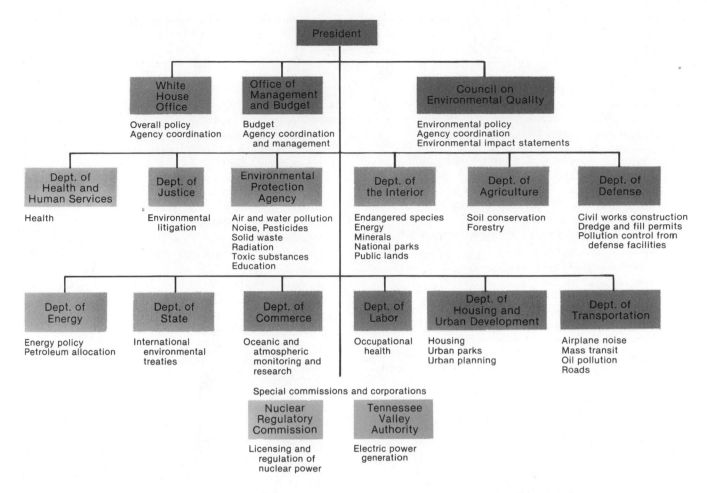

FIGURE 20.3

Major Agencies of the Executive Branch. Major agencies of the executive branch are shown with their environmental responsibility.

U.S. Environmental Policy—Earth Day I

President Teddy Roosevelt declared eighty years ago that nothing short of defending this country in wartime "compares in leaving this land even a better land for our descendants than it is for us." Environmental issues that Roosevelt strongly believed in, however, did not become major political issues until the early 1970s.

April 22, 1970, was the first Earth Day, and in the opinion of many, it was the event that gave birth to the modern environmental movement. In 1970, as a result of mounting public concern over environmental deterioration—cities clouded by smog, rivers on fire, waterways choked by raw sewage—the United States began to address the most obvious, most acute environmental problems. (See table 20.1.)

TABLE 20.1
Major U.S. environmental and resource conservation legislation.

Wildlife conservation
Anadromous Fish Conservation Act of 1965
Fur Seal Act of 1966
National Wildlife Refuge System Act of 1966, 1976, 1978
Species Conservation Act of 1966, 1969
Marine Mammal Protection Act of 1972
Marine Protection, Research, and Sanctuaries Act of 1972
Endangered Species Act of 1973, 1982, 1985, 1988
Fishery Conservation and Management Act of 1976, 1978, 1982
Whale Conservation and Protection Study Act of 1976
Fish and Wildlife Improvement Act of 1978
Fish and Wildlife Conservation Act of 1980 (Nongame Act)

Land use and conservation
Taylor Grazing Act of 1934
Wilderness Act of 1964
Multiple Use Sustained Yield Act of 1968
Wild and Scenic Rivers Act of 1968
National Trails System Act of 1968
National Coastal Zone Management Act of 1972, 1980
Forest Reserves Management Act of 1974, 1976
Forest and Rangeland Renewable Resources Act of 1974, 1978
Federal Land Policy and Management Act of 1976
National Forest Management Act of 1976
Soil and Water Conservation Act of 1977
Surface Mining Control and Reclamation Act of 1977
Antarctic Conservation Act of 1978
Endangered American Wilderness Act of 1978
Alaskan National Interests Lands Conservation Act of 1980
Coastal Barrier Resources Act of 1982
Food Security Act of 1985
Coastal Development Act of 1990

General
National Environmental Policy Act of 1969 (NEPA)
International Environmental Protection Act of 1983
Environmental Education Act of 1990

Energy
National Energy Act of 1978, 1980

Water quality
Water Quality Act of 1965
Water Resources Planning Act of 1965
Federal Water Pollution Control Acts of 1965, 1972
Ocean Dumping Act of 1972
Safe Drinking Water Act of 1974, 1984
Clean Water Act of 1977, 1987
Great Lakes Critical Programs Act of 1990
Oil Spill Prevention and Liability Act of 1990

Air quality
Clean Air Act of 1963, 1965, 1970, 1977, 1990

Noise control
Noise Control Act of 1965
Quiet Communities Act of 1978

Resources and solid-waste management
Solid Waste Disposal Act of 1965
Resources Recovery Act of 1970
Resource Conservation and Recovery Act of 1976
Waste Reduction Act of 1990

Toxic substances
Toxic Substances Control Act of 1976
Resource Conservation and Recovery Act of 1976
Comprehensive Environmental Response, Compensation, and Liability (Superfund) Act of 1980, 1986, 1990
Nuclear Waste Policy Act of 1982

Pesticides
Federal Insecticide, Fungicide, and Rodenticide Control Act of 1972, 1988

Earth Day was the largest organized demonstration in U.S. history. Thousands of schools, colleges, and universities took part, and millions of ordinary citizens demonstrated their desire to work toward environmental goals. Congress adjourned for the day, New York City's Fifth Avenue was closed, and hundreds of ecology fairs were held.

Public opinion polls indicate that a permanent change in national priorities followed Earth Day 1970. When polled in May 1971, 25 percent of the U.S. public declared protecting the environment to be an important goal—a 2,500 percent increase over 1969.

During the early 1970s, Congress tackled many environmental problems. Important pieces of legislation were passed that called for setting air-quality standards, cleaning up rivers and lakes, protecting coastal areas, regulating pesticides, protecting endangered species, and safeguarding drinking water. Many of the identified environmental problems were so immediate, so obvious, that it was relatively easy to see what had to be done and to summon the political will to do it.

When Thomas Jefferson, John Hancock, Ben Franklin, and fifty-two other delegates wrote the U.S. Constitution in 1787, they never considered addressing the right to a clean and healthy environment. They did not include an environmental guarantee because they never envisioned that the land and water would ever be so extensively utilized; nor could they foresee the environmentally destructive by-products of modern industrialized society.

There were eight attempts between 1968 and 1972 to amend the U.S. Constitution to include environmental-quality rights. Although none of these proposals survived congressional committees, their introduction indicated the popularity of such an idea.

A total of twenty-one states already have adopted language in their constitutions that guarantees certain environmental rights. In addition, the constitutions of eight countries—including China, Canada, and Switzerland—also address environmental issues.

In 1989, the National Wildlife Federation passed a resolution calling for the U.S. Constitution to be amended to include an Environmental Bill of Rights. According to the National Wildlife Federation, the principles that an Environmental Quality Amendment should embody are:

> The people have a right to clean air, pure water, productive soils, and to the conservation of the natural, scenic, historic, recreational, esthetic, and economic values of the environment. America's natural resources are the common

Should the U.S. Constitution be amended to include environmental rights?

property of all people, including generations yet to come. As trustee of these resources, the United States Government shall conserve and maintain them for the benefit of all people.

In spite of support for an Environmental Quality Amendment, considerable opposition also exists. Other nations also experienced opposition, and yet the environmental quality conditions were incorporated into similar constitutions. Perhaps the U.S. Constitution will also incorporate environmental rights in the future.

Just as it was beginning to gain momentum, however, the environmental movement began to decline. When the energy crisis threatened to stall the U.S. economy in the early 1970s, environmental concerns quickly faded. By 1974, President Ford had proposed an acceleration of his administration's leasing program for offshore gas and oil drilling. A turnaround in environmental policy was even more pronounced during the 1980s. Former Vice President Walter Mondale was fond of noting that President Reagan "would rather take a polluter to lunch than to court."

During the mid-1980s, the Reagan administration embarked on a policy course that radically altered the environmental agenda in the United States. The guiding principles, set out by the President's Council on Environmental Quality, were: (1) regulatory reform, including the use of cost-benefit analysis to help determine the value of environmental regulations and programs; (2) reliance, as much as possible, on the free market to allocate resources; and (3) decentralization—shifting responsibility for environmental protection to state and local governments whenever feasible. These goals reversed much of the environmental policy developed in the early 1970s.

BOX 20.3
Earth Day 1970–1990

On 22 April 1970, citizens of all ages celebrated Earth Day by planting trees, cleaning up vacant lots, starting gardens, and learning about the delicacy of the natural world.

The spontaneity of Earth Day 1970 and the grassroots-powered energy it developed were felt in the political arena for many years. Plans for an American SST (a controversial supersonic transport plane) were scrapped, Congress enacted the Clean Air Act, new in-plant pollution laws were passed, and the U.S. government banned the military's controversial use of toxic defoliants in Indochina. Earth Day 1970 marked the beginning of political activism for many individuals and awakened public sentiment for the health of our environment.

Many saw Earth Day 1970 as a chance to reflect on the environment of the twenty-first century. While many of the issues are the same today as they were twenty years ago, scientific knowledge and public awareness have grown. We are also beginning to recognize that environmental problems in one country cross national borders, affecting the planet on which all of us must live. While Earth Day 1970 was primarily a North American event, Earth Day 1990 was international in scope.

The organizers of Earth Day 1990 were working toward specific accomplishments, including:

A worldwide ban on emissions of chlorofluorocarbons, which destroy the ozone layer and contribute to global warming, to be fully implemented within five years

Slowing the rate of global warming through dramatic, sustained reductions in carbon dioxide emissions, including higher standards for automobile fuel efficiency and the rapid adoption of a transportation system not powered by fossil fuels

Preservation of old-growth forests, in both temperate and tropical areas

A ban on packaging that is neither recyclable nor biodegradable, and the implementation of strong, effective recycling programs in every community

A swift transition to renewable energy resources

Dramatic increases in residential and industrial energy efficiency

A comprehensive hazardous-waste minimization program, emphasizing source reduction

Heightened protection for endangered species and habitats

New protections for marine resources, including marine mammals and fisheries

Public concern about the environment took many forms on Earth Day, both in 1970 and in 1990.

A powerful international agency with authority to safeguard the atmosphere, the oceans, and other global commons from international threats

A new sense of responsibility for the protection of the planet by individuals, communities, and nations

The adoption by all countries of strategies to stabilize their populations within limits sustainable using environmentally available agricultural and industrial processes

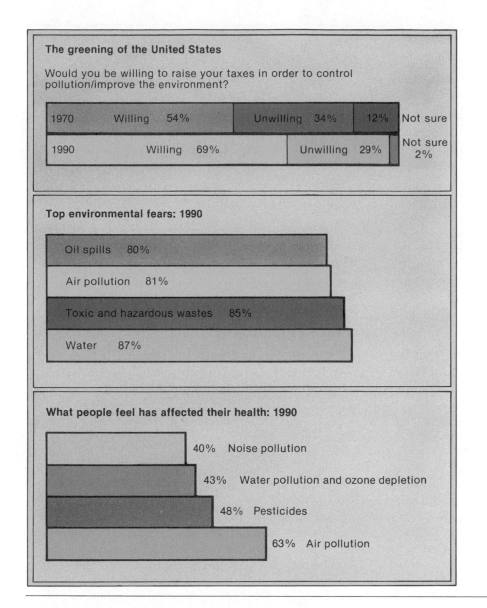

The greening of the United States

Would you be willing to raise your taxes in order to control pollution/improve the environment?

1970 | Willing 54% | Unwilling 34% | 12% Not sure
1990 | Willing 69% | Unwilling 29% | Not sure 2%

Top environmental fears: 1990

Oil spills 80%
Air pollution 81%
Toxic and hazardous wastes 85%
Water 87%

What people feel has affected their health: 1990

40% Noise pollution
43% Water pollution and ozone depletion
48% Pesticides
63% Air pollution

FIGURE 20.4

Greening of the United States. By 1990, concern about the environment was growing. These 1990 polls indicate that a strong majority of the U.S. population favored increased taxes to control pollution and that the public is concerned about the health effects of environmental pollution.

Despite the wavering of political will toward environmental concerns in the period from 1970–1990, there were some very tangible accomplishments. Among the most visible and quantifiable is the expansion of protected areas. Since 1970, federal parklands in the United States—excluding Alaska—increased 800,000 hectares, to 10.5 million. In Alaska, 18.3 million additional hectares have been protected, bringing the state's total to 21.8 million hectares. Also, the extent of the waterways included in the National Wild and Scenic Rivers System increased by more than twelve times, to about 15,000 kilometers.

By the late 1980s, however, a new environmental awareness and concern began to surface as a major political issue. Once again, the public reacted by organizing and putting pressure on the political system, and as in 1970, the politicians began to respond. For the first time in the history of the United States, the environment became a key issue in a presidential campaign. In 1988, the environmental records of the two major candidates were actively debated. Environmentalism was evolving as a major public issue. Polls indicated that the public was even willing to raise taxes to improve the environment. (See figure 20.4.)

TABLE 20.2
Major environmental advocacy groups.

Organization	Membership
Nature Conservancy	436,407
National Wildlife Federation	5,800,000
National Audubon Society	550,000
Sierra Club	426,000
Natural Resources Defense Council	95,000
Wilderness Society	250,000
Environmental Defense Fund	100,000
Environmental Policy Institute/Friends of the Earth	42,000
Izaak Walton League	50,000
Environmental Action	20,000
Total	**7,769,407**

Source: Capital Research Center, Washington, D.C.

U.S. Environmental Policy—Earth Day II

April 22, 1990, was the twentieth anniversary of Earth Day and, in the opinion of many, the beginning of a new decade of environmental concern in the United States. In many respects, the environmental movement of the 1970s and 1980s has come of age in the 1990s. Putting the bands of politics and science, or emotionalism and logic, together in a new environmental movement represents a significant integration of human thinking.

It has been said that politics have always forged science. Prioritization of issues and political will determine where money will be spent. By the early 1990s, it appeared that the United States political will to address environmental concerns was on the rise. A 1990 Gallup Poll indicated that 79 percent of people ages thirty to forty-nine called themselves environmentalists. Two out of three Americans said that pollution was a "very serious threat," according to the poll. A National Wildlife Federation study in 1990 documented a heightened public concern for the environment. In that study, 84 percent of the population called the threat of freshwater pollution "very serious," up from 48 percent in 1970; 73 percent called air pollution a "very serious" threat, up from 46 percent in 1970.

The National Wildlife Federation study also revealed that most Americans would prefer a federal government more actively involved in environmental protection. Ninety percent desired stronger active participation by leaders in government and business in environmental concerns. Eighty-two percent believed that it was the federal government's role to lead the way, while 79 percent thought that the government too often put the interests of business ahead of the environment. By 1991, the rising figures in the polls were being matched by rising membership in nongovernmental environmental organizations. (See table 20.2.)

The dramatic increase in concern about environmental issues by Americans led pollster Louis Harris to predict that, within the decade of the 1990s, a president would be chosen and elected with a pro-environmental stance as a primary identification factor. Harris further stated that the "issue of the environment has become an explosive and decisive cutting edge in mainstream politics."

The United States is not unique in witnessing the growth of the environment as a political issue. Environmental or "green" politics has emerged from minority status and become a political movement in many nations.

Pollution and Policy

BOX 20.4
The Greens

In all Western countries, the growing environmental movement has been an important political development over the past twenty years. Numerous environmental organizations have been created. Indeed, not just Western nations but virtually all countries now have environmental organizations that are helping to shape the perspective of governments on environmental issues.

As environmental organizations have developed, they have faced the question whether to work with existing political parties or to create their own. While organizations in the United States generally work through existing structures, in West Germany, a new political party—the **Greens**—is closely identified with the environmental agenda. As distinguished from environmental organizations, the need for a "Green" party is not evident in every country. Nevertheless, the German Greens has been modeled in many countries, so that it is possible to speak of a "Green phenomenon."

In the West German elections in April 1983, the Green Party gained international attention by breaking through the 5 percent vote barrier of the nation's proportional representation system and placing twenty-seven people in the national parliament. Since then, Greens in seven other Western European countries have entered parliament with between 5 and 8 percent of the vote, and Greens have a solid voice in the all-European parliament that meets in Strasbourg.

In the early summer of 1989, Greens in Great Britain astounded the experts, and themselves, by getting 15 percent of the vote, a figure that unfortunately for them does not translate into parliamentary representation, given the simple majority, winner-take-all system in that country. There are organized Green political tendencies in Poland, Hungary, and the USSR. Japan has a small but very vigorous Green Party, as do Australia, Brazil, and Canada.

WIR HABEN DIE ERDE VON UNSEREN KINDERN NUR GEBORGT.
DIE GRÜNEN

While Greens exist in the United States and Canada, there is little sign that they will be able to elect representatives in significant numbers. They are ultimately victims of the political system within which they work. Also, the need for a specifically Green Party is arguably less in North America than elsewhere because of a strong tradition of freedom of association and the corresponding characteristic for forming political interest and pressure groups. Thus, environmental issues are pushed by a long list of local pressure groups and by strong national environmental lobbies. If the Greens continue to grow in Western Europe, there may come a time when they will also become a significant political force in North America.

Source: U.S.E.P.A., and World Bank.

Such events as the massive destruction that resulted from a chemical spill on the Rhine River and the nuclear disaster at Chernobyl have served to bolster the emergence of green politics. Even in the Soviet Union and the rest of newly enfranchised Eastern Europe, the public has demanded more environmental protection, and the leaders are beginning to respond.

The Greening of Geopolitics

In October 1988, three hundred thousand Lithuanians signed a petition against a nuclear power plant. In Poland, where 80 percent of the Vistula River is too polluted for even industrial use, concern about the environment helped spark the 1989 Polish revolution. Norway has volunteered to contribute 0.1 percent of its gross national product, about $65 million, to an international climate fund if other countries are willing to join. At the economic summit in Paris in 1989, the leaders of the seven largest industrial democracies devoted a third of their final statement to an appeal for "decisive action" to "understand and protect the earth's ecological balance." A month prior to the summit, the Organization for Economic Cooperation and Development, representing the major economic powers, had called upon "all relevant national, regional, and international organizations" of its twenty-four member states to take a "vigilant, serious, and realistic" look at "balancing long-term environmental costs and benefits against near-term economic growth."

Concern about the environment is not limited to developed nations. A 1989 treaty signed in Switzerland limits what poorer nations call toxic terrorism: use of their lands by richer countries as dumping grounds for industrial waste. In 1990, more than one hundred nonaligned nations called for a "productive dialogue with the developed world" on "protection of the environment." Shortly after this statement, Japan pledged $2.25 billion to tackle pollution in the Third World.

Environmentalism is also a growing factor in international relations. Many world leaders see the concern for environment, health, and natural resources as entering the policy mainstream. A new sense of urgency and common cause about the environment is leading to cooperation in some areas. Ecological degradation in any nation is now understood almost inevitably to impinge on the quality of life in others. Drought in Africa and deforestation in Haiti have resulted in large numbers of refugees, whose migrations generate tensions both within and between nations. From the Nile to the Rio Grande, conflicts flare over water rights. The growing megacities of the Third World are areas of potential civil unrest. Sheer numbers of people overwhelm social services and natural resources. The government of the Maldives has pleaded with the industrialized nations to reduce their production of greenhouse gases, fearing that the polar ice caps may melt and inundate the island nation.

Economic progress in the Third World also could bring the possibility for environmental peril and international tension. China, which accounts for 21 percent of the world's population, has the world's third largest recoverable coal reserves. If China's current "modernization" campaign succeeds, the boom will be fueled by that coal, to the detriment of the planet as a whole. Some experts estimate that the developing world, which today produces one-fourth of all greenhouse-gas emissions, could be responsible for nearly two-thirds by the middle of the next century. Third World nations have repeatedly indicated that they are not prepared to slow down their own already weak economic growth to help compensate for decades of environmental problems caused largely by the industrialized world.

Some developing countries may resist environmental action because they see a chance to improve their bargaining leverage with foreign aid donors and international bankers. Where before the poor nations never had a strategic advantage, they now may have an ecological edge. Ecologically, there could be more parity than there ever was economically or militarily.

In 1990, the U.S. Ambassador to the United Nations stated that, just as the cold war between the East and West seems to be winding down, "eco-conflicts" between the industrialized North and the developing South may pose a comparable challenge to world peace. National security may no longer be about fighting forces and weaponry alone. It relates increasingly

BOX 20.5
Eco-Terrorism

The effects of war on the natural environment are frequently catastrophic. By its nature, war is destructive, and the environment is an innocent victim. The 1990–91 war in the Persian Gulf, however, witnessed a new role for the environment in war; its degradation was used as a weapon.

The phrase "eco-terrorism" was applied to Iraqi leader Saddam Hussein's use of oil as a weapon in the war. Iraq dumped an estimated 1.1 billion liters of crude oil into the Persian Gulf from Kuwait's Sea Island terminal before the United States made bombing raids on the pumps feeding the facility. The oil spill that resulted from the dumping was the world's largest—over twenty-five times the size of the 1989 *Exxon Valdez* disaster in Alaska.

Saddam Hussein may have engineered the spill to block U.S. plans for an amphibious invasion of Kuwait, but he was also probably trying to shut down seaside desalination plants that provide much of the fresh water for Saudi Arabia's Eastern Province. Whatever the military or political effect, the environmental effect was much greater.

The Persian Gulf is a sensitive ecosystem. The water is very salty, and temperatures vary widely from one season to another. As a result, indigenous animals and plants are finely attuned to specialized conditions. The gulf is also rather isolated, with only one narrow outlet—the Strait of Hormuz, just 55 kilometers across. The Gulf takes up to five years to flush out.

The Persian Gulf waters, shores, and islands are dotted with coral reefs, mangrove swamps, and beds of sea grass, alive with many unique species of birds, fish, and marine mammals. Many of the region's species, such as the bottlenose dolphin, dugong, green turtle, and caspian tern were already classified as threatened. This complex ecosystem, already pushed to the limits of survival by years of pollution, was again being threatened by a deliberate human act.

Another act of eco-terrorism that took place during the Iraq-Kuwait war was the intentional burning of

hundreds of oil wells in Kuwait. The air pollution resulting from the burning oil covered a large section of the area. It is estimated that it could be more than a year before all the fires are extinguished.

Using environmental damage as a weapon of war is a frightening concept. The environmental consequences of such acts of eco-terrorism will not be fully understood for years. Though the war in the Persian Gulf lasted less than one year, the environmental effects on some species could be irreversible.

to watersheds, croplands, forests, climate, and other factors rarely considered by military experts and political leaders, but that, when taken together, deserve to be viewed as equally crucial to a nation's security as are military factors.

The increased attention to the environment as a foreign policy and national security issue is only the beginning of what will be necessary to avert problems in the future. The most formidable obstacle may be the entrenched economic and political interests of the world's most advanced

BOX 20.6
Environmental Disaster in Eastern Europe

The environmental legacy of four decades of Communist rule is one of stark environmental degradation. While the causes of such degradation are many, four appear to be primary: (1) burning of high-sulfur brown coal, (2) uncontrolled auto tail-pipe emissions, (3) little or no control equipment for industrial smokestacks, and (4) intensive use of pesticides in a monoculture-based agriculture.

In many of the countries, environmental regulations are thought to be adequate but poorly enforced. Environmental institutions, official and nonofficial, exist. Environmental knowledge is believed considerable, but more theoretical than practical. Many environmental activists view their efforts as part and parcel of greater political and economic reform. For over a decade, West Germany paid East Germany to take its toxic wastes and other garbage. The new Germany will bear the responsibility of cleaning them up.

Cleaning up the problems in Eastern Europe will not be easy. Some suggest that the United States demand all loans to the area from the World Bank and other such institutions be conditioned upon tight environmental standards and that private loans from banks uphold basic environmental standards. As Eastern European leaders find that foreign investment is scarce and prospects for growth are dimming, however, they may be tempted to accept dirty technologies from the West: plants and production lines that fail Western environmental standards but are a bit above those now in the East.

Overview of Environmental Problems in Four Countries

Czechoslovakia
Major air pollutant is sulfur dioxide from uncontrolled industries and power plants. Sixty percent of energy comes from burning brown coal. About 70 percent of the rivers are polluted from mining wastes and agricultural runoff, and 40 percent of all sewage is untreated. Seventy-five percent of toxic waste is dangerously stored. Poor agricultural practices and heavy metals from open-cast mining have contaminated soil.

East Germany
Produces most sulfur dioxide in East bloc. Uncontrolled burning of brown coal provides 70 percent of energy. Sixty-six percent of rivers are polluted, and sewage is poorly treated. Toxic waste, stored for the West, is poorly controlled. Agricultural land is reduced as a result of heavy fertilizer use and monoculture agriculture.

Hungary
One-third of the population lives in areas where air pollution exceeds international standards. Primary pollutant is sulfur dioxide. Burning of brown coal produces 20 percent of energy; oil, gas, and power from one nuclear plant provide another 60 percent of energy needs. Hundreds of villages and towns have water that is unfit for drinking. Adequate sewage treatment exists for less than half of the population. Toxic waste, transported from Western Europe, is stored inadequately. Soil erosion is a problem in some areas.

Poland
Twenty-seven areas have been declared ecological danger zones; five regions have been declared disaster areas. Sulfur dioxide is the major air pollutant, especially in Upper Silesia and Krakow. The burning of brown coal supplies almost all energy needs. Nearly all rivers are polluted; water from one-third is unfit for industrial use. Seventy percent of sewage is untreated. Of the food produced in the Krakow area, 60 percent may be unfit for human consumption because of heavy metal contamination.

Sources: U.S. Environmental Protection Agency, World Bank.

nations. If the United States, for example, asks others not to cut their forests, then it will have to be more judicious about cutting its own. If North Americans wish to stem the supply of hardwood from a fragile jungle or furs from endangered species, then they will have to stem demand for fancy furniture and fur coats. If they wish to preserve wildernesses from the intrusions of the oil industry, then they will have to find alternative sources of energy and use all fuels more efficiently.

What may be needed is self-discipline on the part of the world's haves and increased assistance to the have-nots. In the world today, a billion people live in a degree of poverty that forces them to deplete the environ-

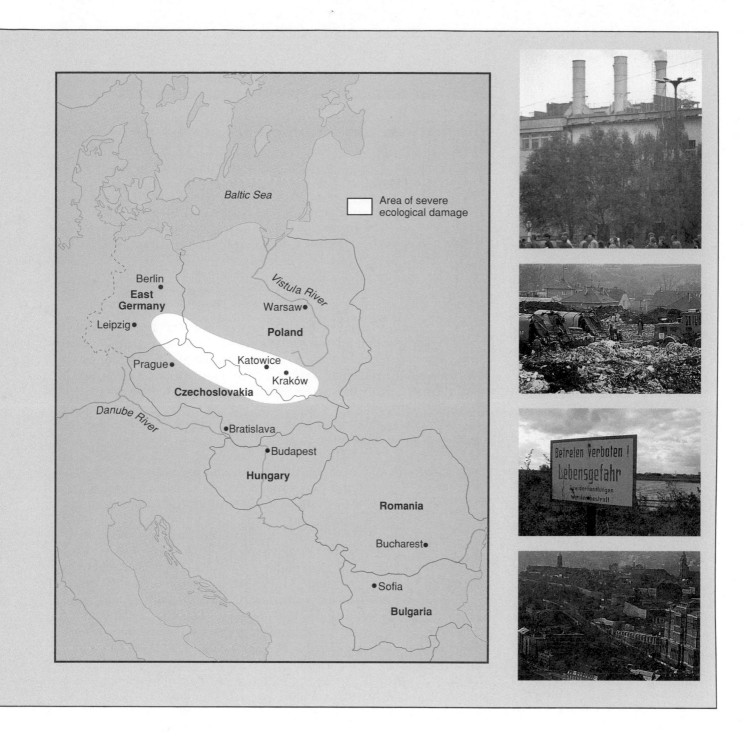

ment without regard to its future. Similarly, their governments often are too crippled by international debt to afford the short-term costs of environmental safeguards.

William Ruckelshaus, the former administrator of the U.S. Environmental Protection Agency, believes a historical watershed may be at hand. If the industrialized and developing countries did everything they should, he says, the resulting change would represent "a modification of society comparable in scale to the agricultural revolution of the Neolithic age and to the Industrial Revolution of the past two centuries."

Just as environmentalism began as a local movement, it needs to continue to grow at the local level, even as it gathers force in the parliaments and legislative bodies of the world. Individuals and nations alike should learn to think on a far broader scale about themselves, their needs, and their interests before a global catastrophe forces them to do so.

International Environmental Policy

If there must be a war, let it be against environmental contamination, nuclear contamination, chemical contamination, against the bankruptcy of soil and water systems; against the driving of people away from the lands as environmental refugees. If there must be war, let it be against those who assault people and other forms of life by profiteering at the expense of nature's capacity to support life. If there must be war, let the weapons be your healing hands, the hands of the world's [youth] in defence of the environment.

Mustafa Tolba
Secretary General
United Nations Environmental Program

Examples of international cooperation on issues other than war are unfortunately somewhat limited. Institutional coordination and political resolve, however, are necessary to preserve and protect the global environment. A major step in that direction was the 1972 United Nations Conference in Stockholm, Sweden. This was the first international conference specifically dealing with environmental concerns. Out of that conference was born the U.N. Environment Program, a separate department of the United Nations that deals with environmental issues.

In 1982, the Third United Nations Conference on the Law of the Sea produced a comprehensive convention that addressed many of the issues concerning jurisdiction over ocean waters and use of ocean resources. This treaty is viewed by many as a model for international environmental protection. Positive results are already evident from application of the agreements on pollution control, marine mammal protection, navigation safety, and other aspects of the marine environment. The issue of deep ocean mining, primarily of manganese nodules, has to date kept many industrial nations, such as Germany and the United States, from ratifying the treaty.

There have been several other successful international conventions and treaties. The Antarctic Treaty of 1961 reserves the Antarctic continent for peaceful scientific research and bans all military activities in the region. The 1979 Convention on Long-Range Transboundary Air Pollution was the first multilateral agreement on air pollution and the first environmental accord involving all the nations of Eastern and Western Europe and North America. The 1989 Accord on Chlorofluorocarbon Emissions addresses the threat to the stratospheric ozone shield.

There is no international legislature with authority to pass laws; nor are there international agencies with power to regulate resources on a global scale. An international court at the Hague in the Netherlands has no power

TABLE 20.3
Factors affecting international environmental laws.

1. **Severity of the problem.** Once a problem is widely acknowledged as critical, it is easier to make progress.
2. **Science.** Good data are needed on the extent of the problem and possible solutions.
3. **Geography.** The extent to which the problem is transboundary and where its effects are felt.
4. **Law.** Whether countries have laws protecting the environment and whether foreigners have access to court and administrative proceedings to enforce those laws.
5. **Domestic interests and pressures.** Who favors and who opposes action on the issue in each country.
6. **Formal institutions and policies.** Whether there is a mechanism in place for cooperative action among the interested countries.
7. **History.** Whether there is a tradition of cooperation or conflict among the countries.
8. **Relative economic strength, military power, and population.** Large disparities in size or strength may hinder agreement unless the stronger country depends on the weaker for the resource.
9. **Outside influences.** Third parties can influence negotiations positively or negatively.
10. **Timing.** It may be easier to reach agreement before various interests are entrenched or at times when other changes in societies or economies are occurring.

Figure from *World Resources 1987* by The World Resources Institute. Copyright © 1987 by The International Institute for Environment and Development and the World Resources Institute. Reprinted by permission of Basic Books, Inc., a division of HarperCollins Publishers.

to enforce its decisions. Nations can simply ignore the court if they wish. However, a growing network of multilateral environmental organizations have a greater sense of their roles and a greater incentive to work together. These include not only the United Nations Environment Program but also the Environment Committee of the Organization for Economic Cooperation and Development and the Senior Advisors on Environmental Problems of the Economic Commission for Europe. Such institutions perform unique functions that cannot be carried out by governments acting alone or bilaterally. Table 20.3 lists factors that affect international environmental laws.

Is the goal of an environmentally healthy world realistic? There is growing optimism that the community of nations is slowly maturing with regard to our common environment. We have all suffered a loss of innocence about "earth management." Laissez-faire may be good economics, but it can be a prescription for disaster in ecology.

This environmental "coming of age" is reflected in the broadening of intellectual perspective. Governments used to be preoccupied with domestic environmental affairs. Now, they are beginning to broaden their scope to confront problems that cross international borders, such as transboundary air and water pollution, and threats of a planetary nature, such as stratospheric ozone depletion and climatic warming. It is becoming increasingly evident that only decisive mutual action can secure the kind of world we seek.

BOX 20.7
What *You* Can Do to Make the World a Better Place to Live In

It is easy to complain; in fact, we all tend to do so almost on a daily basis. Complaining, however, never corrects a problem or helps to resolve a dispute. There are actions that you can take that *will* help in improving the environment. The following list offers a few suggestions, and others will probably come to mind once you have thought about it. Try to expand the list with others in your class, and give your suggestions to your instructor. Have your instructor send your suggestions to the authors at Delta College, University Center, Michigan 48710, and we can expand the list in the next edition of the textbook. Remember, you do not have to start big to help!

1. **Continue your education.** Learning should not stop when you leave school. Become informed about issues; then you can begin to bring about change.

2. **Do not feel responsible for every problem in the world.** You cannot do everything. Concentrate on issues that you feel strongly about and that you can do something about. Focus your energy.

3. **Think about the consequences of your profession and your life-style.** If they are damaging to other people or to the environment, adjust your behavior accordingly. Try to persuade friends, family, and coworkers to do the same. Make environmental awareness a family affair.

4. **Work with others.** Attend meetings of your local government and ask officials about their plans to prevent pollution. Often, officials are very responsive to visits of this kind. Being part of a group of people with similar interests gives you support and increases your effectiveness. If you cannot find an appropriate group, start one of your own.

5. **Become active in your community.** Organize a community conference to discuss positive approaches to pollution prevention. Invite public officials, industry and labor representatives, other interested groups, and individual citizens. Get all the facts, and then try to get the appropriate action programs initiated.

6. **Learn about the ecology of your bioregion.** Develop a sense of place that puts you in contact with your local physical environment. Learn about the unique environmental features of your area. What are the most urgent environmental problems?

7. **Vote.** You cannot improve your world by not voting. If you do not like the choices available, work to get individuals on the ballot who represent your interests.

8. **Think globally and act locally.** You need to be aware of global conditions, but you should also work to improve your own particular place.

9. **Do not be discouraged.** It is important to face facts honestly and to be realistic about the state of the world, but it does not help to wallow in despair. Do not dwell on negatives. Do what you can to improve the world, and take pleasure and pride in the small victories and elements of success.

10. **Try to leave things better than you found them.** Pick up a piece of litter on a beach or on your street, plant a tree, recycle your papers . . . the list goes on. Don't wait for the next person to begin—you can make a difference!

CONSIDER THIS CASE STUDY
What Should We Expect of the EPA?

In the two decades of its existence, the Environmental Protection Agency (EPA) has accomplished a great deal. Yet, many environmental activists feel that the EPA has lost its way. In its early years, the EPA was full of zeal to achieve its mission. Today, many view the EPA as middle-aged and cautiously balancing its objectives. Some observers even detect in the EPA the early stages of senility, in which an agency just "goes through the motions."

For the past twenty years, the EPA has been basically a "reactive" agency. As environmental problems were identified, the public conveyed its concern to Congress, and Congress passed laws to try to solve the problems, often within a well-defined time frame. The EPA then implemented the laws, using the resources, budget, and staff allocated by Congress. Consequently, the EPA has seen its mission largely as managing the reduction of pollution, particularly that pollution defined in the environmental legislation the EPA is required to implement. Moreover, the tools the EPA traditionally has used to reduce pollution have been limited to the emissions controls it could force polluters to comply with through regulatory action.

This reactive mode, although understandable when seen in its historical context, has limited the efficiency and effectiveness of the EPA's environmental protection efforts. Because of the EPA's tendency to react to environmental problems defined in specific environmental laws, the agency has made little effort to compare the relative seriousness of different problems, to anticipate environmental problems, or to take preemptive actions that reduce the likelihood of an environmental problem occurring.

Because most of the EPA's separate offices have been responsible for implementing specific laws, they have tended to view environmental problems separately; each office has been concerned primarily with those problems that it has been mandated to remediate, and questions of relative seriousness or urgency generally have remained unasked. Consequently, at the EPA, there has been little correlation between the relative resources dedicated to different environmental problems and the relative risks posed by those problems.

As the EPA enters its third decade, it is facing tough questions, such as: How does the EPA communicate with the public? How does the EPA relate to the worldwide environmental movement? Is the EPA accountable? The EPA of the 1990s appears to have an expanded vision that includes more research, public education, technical assistance, and market incentives to prevent as well as clean up pollution. Changing the basic agenda of the EPA, however, will not be easy or happen quickly.

What should the role of the EPA be? Research? Enforcement? Education? Advocacy? Pollution prevention? Can the EPA be all five? Should it?

If you were drafting the budget for the EPA, how would you allocate funding?

Summary

Politics and the environment cannot be separated. In the United States, the government is structured into three separate branches, each of which impacts environmental policy.

April 22, 1970 was the first Earth Day, and in the opinion of many, it was the event that gave birth to the modern environmental movement. By 1990, it appeared that the United States political will to address environmental concerns was increasing. Earth Day II on 22 April 1990 was seen as the beginning of a new era of environmental concern.

The late 1980s and early 1990s witnessed a new international concern about the environment, both in the developed and developing nations of the world. Environmentalism is also seen as a growing factor in international relations. This concern is leading to international cooperation where only tension has existed before.

While there exists no world political body that can enforce international environmental protection, the list of multilateral environmental organizations is growing.

It remains too early to tell what the ultimate outcome will be, but progress is being made in protecting our common resources for future generations. Several international conventions and treaties have been successful. In the final analysis, however, each of us has to adjust our life-style to clean up our own small part of the world.

Review Questions

1. What are the major responsibilities of each of the three branches of the U.S. government?
2. In which branch of the government would a group find it most effective to apply pressure to develop programs and policies favorable to that group?
3. How has public opinion in the United States changed concerning the protection of the environment in the past ten years?
4. Why is environmentalism a growing factor in international relations?
5. Give some examples of some international environmental conventions and treaties.

Pollution and Policy

APPENDIX ONE
Active Environmental Organizations

Alliance for Environmental Education, Inc.
Box 1040, 3421 M Street NW
Washington, DC 20037
(Works to further formal and informal educational
activities at all levels.)

Common Cause
2030 M Street NW
Washington, DC 20036
(An active citizens' lobby that covers a broad range
of political issues.)

Conservation Council of Ontario
Suite 202, 74 Victoria Street
Toronto M5C 2A5
(Coordinates provincial conservation associations.)

The Conservation Foundation
1255 Twenty-third Street NW
Washington, DC 20037
(Nonprofit conservation research and educational
association.)

Ducks Unlimited
One Waterford Way
Long Grove, IL 60047
(Primarily dedicated to the protection of migrating
waterfowl through acquisition of vital breeding
habitats.)

Environmental Defense Fund, Inc.
257 Park Avenue S
New York, NY 10010
(Works to link laws and science in defense of the
environment before courts and regulatory agencies.)

Environmental Policy Institute
218 D Street SE
Washington, DC 20003
(Lobbying for all aspects of energy development.)

Friends of the Earth
530 Seventh Street SE
Washington, DC 20003
(Active organization committed to the preservation,
restoration, and wise use of the earth.)

Great Lakes United
24 Agassiz Circle
Buffalo, NY 14214
(Focuses on Great Lakes issues.)

The Izaak Walton League of America
1701 North Fort Myer Drive, Suite 1100
Arlington, VA 22209
(Promotes conservation of renewable natural
resources and the development of outdoor
recreation.)

League of Women Voters of the United States
1730 M Street NW
Washington, DC 20003
(Works for political responsibility through an
informed and active citizenry.)

National Audubon Society
950 Third Avenue
New York, NY 10011
(Operates over forty wildlife sanctuaries across the
United States and provides a wide array of
environmental education services.)

National Parks and Conservation Association
1701 Eighteenth Street NW
Washington, DC 20009
(Works for the acquisition and protection of public
parklands.)

National Wildlife Federation
1412 Sixteenth Street NW
Washington, DC 20036
(Promotes citizen and governmental action for
conservation.)

The Nature Conservancy
Suite 800
1800 North Kent Street
Arlington, VA 22209
(A membership organization dedicated to protecting natural areas.)

Ontario Federation of Anglers and Hunters, Inc.
Box 28
Peterborough, Ontario K9J 6Y5
(Promotes ethical angling and hunting practices.)

Population Reference Bureau
2213 M Street NW
Washington, DC 20037
(Clearinghouse for data concerning the effects of a worldwide population explosion.)

Sierra Club
730 Polk Street
San Francisco, CA 94109
(Works to protect and conserve the world's natural resources.)

Smithsonian Institution
1000 Jefferson Drive SW
Washington, DC 20560
(Promotes environmental education through a wide variety of programs.)

Student Conservation Association, Inc.
Box 550
Charlestown, NH 03603
(Helps provide opportunities for conservation-minded students to work and learn during summer vacations.)

The Wilderness Society
1400 I Street NW, 10th floor
Washington, DC 20005

Wildlife Society
5410 Grosvenor Lane
Bethesda, MD 20814
(Professional society for the preservation of wildlife.)

Worldwatch Institute
1776 Massachusetts Avenue NW
Washington, DC 20036
(Research and education on major worldwide environmental problems.)

World Wildlife Fund
1255 Twenty-third Street NW
Washington, DC 20037
(Research and education on endangered species.)

Zero Population Growth
1601 Connecticut Avenue NW
Washington, DC 20009
(Primary emphasis is on family planning, population stabilization, and voluntary sterilization.)

For a more detailed list of national, state, and local environmental organizations, see *Conservation Directory,* published annually by the National Wildlife Federation, 1412 Sixteenth Street NW, Washington, DC 20036. For international organizations, see Mark Baker et al., *The World Environment Handbook,* New York: World Environment Center Books, 1985.

APPENDIX TWO
Metric Unit Conversion Tables

The metric system.		
Standard metric units		**Abbreviations**
Standard unit of mass	Gram	g
Standard unit of length	Meter	m
Standard unit of volume	Liter	l
Common prefixes		**Examples**
Kilo	1,000	A kilogram is 1,000 grams.
Centi	0.01	A centimeter is 0.01 meter.
Milli	0.001	A milliliter is 0.001 liter.
Micro (μ)	One-millionth	A micrometer is 0.000001 (one-millionth) of a meter.
Nano (n)	One-billionth	A nanogram is 10^{-9} (one-billionth) of a gram.
Pico (p)	One-trillionth	A picogram is 10^{-12} (one-trillionth) of a gram.

Units of length.		
Unit	**Abbreviation**	**Equivalent**
Meter	m	Approximately 39 in
Centimeter	cm	10^{-2} m
Millimeter	mm	10^{-3} m
Micrometer	μm	10^{-6} m
Nanometer	nm	10^{-9} m
Angstrom	Å	10^{-10} m

Length conversions

1 in = 2.5 cm		1 mm = 0.039 in	
1 ft = 30 cm		1 cm = 0.39 in	
1 yd = 0.9 m		1 m = 39 in	
1 mi = 1.6 km		1 m = 1.094 yd	
		1 km = 0.6 mi	

To convert	**Multiply by**	**To obtain**
Inches	2.54	Centimeters
Feet	30	Centimeters
Centimeters	0.39	Inches
Millimeters	0.039	Inches

Units of volume.		
Unit	**Abbreviation**	**Equivalent**
Liter	l	Approximately 1.06 qt
Milliliter	ml	10^{-3} l (1 ml = 1 cm^3 = 1 cc)
Microliter	μl	10^{-6} l
Volume conversions		

1 tsp	= 5 ml	1 ml	= 0.03 fl oz
1 tbsp	= 15 ml	1 l	= 2.1 pt
1 fl oz	= 30 ml	1 l	= 1.06 qt
1 cup	= 0.24 l	1 l	= 0.26 gal
1 pt	= 0.47 l		
1 qt	= 0.95 l		
1 gal	= 3.8 l		

To convert	**Multiply by**	**To obtain**
Fluid ounces	30	Milliliters
Quarts	0.95	Liters
Milliliters	0.03	Fluid ounces
Liters	1.06	Quarts

Units of weight.		
Unit	**Abbreviation**	**Equivalent**
Kilogram	kg	10^3 g (approximately 2.2 lb)
Gram	g	Approximately 0.035 oz
Milligram	mg	10^{-3} g
Microgram	μg	10^{-6} g
Nanogram	ng	10^{-9} g
Picogram	pg	10^{-12} g
Weight conversions		

1 oz	= 28.3 g	1 g	= 0.035 oz
1 lb	= 453.6 g	1 kg	= 2.2 lb
1 lb	= 0.45 kg		

To convert	**Multiply by**	**To obtain**
Ounces	28.3	Grams
Pounds	453.6	Grams
Pounds	0.45	Kilograms
Grams	0.035	Ounces
Kilograms	2.2	Pounds

Temperature conversions   **Some equivalents**

$$°C = \frac{(°F - 32) \times 5}{9}$$

$$°F = \frac{°C \times 9}{5} + 32$$

0° C = 32° F
37° C = 98.6° F
100° C = 212° F

SUGGESTED READINGS

General References That Publish Regularly on Environmental Issues

BP Statistical Review of World Energy. New York, BP America, Inc.

Council on Environmental Quality. *Annual Report.* Washington, D.C.: U.S. Government Printing Office.

National Wildlife Federation, 1400 Sixteenth Street NW, Washington, D.C. 20036. (Publishes annual directory of conservation organizations.)

Population Reference Bureau, Inc., 2213 M Street NW, Washington, D.C. 20037.

U.S. Bureau of the Census. *Statistical Abstract of the United States.* Washington, D.C.: U.S. Bureau of the Census.

World Resources Institute. *World Resources 1990–1991.* New York: Oxford University Press, 1990.

Worldwatch Institute, 1776 Massachusetts Avenue NW, Washington, D.C. 20036. (Publishes annual *State of the World.*)

Part One

Chapter One

Ashworth, W. *The Late, Great Lakes: An Environmental History.* New York: Alfred A. Knopf, 1986.

Great Lakes Diversions and Consumptive Uses. Ottawa, Ontario, and Washington, D.C.: International Joint Commission, 1985.

The Great Lakes: An Environmental Atlas and Resource Book. Chicago: U.S. Environmental Protection Agency; Toronto, Ontario: Environment Canada, 1987.

The Great Lakes Primer. Toronto, Ontario: Pollution Probe.

National Research Council of the United States and the Royal Society of Canada. *The Great Lakes Water Quality Agreement: An Evolving Instrument for Ecosystem Management.* Washington, D.C.: National Academy Press, 1985.

"A Tribute to the Great Lakes." *Michigan Natural Resources Magazine,* 55.3 (1986). Lansing, Mich.: Michigan Department of Natural Resources.

Chapter Two

Carson, Rachel. *Silent Spring.* Boston: Houghton Mifflin, 1962.

Daly, Herman E., ed. *Economics, Ecology, and Ethics.* San Francisco: W. H. Freeman, 1980.

Devall, Bill, and George Sessions. *Deep Ecology: Living As If Nature Mattered.* Salt Lake City, Utah: Gibbs M. Smith, 1985.

Emerson, Ralph W. *Society and Solitude.* Boston: Houghton Mifflin, 1883.

Environmental Ethics Journal. Athens, Ga.: University of Georgia, Department of Philosophy.

Leopold, Aldo. *A Sand County Almanac.* New York: Sierra Club/Ballantine Books, 1949.

Lovelock, James E. *The Ages of Gaia: A Biography of Our Living Earth.* New York: Norton, 1988.

Naess, Arne. *Ecology, Community, and Lifestyle.* New York: Cambridge University Press, 1989.

Nash, Roderick. *The Rights of Nature: A History of Environmental Ethics.* Madison: University of Wisconsin Press, 1988.

Regan, Tom. *Earthbound: New Introductory Essays in Environmental Ethics.* New York: Random House, 1984.

Rifkin, Jeremy. *Declaration of a Heretic.* Boston: Routledge & Kegan Paul, 1985.

Schumacher, E. F. *Small Is Beautiful.* New York: Harper & Row, 1973.

Shrader-Frechette, Kristin. "Environmental Ethics and Global Imperatives." In *The Global Possible: Resources, Development, and the New Century,* edited by Robert Repetto. New Haven, Conn.: Yale University Press, 1986.

Taylor, Paul W. *Respect for Nature: A Theory of Environmental Ethics.* Lawrenceville, N.J.: Princeton University Press, 1986.

Thoreau, Henry D. *Walden.* New York: W. W. Norton, 1966.

Part Two

Chapter Three

Brady, James, and Gerard E. Humiston. *General Chemistry: Principles and Structure.* 4th ed. New York: John Wiley & Sons, 1986.

Hill, J. W. *Chemistry for Changing Times.* 5th ed. New York: Macmillan, 1988.

Kotz, J. C., and K. F. Purcell. *Chemistry and Chemical Reactivity.* Philadelphia: Saunders College Publishing, 1987.

von Frisch, Karl. *Bees: Their Vision, Chemical Senses and Language.* Rev. ed. Ithaca, N.Y.: Cornell University Press, 1971.

Chapter Four

Brewer, Richard. *The Science of Ecology.* New York: Saunders, 1988.

Colinvaux, Paul. *Ecology.* New York: John Wiley & Sons, 1986.

Ehrlich, P. R., and J. Roughgarden. *The Science of Ecology.* New York: Macmillan, 1987.

Gates, David M. *Energy and Ecology.* Sunderland, Mass.: Sinauer, 1985.

Odum, Eugene P. *Basic Ecology.* New York: Saunders College Publishing, 1983.

Odum, Eugene P. *Ecology and Our Endangered Life-Support Systems.* Sunderland, Mass.: Sinauer, 1989.

Odum, Howard T. *Systems Ecology: An Introduction.* New York: John Wiley & Sons, 1983.

Smith, Robert Leo. *Ecology and Field Biology.* 4th ed. New York: Harper & Row, 1990.

Chapter Five

(The references in chapter four would also be useful.)

Attenborough, David. *The Living Planet.* Boston: Little, Brown, 1984.

Calder, Nigel. *The Weather Machine: How Our Weather Works and Why It Is Changing.* New York: Viking Press, 1974.

Clapham, W. B., Jr. *Natural Ecosystems.* 2d ed. New York: Macmillan, 1989.

Hynes, H. B. N. *The Biology of Running Waters.* Toronto: University of Toronto Press, 1970.

McArthur, R. H. *Geographical Ecology.* New York: Harper & Row, 1972.

Mitsch, W. J., and J. G. Gosselink. *Wetlands.* New York: Van Nostrand Reinhold, 1986.

Whittaker, R. H. *Communities and Ecosystems.* 2d ed. New York: Macmillan, 1975.

Chapter Six

(The references in chapter four would also be useful.)

Andrewartha, H. G., and L. C. Birch. *The Ecological Web: More on the Distribution and Abundance of Animals.* Chicago: University of Chicago Press, 1986.

Begon, Michael, and Martin Mortimer. *Population Ecology: A Unified Study of Animals and Plants.* Sunderland, Mass.: Sinauer, 1986.

Haupt, Arthur, and Thomas T. Kane. *Population Handbook: A Quick Guide to Population Dynamics for Journalists, Policymakers, Teachers, Students, and Other People Interested in People.* Updated ed. Washington, D.C.: Population Reference Bureau, 1982.

Slobodkin, Laurence B. *Growth and Regulation of Animal Populations.* New York: Dover Press, 1980.

Wilson, E. O., ed. *Biodiversity.* Washington, D.C.: National Academy Press, 1988.

Chapter Seven

(The references in chapter six would also be useful.)

Ehrlich, P. R. *The Population Bomb.* New York: Ballantine Books, 1968.

Gupte, Pranay. *The Crowded Earth: People and the Politics of Population.* New York: W. W. Norton, 1984.

Hardin, Garrett. *Mandatory Motherhood: The True Meaning of "Right to Life."* Boston: Beacon Press, 1974.

Hardin, Garrett. *Naked Emperors, Essays of a Taboo Stalker.* San Francisco: William Kaufman, 1982.

Hardin, Garrett, ed. *Population, Evolution, and Birth Control.* New York: W. H. Freeman, 1977.

Population Reference Bureau, Inc., 2213 M Street NW, Washington, D.C. 20037. (A good source of up-to-date information on a variety of human population issues.)

Part Three

Chapter Eight

BP Statistical Review of World Energy. New York: BP America, Inc., 1990.

Edmonds, Jae, and John M. Reilly. *Global Energy: Assessing the Future.* New York: Oxford University Press, 1985.

Flavin, Christopher. *Electricity for a Developing World.* Washington, D.C.: Worldwatch Institute, 1986.

Gever, John. *Beyond Oil.* Washington, D.C.: Carrying Capacity, 1986.

Hughes, Barry B. *Energy in the Global Arena: Values, Policies, and Futures.* Durham, N.C.: Duke University Press, 1985.

Meadows, Donella H. et al. *The Limits to Growth.* New York: Universe Books, 1972.

Odum, Howard T., and Elisabeth C. Odum. *Energy Basis for Man and Nature.* New York: McGraw-Hill, 1980.

Simmons, I. G. *Changing the Face of the Earth: Culture, Environment, and History.* London: Basil Blackwell, 1989.

World Resources Institute. *World Resources 1990–1991.* New York: Oxford University Press, 1990.

Chapter Nine

(The references in chapter eight would also be useful.)

American Institute of Physics. *Energy Efficiency and Renewable Resources.* New York: American Institute of Physics, 1985.

Blackburn, John O. *The Renewable Energy Alternative: How the United States and the World Can Prosper without Nuclear Energy or Coal.* Durham, N.C.: Duke University Press, 1987.

Bleviss, Deborah Lynn. *The New Oil Crisis and Fuel Economy Technologies: Preparing the Light Transportation Industry for the 1990s.* Westport, Conn.: Quorum Press, 1988.

Chandler, William U. et al. *Energy Efficiency: A New Agenda.* Washington, D.C.: American Council for an Energy-Efficient Economy, 1988.

Charlier, Roger H. *Tidal Energy.* New York: Van Nostrand Reinhold, 1982.

Glasner, David. *Politics, Prices, and Petroleum: The Political Economy of Energy.* San Francisco: Pacific Institute of Public Policy Analysis, 1986.

Lovins, Amory B. *Soft Energy Paths.* Cambridge, Mass.: Ballinger, 1977.

Miller, Alan S. *Growing Power: Bioenergy for Development and Industry.* Washington, D.C.: World Resources Institute, 1986.

Sawyer, Stephen W. *Renewable Energy: Progress, Prospects.* Washington, D.C.: Association of American Geographers, 1986.

Swan, Christopher C. *Suncell: Energy, Economy, Photovoltaics.* New York: Random House, 1986.

Chapter Ten

Campbell, John L. *Collapse of an Industry: Nuclear Power and the Contradictions of U.S. Policy.* Ithaca, N.Y.: Cornell University Press, 1988.

Cohen, Bernard L. *Before It's Too Late: A Scientist's Case for Nuclear Power.* New York: Plenum Press, 1983.

Flavin, Christopher. *Reassessing Nuclear Power: The Fallout from Chernobyl.* Washington, D.C.: Worldwatch Institute, 1987.

Ford, Daniel F. *Meltdown.* New York: Simon and Schuster, 1986.

Loeb, Paul. *Nuclear Culture: Living and Working in the World's Largest Atomic Complex.* Philadelphia: New Society, 1986.

Marples, David R. *Chernobyl and Nuclear Power in the USSR.* New York: St. Martin's Press, 1986.

Mould, Richard F. *Chernobyl: The Real Story.* New York: Pergamon, 1988.

Pochin, Edward. *Nuclear Radiation: Risks and Benefits.* New York: Oxford University Press, 1985.

Pollack, Cynthia. *Decommissioning: Nuclear Power's Missing Link.* Washington, D.C.: Worldwatch Institute, 1986.

President's Commission on the Accident at Three Mile Island. *Report of the President's Commission on the Accident at Three Mile Island.* Washington, D.C.: U.S. Government Printing Office.

Part Four

Chapter Eleven

Bailey, J. A. *Priciples of Wildlife Management.* New York: Wiley, 1984.

Bronowski, Jacob, Jr. *The Ascent of Man.* Boston: Little, Brown, 1974.

Brundtland, G. H. *Our Common Future: World Commission on Environment and Development.* New York: Oxford University Press, 1987.

Cairns, John, ed. *Rehabilitating Damaged Ecosystems.* Toledo, Ohio: CRC Press, 1988.

Carson, Rachel. *Silent Spring.* Boston: Houghton Mifflin, 1962.

Dana, Samuel T., and Salley K. Fairfax. *Forest and Range Policy: Its Development in the United States.* 2d ed. New York: McGraw Hill, 1980.

Erhlich, Paul R., and Anne H. Erhlich. *Extinction: The Causes and Consequences of the Disappearance of Species.* New York: Ballantine, 1983.

Harris, L. D. *The Fragmented Forest: Island Biogeography Theory and the Preservation of Biotic Diversity.* Chicago: Chicago University Press.

Kimmins, Peter H., and Dennis D. DiPietre. *Forest Ecology: Successful Decisions in a Changing Environment.* New York: Macmillan, 1983.

Leopold, Aldo. *Game Management.* New York: Charles Scribner's Sons, 1933.

Leopold, Aldo. *A Sand County Almanac with Essays on Conservation from Round River.* New York: Oxford University Press, 1966.

Smith, Robert Leo. *Ecology and Field Biology.* 4th ed. New York: Harper & Row, 1990.

Soulé, Michael E., ed. *Viable Populations for Conservation.* New York: Cambridge University Press, 1987.

Spurr, Stephen H., and Buron V. Barnes. *Forest Ecology.* 3d ed. New York: Ronald Press, 1980.

Stoddard, Charles H., and Glenn M. Stoddard. *Essentials of Forestry Practice.* 4th ed. New York: John Wiley, 1987.

Chapter Twelve

Agriculture and the Environment in a Changing World Economy. Washington, D.C.: Conservation Foundation, 1986.

Baker, R. Lisle, and Norman H. Wolfe. *Negotiated Development and Open Space Preservation.* Cambridge, Mass.: Lincoln Institute of Land Policy, 1984.

Beewer, Wm. E., and Charles P. Alter. *The Complete Manual of Land Planning and Development.* Englewood Cliffs, N.J.: Prentice-Hall, 1988.

Black, Peter E. *Conservation of Water and Related Land Resources.* Totowa, N.J.: Rowman and Littlefield, 1987.

Fabos, J. G. *Land-Use Planning.* New York: Chapman and Hall, 1985.

Horowitz, Alan J. et al. *Assessment of Land-Use Impacts of Highways in Small Urban Areas.* Milwaukee: University of Wisconsin–Milwaukee, Center for Architecture and Urban Planning Research, 1985.

McHarg, Ian L. *Design with Nature.* Garden City, N.Y.: Natural History Press, 1969.

Chapter Thirteen

Courtney, F. M., and S. T. Trudgill. *Soil: An Introduction to Soil Study.* 2d ed. Baltimore: E. Arnold, 1984.

Dale, Tom, and V. G. Carter. *Topsoil and Civilization.* Norman, Okla.: University of Oklahoma Press, 1955.

Dregnue, Harold E. *Desertification of Arid Lands.* New York: Academic Press, 1983.

Hudson, Norman. *Soil Conservation.* 2d ed. Ithaca, N.Y.: Cornell University Press, 1985.

Morgan, R. P. *Soil Erosion and Its Control.* New York: Van Nostrand Reinhold, 1986.

National Academy of Sciences. *Soil Conservation.* Washington, D.C.: National Academy Press, 1986.

Paddock, Joe et al. *Soil and Survival: Land Stewardship and the Future of American Agriculture.* San Francisco: Sierra Club Books, 1987.

Sophen, C. D., and J. V. Baird. *Soils and Soil Management.* Reston, Va.: Reston Publishing, 1982.

Tompkins, Peter, and Christopher Bird. *Secrets of the Soil.* New York: Harper & Row, 1989.

U.S. Department of Agriculture. *Analysis of Policies to Conserve Soil and Reduce Surplus Crop Production.* Washington, D.C.: U.S. Department of Agriculture Economic Report 534, 1985.

Wilson, G. F. et al. *The Soul of the Soil: A Guide to Ecological Soil Management.* 2d ed. Quebec, Canada: Gaia Services, 1986.

Chapter Fourteen

Aliteri, Miguel A. *Agroecology: The Scientific Basis of Alternative Agriculture.* Berkeley, Calif.: Division of Biological Control, University of California, Berkeley, 1983.

Carson, Rachel. *Silent Spring.* New York: Fawcett Crest, 1966.

Georghiou, G. P., and Tetsuo Saito. *Pest Resistance and Pesticides.* New York: Plenum Publishers, 1983.

Gips, Terry. *Breaking the Pesticide Habit.* Minneapolis, Minn.: IASA, 1987.

Jackson, Wes. *New Roots for Agriculture.* San Francisco: Friends of the Earth, 1980.

Jackson, Wes et al., eds. *Meeting the Expectations of Land: Essays in Sustainable Agriculture and Stewardship.* Berkeley, Calif.: North Point Press, 1985.

Lockeretz, William G., ed. *Sustaining Agriculture Near Cities.* Ankeny, Iowa: Soil and Water Conservation Society, 1987.

Lowrance, Richard, ed. *Agricultural Ecosystems: Unifying Concepts.* Somerset, N.J.: John Wiley, 1984.

Matthews, G. A. *Pest Management.* John Wiley & Sons, 1985.

Pimentel, David, ed. *Handbook of Pest Management.* New York: CRC Press, 1981.

Pimentel, David, and Carl W. Hall. *Food and Energy Resources.* Orlando, Fla.: Academic Press, 1984.

Pimentel, David, and Carl W. Hall. *Food and Natural Resources.* Orlando, Fla.: Academic Press, 1989.

van den Bosch, Robert, and Mary L. Flint. *Introduction to Integrated Pest Management.* New York: Plenum Press, 1981.

Ware, George W. *Complete Guide to Pest Control.* 2d ed. Fresno, Calif.: Thomson Publications, 1988.

Ware, George W. *The Pesticide Book.* 3d ed. Fresno, Calif.: Thomson Publications, 1989.

Chapter Fifteen

Anderson, T. L. *Water Rights: Scarce Resource Allocation, Bureaucracy, and the Environment.* San Francisco: Pacific Institute for Public Policy, 1986.

Carmichael, J., ed. *Industrial Water Use and Treatment.* New York: Tayler and Francis, 1986.

Clark, Edwin H. II et al. *Eroding Soils: The Off-Farm Impacts.* Washington, D.C.: Conservation Foundation, 1985.

Consumer Reports Book Editors and Raymond Gabler. *Is Your Water Safe to Drink?* New York: Consumer Reports Books, 1987.

DeKruij, H. A., and H. J. Kool, eds. *Organic Micropollutants in Drinking Water and Health.* New York: Elsevier Science, 1986.

Evaluation of Hydrocarbon in Runoff to San Francisco Bay. Oakland, Calif.: Association of Bay Area Governments, 1985.

Franco, David A., and Robert G. Wetel. *To Quench Our Thirst: The Present and Future Status of Freshwater Resources of the United States.* Ann Arbor, Mich.: University of Michigan Press, 1983.

Giorgini, A. et al., eds. *Agricultural Nonpoint Source Pollution: Model Selection and Application.* New York: Elsevier Science, 1986.

Golfarb, William. *Water Law.* 2d ed. New York: Lewis Publishers, 1988.

Gottlieb, Robert. *A Life of Its Own: The Politics and Power of Water.* New York: Harcourt Brace Jovanovich, 1989.

Murty, A. S., ed. *Toxicity of Pesticides to Fish.* Vol. 1. Boca Raton, Fla.: CRC Press, 1986.

National Research Council Groundwater Contamination, Oil in the Sea: Inputs, Fates and Effects. Washington, D.C.: National Academy Press, 1985.

Nemerow, Nelson L. *Stream, Lake, Estuary and Ocean Pollution.* New York: Van Nostrand Reinhold, 1985.

Patrick, R., E. Ford, and J. Quarles, eds. *Groundwater Contamination in the United States.* Philadelphia: University of Pennsylvania Press, 1987.

Pettyjohn, Wayne A., ed. *Protection of Public Water Supplies from Groundwater Contamination.* Park Ridge, N.J.: Noyes Data Corporation, 1987.

Rail, Chester D. *Groundwater Contamination: Sources, Control, and Preventive Measures.* Lancaster, Penn.: Technomic Publishing, 1989.

Shainberg, I., and J. Shalhevet, eds. *Soil Salinity under Irrigation.* New York: Springer-Verlag, 1984.

Part Five

Chapter Sixteen

Boulding, Kenneth E. *The World As a Total System.* Beverly Hills, Calif.: Sage Publications, 1985.

Bowden, Elbert V. *Principles of Economics: Theory, Problems, Policies.* 4th ed. Cincinnati, Ohio: South-Western, 1987.

Collard, David et al., eds. *Economics, Growth, and Sustainable Environments.* New York: St. Martin's Press, 1988.

Covello, V. T. et al. *Risk Evaluation and Management.* New York: Plenum, 1986.

Daly, Herman E. *Steady-State Economics.* San Francisco: W. H. Freeman, 1977.

Freeman, A. Myrick III. *Air and Water Pollution Control: A Benefit-Cost Assessment.* New York: John Wiley, 1982.

Galbraith, John Kenneth. *Economics in Perspective: A Critical History.* New York: Houghton Mifflin, 1988.

Hardin, Garrett. "The Tragedy of the Commons." *Science* 162 (1968): 1243–48.

Lave, Lester B. *Risk Assessment and Management.* New York: Plenum, 1987.

McConnell, Campbell R. *Economics: Principles, Problems, and Policies.* 10th ed. New York: McGraw-Hill, 1987.

Meeker-Lowry, Susan. *Economics As If the Earth Mattered: A Catalyst Guide to Socially Conscious Investing.* Philadelphia, Penn.: New Society Publishers, 1988.

Pearson, C. S. *Multinational Corporations, Environment, and the Third World.* Washington, D.C.: World Resources Institute, Duke University Press, 1987.

Schumacher, E. F. *Small Is Beautiful: Economics As If People Mattered.* New York: Harper & Row, 1973.

Chapter Seventeen

Abrahamson, Dean E., ed. *The Challenge of Global Warming.* Covelo, Calif.: Island Press, 1989.

Air Pollution and Acid Rain. Bowling Green Station, N.Y.: Gordon Press, 1986.

Brenner, David J. *Radon: Risk and Remedy.* Salt Lake City, Utah: W. H. Freeman, 1989.

Brown, Michael H. *The Toxic Cloud: A Cross-Country Report of the Poisoning of America's Air.* New York: Harper & Row, 1987.

Flagan, Richard C., and John H. Seinfeld. *Fundamentals of Air Pollution Engineering.* Englewood Cliffs, N.J.: Prentice-Hall, 1988.

Freedman, Bill. *Environmental Ecology: The Impacts of Pollution and Other Stresses on Ecosystem Structure and Function.* San Diego, Calif.: Academic Press, 1989.

Geller, H. et al. *Acid Rain and Energy Conservation.* Washington, D.C.: American Council for an Energy-Efficient America, 1986.

Kryter, Karl D. *The Effects of Noise.* 2d ed. Orlando, Fla.: Academic Press, 1985.

Loiy, Paul J., and Joan M. Daisey. *Toxic Air Pollution.* Chelsea, Mich.: Lewis Publishing, 1987.

Luoma, Jon R. *The Air around Us, and Air Pollution Primer.* Edited by Harriett S. Stubbs. Washington, D.C.: Acid Rain Foundation, 1987.

Mintzer, Irving M. et al. *Protecting the Ozone Shield: Strategies for Phasing Out CFCs during the 1990s.* Washington, D.C.: World Resources Institute, 1989.

Postel, Sandra. *Air Pollution, Acid Rain, and the Future of Forests.* Washington, D.C.: Worldwatch Institute, 1984.

Seinfeld, John H. *Atmospheric Chemistry and Physics of Air Pollution.* New York: John Wiley & Sons, 1986.

Chapter Eighteen

Abernathy, William et al. *Industrial Renaissance.* New York: Basic Books, 1983.

Berger, John. *Restoring the Earth: How Americans Are Working to Renew Our Damaged Environment.* New York: Alfred A. Knopf, 1986.

Chandler, W. U. *Materials Recycling: The Virtue of Necessity.* Washington, D.C.: Worldwatch Institute, 1983.

Environmental Protection Agency. *Solid Waste Disposal in the United States.* Washington, D.C.: U.S. Government Printing Office, 1989.

Hershkowitz, Allen, and Eugene Salermi. *Garbage Management in Japan: Leading the Way.* New York: INFORM, 1987.

Neal, Homer A., and J. R. Schubel. *Solid Waste Management and the Environment: The Mounting Garbage and Trash Crisis.* Englewood Cliffs, N.J.: Prentice-Hall, 1987.

Office of Technology Assessment. *Facing America's Trash: What's Next for Municipal Solid Waste.* Washington, D.C.: U.S. Government Printing Office, 1989.

Pollack, Cynthia. *Mining Urban Wastes: The Potential for Recycling.* Washington, D.C.: Worldwatch Institute, 1987.

Polprasert, Chorgrah. *Organic Waste Recycling.* New York: John Wiley, 1989.

Rosenbaum, Walter A. *Environment, Politics, and Policy.* Washington, D.C.: Congressional Quarterly, 1991.

Seldman, Neil, and Bill Perkins. *Designing the Waste Stream.* Washington, D.C.: Institute for Local Self-Reliance, 1988.

Stokes, Bruce. *Helping Ourselves: Local Solutions to Global Problems.* New York: W. W. Norton, 1981.

Tietenberg, Tom. *Environmental and Natural Resource Economics.* Glenview, Ill.: Scott, Foresman, 1984.

Chapter Nineteen

Clarke, Lee. *Acceptable Risk? Making Decisions in a Toxic Environment.* Berkeley, Calif.: University of California Press, 1989.

Dowling, Michael. "Defining and Classifying Hazardous Wastes." *Environment* 27 (1985): 18–20, 36–41.

Environmental Protection Agency. *Damages and Threats Caused by Hazardous Material.* Washington, D.C.: U.S. Environmental Protection Agency, 1980.

Fortuna, Richard C., and David J. Lennett. *Hazardous Waste Regulation: The New Era.* New York: McGraw-Hill, 1987.

League of Women Voters Education Fund. *A Hazardous Waste Primer.* Washington, D.C.: League of Women Voters, 1981.

Lester, James P., and Ann O. M. Bowman, eds. *Politics of Hazardous Waste Management.* Durham, N.C.: Duke University Press, 1983.

Office of Technology Assessment. *Serious Reduction of Hazardous Waste.* Washington, D.C.: U.S. Government Printing Office, 1986.

Piasecki, Bruce, ed. *Beyond Dumping: New Strategies for Controlling Toxic Contamination.* Westport, Conn.: Quorum Books, Greenwood Press, 1984.

Robinson, William D., ed. *The Solid Waste Handbook.* New York: Wiley, 1986.

World Health Organization. *1989 World Health Statistics Annual.* Geneva, Switzerland: World Health Organization, 1989.

Chapter Twenty

Adams, Jane. *Democracy and Societal Ethics.* Raleigh, N.C.: American Biographical Institute, 1985.

Bahro, Rudolf. *Building the Green Movement.* London: Heretic Books, 1986.

Borrelli, Peter, ed. *Crossroads: Environmental Priorities for the Future.* Covelo, Calif.: Island Press, 1988.

Bureau of National Affairs. *U.S. Environmental Laws: 1989 Edition.* Washington, D.C.: Bureau of National Affairs Books, 1989.

Durning, Alan B. *Action at the Grassroots: Fighting Poverty and Environmental Decline*. Washington, D.C.: Worldwatch Institute, 1989.

Gruber, Judith E. *Controlling Bureaucracies, Dilemmas in Democratic Governance*. Berkeley, Calif.: University of California Press, 1986.

Lemos, Ramon M. *Rights, Goods, and Democracy*. Cranbury, N.J.: University of Delaware Press, 1986.

Ophuls, William. *Ecology and the Politics of Scarcity*. San Francisco: W. H. Freeman, 1977.

Paehilke, Robert C. *Environmentalism and the Future of Progressive Politics*. New Haven, Conn.: Yale University Press, 1989.

Renner, Michael. *National Security: The Economic and Environmental Dimensions*. Washington, D.C.: Worldwatch Institute, 1989.

Yandle, Bruce. *The Political Limits of Environmental Regulation*. Westport, Conn.: Quorum Books, 1989.

GLOSSARY

A

abiotic factors
Nonliving factors that influence the life and activities of an organism.

abyssal ecosystem
The collection of organisms and the conditions that exist in the deep portions of the ocean.

acid
Any substance that, when dissolved in water, releases hydrogen ions.

acid deposition
The accumulation of potential acid-forming particles on a surface.

acid mine drainage
A kind of pollution, associated with coal mines, in which bacteria convert the sulfur in coal into compounds that form sulfuric acid.

acid rain (acid precipitation)
The deposition of wet acidic solutions or dry acidic particles from air.

activated sludge sewage treatment
Method of treating sewage in which some of the sludge is returned to aeration tanks, where it is mixed with incoming wastewater to encourage degradation of the wastes in the sewage.

activation energy
The initial energy input required to start a reaction.

active solar system
A system that traps sunlight energy as heat energy and uses mechanical means to move it to another location.

acute toxicity
A serious effect, such as a burn, illness, or death, that occurs shortly after exposure to a hazardous-waste substance.

age distribution
The comparative percentages of different age groups within a population.

agricultural products
Any output from farming: milk, grain, meat, etc.

agricultural runoff
Surface water that carries nutrients, such as phosphate and nitrates, as it runs off agricultural land to lakes and streams.

air stripping
The process of pumping air through water to remove volatile materials dissolved in the water.

alpha radiation
A type of radiation consisting of a particle with two neutrons and two protons.

alpine tundra
The biome that exists above the tree line in mountainous regions.

aquiclude
An impermeable layer in an artesian aquifer.

aquifer
A layer of earth material that can transmit water sufficient for water supply purposes.

aquitard
A permeable layer in an artesian aquifer.

artesian aquifer
The result of a pressurized aquifer intersecting the surface or being penetrated by a pipe or conduit, from which water gushes without being pumped.

atom
The basic subunit of elements, composed of protons, neutrons, and electrons.

atomic fission
The decomposition of an atom's nucleus, with the release of particles and energy.

auxin
A plant hormone that stimulates growth.

B

base
Any substance that, when dissolved in water, removes hydrogen ions from solution; forms a salt when combined with an acid.

benthic
Describes organisms that live on the bottom of marine and freshwater ecosystems.

benthic ecosystem
A type of marine or freshwater ecosystem consisting of organisms that live on the bottom.

beta radiation
A type of radiation consisting of electrons released from the nuclei of many fissionable atoms.

bioaccumulation
The buildup of a material in the body of an organism.

biochemical oxygen demand (BOD)
The amount of oxygen required to destroy organic molecules in aquatic ecosystems.

biocide
A kind of chemical that kills many different types of living things.

biodegradable
Able to be broken down by natural biological processes.

biological amplification
The increases in the amount of a material at successively higher trophic levels.

biomass
The weight of a certain amount of living material.

biome
A kind of plant and animal community that covers major geographic areas. Climate is a major determiner of the biome found in a particular area.

biotic factors
Living portions of the environment.

biotic potential
The inherent reproductive capacity.

birthrate (natality)
The number of individuals added to the population through reproduction per thousand individuals per year.

black lung disease
A respiratory condition resulting from the accumulation of large amounts of fine coal dust particles in miners' lungs.

boiling-water reactor (BWR)
A type of light-water reactor in which steam is formed directly in the reactor, which is used to generate electricity.

boreal forest
A broad band of mixed coniferous and deciduous trees that stretches across northern North America (and also Europe and Asia); its northernmost edge is integrated with the arctic tundra.

C

carbamate
A class of soft pesticides that work by interfering with normal nerve impulses.

carbon absorption
The use of carbon particles to treat chemicals by having the chemicals attach to the carbon particles.

carbon cycle
The cyclic flow of carbon from the atmosphere to living organisms and back to the atmospheric reservoir.

carbon dioxide (CO_2)
A normal component of the earth's atmosphere, which, when in elevated concentrations, may interfere with the earth's heat budget.

carbon monoxide (CO)
A primary air pollutant produced when organic materials, such as gasoline, coal, wood, and trash, are incompletely burned.

carcinogen
A substance that causes cancer.

carnivores
Animals that eat other animals.

carrying capacity
The optimum number of individuals of a species that can be supported in an area over an extended period of time.

catalyst
A substance that alters the rate of a reaction but is not itself changed.

cellular respiration
The process that organisms use to release chemical bond energy from food.

chain reaction
A nuclear reaction in which the products of the disintegration of one nucleus cause the disintegration of other nuclei, which, in turn, cause other nuclei to disintegrate.

chemical bond
The physical attraction between atoms that results from the interaction of their electrons.

chlorinated hydrocarbon
A class of pesticide consisting of carbon, hydrogen, and chlorine, which are very stable.

chronic toxicity
A serious effect, such as an illness or death, that occurs after prolonged exposure to small doses of a toxic substance.

clear-cutting
A forest harvesting method in which all the trees in a large area are cut and removed.

climax community
Last stage of succession; a relatively stable, long-lasting, complex, and interrelated community of plants, animals, fungi, and bacteria.

combustion
The process of releasing chemical bond energy from fuel.

commensalism
The relationship between organisms in which one organism benefits while the other is not affected.

community
Interacting groups of different species.

competition
An interaction between two organisms in which both require the same limited resource, which results in harm to both.

composting
A waste-disposal system whereby organic matter is allowed to decay to a usable product.

compound
A kind of matter composed of two or more different kinds of atoms.

Comprehensive Environmental Response, Compensation, and Liability Act (CERCLA)
The 1980 U.S. law that addressed the issue of cleanup of hazardous-waste sites.

comprehensive water management
The process of protecting and developing water resources in a broad, integrated, and foresighted manner.

comprehensive water planning
Process used to integrate the diverse functions necessary for comprehensive water management.

confined aquifer
An aquifer that is bounded on the top and bottom by confining layers.

conservation
To use in the best possible way so that the greatest long-term benefit is realized by society.

conservation ethic
An environmental ethic that stresses a balance between total development and absolute preservation.

consumers
Organisms that rely on other organisms for food.

contour farming
A method of tilling and planting at right angles to the slope, which reduces soil erosion by runoff.

controlled experiment
An experiment in which two groups are compared. One, the control, is used as a basis of comparison and the other, the experimental, has one factor different from the control.

coral reef ecosystem
A tropical, shallow-water, marine ecosystem dominated by coral organisms that produce external skeletons.

corporation
A business structure that has a particular legal status.

corrosiveness
Ability to degrade standard materials.

cost-benefit analysis
A method used to determine the feasibility of pursuing a particular project by balancing estimated costs against expected benefits.

cover
A term used to refer to any set of physical features that conceals or protects animals from the elements or their enemies.

D

death phase
The portion of the population growth curve that shows the population declining.

deathrate (mortality)
The number of deaths per thousand individuals per year.

decommissioning
Decontaminating and disassembling a nuclear power plant and safely disposing of the radioactive materials.

decomposers
Small organisms, like bacteria and fungi, that cause the decay of dead organic matter and recycle nutrients.

demand
Amount of a product that consumers are willing and able to buy at various prices.

demographic transition
The hypothesis that economies proceed through a series of stages, resulting in stable populations and high economic development.

demography
The study of human populations, their characteristics, and changes.

denitrifying bacteria
Bacteria that convert nitrogen compounds in the soil into nitrogen gas.

density-dependent factors
Those limiting factors that become more severe as the size of the population increases.

density-independent factors
Those limiting factors that are not affected by population size.

desert
A biome that receives less than 25 centimeters of precipitation per year.

desertification
The conversion of arid and semiarid lands into deserts by inappropriate farming practices or overgrazing.

detritus
Organic material that results from fecal waste material or the decomposition of plants and animals.

development ethic
Philosophy that states that the human race should be the master of nature and that the earth and its resources exist for human benefit and pleasure.

dispersal
Migration of organisms from a concentrated population into areas with lower population densities.

domestic water
Water used for domestic activities, such as drinking, air conditioning, bathing, washing clothes, washing dishes, flushing toilets, and watering lawns and gardens.

dredging
The removal of accumulated sediments from a water course.

E

ecology
A branch of science that deals with the interrelationship between organisms and their environment.

economic costs
Those monetary costs that are necessary to exploit a natural resource.

economic growth
The perceived increase in monetary growth within a society.

ecosystem
A group of interacting species combined with the physical environment.

ectoparasite
A parasite that is adapted to live on the outside of its host.

electron
The lightweight, negatively charged particle that moves around the nucleus of an atom.

element
A form of matter consisting of a specific kind of atom.

emergent plants
Aquatic vegetation that is rooted on the bottom but has leaves that float on the surface or protrude above the water.

emigration
Movement out of an area that was once one's place of residence.

endangered species
Those species that are present in such small numbers that they are in immediate jeopardy of becoming extinct.

endoparasite
A parasite that is adapted to live within a host.

energy
The ability to do work.

energy cost
The amount of energy required to exploit a resource.

environment
Everything that affects an organism during its lifetime.

environmental costs
Damage done to the environment as a resource is exploited.

Environmental Protection Agency (EPA)
U.S. government organization responsible for the establishment and enforcement of regulations concerning the environment.

environmental resistance
The combination of all environmental influences that tend to keep populations stable.

environmental science
An interdisciplinary area of study that includes both applied and theoretical aspects of human impact on the world.

erosion
The loss of soil because it is carried away by running water or wind.

estuaries
Marine ecosystems that consist of shallow, partially enclosed areas where fresh water enters the ocean.

ethics
A discipline that seeks to define what is fundamentally right and wrong.

euphotic zone
The upper layer in the ocean where the sun's rays penetrate.

eutrophication
The enrichment of water (either natural or cultural) with nutrients.

eutrophic lake
A usually shallow, warm-water lake that is nutrient rich.

evolution
A change in the structure, behavior, or physiology of a population of organisms as a result of some organisms with favorable characteristics having greater reproductive success than those organisms with less favorable characteristics.

executive branch
The office of the President of the United States.

experiment
An artificial situation designed to test the validity of a hypothesis.

exponential growth
The period during population growth when the population increases at an ever-increasing rate.

external costs
Expenses, monetary or otherwise, borne by someone other than the individuals or groups who use a resource.

extinction
The death of a species; the elimination of all the individuals of a particular kind.

F

first law of thermodynamics
A statement about energy that says that under normal physical conditions, energy is neither created nor destroyed.

fixation
A form of waste immobilization in which materials, such as fly ash or cement, are mixed with hazardous waste.

floodplain
Lowland area on either side of a river.

floodplain zoning ordinances
Designations that restrict future building in floodplains.

food chain
The series of organisms involved in the passage of energy from one trophic level to the next.

food web
Intersecting and overlapping food chains.

free-living, nitrogen-fixing bacteria
Bacteria that live in the soil and can convert nitrogen gas (N_2) in the atmosphere into forms that plants can use.

freshwater ecosystem
Aquatic ecosystems that have low salt content.

friability
The ability of a soil to crumble.

fungicide
A pesticide designed to kill or control fungi.

G

gamma radiation
A type of electromagnetic radiation that comes from disintegrating atomic nuclei.

gas-cooled reactor (GCR)
A type of reactor that uses graphite as a moderator and carbon dioxide or helium as a coolant.

geothermal energy
The heat energy from the earth's molten core.

grasslands
Areas receiving between 25 and 75 centimeters of precipitation per year. Grasses are the dominant vegetation, and trees are rare.

greenhouse effect
The property of carbon dioxide (CO_2) that allows light energy to pass through the atmosphere but prevents heat from leaving; similar to the action of glass in a greenhouse.

gross national product (GNP)
An index that measures the total goods and services generated annually within a country.

groundwater
Water that infiltrates the soil and is stored for long periods in underground reservoirs.

groundwater mining
Removal of water from an aquifer faster than it is replaced.

H

habitat
An identifiable region in which a particular kind of organism lives.

habitat management
The process of changing the natural community to encourage the increase in populations of certain desirable species.

hard pesticide
A pesticide that persists for long periods of time; a persistent pesticide.

hazardous-waste dump
A disposal site for hazardous waste in a dump, landfill, or surface impoundment without any concern for potential environmental or health risks.

hazardous wastes
Substances that could endanger life if released into the environment.

heavy-water reactor (HWR)
A type of reactor that uses the hydrogen isotope deuterium in the molecular structure of the coolant water.

herbicide
A pesticide designed to kill or control plants.

herbivores
Primary consumers; animals that eat plants.

high-temperature, gas-cooled reactor (HTGCR)
A type of reactor that uses graphite as a moderator and helium as a coolant.

horizon
A horizontal layer in the soil. The top layer (*A* horizon) has organic matter. The lower layer (*B* horizon) receives nutrients by leaching. The *C* horizon is partially weathered parent material.

host
The organism a parasite uses for its source of food.

humus
Soil organic matter.

hydrocarbons (HC)
Group of organic compounds consisting of carbon and hydrogen atoms that are evaporated from fuel supplies or are remnants of the fuel that did not burn completely, and that act as a primary air pollutant.

hydrologic cycle
Constant movement of water from surface water to air and back to surface water.

hypothesis
A logical guess that explains an event or answers a question.

I

ignitability
Characteristic of materials that results in combustion.

immigration
Movement into an area in which one has not previously resided.

incineration
Method of disposing of solid waste by burning.

Industrial Revolution
A period of history during which machinery replaced human labor.

industrial uses
Uses of water for cooling and for dissipating and transporting waste materials.

insecticide
A pesticide designed to kill or control insects.

in-stream uses
Use of a stream's water flow for such purposes as hydroelectric power, recreation, and navigation.

integrated pest management
A method of pest management in which many aspects of the pest's biology are exploited to control its numbers.

interspecific competition
Competition between members of different species for a limited resource.

intraspecific competition
Competition among members of the same species for a limited resource.

irrigation
Adding water to an agricultural field to allow certain crops to grow where the lack of water would normally prevent their cultivation.

isotope
Atoms of the same element that have different numbers of neutrons.

J

judicial branch
That portion of the U.S. government that includes the court system.

K

kinetic energy
Energy contained by moving objects.

K-strategists
Large organisms that have relatively long lives, produce few offspring, provide care for their offspring, and typically have populations that stabilize at the carrying capacity.

L

lag phase
The initial stage of population growth during which growth occurs very slowly.

land
The surface of the earth not covered by water.

landfill
A method of disposing of solid wastes that involves burying the wastes in specially constructed sites.

land-use planning
The construction of an orderly list of priorities for the use of available land.

law
A hypothesis that has survived repeated examination by many investigators.

LD$_{50}$
A measure of toxicity; the dosage of a substance that will kill (lethal dose) 50 percent of a test population.

leaching
The downward movement of minerals from the *A* horizon to the *B* horizon by the downward movement of soil water.

legislative branch
That portion of the U.S. government that is responsible for the development of laws.

light-water reactor
A reactor that uses ordinary water as a coolant.

limiting factor
The one primary condition of the environment that determines the success of an organism.

limnetic zone
Region that does not have rooted vegetation in a freshwater ecosystem.

liquefied natural gas
Natural gas that has been converted to a liquid by cooling to −162° C.

liquid metal fast-breeder reactor (LMFBR)
Nuclear fission reactor using liquid sodium as the moderator and heat transfer medium; produces radioactive plutonium-235, which can be used as a nuclear fuel.

littoral zone
Region with rooted vegetation in a freshwater ecosystem.

loam
An ideal soil type with good drainage and good texture, and that is ideal for growing crops.

M

macronutrient
A nutrient, such as nitrogen, phosphorus, and potassium, that is required by plants in relatively large amounts.

mangrove swamp ecosystems
Marine shoreline ecosystems dominated by trees that can tolerate high salt concentrations.

marine ecosystems
Aquatic ecosystems that have high salt content.

marsh
Areas of grasses and reeds that are either permanently flooded or are flooded for a major part of the year.

mass burn
A method of incineration of solid waste.

matter
Substance with measurable mass and volume.

megalopolis
A large, regional urban center.

micronutrient
A nutrient needed in extremely small amounts for proper plant growth; examples are boron, zinc, and magnesium.

migratory birds
Birds that fly considerable distances between their summer breeding areas and their wintering areas.

moderator
Material that absorbs the energy from neutrons released by fission.

molecule
Two or more atoms combined to form a stable unit.

monoculture
A system of agriculture in which large tracts of land are planted to the same crop.

morals
Predominant feeling of a culture about ethical issues.

multiple land use
Land uses that do not have to be exclusionary, so that two or more uses of land may occur at the same time.

municipal solid waste
All the waste produced by the residents of a community.

mutualism
The association between organisms in which both benefit.

N

National Priority List
A listing of hazardous-waste dump sites requiring urgent attention as identified by the Superfund.

natural resources
Those structures and processes that can be used by humans for their own purposes, but cannot be created by them.

natural selection
A process that determines which individuals within a species will reproduce most effectively.

nature centers
Teaching institutions that provide a variety of methods for people to learn about and appreciate the natural world.

negligible risk
A point at which there is no significant health or environmental risk.

neutralization
Reacting acids with bases to produce relatively safe end products.

neutron
Neutrally charged particle located in the nucleus of an atom.

niche
The total role an organism plays in a habitat.

nitrogen cycle
The series of stages in the flow of nitrogen in ecosystems.

nitrogen-fixing bacteria
Bacteria that are able to convert the nitrogen gas (N_2) in the atmosphere into forms that plants can use.

nonpersistent pollutants
Those pollutants that do not remain in the environment for long periods and are biodegradable.

nonpoint source
Diffuse pollutants, such as agricultural runoff, road salt, and acid rain, that are not from a single, confined source.

nonrenewable energy
Those energy sources that are not replaced by natural processes within a reasonable length of time.

nonrenewable resources
Those resources that are not replaced by natural processes, or those whose rate of replacement is so slow as to be noneffective.

nontarget organism
An organism whose elimination is not the purpose of pesticide application.

northern coniferous forest
See boreal forest.

nuclear breeder reactor
Nuclear fission reactor designed to produce radioactive fuel from nonradioactive uranium and at the same time release energy to use in the generation of electricity.

nuclear fusion
The union of smaller nuclei to form a heavier nucleus accompanied with the release of energy.

nuclear reactor
A device that permits a controlled nuclear fission chain reaction.

nucleus
The central region of an atom that contains protons and neutrons.

O

observation
Ability to detect events by the senses or machines that extend the senses.

oligotrophic lakes
Deep, cold, nutrient-poor lakes that are low in productivity.

omnivores
Organisms that eat both plants and animals.

organophosphate
A class of soft pesticides that work by interfering with normal nerve impulses.

outdoor recreation
Recreation that uses the natural out-of-doors for leisure-time activities.

overburden
The layer of soil and rock that covers deposits of desirable minerals.

oxides of nitrogen (NO and NO$_2$)
Primary air pollutants consisting of a variety of different compounds containing nitrogen and oxygen.

ozone (O$_3$)
A molecule consisting of three atoms of oxygen, which absorb much of the sun's ultraviolet energy before it reaches the earth's surface.

P

parasite
An organism adapted to survival by using another living organism (host) for nourishment.

parasitism
A relationship between organisms in which one, known as the parasite, lives in or on the host and derives benefit from the relationship while the host is harmed.

parent material
Material that is weathered to become the mineral part of the soil.

particulates
Small pieces of solid materials, such as smoke particles from fires, bits of asbestos from brake linings and insulation, dust particles, or ash from industrial plants, that are dispersed into the atmosphere.

passive solar system
A design that allows for the entrapment and transfer of heat from the sun to a building without the use of moving parts or machinery.

patchwork clear-cutting
A forest harvest method in which patches of trees are clear-cut among patches of timber that are left untouched.

peat
The first stage in the conversion of organic material into coal.

pelagic
Those organisms that swim in open water.

pelagic ecosystem
A portion of a marine or freshwater ecosystem that occurs in open water away from the shore.

periphyton
Attached organisms in freshwater streams and rivers, including algae, animals, and fungi.

permafrost
Permanently frozen ground.

persistent pesticide
A pesticide that remains unchanged for a long period of time; a hard pesticide.

persistent pollutant
A pollutant that remains in the environment for many years in an unchanged condition.

pest
An unwanted plant or animal that interferes with human activity.

pesticide
A chemical used to eliminate pests; a general term used to describe a variety of different kinds of pest killers, such as insecticides, fungicides, rodenticides, and herbicides.

pH
The negative logarithm of the hydrogen ion concentration; a measure of the number of hydrogen ions present.

pheromone
A chemical produced by one animal that changes the behavior of another.

photochemical smog
A yellowish-brown haze that is the result of the interaction of hydrocarbons, oxides of nitrogen, and sunlight.

photosynthesis
The process by which plants manufacture food. Light energy is used to convert carbon dioxide and water to sugar and oxygen.

photovoltaic cell
A means of directly converting light energy into electricity.

phytoplankton
Free-floating, microscopic, chlorophyll-containing organisms.

pioneer community
The early stages of succession that begin the soil-building process.

plutonium-239 (Pu-239)
A radioactive isotope produced in a breeder reactor and used as a nuclear fuel.

point source
Pollution from a single pipe or series of effluent pipes.

policy
Planned course of action on a question or a topic.

pollution
Waste material that people produce in such large quantities that it interferes with their health or well-being.

pollution costs
The private or public expenditures undertaken to avoid pollution damage once pollution has occurred and the increased health costs and loss of the use of public resources because of pollution.

pollution prevention
To prevent either entirely or partially the pollution that would otherwise result from some production or consumption activity.

postwar "baby boom"
A large increase in the birthrate immediately following World War II.

potable waters
Unpolluted freshwater supplies.

potential energy
The energy of position.

prairies
Grasslands.

precipitation
Removal of materials by mixing with chemicals that cause the materials to settle out of the mixture.

predator
An organism that kills and eats another organism.

preservation
To keep from harm or damage; to maintain in its original condition.

preservation ethic
Philosophy that considers nature to be so special that it should remain intact.

pressurized-water reactor (PWR)
A type of light-water reactor in which the water in the reactor is kept at high pressure and steam is formed in a secondary loop.

prey
An organism that is killed and eaten by a predator.

primary consumer
An organism that eats plants (producers) directly.

primary pollutants
Types of unmodified materials that, when released into the environment in sufficient quantities, are considered hazardous.

primary sewage treatment
Process that removes larger particles by filtering raw sewage through large screens into ponds or lagoons.

primary succession
Succession that begins with bare mineral surfaces or water.

probability
A mathematical statement about how likely it is that something will happen.

producer
An organism that can manufacture food from inorganic compounds and light energy.

profitability
The extent to which economic benefits exceed the economic costs of doing business.

proton
The positively charged particle located in the nucleus of an atom.

public resources
Those parts of the environment that are owned by everyone.

R

radiation
Energy that travels through space in the form of waves or particles.

radioactive
Unstable nuclei that release particles and energy as they disintegrate.

radioactive half-life
The time it takes for half of the radioactive material to spontaneously decompose.

radon
Radioactive gas emitted from certain kinds of rock; can accumulate in very tightly sealed buildings.

range of tolerance
The ability that organisms have to succeed under a variety of environmental conditions. The breadth of this tolerance is an important ecological characteristic of a population.

reactivity
The property of materials that indicates the degree to which a material is likely to react vigorously to water or air, or to become unstable or explode.

recycling
The process of reclaiming a resource and reusing it for another or the same structure or purpose.

reforestation
The process of replanting areas after the original trees were removed.

rem
A measure of the biological damage to tissue caused by certain amounts of radiation.

renewable energy
Those energy sources that can be regenerated by natural processes.

renewable resources
Those resources that can be formed or regenerated by natural processes.

repeatability
An important criterion in the scientific method, it requires that independent investigators be able to repeat an experiment and get the same results.

replacement fertility
The number of children per woman needed to replace just the parents.

reserves
The known deposits from which materials can be extracted profitably with existing technology under present economic conditions.

Resource Conservation and Recovery Act (RCRA)
The 1976 U.S. law that specifically addressed the issue of hazardous waste.

resource exploitation
The use of natural resources by society.

resources
Naturally occurring substances that are potentially feasible to extract under prevailing conditions.

ribbon sprawl
Development along transportation routes that usually consists of commercial and industrial building.

risk assessment
The use of facts and assumptions to estimate the probability of harm to human health or the environment that may result from exposures to specific pollutants, toxic agents, or management decisions.

risk management
Decision-making process that uses input such as risk assessment, technological feasibility, economic impacts, public concerns, and legal requirements.

rodenticide
A pesticide designed to kill rodents.

r-strategist
Typically, a small organism that has a short life span, produces a large number of offspring, and does not reach a carrying capacity.

runoff
The surface water that enters a river system.

S

sanitary landfill
An area used for the containment of solid wastes.

savanna
Tropical biome having seasonal rainfall of 50 to 150 centimeters per year. The dominant plants are grasses, with some scattered fire- and drought-resistant trees.

science
A method for gathering and organizing information that involves observation, hypothesis formation, and experimentation.

scientific method
A way of gathering and evaluating information. It involves observation, hypothesis formation, hypothesis testing, critical evaluation of results, and the publishing of findings.

secondary consumers
Organisms that eat animals that have eaten plants.

secondary pollutants
Pollutants that may interact with one another to form new pollutants in the presence of an appropriate energy source.

secondary recovery
Techniques used to obtain the maximum amount of oil or natural gas from a well.

secondary sewage treatment
Involves holding the wastewater until the organic material has been degraded by bacteria and other microorganisms.

secondary succession
Succession that begins with the destruction or disturbance of an existing ecosystem.

second law of thermodynamics
A statement about energy conversion that says that, whenever energy is converted from one form to another, some of the useful energy is lost.

selective harvest
A forest harvesting method in which individual high-value trees are removed from the forest, leaving the majority of the forest undisturbed.

septic tank
Underground holding tank into which sewage is pumped and where biological degradation of organic material takes place; used in places where sewers are not available.

seral stage
A stage in the successional process.

sere
A stage in succession.

sex ratio
Comparison between the number of males and females in a population.

soft pesticide
A nonpersistent pesticide that breaks down into harmless products in a few hours or days.

soil
A mixture of mineral material, organic matter, air, water, and living organisms; capable of supporting plant growth.

soil profile
The series of layers (horizons) seen as one digs down into the soil.

soil structure
Refers to the way that soil particles clump together. Sand has little structure because the particles do not stick to one another.

soil texture
Refers to the size of the particles that make up the soil. Sandy soil has large particles, and clay soil has small particles.

solidification
The conversion of liquid wastes to a solid form to allow for more safe storage or transport.

solid waste
Unusable or unwanted solid products that result from human activity.

source reduction
Reducing the amount of solid waste generated by using less, or converting from heavy packaging materials to lightweight ones.

speciation
The process of developing a new species.

species
A group of organisms that can interbreed and produce offspring capable of reproduction.

stable equilibrium phase
The phase in a population growth pattern in which the deathrate and birthrate become equal.

standard of living
The necessities and luxuries essential to a level of existence that is customary within a society.

steam stripping
The use of heated air to drive volatile compounds from liquids.

steppe
A grassland.

storm-water runoff
Stormwater that runs off of streets and buildings and is often added directly to the sewer system and sent to the municipal waste-water treatment facility.

strip farming
The planting of crops in strips that alternate with other crops. The primary purpose is to reduce erosion.

submerged plants
Aquatic vegetation that is rooted on the bottom and has leaves that stay submerged below the surface of the water.

subsidy
A gift given to private enterprise by government when the enterprise is in temporary economic difficulty but is viewed as being important to the public.

succession
Regular and predictable changes in the structure of a community, ultimately leading to a climax community.

succession stage
A stage in succession.

sulfur dioxide (SO$_2$)
A compound containing sulfur and oxygen produced when sulfur-containing fossil fuels are burned. When released into the atmosphere, it acts as a primary pollutant.

Superfund
The common name given to the U.S. 1980 Comprehensive Environmental Response, Compensation, and Liability Act, which was designed to address hazardous-waste sites.

supply
Amount of a good or service available to be purchased.

supply/demand curve
The relationship between available supply of a commodity or service and its price.

surface impoundment
Pond created to hold liquid materials. Some may hold only water, while others may be used to contain polluted water or liquid contaminants.

surface mining (strip mining)
A type of mining in which the overburden is removed to procure the underlying deposit.

sustainable development
Using renewable resources in harmony with ecological systems to produce a rise in real income per person and an improved standard of living for everyone.

swamp
Areas of trees that are either permanently flooded or are flooded for a major part of the year.

symbiosis
A close, long-lasting physical relationship between members of two different species.

symbiotic nitrogen-fixing bacteria
Bacteria that grow within a plant's root system and that can convert nitrogen gas (N$_2$) from the atmosphere to nitrogen compounds that the plant can use.

synergism
The interaction of materials or energy that increases the potential for harm.

T

taiga
Biome having short, cool summers and long winters with abundant snowfall. The trees are adapted to winter conditions.

target organism
The organism a pesticide is designed to eliminate.

technological advances
Increasing use of machines to replace human labor.

temperate deciduous forest
Biome that has a winter-summer change of seasons and that receives 100 centimeters or more of relatively evenly distributed precipitation throughout the year.

terrace
A level area constructed on steep slopes to allow agriculture without extensive erosion.

tertiary sewage treatment
Involves a variety of different techniques designed to remove dissolved pollutants left after primary and secondary treatments.

theory
A unifying principle that binds together large areas of scientific knowledge.

thermal inversion
The condition in which warm air in a valley is sandwiched between two layers of cold air and acts like a lid on the valley.

thermal pollution
Waste heat that industries release into the environment.

thermal treatment
A form of hazardous-waste destruction involving heating waste. Such incineration can destroy 99 percent of the organic wastes.

threatened species
Those species that could become extinct if a critical factor in their environment were changed.

threshold levels
The minimum amount of something required to cause measurable effects.

toxicity
A property of materials which, in sufficient quantity, can cause danger to human health or the environment.

toxic waste
Substances that are poisonous and cause death or serious injury to humans and animals when released into the environment.

tract development
The construction of similar residential units over large areas.

transpiration
Evaporation of water from the surfaces of plants.

trophic level
A stage in the energy flow through ecosystems.

tropical rain forest
A biome with warm, relatively constant temperatures where there is no frost. These areas receive more than 200 centimeters of rain per year in rains that fall nearly every day.

tundra
A biome that lacks trees and has permanently frozen soil.

U

unconfined aquifer
An aquifer that usually occurs near the land's surface and may be called a water table aquifer.

underground mining
A type of mining in which the deposited material is removed without disturbing the overburden.

underground storage tank
Tank located below ground level for the storage of materials, such as oil, gasoline, or other chemicals.

uranium-235 (U-235)
A naturally occurring radioactive isotope of uranium used as fuel in nuclear reactors.

urban sprawl
Unplanned suburban growth.

W

waste destruction
Destruction of a portion of hazardous waste with harmful residues still left behind.

waste immobilization
Putting hazardous wastes into a solid form that is easier to handle and less likely to enter the surrounding environment.

waste separation
Either separating one hazardous waste from another, or separating hazardous waste from nonhazardous material that it has contaminated.

water diversion
The physical process of transferring water from one area to another.

water table
The top of the layer of water in an aquifer.

waterways
Low areas that water normally flows through.

weathering
The physical and chemical breakdown of materials; involved in the breakdown of parent material in soil formation.

weed
An unwanted plant.

wetlands
Areas that include swamps, tidal marshes, coastal wetlands, and estuaries.

wilderness
Designation of land use for the exclusive protection of the area's natural wildlife; thus, no human development is allowed.

windbreak
The planting of trees or strips of grasses at right angles to the prevailing wind to reduce erosion of soil by wind.

Z

zero population growth
The stabilized growth stage of human population during which births equal deaths and equilibrium is reached.

zoning
Type of land-use regulation in which land is designated for specific potential uses, such as agricultural, commercial, residential, recreational, and industrial.

zooplankton
Weakly swimming microscopic animals.

CREDITS

Photographs

Part Openers
1: © Mark Segal/Tony Stone Worldwide; 2: © Thomas D. Mangelsen/Peter Arnold, Inc.; 3: © F. Stuart Westmorland/Tom Stack and Associates; 4: © Lionel Delevingne/Stock, Boston; 5: © Will McIntyre/Photo Researchers, Inc.

Chapter 1
1.3, 1.6B: © William E. Ferguson; 1.8B: U.S. Fish and Wildlife Service; Ann Arbor, MI; 1.9A, 1.11: © Michael Douglas/The Image Works; p. 21: Credit: Michigan Department of Natural Resources/ Fishery Division

Chapter 2
2.1: NASA Suquamish Museum; p. 31TL: A: The Granger Collection B: The Bettmann Archive C: The Bettmann Archive; 2.2B: © Thomas Kitchin/ Tom Stack and Associates; p. 31BM, p. 31BR: AP/ Wide World Photos; p. 34 L: © Bob Daemmrich/ The Image Works; p. 34 R: © Wendy Shattil/Tom Stack and Associates; 2.3A: © Toni Michaels; 2.3B: © Stephen Trisch/Stock, Boston Science Lab Experiment; 2.3C: © Toni Michaels; 2.4A: © Thomas Kitchin/Tom Stack and Associates; 2.4B: © Paul Fusco/Magnum Photos; p. 41: © Anna E. Zuckerman/Tom Stack and Associates

Chapter 3
p. 54: © Toni Michaels; 3.8: © Carl Purcell/Photo Researchers, Inc.

Chapter 4
4.4: © William E. Ferguson. Photo by Richard E. Ferguson; 4.6: © M. W. Fweedie/Photo Researchers, Inc.; 4.7: © Gregory G. Demijian, M.D./Photo Researchers, Inc.; 4.8: © Fritz Poiking GDT/Peter Arnold, Inc.; 4.10: © Tom McHugh/ Photo Researchers, Inc.; 4.11: © Dr. J. Burgess/ Photo Researchers, Inc.

Chapter 5
5.1: © William E. Ferguson; 5.4: © Peter Arnold; 5.9B: © Len Rue, Jr./Photo Researchers, Inc.; 5.10B: © William E. Ferguson. Photo by Stephanie Ferguson; 5.11B: © Jacques Jangoux/Peter Arnold, Inc.; 5.12B: © Eastcott/Momatiuk/The Image Works; 5.13B: © William E. Ferguson; 5.14B: © Brian Parker/Tom Stack and Associates; 5.15B: © John Shaw/Tom Stack and Associates; 5.19: © David Hall/Photo Researchers, Inc.; 5.20: © William E. Ferguson

Chapter 6
6.3: Carl Houb, Population Bulletin, 1987; 6.4A: © Betty Derig/Photo Researchers, Inc.; 6.4B: © John Colwell/Grant Heilman Photography; 6.4C: © Kees Van Den Berg/Photo Researchers, Inc.; p. 125L: © Toni Michaels; p. 125T: © James Stevenson/Science Photo Library/Photo Researchers, Inc.

Chapter 7
p. 134: © Robert E. Murowchick/Photo Researchers, Inc.

Chapter 8
8.1: © Irven Devore, Anthro—Photo; 8.2: /Art Resource, N.Y.; 8.4B: Field Museum of Natural History; 8.7: Grant Heilman Photography; 8.8: © Steve McCurry/Magnum Photos; 8.9: © Margot Granitsas/The Image Works

Chapter 9
9.5: © Grapes Michand/Photo Researchers, Inc.; 9.7A: © Matt Meadows/Peter Arnold, Inc.; 9.7B: © David J. Cross/Peter Arnold, Inc.; 9.8A: © Sam Pierson Jr./Photo Researchers, Inc.; 9.10: © Toni Michaels; 9.11A: © Paul Fusco/Magnum Photos; 9.13: © John Running/Stock, Boston; 9.14: © Michelangelo Durrazzo/Magnum Photos, Inc.; 9.15: © Paolo Koch/Photo Researchers, Inc.; 9.16: © Stephen Krasemann/Photo Researchers, Inc.; 9.18B, 9.19: © Tom McHugh/Photo Researchers, Inc.; 9.20: Hank Morgan/Science Source/Photo Researchers, Inc.; 9.21: © Susan Meiselas/Magnum Photos, Inc.; 9.23: © Hank Morgan/Science Source/Photo Researchers, Inc.; 9.24A-B: © Toni Michaels; 9.24C: COURTESY ANDERSON CORPORATION; p. 193L: © Greenlar/The Image Works; p. 193R: © Francis Lepine/Valan Photos

Chapter 10
10.6: © Alexander Tscaras/Science Source/Photo Researchers, Inc.; 10.8: © Y. Arthus-Bertrand/Peter Arnold, Inc.; p. 211: © William E. Ferguson; 10.11: © TASS/Sovfoto/Eastfoto; 10.13: © W. Marc Bernsau/The Image Works

Chapter 11
11.1A: © Ulrike Welsch/Photo Researchers, Inc.; 11.1B: © Toni Michaels; 11.1C: © Mark Antman/ The Image Works; 11.1D: © Bob Daemmrich/The Image Works; 11.1E: © C. J. Collins/Photo Researchers, Inc.; 11.1F: © Joe Sohm/The Image Works; 11.1G: © J. C. Allen and Son; 11.1H: © Toni Michaels; 11.2: © Grant Heilman Photography; 11.4: © Grant Heilman Photography; 11.7: © John Canoalos/Peter Arnold, Inc.; p. 236: © Kevin Schafer/Tom Stack and Associates; 11.8A: © G. Brad Lewis/The Image Works; 11.8B: © Bob Evans/Peter Arnold, Inc.; 11.9: © Julia Sims/Peter Arnold, Inc.; 11.11B: © Eastcott/Momatiuck/The Image Works; 11.12A: © Tom Stack and Associates; 11.12B: © Jack Fields/Photo Researchers, Inc.; 11.12C: © Tom Stack and Associates; 11.12D: Department of Fisheries, Olympia, WA; 11.14B: © Thomas Kitchin/Tom Stack and Associates; 11.15: © Grant Heilman Photography; 11.17B: © Gary Melburn/Tom Stack and Associates; 11.17C: © Richard H. Smith/Photo Researchers, Inc.; 11.17D: © Tom Davis/Photo Researchers, Inc.; 11.17E: © Ken M. Highfill; p. 253: © F.

Gohier/Photo Researchers, Inc.; 11.20A: © Tom McHugh/Photo Researchers, Inc.; 11.20B: © Steven C. Kaufman/Peter Arnold, Inc.; 11.20C: © Edward S. Ross; 11.20D: © Tom and Pat Leeson/ Photo Researchers, Inc.; 11.20E: © Douglas Faulkner/Photo Researchers, Inc.; 11.20F: © Leonard Lee Rue, III/Photo Researchers, Inc.

Chapter 12
12.3A: © Howard Dratch/Beringer/The Image Works; 12.3B: © Toni Michaels; 12.4: © Bob Daemmrich/The Image Works; 12.5A: © Michael Dkoniewski/The Image Works; 12.5B: © Mark D. Phillips/Photo Researchers, Inc.; 12.5C: © S. J. Krasemann/Peter Arnold, Inc.; 12.8: © Thomas Hollyman/Photo Researchers, Inc.; 12.9: © Toni Michaels; 12.10A: © Sharon Gerig/Tom Stack and Associates; 12.10B: © Robert Wenslow/Tom Stack and Associates; 12.12 A&B: © Toni Michaels; 12.13: USDA; p. 279T: © George Gardner/The Image Works; p. 282: © Erick Hartmann/Magnum Photos

Chapter 13
13.2: © Spencer Swanger/Tom Stack and Associates; 13.8: © John D. Cunningham/Visuals Unlimited; 13.9: © Lou Jacobs, Jr./Grant Heilman; 13.10: © Grant Heilman Photography; 13.11: © Steve McCurry/Magnum Photos; 13.13A: © Eastcott/Momatiuk/The Image Works; 13.13B, 13.15: © Grant Heilman Photography; p. 298: © Earl Roberge/Photo Researchers, Inc.; 13.16: © Joe Munroe/Photo Researchers, Inc.; 13.17A: © Bruno Zehnder/Peter Arnold, Inc.; 13.17B: USDA; 13.18A: © Grant Heilman Photography; 13.18B: © Larry Lefever/Grant Heilman Photography; 13.19A: © Grant Heilman Photography; 13.19B: © Link/Visuals Unlimited; 13.20: © John Griffin/The Image Works; 13.21: © Francis de Richemond/The Image Works

Chapter 14
14.1: © Jacques Jangou/Peter Arnold, Inc.; 14.2: © Walter H. Hodge/Peter Arnold, Inc.; 14.3: © Earl Roberge/Photo Researchers, Inc.; 14.7: © Larry Lefever/Grant Heilman Photography; 14.8: © Toni Michaels; 14.9: © J. E. Nugent/N.W. Michigan Horticulture Res. Station; 14.12: © Dr. Jeremy Burgess/Science Photo Library/Photo Researchers, Inc.; p. 323: The National Archives

Chapter 15
15.6: © Pat and Tom Leeson/Photo Researchers, Inc.; 15.7A: © Thomas Kitchin/Tom Stack and Associates; 15.7B: © John Colwell/Grant Heilman Photography; 15.8: © Ted Clutter/Photo Researchers, Inc.; 15.9: © Francis Gohier/Photo Researchers, Inc.; 15.11A, 15.11B: Ray Pfortner/ Peter Arnold, Inc.; p. 359: © Ray Pfortner/Peter Arnold, Inc.; p. 362: © Randy Brandon/Peter Arnold, Inc.; p. 365T: © Wendell Metzen/Southern Stock Photos; p. 365B: © Tom and Pat Leeson/

Photo Researchers, Inc.; **15.13:** © M. Timothy O'Keefe/Tom Stack and Associates; **15.15:** © Grant Heilman Photography; **p. 368:** © Tass/Sovfoto/Eastfoto; **p. 373T:** © Ken W. Davis/Tom Stack and Associates; **p. 373:** © Gene Marshall/Tom Stack and Associates; **p. 373:** © William E. Ferguson; **p. 373:** © Toni Michaels

Chapter 16
16.5B: © William E. Ferguson; **16.5C:** © Mark Antman/The Image Works; **16.5D:** © Thomas Kitchin/Tom Stack and Associates; **16.5E, 16.5F:** © Toni Michaels; **16.5G:** © Joe Sohm/The Image Works; **16.5H:** © Scott Blackman/Tom Stack and Associates; **16.5I:** © Jon Feengersh/Tom Stack and Associates; **16.7A:** © C. Max Dunham/Photo Researchers, Inc.; **16.7B:** © Chris Caswell/Photo Researchers, Inc.; **16.7C:** © Jan Halaska/The Image Works; **16.7D:** © Dedeer Gevois Agence Vandystadt/Photo Researchers, Inc.; **16.7E:** © Clyde H. Smith/Peter Arnold, Inc.; **16.7F:** © George E. Jones III/Photo Researchers, Inc.; **16.9A:** © Runk/Schoenberger/Grant Heilman Photography; **16.9B:** © Bob Daemmrich/The Image Works; **p. 395T:** © C. Allan Morgan/Peter Arnold, Inc.; **p. 395B:** © John R. MacGregor/Peter Arnold, Inc.

Chapter 17
17.4A: © John D. Cunningham/Visuals Unlimited; **17.4B:** © The Image Works; **17.5 A&B:** © Peggy Kahana/Peter Arnold, Inc.; **17.7:** Both: © Toni Michaels; **17.9:** © Don and Pat Valenti/Tom Stack and Associates; **17.10:** © John Shaw/Tom Stack and Associates; **17.11:** Canapress Photo Service, Inc.; **p. 419:** © Toni Michaels; **p. 421:** © Bruce M. Wellman/Stock, Boston

Chapter 18
18.1B: © R. Maiman/Sygma; **18.2:** © Ray Pfortner/Peter Arnold, Inc.; **18.4:** © Rapho Division/Photo Researchers, Inc.; **18.9:** © David M. Dennis/Tom Stack and Associates; **18.11:** © Mark Antman/The Image Works; **18.12B:** © Steve Elmore/Tom Stack and Associates; **p. 439A; All, p. 441:** © Toni Michaels

Chapter 19
19.1B: © John Colwell/Grant Heilman Photography; **19.1C:** © Toni Michaels; **19.1D:** © Larry Kolvoord/The Image Works; **19.1E, 19.1F:** © Toni Michaels; **19.1G:** © Benelux/Photo Researchers, Inc.; **19.1H:** © Bob Daemmrich/The Image Works; **19.1I:** © Joseph Nettis/Photo Researchers, Inc.; **p. 449:** © Kerry T. Givens/Tom Stack and Associates; **19.2:** © Gary Melburn/Tom Stack and Associates; **p. 452:** © Lowell Georgia/Photo Researchers, Inc.; **p. 456:** © Ray Pfortner/Peter Arnold, Inc.; **p. 458:** © William Campbell/Peter Arnold, Inc.

Chapter 20
p. 473: © Joe Sohm/The Image Works; **p. 474T:** © Paul Dusco/Magnum Photos; **p. 474B:** © Robert Winslow/Tom Stack and Associates; **p. 477:** © Craig Newbauer/Peter Arnold, Inc.; **BOX 20.1: A:** Peter Turnley/Black Star **B:** P. Durand/Sygma; **p. 481T:** © Mat Jacob/The Image Works; **p. 481:** © G. Giansanti/Sygma; **p. 481:** © Craig Newbauer/Peter Arnold, Inc.; **p. 481B:** © Leonard Freed/Magnum Photos; **p. 484:** © Frank Pedrick/The Image Works

Line Art

Chapter 1
1.4, 1.6 top, 1.7, 1.10: Environment Canada and the Ontario Ministry of the Environment. **1.5b:** Environmental Remote Sensing Center, University of Wisconsin, Madison. **1.8a:** Source: Data from National Wildlife Federation. **1.9b:** Source: U.S. Department of Commerce.

Chapter 5
5.7: Source: Map based on Robinson Projection. **5.9a:** From Arthur Getis, et al.; *Introduction to Geography*; 3d ed. Copyright © 1991 Wm. C. Brown Publishers, Dubuque, Iowa. All Rights Reserved. Reprinted by permission. **5.10a:** From Arthur Getis, et al.; *Introduction to Geography*; 3d ed. Copyright © 1991 Wm. C. Brown Publishers, Dubuque, Iowa. All Rights Reserved. Reprinted by permission. **5.11a:** From Arthur Getis, et al.; *Introduction to Geography*; 3d ed. Copyright © 1991 Wm. C. Brown Publishers, Dubuque, Iowa. All Rights Reserved. Reprinted by permission. **5.12a:** From Arthur Getis, et al.; *Introduction to Geography*; 3d ed. Copyright © 1991 Wm. C. Brown Publishers, Dubuque, Iowa. All Rights Reserved. Reprinted by permission. **5.13a:** From Arthur Getis, et al.; *Introduction to Geography*; 3d ed. Copyright © 1991 Wm. C. Brown Publishers, Dubuque, Iowa. All Rights Reserved. Reprinted by permission. **5.14a:** From Arthur Getis, et al.; *Introduction to Geography*; 3d ed. Copyright © 1991 Wm. C. Brown Publishers, Dubuque, Iowa. All Rights Reserved. Reprinted by permission. **5.15a:** From Arthur Getis, et al.; *Introduction to Geography*; 3d ed. Copyright © 1991 Wm. C. Brown Publishers, Dubuque, Iowa. All Rights Reserved. Reprinted by permission.

Chapter 6
6.3: Arthur Haupt and Thomas T. Kane, *Population Handbook*, 2nd ed. (Washington, D.C.: Population Reference Bureau, Inc., 1988) p. 14. Reprinted by permission. **6.9:** From Jean Van der Tak, et al.; "Our Population Predicament: A New Look" in *Population Bulletin*, Vol. 34, no. 5, December 1979. Used by permission of Population Reference Bureau, Washington, DC.

Chapter 7
7.3: Source: Map based on Robinson Projection. **7.4:** Source: Data from *1990 World Population Data Sheet.*

Chapter 8
8.5, illustration, p. 161: Source: Map based on Robinson Projection.

Chapter 9
9.2: Copyright © 1971 by Arthur N. Strahler. **9.3:** Copyright © 1972 by Arthur N. Strahler. **9.9:** "From MAN, ENERGY, SOCIETY by Earl Cook. W.H. Freeman and Company. Copyright © 1976." **9.18 top:** *Solar Energy: A Biased Guide International Library of Ecology Series* by Domus Books 1977. **illustration, p. 195:** Copyright, June 20, 1977, *U.S. News & World Report.*

Chapter 10
10.4: Source: Atomic Industrial Forum, U.S. Council for Energy Awareness. **10.7:** © 1980 National Wildlife Federation June-July issue of National Wildlife Magazine. Data from Environmental Protection Agency. **10.12:** Based on US Air Force weather data and computer simulation by Lawrence Livermore National Laboratory in California.

Chapter 11
11.3, 11.10, 11.11 top: Source: Map based on Robinson Projection. **11.6:** International Union for Conservation of Nature and Natural Resources (IUCN). *The United Nations List of National Parks and Protected Areas* (IUCN Gland, Switzerland, 1985). **11.16:** Reprinted with permission of Charles Scribner's Sons, an imprint of Macmillan Publishing Company, from *Game Management* by Aldo Leopold. Copyright 1933 by Charles Scribner's Sons, renewed © 1961 by Estella B. Leopold.

Chapter 13
13.5: Figure from *Ecology and Field Biology*, Third Edition, by Robert Leo Smith. © 1980 by Robert Leo Smith. Harper & Row, Publishers, HarperCollins Publishers, New York, NY. **13.12:** R. F. Dasmann *Environmental Conservation* © 1968 John Wiley.

Chapter 14
14.11: D. C. Gunn and J. G. H. Stevens editors, *Pesticides and Human Welfare* Oxford University Press, 1976, 1980 data from *Science* 226:1293 12/14/84. **14.13:** Reprinted with permission of Macmillan Publishing Company from *Man and the Environment,* 2/e by Arthur S. Boughey. Copyright © 1975 by Arthur S. Boughey.

Chapter 15
Illustration, page 363 (left): Source: National Oceanic and Atmospheric Administration. **Illustration, page 365:** Copyright, April 2, 1990, *U.S. News & World Report.* **15.2:** Source: Map based on Robinson Projection. **illustration, p. 370:** Source: Data from Environmental Protection Agency.

Chapter 16
16.2: "Great Lakes Fish Consumption Study," National Wildlife Federation, 1989.

Chapter 17
17.8: Steven L. Rhodes and Paulette Middleton "Public Pressures, Technical Options: The Complex Challenge of Controlling Acid Rain" *Environment* 25, no. 4 1983, 6–9. **17.14:** Source: Data from Environmental Protection Agency.

Chapter 18
18.1: "Reprinted with permission of *The Miami Herald.*" **18.13:** Source: Data from Franklin Associates, Ltd.

Chapter 20
20.4: Source: Data from *U.S. News & World Report,* April 23, 1990; Warwick Baker & Fiore; and *Adweek's Marketing Week.*

INDEX

A

Abiotic factors, 62
Abstinence, 125
Abyssal ecosystem, 104
Accumulation, 316
    bio-. *See* Bioaccumulation
        of pollutants, 401
Acetanilides, 324
Acid deposition, 411–14
Acid mine drainage, 175
Acid rain, 411, 412, 415, 424
Acids, 54–55, 324, 325, 326
Activated sludge sewage treatment,
    357
Activation energy, 52
Active solar systems, 185–86
Acute toxicity, 447
Additives, and food, 333
Advisory, consumption. *See*
    Consumption advisory
Advocacy groups, 476, 487–90
Aeration, zone of, 347
Aesthetic pollution, 419
Africa, 258
Age distribution, 115, 116, 135
Agent Orange, 323
Agriculture, 269, 297, 470
    and fertilizers, 313–15
    and Great Lakes Basin, 17–18
    and groundwater pollution,
        360–61
    labor-intensive, 311
    methods, 308, 310–13
    and pest management, 308–40
    and runoff, 355
    slash-and-burn, 310
    and technology, 313
    use of water, 349–51
Air pollutants, primary and
    secondary, 403–6
Air pollution, 67, 398–425
    and acid deposition, 411–14
    and aesthetic pollution, 419
    and atmosphere, 400–402
    control of, 408–10
    and global warming, 414–18
    indoor air pollution, 420–24
    and Mexico City, 404
    and ozone depletion, 418–19
    and photochemical smog, 406–8
    primary air pollutants, 403–6
Air-pollution control (APC) valves,
    408
Air quality, and cost-benefit analysis,
    388
Air stripping, 461
Alar controversy, 384
Alaska, 100, 110, 194–95, 362–63
Alaskan Arctic National Wildlife
    Range, 233
Alaskan pipeline, 194–95
Alcohol, 54, 58

Aleuts, 194
Algae, 7
Aliphatic acid, 324
Alpha radiation, 200
Alpine tundra, 100
Altitude, and latitude and vegetation,
    101
Altruistic preservationists, 40
Aluminum, 230
Amazon, 37
Amendments, and environment, 473
American Game Association, 30
Amides, 324
Amino acids, 325
Ammodramus maratimus mirabilis,
    258
*Ammodramus maratimus
    nigrescens,* 258
Ammodramus maratimus pelonata,
    258
*Ammodramus maratimus
    peninsulae,* 258
Ammonia, 54
Amoco Cadiz, 362–63
Amplification, biological, 327, 329
Anacapa Island, 339
Analysis, cost-benefit. *See* Cost-
    benefit analysis
Animal power, 151
Animals, that benefit from humans,
    256
"Animal Use Months," 276
Antarctica, resource or refuge, 41
Antarctic World Park, 41
Anthropocentric, 40
AOC. *See* Areas of concern (AOC)
APC. *See* Air-pollution control (APC)
    valves
Apocalypse, 224
Apples, 384
Aquatic ecosystems, 102–10, 240–45
Aquatic primary succession, 87–89
Aquiclude, 347
Aquifers, 347, 366
Aquitard, 347
Areas of concern (AOC), Great
    Lakes, 20
Argentina, 138–39
Army Corps of Engineers. *See* United
    States Army Corps of
    Engineers
Arsenic, 15, 324, 452, 453
Assessment
    OTA. *See* United States Office of
        Technology Assessment
        (OTA)
    risk. *See* Risk assessment
Atmosphere, 400–402
Atmospheric deposition, 15
Atomic fission, 200, 201
Atomic structure, 49–50
Atoms, 49
"Atoms for Peace" speech, 200

Attitudes, environmental. *See*
    Environmental attitudes
Australia, 142
Automobile, 155–56
Auxins, 321
Aztecs, 150

B

Baby boom, 115, 134
Bacillus thuringiensis, 336, 337
Bacteria, 336, 337
    growth curve, 120
    nitrogen-fixing, 78–80
Baking soda, 54
Bald eagle, 257
Baltimore Gas and Electric Company,
    191
Bangladesh, 391
Bankers Trust, 394
Barium, 201
Bases, 54–55
Basin, Great Lakes. *See* Great Lakes
    Basin
Bauxite, 228
Bay of Fundy, 181
Beaver pond, succession, 90
Beetle, 335
Benefits, and cost-benefit analysis,
    387–89
Benthic marine ecosystems, 104–6
Benthic organisms, 104
Benzonitriles, 324
Benzothiadiazoles, 325
Beryllium, 201
Beta radiation, 200
Bicycling, 275
Bill of Rights, Environmental, 473
Bioaccumulation, 327
Biocentric, 40
Biochemical oxygen demand
    (BOD), 108, 109, 354, 355
Biocides, 315, 453
Biodegradable, defined, 225
Biological amplification, 327, 329
Biology, and energy requirements,
    150–51
Biomass, 74, 188–89
Biomass conversion, 188–89
Biomass Users Network (BUN), 189
Biomes, 90–101
Biotic community, 40
Biotic factors, 62
Biotic potential, 117
Bipyridyliums, 324
Birds, migratory, 248, 249
Birth control, 125
Birthrate, 114, 137
Bison, history of, 253
Blackfooted ferret, 257
Black lung disease, 174
Black rhinoceros, 257

Bleach, 54
BOD. *See* Biochemical oxygen
    demand (BOD)
Bog, floating, 89
Boiling-water reactor (BWR), 202
Bonds, chemical. *See* Chemical
    bonds
Boreal forest, 99–100
Boundary Waters Treaty, 19
Brazil, 37, 58
Breeder reactors, 216–19
Brown, Noel, 38
Bulletin, of American Game
    Association, 30
BUN. *See* Biomass Users Network
    (BUN)
Bureau of Land Management. *See*
    United States Bureau of Land
    Management
Burma, 95
Burning
    landfills, 431
    and mass burn, 434
Bush, George, 407
Butterfly, 257
BWR. *See* Boiling-water reactor
    (BWR)
Bythotrephes cedestroemi, 21

C

Cadmium, 15, 228, 452, 453
California Water Plan, 372–73
Camping, 275
Canada, 142, 192–93
    and acid rain, 412
    and decisions about Great Lakes,
        19–22
Cancer
    and alar, 384
    fish, 14
    and particulates, 405
    risk, 379, 380
Canete Valley, 337
Capability classes, of land, 298–99
Capacity, carrying. *See* Carrying
    capacity
Cape Sable seaside sparrow, 258
Carbamates, 318–19, 321, 324
Carbon, 201
Carbon absorption, 461
Carbon cycle, 76–78
Carbon dioxide, 415, 416
Carboniferous Period, 152
Carbon monoxide (CO), 403–4
Carbon tetrachloride, 452
Carcinogens, 14
    and particulates, 405
    risk, 379, 380
Carnivores, 73, 74, 75, 76
Carrying capacity, 118–20
Carson, Rachel, 30–31